Organic Meat Production and Processing

IFT Press

The *IFT Press* series reflects the mission of the Institute of Food Technologists—to advance the science of food contributing to healthier people everywhere. Developed in partnership with Wiley-Blackwell, *IFT Press* books serve as leading-edge handbooks for industrial application and reference and as essential texts for academic programs. Crafted through rigorous peer review and meticulous research, *IFT Press* publications represent the latest, most significant resources available to food scientists and related agriculture professionals worldwide. Founded in 1939, the Institute of Food Technologists is a nonprofit scientific society with 22,000 individual members working in food science, food technology, and related professions in industry, academia, and government. IFT serves as a conduit for multidisciplinary science thought leadership, championing the use of sound science across the food value chain through knowledge sharing, education, and advocacy.

WILEY-BLACKWELL
A John Wiley & Sons, Ltd., Publication

Organic Meat Production and Processing

Edited by

Steven C. Ricke
Professor and Wray Endowed Chair in Food Safety
University of Arkansas, Fayetteville, USA

Ellen J. Van Loo
Doctoral Researcher
Department of Agricultural Economics, Ghent University, Ghent, Belgium

Michael G. Johnson
Emeritus Professor, Food Microbiology & Safety
University of Arkansas, Fayetteville, USA

Corliss A. O'Bryan
Post Doctoral Research Associate
University of Arkansas, Fayetteville, USA

IFT Press
WILEY-BLACKWELL
A John Wiley & Sons, Ltd., Publication

Wiley-Blackwell is an imprint of John Wiley & Sons, formed by the merger of Wiley's global Scientific, Technical and Medical business with Blackwell Publishing.

Editorial offices: 2121 State Avenue, Ames, Iowa 50014-8300, USA
The Atrium, Southern Gate, Chichester, West Sussex, PO19 8SQ, UK
9600 Garsington Road, Oxford, OX4 2DQ, UK

For details of our global editorial offices, for customer services and for information about how to apply for permission to reuse the copyright material in this book please see our website at www.wiley.com/wiley-blackwell.

Library of Congress Cataloging-in-Publication Data

Organic meat production and processing / edited by Steven C. Ricke. [et al.].
p. cm. – (Institute of food technologist series)
Includes bibliographical references and index.
ISBN 978-0-8138-2126-9 (hard cover : alk. paper) 1. Meat industry and trade. 2. Meat industry and trade – United States. 3. Natural foods industry. 4. Meat hygiene. 5. Natural foods 6. Processed foods. 7. Animal industry. 8. Packing-houses. 9. Animal biotechnology. I. Ricke, Steven C., 1957–
TS1970.O74 2012
363.19′29–dc23

2011037450

A catalogue record for this book is available from the British Library.

Wiley also publishes its books in a variety of electronic formats. Some content that appears in print may not be available in electronic books.

Set in 10/12 pt Times by Aptara® Inc., New Delhi, India
Printed and bound in Malaysia by Vivar Printing Sdn Bhd

1 2012

Titles in the *IFT Press* series

- *Accelerating New Food Product Design and Development* (Jacqueline H. Beckley, Elizabeth J. Topp, M. Michele Foley, J.C. Huang, and Witoon Prinyawiwatkul)
- *Advances in Dairy Ingredients* (Geoffrey W. Smithers and Mary Ann Augustin)
- *Bioactive Proteins and Peptides as Functional Foods and Nutraceuticals* (Yoshinori Mine, Eunice Li-Chan, and Bo Jiang)
- *Biofilms in the Food Environment* (Hans P. Blaschek, Hua H. Wang, and Meredith E. Agle)
- *Calorimetry in Food Processing: Analysis and Design of Food Systems* (Gönül Kaletunç)
- *Coffee: Emerging Health Effects and Disease Prevention* (YiFang Chu)
- *Food Carbohydrate Chemistry* (Ronald E. Wrolstad)
- *Food Irradiation Research and Technology* (Christopher H. Sommers and Xuetong Fan)
- *High Pressure Processing of Foods* (Christopher J. Doona and Florence E. Feeherry)
- *Hydrocolloids in Food Processing* (Thomas R. Laaman)
- *Improving Import Food Safety* (Wayne C. Ellefson, Lorna Zach, and Darryl Sullivan)
- *Innovative Food Processing Technologies: Advances in Multiphysics Simulation* (Kai Knoerzer, Pablo Juliano, Peter Roupas, and Cornelis Versteeg)
- *Microbial Safety of Fresh Produce* (Xuetong Fan, Brendan A. Niemira, Christopher J. Doona, Florence E. Feeherry, and Robert B. Gravani)
- *Microbiology and Technology of Fermented Foods* (Robert W. Hutkins)
- *Multivariate and Probabilistic Analyses of Sensory Science Problems* (Jean-François Meullenet, Rui Xiong, and Christopher J. Findlay)
- *Natural Food Flavors and Colorants* (Mathew Attokaran)
- *Nondestructive Testing of Food Quality* (Joseph Irudayaraj and Christoph Reh)
- *Nondigestible Carbohydrates and Digestive Health* (Teresa M. Paeschke and William R. Aimutis)
- *Nonthermal Processing Technologies for Food* (Howard Q. Zhang, Gustavo V. Barbosa-Cánovas, V.M. Balasubramaniam, C. Patrick Dunne, Daniel F. Farkas, and James T.C. Yuan)
- *Nutraceuticals, Glycemic Health and Type 2 Diabetes* (Vijai K. Pasupuleti and James W. Anderson)
- *Organic Meat Production and Processing* (Steven C. Ricke, Michael G. Johnson, and Corliss A. O'Bryan)
- *Packaging for Nonthermal Processing of Food* (Jung H. Han)
- *Preharvest and Postharvest Food Safety: Contemporary Issues and Future Directions* (Ross C. Beier, Suresh D. Pillai, and Timothy D. Phillips, Editors; Richard L. Ziprin, Associate Editor)
- *Regulation of Functional Foods and Nutraceuticals: A Global Perspective* (Clare M. Hasler)
- *Sensory and Consumer Research in Food Product Design and Development, second edition* (Howard R. Moskowitz, Jacqueline H. Beckley, and Anna V.A. Resurreccion)
- *Sustainability in the Food Industry* (Cheryl J. Baldwin)
- *Thermal Processing of Foods: Control and Automation* (K.P. Sandeep)
- *Water Activity in Foods: Fundamentals and Applications* (Gustavo V. Barbosa-Cánovas, Anthony J. Fontana Jr., Shelly J. Schmidt, and Theodore P. Labuza)
- *Whey Processing, Functionality and Health Benefits* (Charles I. Onwulata and Peter J. Huth)

WILEY-BLACKWELL
A John Wiley & Sons, Ltd., Publication

Contents

List of Contributors

Raúl Avila-Sosa
Departamento de Bioquímica-Alimentos
Benemérita Universidad Autónoma de Puebla
Puebla
México

Maciej Badowski
Warsaw University of Life Sciences
Warszawa
Poland

Debabrata Biswas
Department of Animal and Avian Sciences
University of Maryland
College Park, MD
USA

Antonio Boggia
Department of Economics and Appraisal
University of Perugia
Perugia
Italy

Kerri C. Boling
National Agricultural Law Center
University of Arkansas
Fayetteville, AR
USA

Thelma Calix-Lara
Department of Animal Science
Texas A&M University
College Station, TX
USA

Maurizio Canavari
Department of Agricultural Economics and Engineering
Alma Mater Studiorum
University of Bologna
Bologna
Italy

Vincenzina Caputo
Department of Agricultural Economics and Engineering
Alma Mater Studiorum
University of Bologna
Bologna
Italy

Dave Carter
Crystal Springs Consulting, Inc.
Westminster, CO
USA

Cesare Castellini
Department of Applied Biology
University of Perugia
Perugia
Italy

Vesela I. Chalova
Department of Biochemistry and Molecular Biology
University of Food Technology
Plovdiv
Bulgaria

Mauro Conter
Department of Animal Production
Veterinary Biotechnology
Food Quality and Safety
Parma University
Parma
Italy

Philip G. Crandall
Center for Food Safety
Department of Food Science
University of Arkansas
Fayetteville, AR
USA

Claudia S. Dunkley
Department of Poultry Science
University of Georgia
Tifton, GA 31794
USA

Kingsley Dunkley
School of Science and Mathematics
Abraham Baldwin Agricultural College
Tifton, GA
USA

Paul D. Ebner
Department of Animal Sciences
Purdue University
West Lafayette, IN
USA

J. Trent Fox
JBS Five Rivers Cattle Feeding LLC
Greeley, CO
USA

Sergio Ghidini
Department of Animal Production
Veterinary Biotechnology
Food Quality and Safety
Parma University
Parma
Italy

Kristen E. Gibson
Center for Food Safety
Department of Food Science
University of Arkansas
Fayetteville, AR
USA

Irene B. Hanning
Department of Food Science and Technology
University of Tennessee
Knoxville, TN
USA

Paul Hererra
Food Safety and Inspection Service
US Department of Agriculture
Waco, TX
USA

Adriana Ianieri
Department of Animal Production
Veterinary Biotechnology
Food Quality and Safety
Parma University
Parma
Italy

Jacqueline Jacob
Department of Animal and Food Sciences
University of Kentucky
Lexington, KY
USA

Megan E. Jacob
Department of Population Health and Pathobiology
College of Veterinary Medicine
North Carolina State University
Raleigh, NC
USA

Rolf Joerger
Department of Animal and Food Sciences
University of Delaware
Newark, DE
USA

Michael G. Johnson
Center for Food Safety
Department of Food Science
University of Arkansas
Fayetteville, AR
USA

Dae-soo Kim
Ralph E. Martin Department of Chemical Engineering
University of Arkansas
Fayetteville, AR
USA

Aurelio López-Malo
Departamento de Ingeniería Química y Alimentos
Universidad de las Americas
Puebla
México

Ivan Manolov
Department of Agrochemistry and Soil Science
Agricultural University
Plovdiv
Bulgaria

Sherry N. Melendez
Microbiology Laboratory
Mars Inc.
Hackettstown, NJ
USA

Jean-François Meullenet
Center for Food Safety
Department of Food Science
University of Arkansas
Fayetteville, AR
USA

Shannon J. Mirus
National Agricultural Law Center
University of Arkansas
Fayetteville, AR
USA

T. G. Nagaraja
Department of Diagnostic Medicine/Pathobiology
College of Veterinary Medicine
Kansas State University
Manhattan, KS
USA

Simona Naspetti
Università Politecnica delle Marche
Agricultural Economics
Ancona
Italy

Rodolfo M. Nayga, Jr.
Department of Agricultural Economics and Agribusiness
University of Arkansas
Fayetteville, AR
USA

Corliss A. O'Bryan
Center for Food Safety
Department of Food Science
University of Arkansas
Fayetteville, AR
USA

Enrique Palou
Departamento de Ingeniería Química y Alimentos
Universidad de las Americas
Puebla
México

Luisa Paolotti
Department of Economics and Appraisal
University of Perugia
Perugia
Italy

Anthony Pescatore
Department of Animal and Food Sciences
University of Kentucky
Lexington, KY
USA

Harrison M. Pittman
National Agricultural Law Center
University of Arkansas
Fayetteville, AR
USA

Ewa Rembiałkowska
Warsaw University of Life Sciences
Division of Organic Foodstuffs
Warszawa
Poland

Lydia J. Rice
Department of Food Science
University of Arkansas
Fayetteville, AR
Arkansas
USA

Steven C. Ricke
Center for Food Safety
Department of Food Science
University of Arkansas
Fayetteville, AR
USA

Marcos H. Rostagno
USDA-ARS
Livestock Behavior Research Unit
West Lafayette, IN
USA

Gregory R. Siragusa
Danisco USA, Inc.
Research and Development
Waukesha, WI
USA

Poul Sørensen
Institute of Genetics and Biotechnology
Faculty of Natural Science and Technology
Aarhus University
Denmark

Albert Sundrum
Department of Animal Nutrition and Animal Health
Kassel University
Germany

T. Matthew Taylor
Department of Animal Science
Texas A&M University
College Station, TX
USA

Greg J. Thoma
Ralph E. Martin Department of Chemical Engineering
University of Arkansas
Fayetteville, AR
USA

Ellen J. Van Loo
Department of Agricultural Economics
Faculty of Bioscience Engineering
Ghent University
Ghent
Belgium

Christina Yancheva
Department of Agrochemistry and Soil Science
Agricultural University
Plovdiv
Bulgaria

Emanuela Zanardi
Department of Animal Production
Veterinary Biotechnology
Agricultural Economics
Food Quality and Safety
Parma University
Parma
Italy

Raffaele Zanoli
Università Politecnica delle Marche
Agricultural Economics
Ancona
Italy

1 Historical and Current Perspectives on Organic Meat Production

Ellen J. Van Loo, Steven C. Ricke, Corliss A. O'Bryan, and Michael G. Johnson

Abstract: The public concerns about the environment and the chemicals used in food production have made consumers more sensitive to the potential ecological problems associated with synthetic chemicals used in farming. This has resulted in a growing interest in the production of organic foods. Based on the growth of market demand, organic foods have become much more precisely defined as a regulatory entity, and now these food products often carry an organic label when sold at retail establishments. However, there is still some confusion among consumers about the different organic labels, and some of those reasons are discussed here. In addition to covering the development of organic foods and meats and current perceptions, this chapter also discusses the specific objectives on organic meat production and processing, which is the subject of this book.

Keywords: labeling; definition; objectives

1.1 WHAT IS ORGANIC – DEFINITION

A very simple definition of organic foods is: foods that are produced using methods that do not involve the use of synthetic pesticides or chemical fertilizers, do not contain genetically modified organisms, and are not processed using irradiation, industrial solvents, or chemical food additives (DeSoucey, 2007). Certified organic food production has evolved into a highly regulated industry in the European Union (EU), the United States, Canada, Japan, and many other countries. Consequently, the term "organic food" is now understood to mean food produced in a way that complies with organic standards set by national governments and international organizations. However, variations among the standards may exist and the standards are not internationally uniform.

1.2 HISTORY AND DEVELOPMENT OF THE MODERN ORGANIC FOOD INDUSTRY

Prior to the twentieth century, traditional farming practices, based on current definitions, would be considered "organic." However, as chemically synthesized compounds such as synthetic pesticides, fertilizers, preservatives, and synthetic chemicals were introduced over

Organic Meat Production and Processing, First Edition. Edited by Steven C. Ricke, Ellen J. Van Loo, Michael G. Johnson, and Corliss A. O'Bryan.

the twentieth century to improve the yield and quality and lengthen the shelf-life of foods, they would no longer be considered in the organic food category as it is now defined. In the 1960s and 1970s, a variety of investigative reports and awareness movements raised public concerns about the environment and the chemicals used in food production and consumers became more sensitive to the potential ecological problems associated with synthetic chemicals used in farming. This created a growing demand of more environmentally friendly food production systems and resulted in an organic food movement (DeSoucey, 2007; Baker, 2005).

Initially, locating genuine organic foods was difficult and confusing for consumers. Without strict regulations, the consumer was not able to distinguish "organic" from "natural," "locally grown," and other similar claims and labels. The number of companies attempting to mislead the consumer and use false claims about their products increased. With an organic food market that was skyrocketing in the 1980s and 1990s, the need for regulation, standard, and certification became necessary. In the United States, the Federal Congress passed the Organic Foods Production Act (OFPA) in 1990 to avoid the confusion, prevent misrepresentation, and develop a uniform standard for organic food claims. The United States Department of Agriculture (USDA) established the National Organic Program (NOP) in 2002 (USDA-NOP, 2008). The NOP made efforts to devise uniform organic standards and certification procedures. The NOP established national standards for the production, handling, and processing of organic food.

Since the OFPA initiation in 1990, the organic food sector has developed a robust market. At the end of the nineties (1999), the total US organic food sales were valued at $5,039 million; however, since then they have quintupled and were estimated at $26,708 million in 2010 (OTA, 2009, 2011). However, strong organic markets are emerging in other countries as well. In Denmark, Switzerland, and Austria, the estimated consumption of organic foods in 2009 was €100 per capita (Willer & Kilcher, 2011) (Table 1.1). These are also countries with a high GDP (Table 1.1). In the United States, the per capita organic food consumption in 2009 was estimated at €58 and is similar to other Western countries. In Europe, most countries have a per capita organic food consumption of €20 or more. Residents of Japan spend on average only €8 on organic food purchases. In China, this is only €1 and is not surprising knowing that its GDP per capita is more than ten times smaller than that of the United States (Table 1.1).

Organic foods may not be affordable to most consumers with low incomes. High income and education level are associated with organic food consumers (Tsakiridou *et al.*, 2008). The high price for organic foods is often a barrier (Van Loo *et al.*, 2010), especially for lower income households. Organic buyers often have a higher income (FMI & AMI, 2010) and consumers with a higher income are also more willing to pay a higher price for organic meat products (Van Loo *et al.*, 2011). Mintel's US Organic Food Report (2008) suggested that a strong consumer base across a range of income levels would be important for the success of organic food products. With an increasing organic meat production, organic meat products are expected to be priced more competitively as compared to the conventional products (FMI, 2007).

Historically, organic food has been associated with mostly fruit and vegetables but as the organic food market is expanding, other food items are finding their way into the organic food market (O'Bryan *et al.*, 2008; Winter & Davis, 2006). The organic meat segment has followed the pattern of organic food growth but has grown at a faster rate. In the first decade of the 2000s, the US organic meat industry has climbed over 20-fold from $23 million in sales in 2000 to $470 million in 2010 (OTA, 2011) with growth rates that are substantially outpacing growth of all other organic foods in general. With this remarkable increase of organic meat sales, there is a need for further development of all aspects of organic meat

Table 1.1 Organic food spending in several countries.

Region	Per capita organic food consumption levels in 2009 (€)[a]	Organic food sales in 2009 in (€ million)[a]	GDP per capita in 2010 ($)[b]	Per capita organic food consumption as percentage of GDP[c]
Denmark	139	765	56,147	0.36
Switzerland	132	1,023	67,246	0.28
Austria	104	868	44,987	0.33
Sweden	75	698	48,875	0.22
Germany	71	5,800	40,631	0.25
USA	58	17,835	47,284	0.18
France	47	3,041	41,019	0.16
Canada	38	1,284	46,215	0.12
Netherlands	36	591	47,172	0.11
United Kingdom	34	2,065	36,120	0.14
New Zealand	33	143	32,145	0.15
Belgium	32	350	42,630	0.11
Australia	25	536	55,590	0.06
Italy	25	1,500	34,059	0.11
Norway	24	114	84,444	0.04
Ireland	24	113	45,689	0.08
Spain	20	905	30,639	0.09
Slovenia	17	34	23,706	0.10
Japan	8	1,000	42,820	0.03
Greece	5	58	27,302	0.03
China	1	791	4,382	0.03

[a]From Willer and Kilcher (2011).
[b]From International Monetary Fund (2011)–World Economic Outlook Database, April 2011.
[c]Based on the exchange rate of €1 = $1.4362.

production not only to meet the demands of increasing consumer interest but also to retain product integrity, food safety, and economic value.

The organic meat industry has gained significant importance in the organic food sector. However, limited resources with information are available, and as of now, there are no sources that package all the information into one publication. The purpose of this book is to address specific issues associated with organic meat production and give a comprehensive overview of all the knowledge on organic meat production and processing.

1.3 ORGANIC FOOD LABELS

A product label is a quality signal for the consumer. The organic food label is an important identification means to help the consumer to easily locate organic products. The consumer might not be aware that the product is organic because differentiation between conventional and organic food may not be that discernable. According to Yiridoe *et al.* (2005), product labels help buyers assess product quality by transforming credence characteristics into search attributes. However, mislabeling and product misrepresentation have discouraged consumers from buying organic foods.

It is necessary to have defined rules for production methods and labeling of organic foods. For consumers, it is important that they can trust the organic labels and that the label guarantees that the products are organic. Therefore, it is essential to have uniform organic standards and certification procedures resulting in clear and nonmisleading organic labels. Numerous organic certifying agencies exist internationally with each having their own standards, certification methods, and labeling. Both the United States and Europe have established certification agencies and regulation to protect the consumer and regulate the organic food production. In the United States, the United States Department of Agriculture's (USDA) National Organic Program (NOP) is in charge of the organic foods regulation. In Europe, the production, control, and labeling is monitored by the EU Organic regulation. The EU Organic Regulation ensures that all European agencies work to the same standard. Examples of a few of the different agencies are shown in Table 1.2.

1.3.1 Meat labeling

Only animals that have been raised according to the organic certifying body of each country can be labeled as organic. However, the consumer could still confuse organic meat with other terms that are used to label meat and poultry. In particular, the differences between organic, free range, or natural may not always be clear to the consumer. For a meat product to be labeled "free range" or "free roaming," there is only one requirement: the livestock has been allowed access to an outdoor environment.

In the United States, "natural" can only be used when no artificial ingredients or color has been added and when the product is minimally processed. The label must further explain the use of the term "natural" (for example: "no added artificial ingredients or color ingredients; minimally processed"). Meats labeled as "natural" are not subject to the strict production standards that the USDA sets for organic meats. In January 2009, the USDA established a voluntary "naturally raised" marketing claim that livestock producers can request to have verified by the USDA (USDA & AMS, 2009a, 2009b). To meet this standard, the livestock has to meet following three requirements: (1) be raised entirely without growth promotants, (2) without antibiotics (except for ionophores used as coccidiostats for parasite control), and (3) have never been fed animal (mammalian, avian, or aquatic) by-products. The three requirements must be explicitly mentioned in addition to the "USDA-certified naturally raised" claim. Just like the organic claim, this naturally raised voluntary claim can help to increase the potential for niche marketing of the particular meat product as alternative production methods may distinguish the products and may result in value-added opportunities. A USDA-certified naturally raised claim may also increase the potential for exporting to foreign markets where the livestock is required to be raised without growth promotants (USDA & AMS, 2009a, 2009b).

Other possible claims for meat and poultry products include "no antibiotics added," which may be used on the label if no antibiotics have been used in raising the animals. "No hormones added" can be used in beef products but not in the pork and poultry products since hormones are not allowed in raising swine and poultry under any circumstances (Food Safety and Inspection Service (FSIS), 2006). However, when followed by "Federal regulation prohibits the use of hormones," the claim is allowed for pork and poultry as well. The truthfulness of the labeling of meat and poultry products is overseen by the USDA's Food Safety and Inspection Service (FSIS, 2006). Another label that is frequently used is "pastured poultry," which refers to birds raised on pasture with shelters that can be moved. In contrast to the other labels, "pasture poultry" is an unregulated label and is thus not regulated by the USDA or FSIS (FSIS, 2006; Oberholtzer *et al.*, 2006).

1.3.2 Consumer confusion about labeling

Consumers are confused and are unclear about the differences between "organic" and "natural," which is one of the major challenges for the organic food industry (Mintel, 2008,

Table 1.2 Examples of organic programs.

Name	**Region**	**Label**
International Federation of Organic Agriculture Movements		
North America		
USDA's National Organic Program	United States	
Canada Organic	Canada	
Europe		
EU Organic Farming	EU	
Austria Bio Garantie	Austria	
Biogarantie	Belgium	
Agriculture Biologique	France	
Bio-Siegel	Germany	
Irish Organic Farmers and Growers Association	Ireland	

(continued)

Table 1.2 (*Continued*)

Name	Region	Label
Associazione Italiana per l'agricoltura Biologica	Italy	
EKO Quality Mark	Netherlands	
Debio	Norway	
AgroBio	Portugal	
Bio Suisse	Switzerland	
Soil Association	United Kingdom	
Organic Farmers & Growers	United Kingdom	
Others		
Australian Certified Organic	Australia	

Table 1.2 (*Continued*)

Name	Region	Label
NASAA Certified Organic	Australia	
JAS	Japan	
BioGro	New Zealand	
China Green Food Development Center	China	

EU, European Union; USDA, United States Department of Agriculture. (Logos reproduced with permission of Organic Farmers & Growers Ltd, the Soil Association, Associazione Italiana per l'agricoltura Biologica (AIAB), USDA Organic, Irish Organic Farmers and Growers Association (IOFGA), EKO, NASAA, BIO-SIEGEL, EU Organic Farming, Austria Bio garantie, Biogarantie (Belgium), Debio, Bio Suisse and Australian Certified Organic.)

2010). Since they are able to distinguish between natural and organic, many consumers do not understand the higher price premium for organic foods compared to natural foods. Less than half of the respondents distinguished between organic and natural foods suggesting the need for growers, manufacturers, and retailers to explain the differences (Mintel, 2008). Particularly, for the organic industry, it will be necessary to educate the current and potential future consumers about the different standards. Mintel (2008) reports that the biggest competition between natural and organic is in the meat and egg industries. It will be important to improve the consumer's knowledge about natural and organic meat products so that they can make an educated choice.

As communication has been identified as being highly important for markets such as organic foods (Mintel, 2008), the communication to consumer should be improved to better market the organic products and educate the consumer. Even communicating with the consumers who currently purchase organic meat will be beneficial in maintaining their loyalty (Mintel, 2008). The market needs to continue to appeal to the current strong organic consumer base with product innovation as well as competitive priced products to boost the sales among the current users as well as new consumers.

Another set of competing products is the locally grown products. With the increasing concern about the sustainability and other ecological issues, the food miles corresponding with transportation of food products has received increased awareness across consumer groups. The "locavore movement" is convinced that eating local foods is better (Mintel, 2008). So, buying a food identified as a local food even if it is not produced organically may be a competitor of organic food products: 59% of the respondents agreed to buy locally grown food whether it is organic or not (Mintel, 2008). The increase in farmer's markets across the country as well as Mintel's report indicate that consumers have become likely to prefer local rather than organic.

1.4 ORGANIC MEAT AND OBJECTIVES OF THIS BOOK

This book is an attempt to address the different aspects of organic meat and package the knowledge on organic meat under the appropriate subject areas. In this introductory chapter, the organic food labels and meat labels according to the US regulation are described. The first section of this book, "Economics, Market, and Regulatory Issues," outlays the organic meat industry in the United States and Europe including the regulations. Marketing aspects of organic food and organic meat in particular are discussed.

The second section, "Management Issues for Organically Raised and Processed Meat Animals," covers a wide range of topics including a discussion about the welfare and health of the organic livestock and the environmental impacts of organic meat production. With the consumers' increasing awareness for environment-friendly products and animal welfare issues, these topics will impact the future of the organic meat market. Genetics and details on the feeding options for organic livestock are described in detail. With the increasing interest in finding a useful outlet for the by-products created in food production, there is potential in using organic meat by-products to create other useful organic products such as organic pet food.

The third section, "Processing, Sensory, and Human Health Aspects of Organic Meat," aims to give an overview of the slaughter options for organic meat producers in the United States and gives a review of sensory and nutritional attributes of organic meat.

The fourth section, "The Current Food Safety Status of Organic Meats," gives an overview of the literature of the food-borne pathogen occurrence in organic beef, swine, and poultry. Goat and sheep were not included since there is virtually no research done on livestock raised organically.

The last section, "Preharvest Control Measures for Assuring the Safety of Organic Meats," gives an overview of the potential methods to assure safe organic meats. This includes the use of prebiotics, probiotics, bacteriophages, organic acids, antimicrobial peptides, and botanicals as feed additives.

In the final chapter, each section is briefly discussed. The future of the organic meat industry has limitation and challenges, which are highlighted in the final conclusion chapter. The needs of the organic meat industry are described as well.

ACKNOWLEDGMENT

Support is gratefully acknowledged from a National Integrated Food Safety Initiative (NIFSI) grant (2008–51110-04339) to coauthor Steven C. Ricke.

REFERENCES

Baker, B. 2005. Brief history of organic farming and the National Organic Program. In: B. Baker, S. L. Swezey, D. Granatstein, S. Guldan, and D. Chaney. (eds) *Organic Farming Compliance Handbook: A*

Resource Guide for Western Region Agricultural Professionals. Available at: http://www.sarep.ucdavis.edu/organic/complianceguide/ (accessed October 01, 2011).
DeSoucey, M. 2007. Organic food. In: G. J. Allen and K. Albala. (eds) *The Business of Food: Encyclopedia of the Food and Drink Industries*. ABC-CLIO Greenwood, Santa Barbara, CA.
Food Marketing Institute (FMI) and American Meat Institute (AMI). 2010. The power of meat – an in-depth look at meat through the shoppers' eyes. Joint report AMI/FMI, Arlington, VA. p. 77.
Food Marketing Institute (FMI). 2007. Natural and organic foods. Available at: http://helbreds.info/doc/9.pdf (accessed September 20, 2011).
Food Safety and Inspection Service (FSIS). 2006. Meat and poultry labeling terms. Available at: http://www.fsis.usda.gov/pdf/meat_and_poultry_labeling_terms.pdf (accessed September 23, 2011).
International Monetary Fund. 2011. World economic outlook database. Available at: http://www.imf.org/external/pubs/ft/weo/2011/02/weodata/index.aspx (accessed October 01, 2011).
Mintel. 2008. Organic food – US – October 2008. Available at: http://oxygen.mintel.com/sinatra/oxygen/display/id=297944 (accessed September 24, 2011).
Mintel. 2010. Consumer attitudes toward natural and organic food and beverage – US – March 2010. Available at: http://oxygen.mintel.com/sinatra/oxygen/display/id=482491 (accessed September 24, 2011).
Oberholtzer, L., C. Greene, and E. Lopez. 2006. *Organic Poultry and Eggs Capture High Price Premiums and Growing Share of Specialty Markets*. Economic Research Service/USDA, Washington, DC. Available at: http://www.ers.usda.gov/publications/ldp/2006/12dec/ldpm15001/ldpm15001.pdf (accessed October 03, 2011).
O'Bryan, C. A., P. G. Crandall, and S. C. Ricke. 2008. Organic poultry pathogen control from farm to fork. *Foodborne. Path. Dis*. 5:709–720.
Organic Trade Association (OTA). 2009. *Organic Trade Association's 2009 Organic Industry Survey*.
Organic Trade Association (OTA). 2011. *Organic Trade Association's 2011 Organic Industry Survey*.
Tsakiridou, E., C. Boutsouki, Y. Zotos, and K. Mattas. 2008. Attitudes and behaviour towards organic products: an exploratory study. *Int. J. Retail Distrib. Manag*. 36:158–175.
US Department of Agriculture National Organic Program (USDA-NOP). 2008. Agricultural Marketing Service 7, Code of Federal Regulations (CFR), Title 7, Part 205: the National Organic Program, 2008 ed.
USDA and Agricultural Marketing Service (AMS) (2009a). Naturally Raised Marketing Claim Standards. Available at: http://www.ams.usda.gov/AMSv1.0/ams.fetchTemplateData.do?template=TemplateN&navID=NaturallyRaisedMarketingClaimStandards&rightNav1=NaturallyRaisedMarketingClaimStandards&topNav=&leftNav=GradingCertificationandVerfication&page=NaturallyRaisedMarketingClaims (accessed March 24, 2010).
USDA and AMS (2009b). United States standards for livestock and meat marketing claims, naturally raised claim for livestock and the meat and meat products derived from such livestock. *Fed. Regist*. 74(12):3541–3545.
Van Loo, E., V. Caputo, R. M. Nayga, J. F. Meullenet, P. G. Crandall, and S. C. Ricke. 2010. Effect of organic poultry purchase frequency on consumer attitudes toward organic poultry meat. *J. Food. Sci*. 75:S384–S397.
Van Loo, E. J., V. Caputo, R. M. Nayga, J.-F. Meullenet, and S. C. Ricke. 2011. Consumers' willingness to pay for organic meat: Experimental evidence from chicken breast. *Food. Qual. Prefer*. 22:603–613.
Willer, H. and L. Kilcher (eds). 2011. The world of organic agriculture. Statistics and emerging trends 2011. FiBL-IFOAM Report. IFOAM, Bonn, DE and FiBL, Frick, CH, p. 288. Available at: http://www.organic-world.net/fileadmin/documents/yearbook/2011/world-of-organic-agriculture-2011-page-1-34.pdf (accessed October 01, 2011).
Winter, C. K. and S. F. Davis. 2006. Organic foods. *J. Food. Sci*. 71:R117–R124.
Yiridoe, E. K., S. Bonti-Ankomah, and R. C. Martin. 2005. Comparison of consumer perceptions and preference toward organic versus conventionally produced foods: a review and update of the literature. *Renew. Agric. Food Syst*. 20:193–205.

Section I

Economics, Market, and Regulatory Issues

2 Organic Meat Operations in the United States

Corliss A. O'Bryan, Ellen J. Van Loo, Steven C. Ricke, and Philip G. Crandall

Abstract: Sales of organic meat in the United States have increased more than tenfold in the past 5 years. Consumers concerned about mad cow disease and other health concerns are demanding more organic, natural, and grain-fed alternatives to conventional beef, pork, poultry, and other protein products.

Keywords: organic meat; US meat production; beef; poultry; pork

2.1 INTRODUCTION

The US meat industry was confined to on-farm slaughtering and processing until the development of ammonia refrigeration and the central generation of electricity in the 1800s (Skaggs, 1986). In 1878, businessman Gustavus Swift and engineer Andrew J. Chase developed the insulated refrigerator railroad car to ship dressed beef, fruit, vegetables, and poultry (Skaggs, 1986). In the first years of the twentieth century, five firms were the major providers of meat processing: Swift, Armour, Morris, Wilson, and Cudahy (Arnould, 1971). The development of labor-saving equipment and restructuring of the work process allowed the packers to employ workers with few skills (Corey, 1950). In the 1950s, another transformation occurred as packers migrated to rural areas, where land was cheaper and local communities were willing to provide tax and other incentives to relocating firms (Whitaker, 2006). It was also more economical for growers to move livestock to a local or regional center rather than shipping animals to a big city (Whitaker, 2006).

The poultry industry remained a small farm-type operation until the early 1940s. It began to transform during World War II because of increasing demand with large numbers of small growers available at that time, but 10 or 15 years later some consolidation had begun, and by the late twentieth century, there were only five or six major companies dominating poultry production (Whitaker, 2006). The poultry industry produced more than 8.6 billion broilers in 2009, for cash receipts of $45 billion (Economic Research Service, 2011).

Thus, the use of large-scale irrigation, fertilization, and the wide-scale use of pesticides became common after World War II. The general public in the United States began to perceive organic agriculture as separate from conventional farming after the publication of *Silent Spring* (Carson, 1962) raised public awareness of the ecological problems associated with

Organic Meat Production and Processing, First Edition. Edited by Steven C. Ricke, Ellen J. Van Loo, Michael G. Johnson, and Corliss A. O'Bryan.

agricultural chemicals, particularly synthetic insecticides. Public attention to industrialized farming continued in the 1970s focusing on raising awareness of the importance of buying locally grown food, followed by efforts to persuade government to regulate organically grown food in the 1980s.

The Organic Foods Production Act (OFPA) was passed by the US Congress in 1990 to develop a national standard for organic food and fiber production. This act mandated that United States Department of Agriculture (USDA) develop and write regulations to explain the law to producers, handlers, and certifiers. OFPA also called for an advisory National Organic Standards Board to make recommendations regarding the substances that could be used in organic production and handling, and to help USDA write the regulations. Final rules were written and implemented in the fall of 2002.

Currently, the term "certified organic" refers to the way livestock and agricultural products are raised and processed, avoiding synthetic pesticides and fertilizers. Organic production focuses on animal health and welfare, good environmental practices, and product quality. In contrast, conventional production focuses on reducing costs and maximizing production through weight gain and feed efficiency, among other factors (Sundrum, 2006). On the basis of these requirements, traditional farming practices that were used prior to the twentieth century are generally regarded as "organic."

2.2 THE MARKET FOR ORGANIC MEAT IN THE UNITED STATES

The US organic meat industry is relatively young compared with the conventional meat industry, although it is fast growing. In 2011, the Organic Trade Association (OTA) conducted their annual Organic Industry Survey and determined that the organic industry in the United States had grown from \$3.6 billion in 1997 to \$29 billion in 2010, despite the worst economic downturn in 80 years (Organic Trade Association, 2011a).

Consumers have expressed concerns over antibiotics, growth hormones, and other drugs in meats, as well as animal feeds containing genetically engineered grains and animal parts. In addition, consumers demand that animals are treated humanely and be provided plenty of wide open space, not confined in cages or feedlots. However, in addition to being a lifestyle choice for a small group of consumers, organic products have gained wider acceptance and now 75% of adults purchase natural and/or organic foods (Whole Foods Market, 2010). In addition, consumers are purchasing an increasing number of organic products with 27% of adults claiming that one-fourth of their groceries are natural/organic (Whole Foods Market, 2010).

2.3 PRODUCTION AND SUPPLY OF ORGANIC MEAT IN THE UNITED STATES

One of the fastest growing sectors of agriculture in the United States continues to be organic farming. This has been true for over a decade; when the OFPA was passed by Congress in 1990, there were less than 1 million acres of certified organic cropland, and when the national organic standards were implemented in 2002, this had doubled to more than 2 million acres. In 2008, this figure was nearly 5 million acres of certified organic cropland (Economic Research

Table 2.1 Total US-certified organic livestock in 2008. Data extracted from the 2008 USDA/NASS organic survey.

Organic livestock	Number
Beef cows	63,680
Milk cows	249,766
Other cows	144,817
Hogs and pigs	10,111
Sheep and lambs	7,455
Total cows, pigs, and sheep	475,829
Layer hens	5,538,011
Broilers	9,015,984
Turkeys	398,531
Other	565,549
Total poultry	15,518,075
Other animals	6,860
Total	**16,000,764**

Source: Economic Research Service, 2010.

Service, 2010). Total certified organic livestock has risen from less than 12,000 heads in 1992 to more than 475,000 in 2008 (Table 2.1). Poultry has also seen an increase from just over 3,000,000 to nearly 16,000,000 in the same time period (Economic Research Service, 2010) (Table 2.1). The majority of states had certified organic rangeland and pasture in 2008, with 13 states having more than 100,000 acres, reflecting the strong growth in the organic dairy sector between 2005 and 2008 (Economic Research Service, 2010). The OTA estimated the total sales in organic meat at approximately $470 million in 2010, which is a steep increase compared to the $23 million in organic meat sales in 2000 (Organic Trade Association, 2011b). An overview of the different subcategories is shown in Table 2.2.

2.3.1 Beef

A traditional beef rancher has no control over their respective markets, and one way for them to avoid this is by niche marketing. Organic beef represents such a niche market that has been growing and offers a potential of higher profits as well as being a more sustainable agriculture (Appropriate Technology Transfer for Rural Areas, 2003). Farmers can charge a higher price for organic beef than conventional beef because of the added labor and management that goes into producing and marketing it (Economic Research Service, 2003). One of the primary

Table 2.2 Organic meat sales by subcategory ($ million).

	Sales in 2010 ($ million)	
Poultry	294	62.55%
Beef	102	21.70%
Sausages/Deli meats	53	11.28%
Pork	14	2.98%
Lamb	7	1.49%
Total	**470**	

Source: Organic Trade Association (2011b).

Table 2.3 Organic beef cattle and sales in the United States, December 2008, top ten states.

State	Number of cattle	Sales ($)
California	5,081	432,734
Wyoming	4,615	189,865
Wisconsin	2,943	475,618
Idaho	2,447	249,324
South Dakota	2,391	103,280
Nebraska	2,135	N
North Dakota	1,594	N
New Mexico	1,590	264,840
New York	1,556	368,435
Iowa	1,533	208,865

Source: Data extracted from the 2008 USDA/NASS organic survey (USDA/NASS, 2010).
N, not revealed in order not to identify a particular farm.

limitations to further growth of US organic beef is the low number of certified organic beef slaughter facilities (Seideman *et al.*, 2010).

"Artisan beef" is a niche market that may or may not be a subset of organic and therefore could directly compete for the organic beef market (Seideman *et al.*, 2010). Artisan beef as defined by the Artisan Beef Institute (2010) as a single breed or crossbreed that is deemed appropriate to the growing region, has a preferably all-grass diet, and is "gently" handled on the ranch and through the point of slaughter. No growth hormones or preventative antibiotics are used and the beef must also be traceable from farm to fork. Artisan-produced beef has been embraced by some producers as a means to market directly as a farm-to-market natural product without going through the certified organic process (Seideman *et al.*, 2010).

The 2008 organic survey conducted by USDA/NASS (National Agricultural Statistics Service) (USDA/NASS, 2010) revealed a total of 43,782 head of organic beef cattle in the United States as of December 2008, with total sales of $6,141,084. Table 2.3 presents the top ten states for numbers of cattle and the sales numbers for those states.

Organic beef sales in 2010 were $100 million according to the 2010 Organic Industry Survey conducted by the OTA (2011b). Beef sales are monitored at the retail level via scanner data by the National Cattlemen's Beef Association. This data indicated that the natural/organic beef share of fresh beef sales has increased from 1.1% in 2003 to 4.2% in the first quarter of 2011. In the first quarter of 2011, prices for all beef products offered in retail supermarkets averaged $3.78 while natural/organic beef products averaged $5.48 per pound, indicating that consumers were willing to pay a premium of $1.70 per pound for natural/organic beef (National Cattlemen's Beef Association, 2011).

2.3.2 Poultry and eggs

Poultry is the organic meat product that is in greatest demand, and has shown an average of 39% growth annually from 2000 through 2005. The number of organic broilers, a significant part of overall organic poultry growth, increased from less than 2,000,000 in 2000 to more than 10,000,000 in 2005. Numbers have fallen slightly to just over 9,000,000 in 2008 (Economic

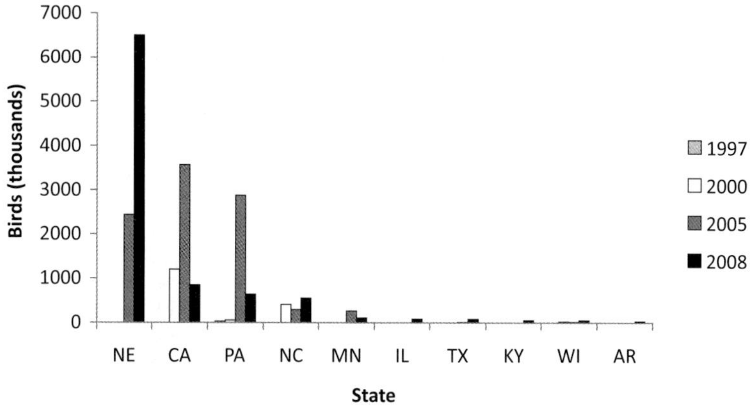

Figure 2.1 Certified organic broilers, top ten states. (Data extracted from the 2008 USDA/NASS organic survey (USDA/NASS, 2010).)

Research Service, 2010). Growth of organic broiler production in the top ten producing states is illustrated in Figure 2.1. The numbers of certified organic broilers produced in Nebraska has increased sharply, while California and Pennsylvania have experienced deep declines.

Numbers of laying hens increased from 1,113,746 in 2000 to almost 5,600,000 in 2008 (Economic Research Service, 2010). In 2008, the 978 certified organic egg farms in the United States produced more than 80.4 million dozen organic eggs valued at $154.8 million. The largest producers of organic eggs in 2008 were located in California (15.8 million dozen), Pennsylvania (14.4 million dozen), and Iowa (4.9 million dozen) (USDA/NASS, 2010). Figure 2.2 illustrates the growth of the organic egg industry in the top ten producing states from 1997 to 2008.

The US sales of organic turkeys topped $8.7 million in 2008, with highest sales reported in California ($2.7 million), Wisconsin (more than $144 thousand), and Washington (more than $35 thousand) (USDA/NASS, 2010). Acreage certified for organic turkey production

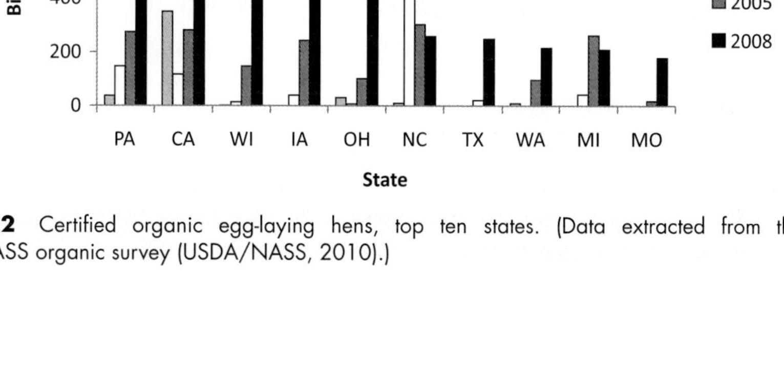

Figure 2.2 Certified organic egg-laying hens, top ten states. (Data extracted from the 2008 USDA/NASS organic survey (USDA/NASS, 2010).)

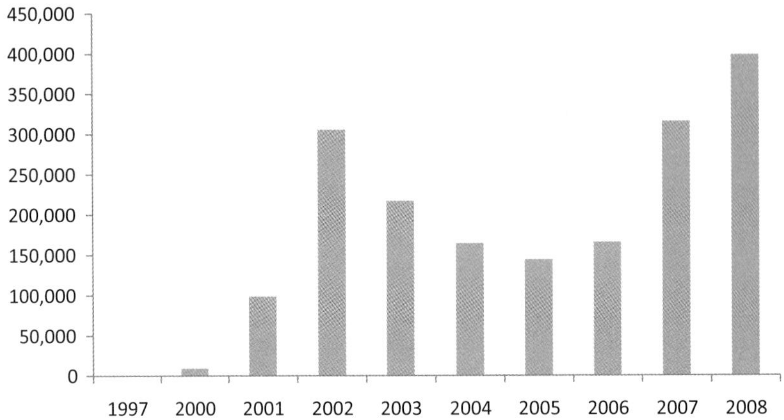

Figure 2.3 Growth of certified organic turkey numbers, 1997–2008. (Data extracted from the 2008 USDA/NASS organic survey (USDA/NASS, 2010).)

was up over 4000% between 2000 and 2008 (USDA/NASS, 2010). Organic turkey acreage was reported as zero in 1997, and although organic turkey still accounts for only 0.15% of overall turkey acreage production in the United States (USDA/NASS, 2010), the growth of this segment has been considerable. Numbers of certified organic turkeys have increased from a mere 750 nationwide in 1997 to 398,531 in 2008. Figure 2.3 illustrates this growth.

It is difficult to predict the extent and the direction that future growth of organic poultry markets will take. However, it has been suggested that a logical pattern would be to follow a market development direction similar to conventional poultry starting with expanding the production base to meet increasing market demands followed by industry integration and developing more value-added organic poultry products (Crandall *et al.*, 2010). There is historical precedent for this as the conventional poultry industry evolved in a similar fashion starting with small farms and subsequently growing into the vertically integrated industry known today (Strausberg, 1995).

2.3.3 Pork

The organic pork market is a relatively new and expanding segment of the industry. In December 2008, there were 8,940 hogs and pigs certified organic on 258 farms in the United States. Iowa had 19 organic hog farms with sales of $1.59 million as compared to Wisconsin with 32 farms and sales of $419,528 (Economic Research Service, 2010). Figure 2.4 illustrates the growth of the organic pork industry since 1997.

2.3.4 Sheep and goats

The demand for organic commodities derived from sheep and goats (meat, milk, and fiber) is increasing. In 2008, the United States had 11,202 organically certified sheep and lambs, up from 4471 in 2005 (Economic Research Service, 2010). Separate records have not been kept for goats, which are included in the "other animals" with buffalo, bison, rabbits, and other specialty meats.

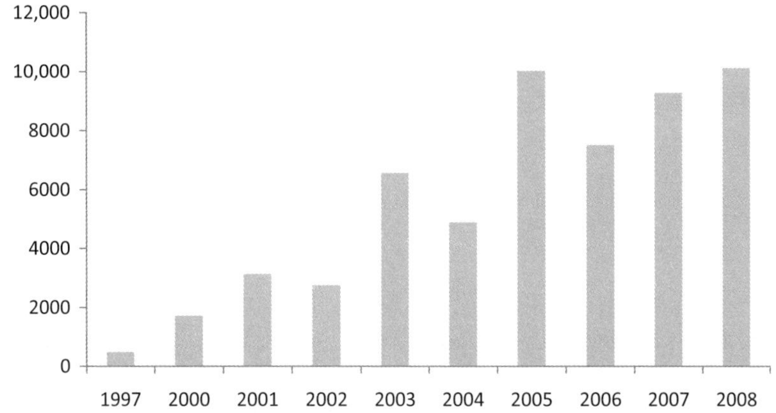

Figure 2.4 Growth of certified organic pork numbers, 1997–2008. (Data extracted from the 2008 USDA/NASS organic survey (USDA/NASS, 2010).)

2.3.5 Dairy

Organic dairy products must be made from the milk of animals raised under organic management, meaning that the cows are kept separately from conventional dairy cows and are not given growth hormones or antibiotics. Dairy products marketed as organic must be made from milk from animals raised organically for at least 1 year prior to producing the milk. The process of bottling milk, making and packing cheese, ice cream, and yogurt also must be certified as organic.

Annual growth rates in the organic dairy sector have ranged from 16% to 34% between 1997 and 2007, making it one of the fastest-growing segments of the organic industry. Figure 2.5 shows the increase in the number of certified dairy cows between 2000 and 2008. However, there have been periodic shortages in milk supplies, which have been detrimental to further growth in this area (Oliver, 2008; Weinraub & Nicholls, 2005).

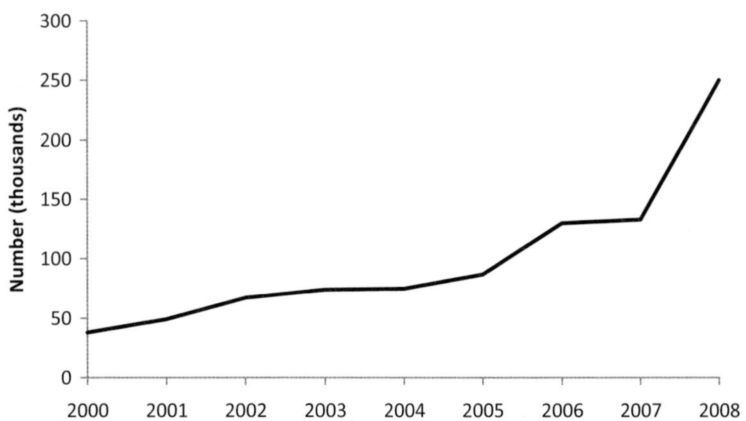

Figure 2.5 Number of organic milk cows on US farms; between 2000 and 2008. (Data extracted from the 2008 USDA/NASS organic survey (USDA/NASS, 2010).)

Table 2.4 Dairy cattle and sales in the United States, December 2008, top ten states.

	Number of cows	Sales ($)
California	35,333	3,984,136
Wisconsin	25,916	3,182,616
Texas	18,854	8,748,974
New York	17,431	N
Oregon	16,290	2,408,238
Vermont	12,653	1,335,064
Pennsylvania	11,078	1,419,022
Washington	8,589	2,111,850
Minnesota	7,826	842,088
Idaho	7,370	1,412,284

Source: Data extracted from the 2008 USDA/NASS organic survey (USDA/NASS, 2010).
N, not revealed in order not to identify a particular farm.

Factors that influence a dairy producer to embrace organic production methods include size and location of the operation and milk production costs. Small producers are more likely than larger producers to consider organic production to increase income, and being located in the Northeast or Upper Midwest also increases the likelihood for a dairy to turn to organic production (McBride & Greene, 2007). Even though organic dairy production costs are higher than for conventional dairies, organic dairy producers get an average premium of almost $7.00 per hundredweight of milk (McBride & Greene, 2007). Table 2.4 details the sales for the top ten organic dairy states for 2008.

2.4 FUTURE OF THE US ORGANIC MEAT INDUSTRY

The US market for organics continues to grow in all food categories, but the markets that show the most potential are dairy and meat. The demand for all organic food products remains strong, including processed foods and prepared organic meals. The single biggest barrier to market growth of organic meats is the lack of certified organic feeds (see Chapter 11). There are shortages of organic dairy farms and organic feed producers. Converting from conventional to organic production is expensive and time consuming. In addition, depending on how organic animals are raised, some of the more open free-range operations represent new biosecurity and food safety concerns that will require development of new management approaches to retain safety of these products (Ricke, 2010). Also, the rapid growth of the ethanol industry, which does not require organic production practices, may also slow the growth of organic production.

REFERENCES

Appropriate Technology Transfer for Rural Areas. 2003. Available at: http://attra.ncat.org/organic.html.
Arnould, R. J. 1971. Changing patterns of concentration in American meat packing, 1880–1963. *Bus. Hist. Rev.*, Spring 1971, pp. 20–22.
Artisan Beef Institute. 2010. Discover the world of beef. Available at: http://artisanbeefinstitute.com/about/.

Carson, R. 1962. *Silent Spring*. Houghton Mifflin Company, Boston, MA. p. 400.
Corey, L. 1950. *Meat and Man: A Study of Monopoly, Unionism, and Food Policy*. The Viking Press, New York. p. 37.
Crandall, P. G., C. A. O'Bryan, S. C. Ricke, F. T. Jones, S. C. Seideman, R. Rainey, E. A. Bihn, T. Maurer, and A. C. Fanatico. 2010. Food safety of natural and organic poultry. In: S. C. Ricke and F. T. Jones (eds) *Perspectives on Food-Safety Issues of Animal-Derived Foods*. University of Arkansas Press, Fayetteville, AR. pp. 289–305.
Economic Research Services. 2003. Organic farming and marketing. Available at: http://www.ers.usda.gov/.
Economic Research Service. 2010. Organic production. Available at: http://www.ers.usda.gov/Data/Organic/.
Economic Research Service. 2011. U.S. broiler industry: background statistics and information. Available at: http://www.ers.usda.gov/News/broilercoverage.htm.
McBride, W. and C. Greene. 2007. A comparison of conventional and organic milk production systems in the U.S. *Selected Paper at the American Agricultural Economics Association Annual Meeting*, Portland, OR, July 29th–August 1st. Available at: http://ageconsearch.umn.edu/bitstream/9680/1/sp07mc01.pdf.
National Cattlemen's Beef Association. 2011. Fresh data shows trend in natural/organic beef category. Available at: http://www.beefretail.org/NaturalOrganicCategory.aspx.
Oliver, H. 2008. Organic dairy demand exceeds supply. *NFM* 27:1, 4.
Organic Trade Association. 2011a. U.S. organic industry valued at nearly $29 billion in 2010. Available at: http://www.organicnewsroom.com/2011/04/us_organic_industry_valued_at.html.
Organic Trade Association. 2011b. *Organic Trade Association's 2011 Organic Industry Survey.*
Ricke, S. C. 2010. Future prospects for advancing food – safety research in food animals. In: S. C. Ricke and F. T. Jones (eds) *Perspectives on Food-Safety Issues of Animal-Derived Foods*. University of Arkansas Press, Fayetteville, AR. pp. 335–350.
Seideman, S. C., T. R. Callaway, P. G. Crandall, S. C. Ricke, and D J. Nisbet. 2010. Alternative and organic beef production: food-safety issues. In: S. C. Ricke and F. T. Jones (eds) *Perspectives on Food-Safety Issues of Animal-Derived Foods*. University of Arkansas Press, Fayetteville, AR. pp. 307–321.
Skaggs, J. 1986. *Prime Cut: Livestock Raising and Meatpacking in the United States, 1607–1983*. TAMU Press, College Station, TX, p. 280.
Sundrum, A. 2006. Protein supply in organic poultry and pig production. *Proc. 1st IFOAM Int. Conf. Anim. Org. Prod.*, St. Paul, MN, August 23rd–25th August. pp. 195–199.
Strausberg, S. F. 1995. *From Hills and Hollers: Rise of the Poultry Industry in Arkansas*. Arkansas Agricultural Experimental Station, Fayetteville, AR. p. 221.
USDA/NASS. 2010. 2008 organic survey. Available at: http://www.agcensus.usda.gov/Publications/2007/Online_Highlights/Organics/.
Weinraub, J., W. Nicholls. 2005. Organic milk supply falls short, Washington Post. June 1. p. F01.
Whitaker, W. G. 2006. Labor practices in the meat packing and poultry processing industry: an overview. Available at: http://www.ansarilawfirm.com/docs/Labor-Practices-in-the-Meat-Packing-and-Poultry-Processing-Industry.pdf.
Whole Foods Market. 2010. National survey shows organic foods now represent a larger part of total food purchases. Available at: http://wholefoodsmarket.com/pressroom/blog/2010/08/16/national-survey-shows-organic-foods-now-represent-larger-part-of-total-food-purchases/.

3 Regulatory Issues in Domestically Raised and Imported Organic Meats in the United States

Harrison M. Pittman, Kerri C. Boling, and Shannon J. Mirus

Abstract: In 1990, Congress enacted the Organic Foods Production Act (OFPA), setting the legal foundation for the National Organic Program (NOP). NOP governs virtually every aspect of the production, handling, distribution, and retail sale of organic products sold in the United States, in addition to standards applied to imported organic products. Following enactment of OFPA, the United States Department of Agriculture (USDA), Agricultural Marketing Service promulgated the regulations implementing OFPA that serve as the NOP governing rules. This chapter provides an overview of the NOP, with a focus on those aspects of the rule pertinent to organic meat production and production issues.

Keywords: organic; organic meat; National Organic Program (NOP); Organic Foods Production Act (OFPA); organic program

3.1 INTRODUCTION

Congress enacted the Organic Foods Production Act (OFPA) of 1990 to (1) create "national standards governing the marketing of certain agricultural products as organically produced products," (2) assure consumers that "organically produced products meet a consistent standard," and (3) facilitate "interstate commerce in fresh and processed food that is organically produced."[1] OFPA requires the United States Department of Agriculture (USDA) Secretary to establish national standards for organic production and handling consistent with OFPA. On December 21, 2000, the USDA published a final rule that created these national standards. The combination of OFPA and the final rule created the National Organic Program (NOP). As of October 21, 2002, for a "producer" or "handler" to sell, label, or represent agricultural products as "organic," that producer or handler must comply with all applicable requirements set forth in the OFPA and the final rule.

The NOP operates through a system in which USDA-accredited agents certify that producers and handlers comply with all applicable NOP requirements. It sets forth standards with which entities and individuals must comply to be accredited as certifying agents, as well as requirements with which producers and handlers of production and handling operations

[1] 7 U.S.C. §§ 6501-6522.

Organic Meat Production and Processing, First Edition. Edited by Steven C. Ricke, Ellen J. Van Loo, Michael G. Johnson, and Corliss A. O'Bryan.

must comply to be certified. The NOP also establishes organic standards that govern crop and livestock production and the handling of these products. These standards require, among other things, that producers and handlers submit an organic system plan that describes all aspects of production and handling.

NOP publishes a national list of substances that are allowed or prohibited for use in organic production and handling. NOP also established requirements that govern the labels, labeling, and market information for organic products. NOP permits states to create their own organic programs, subject to conditions and requirements set forth in NOP. The NOP enforcement mechanisms ensure that certifying agents, producers, and handlers comply with NOP requirements. The NOP provisions also govern the importation and exportation of organic products, and establish mediation and adverse action appeals processes.

It is useful to understand how and why NOP was created, in order to place the NOP livestock provisions in perspective. The practice of producing organically grown agricultural products has existed for several decades in the United States, evolving from a small-scale and localized system into a highly organized and global production and marketing system (Dimitri & Greene, 2002; Greene & Kremen, 2003). These changes spurred the organic industry to establish uniform standards for organic production and marketing.

As a result, private organizations and some states developed third-party certification systems. These systems required a third-party to determine that the producer or processor had complied with its particular requirements for organic production, in order to be certified. This allowed the producer or processor to represent to consumers that its product had been produced and processed in accordance with that third-party's standards for organic production. Although this was a positive development for the organic industry, the goal of creating uniform standards for organic production and marketing remained unattained.

This goal remained unaccomplished because the standards for organic production often differed from one third-party certifier to another, as well from one state to another. Many third parties refused to recognize products that were certified by other certifying organizations because of differing standards, thereby impeding the flow of organic products into the marketplace and across state lines. This lack of uniformity in organic production standards was a significant problem. For example, organic livestock producers that needed organic feed for their operations in order to be certified and processors that needed to purchase organic ingredients for their processing operations were often unable to purchase the feed or ingredients that they needed because the feed ingredients were certified under standards not recognized by the third-party that certified the producer or processor. Another problem was the lack of effective enforcement mechanisms to combat fraud and other violations.

Recognizing these problems, Congress enacted OFPA and authorized the USDA Secretary to promulgate regulations. The result was NOP, a comprehensive statutory and regulatory framework governing all stages of organic production and handling. For a producer or handler to sell, label, or represent agricultural products as "organic," the producer or handler must comply with all applicable NOP requirements.

3.2 THE NATIONAL ORGANIC PROGRAM

3.2.1 Definitions—7 C.F.R. §§ 205.1-205.2

The definitions provided in this section are basic to understanding NOP and are used frequently throughout its provisions, specifically including the livestock provisions.

A "producer" is any person engaged in the business of growing or producing food or feed, but does not include the selling, transporting, or delivering of crops or livestock by a producer of the crops or livestock to a handler.[2] The term "handle" is defined as the selling, processing, or packaging of agricultural products.[3] A "handler" is any person engaged in the business of selling, processing, or packaging agricultural products, but does not include final retailers that do not process agricultural products.[4] A "handling operation" is "[a]ny operation or portion of an operation (except final retailers of agricultural products) that receives or otherwise acquires agricultural products and processes, packages, or stores such products."[5]

The OFPA defines "livestock" as "any cattle, sheep, goats, swine, poultry, equine animals used for food or in the production of food, fish used for food, wild or domesticated game, or other non-plant life."[6] The final rule defines "livestock" as

> [a]ny cattle, sheep, goat, swine, poultry, or equine animals used for food or in the production of food, fiber, feed, or other agricultural-based consumer products; wild or domesticated game; or other non-plant life, except such term shall not include aquatic animals for the production of food, fiber, feed, or other agricultural-based consumer products.[7]

An "agricultural product" is any agricultural commodity or product in raw or processed form and includes any commodity or product derived from livestock that is marketed for either human or livestock consumption.[8] A "person" can be an individual, group of individuals, or any type of business entity.[9] "Administrator" means the "Administrator for the Agricultural Marketing Service, United States Department of Agriculture, or the representative to whom authority has been delegated to act in the stead of the Administrator."[10]

3.2.2 Applicability—7 C.F.R. §§ 205.100-205.105

The NOP establishes certification requirements and specifies operations exempt or excluded from the certification requirements. The NOP also established recordkeeping requirements for certified and exempted operations. Although exempted and excluded operations are not required to comply with certification requirements, they must still comply with other NOP requirements.

3.2.2.1 *Certification, exemptions, and exclusions—7 C.F.R. §§ 205.100-205.101*

Certification

A production or handling operation or part thereof "that produces or handles crops, livestock, livestock products, or other agricultural products that are intended to be sold, labeled,

[2] *See* 7 U.S.C. § 6502(18). *See also* 7 C.F.R. § 205.2.
[3] *See* 7 U.S.C. § 6502(8). *See also* 7 C.F.R. § 205.2.
[4] *See* 7 U.S.C. § 6502(9). *See also* 7 C.F.R. § 205.2.
[5] *See* 7 C.F.R. at 205.2. There is no definition of "production operation" contained in the statute or the regulations.
[6] 7 U.S.C. § 6502(11).
[7] 7 C.F.R. § 205.2.
[8] *See* 7 U.S.C. § 6502(1). *See also* 7 C.F.R. § 205.2.
[9] *See* 7 U.S.C. § 6502(15). *See also* 7 C.F.R. § 205.2.
[10] 7 C.F.R. § 205.2.

or represented" as "organic" must comply with the certification requirements and all other applicable NOP requirements, unless exempted or excluded.[11] "Certification" is a determination made by an accredited certifying agent that an operation has complied with all applicable requirements in the OFPA and the final rule.[12]

Exemptions

Four types of operations are exempted from certification requirements. The first is a production or handling operation that has a gross annual income from sales of organic products totaling $5000.00 or less.[13] This exemption is intended primarily for the benefit of producers who market their product(s) directly to consumers.[14] This would include, for example, producers who sell their product(s) at farmers' markets or directly to "retail food establishments" for resale to consumers.[15] A "retail food establishment" is "[a] restaurant; delicatessen; bakery; grocery store; or any retail outlet with an in-store restaurant, delicatessen, bakery, salad bar, or other eat-in or carry-out service of processed or prepared raw and ready-to-eat-food."[16]

An operation exemption under the "$5000.00 or less" standard must comply, however, with the applicable organic production and handling requirements, except for the requirement that it submit an organic system plan.[17] The operation must also comply with the labeling requirements for exempted and excluded operations.[18] In addition, products derived from an operation exempted under the "$5000.00 or less" standard cannot be used as ingredients in processed products identified as "organic" produced by another handling operation.[19]

Any retail food establishment or portion thereof that handles organic products but does not process them is also exempted from the certification requirements.[20] There are no other requirements in either the OFPA or the final rule with which a retail food establish falling under this exemption must comply.

A handling operation that only handles agricultural products containing less than 70% organic ingredients by total weight of the finished product is also exempted from the requirements for certification.[21] This type of operation, however, must comply with (1) the standards for avoiding the commingling and contact of organic products with prohibited substances[22]; (2) the labeling standards applicable to exempt or excluded operations; (3) the labeling standards for multi-ingredient packaged products with less than 70% organic ingredients; and (4) the recordkeeping requirements for exempted operations.[22]

[11] 7 C.F.R. § 205.100(a).

[12] *See id.* at § 205.2 (defining "certification" or "certified"). The certification requirements are addressed in Part E of this article.

[13] *See id.* at § 205.101(a)(1).

[14] 65 Fed. Reg. 80,547, 80,552 (prefatory comments to final rule with request for comments).

[15] *See id.*

[16] 7 C.F.R. at § 205.2.

[17] *See id.* at § 205.101(a)(1).

[18] *See id.*

[19] *See id.*

[20] *See id.* at § 205.101(a)(2).

[21] *See id.* at § 205.101(a)(3). The total weight of the finished product excludes the weight of water and salt in the product. The standards for calculating the organic composition of a product are discussed in Part D of this article and are found at 7 C.F.R. § 205.302.

[22] *See id.* at § 205.101(a)(3)(i)-(iii).

The final type of exempted operation is a handling operation that identifies ingredients as "organic" only on the information panel.[23] An "information panel" is:

> [t]hat part of the label of a packaged product that is immediately contiguous to and to the right of the principle display panel, unless another section of the label is designated as the information panel because of package size or other package attributes (e.g., irregular shape with one useable surface).[24]

A "principal display panel" is the "part of a label that is most likely to be displayed, presented, shown, or examined under customary conditions of display for sale."[25]

This type of exempted operation must comply with (1) the provisions for the prevention of contact of organic products with prohibited substances; (2) the labeling requirements for multi-ingredient products containing less than 70% organically produced ingredients; (3) the labeling requirements for exempted and excluded operations; and (4) the recordkeeping requirements for exempted operations.[26]

Exclusions

There are two types of operations that are excluded from the requirements for certification.[27] The first are handling operations that sell only "organic" products that are packaged or otherwise enclosed in a container before the operation receives or acquires the product and "remain in the same package or container and are not otherwise processed while in the control of the handling operation."[28] The prefatory comments to the final rule explain that this exclusion is designed to avoid creating an unnecessary barrier for handlers who distribute nonorganic products and also want to offer a selection of organic products.[29] These operations must comply with the standards for preventing the commingling and contact of organic products with prohibited substances.

The second type of excluded operations is retail food establishments or portions thereof that process on their premises "raw and ready-to-eat food from products that were previously labeled" as "organic."[30] These operations must comply with the labeling requirements specifically applicable to agricultural products produced on exempted and excluded operations and the standards for preventing the commingling and contact of organic agricultural products with prohibited substances.[31]

With respect to exempted and excluded retail food establishments, the prefatory comments to the final rule state the following:

> [T]here is clearly a great deal of public concern regarding the handling of organic products by retail food establishments. We have not required certification of retail food establishments at this time because of a lack of consensus as to whether retail food establishments should be certified, a

[23] *See id.* at § 205.101(a)(4).
[24] *Id.* at § 205.2.
[25] *Id.*
[26] *See id.* at § 205.101(a)(4)(i)-(iii).
[27] *See id.* at § 205.101(b)(1)-(2).
[28] *Id.* at § 205.101(b)(1).
[29] 65 Fed. Reg. 80,547, 80,552 (prefatory comments to final rule with request for comments).
[30] 7 C.F.R. § 205.101(2).
[31] *See id.* at § 205.101(b)(i)-(ii).

lack of consensus on retailer certification standards, and a concern about the capacity of existing certifying agents to certify the sheer volume of such businesses. Retail food establishments, not exempt under the Act, could at some future date be subject to regulation under the NOP. Any such regulation would be preceded by rulemaking with an opportunity for public comment.[32]

3.2.2.2 *Recordkeeping requirements for certified, exempt, and excluded operations—7 C.F.R. § 103*

Certified operations

A certified operation must maintain records that relate to the "production or handling of agricultural products sold or labeled as organically produced" for 5 years beyond the date the records are created.[33] The records must include a detailed history of the substances applied to the operation's fields or agricultural products, the names and addresses of the person who applied the substances, the dates and rates the substances were applied, and the method used to apply the substances.[34] The records must "[b]e adapted to the particular business that the certified operation is conducting" and must provide full disclosure of all "activities and transactions" of the operation "in sufficient detail as to be readily understood and audited."[35] They must also be sufficient to demonstrate that the operation has complied with all applicable NOP requirements.[36] The records must be available during normal business hours to representatives of the Secretary, the State official governing a State's organic program, if applicable, and the certifying agent for inspection and copying.[37]

Exempted operations

An exempted operation must maintain records sufficient to demonstrate that ingredients identified as "organic" were organically produced from those ingredients.[38] The records must be kept for at least 3 years after they are created.[39] Representatives of the USDA Secretary of Agriculture and the state official governing a state organic program, if applicable, must be allowed access to the records for inspection and copying during normal business hours.[40]

Excluded operations

Recordkeeping requirements for excluded operations are confusing and unsettled. There are no recordkeeping requirements in the OFPA or the final rule specifically applicable to excluded operations. In describing the qualification for excluded operations, the final rule expressly states that the first type of excluded operation—a handling operation that sells "organic" products that are received and maintained in an enclosed package or container and are not processed by the operation—is "excluded from the requirements of this part, except for the requirements for the prevention of commingling and contact with prohibited substances as set forth in § 205.272. . . ."[41] The final rule states that the second type of excluded

[32] 65 Fed. Reg. 80,547, 80,553 (prefatory comments to final rule with request for comments).
[33] 7 U.S.C. § 6511(d).
[34] *See id.* at § 6511(d)(1)-(2).
[35] 7 C.F.R. § 205.103(a)(1)-(2).
[36] *See id.* at § 205.103(a)(3)-(4).
[37] *See id.* at § 205.103(c).
[38] *See id.* at § 205.101(c)(1)(i)-(ii).
[39] *See id.* at § 205.101(c)(2).
[40] *See id.*
[41] *Id.* at § 205.101(b)(1).

operation—a retail food establishment or portion thereof that processes raw and ready-to-eat food from products certified as "organic"—is "excluded from the requirements in this part, except . . . the requirements for the prevention of contact with prohibited substances as set forth in § 205.272 . . . and . . . the labeling provisions of § 205.310."[42]

In neither of these provisions does the final rule reference recordkeeping requirements for either type of excluded operation, leading one to conclude perhaps that no such recordkeeping requirements exist. The Agricultural Marketing Service (AMS), however, signals in its prefatory comments to the final rule that excluded operations are subject to recordkeeping requirements. Specifically, the comments state, in pertinent part, that "[e]ach exempt, excluded, and certified operation should maintain the records which demonstrate compliance with the Act and the regulations applicable to it and . . . establish . . . that the exempt, excluded, or certified operation is and has been in compliance with the Act and the regulations."[43]

In response to comments made concerning the recordkeeping requirements for excluded operations, the Agricultural Marketing Service also stated in its prefatory comments to the final rule the following:

> [S]everal commenters argued that excluded operations should be required to comply *with the same recordkeeping requirements as exempt operations*. Some commenters expressed concern over the inability to verify compliance for either exempt or excluded operations and asked that exempt or excluded operations be subject to additional recordkeeping requirements. *We disagree with these commenters* and have retained the provisions from the proposed rule on recordkeeping for excluded operations. Given the nature of these excluded operations . . . we believe that extensive recordkeeping requirements would be an unwarranted regulatory burden.[44]

The AMS's acknowledgment that it rejected the argument that "excluded operations should be required to comply with the same recordkeeping requirements as exempt operations" indicates that it believes there are separate recordkeeping requirements for exempt and excluded operations. However, the only recordkeeping requirements contained in the final rule apply to certified and exempted operations. There is no reference to excluded operations in either set of recordkeeping requirements.

Moreover, although the AMS states that it "retained the provisions from the proposed rule on recordkeeping for excluded operations," a review of the proposed rule reveals that there were no such requirements contained in the proposed rule. Thus, it is not entirely clear whether excluded operations must comply with any recordkeeping requirements and, if so, what those requirements would specifically be.[45]

3.2.3 Organic production and handling standards—7 C.F.R. §§ 205.200-205.290

For a producer or handler of a production or handling operation to sell, label, or represent products as "organic," the producer or handler must comply with all applicable organic

[42] *Id.* at § 205.101(b)(2).

[43] 65 Fed. Reg. 80,547, 80,553 (prefatory comments to the final rule with request for comments).

[44] 65 Fed. Reg. 80,555 (prefatory comments to the final rule with request for comments – emphasis added).

[45] At § 205.307(c), the final rule references "recordkeeping requirements for exempt and excluded operations under § 205.101." This point is not discussed in this part of this article but is addressed in Part D.

production and handling requirements.[46] The organic production and handling standards contain requirements for organic system plans, livestock productions, and organic handling.

3.2.3.1 *Organic system plan—7 C.F.R. §§ 205.200-205.201*

A producer or handler of a production or handling operation seeking certification, unless exempted or excluded from the certification requirements, must submit an "organic system plan" to its certifying agent or, if applicable, to the governing official of its state's organic program.[47] An "organic system plan" is "[a] plan of management of an organic production or handling operation that has been agreed to by the producer or handler and the certifying agent and that includes written plans concerning all aspects of agricultural production or handling" as set forth in OFPA and the final rule's organic production and handling requirements.[48] The prefatory comments to the final rule explain that the organic system plan "is the forum through which the producer or handler and certifying agent collaborate to define, on a site-specific basis, how to achieve and document compliance with the requirements of certification."[49]

An organic system plan must describe the practices and procedures that the producer or handler will implement and maintain in its operation.[50] It must also explain how often the practices and procedures will be performed.[51] The requirement that the organic system plan explain how often the practices and procedures will be performed requires, according to the prefatory comments, a plan to include an implementation schedule that describes the timing and sequence of activities such as crop rotation, timing and location of soil tests, or the addition of feed supplements to livestock feed.[52]

The prefatory comments describe "practices" as tangible techniques, "such as the method of applying manure, the mechanical and biological methods used to prepare and combine ingredients and package finished products, and the measures taken to exclude pests from a facility."[53] The comments describe "procedures" as the decision-making process for locating commercially available, organically produced seed.[54]

The organic system plan must also describe the monitoring practices and procedures that will be used in the operation to verify that the plan is implemented effectively, including a description of how often the monitoring practices and procedure will be performed.[55] Elaborating on this requirement, the prefatory comments explain that procedures and handlers are required to identify "measurable indicators" that will be used to monitor the degree to which the production objectives for the operation are being met.[56] For example, if an

[46] *See* 7 C.F.R. § 205.200. *See also* 7 C.F.R. § 205.200 (requiring that production practices implemented pursuant to the organic production and handling requirements "maintain or improve the natural resources of the operation, including soil and water quality").

[47] 7 U.S.C. § 6513. *See also* 7 C.F.R. § 205.201.

[48] 7 U.S.C. § 6502(13). *See also* 7 C.F.R. § 205.2.

[49] 65 Fed. Reg. 80,547, 80,558 (prefatory comments to final rule with request for comments).

[50] 7 C.F.R. § 205.201(a)(1).

[51] *See id.*

[52] 65 Fed. Reg. 80,547, 80,558 (prefatory comments to final rule with request for comments).

[53] *Id.*

[54] *See id.*

[55] *See id.* at § 205.201(a)(3).

[56] 65 Fed. Reg. 80,547, 80,559 (prefatory comments to final rule with request for comments).

organic system plan calls for improvements in soil organic system plan calls for improvements in soil organic matter content in a particular field, it would include provisions for analyzing soil organic matter levels at periodic intervals.[57] Also, if herd health improvement is an objective stated in the plan, factors such as somatic cell count or observation about changes in reproductive patterns might be used as indicators.[58] The indicators used to monitor a particular organic system plan are determined by the producer or handler through consultation with the certifying agent.[59]

The organic system plan must include a description of the recordkeeping system that a producer or handler will use in its operation to ensure compliance with the recordkeeping requirements for certified operations.[60] The prefatory comments state that this description must be sufficient to verify and document an audit trail for the operation and that, with respect to "each crop or wild-crop harvested, the audit trail must trace the product from the field, farm parcel, or area where it is harvested through the transfer of legal title."[61] With respect to livestock operations, the description must be sufficient to trace each animal from the time it enters the operation until the animal is removed from the operation.[62] For handling operations, the operation must trace each product sold, labeled, or represented as "organic" "from the receipt of its constituent ingredients to the sale of the processed product."[63]

The organic system plan must also include a description of the management practices that a producer or handler will implement "to prevent commingling or organic and nonorganic products on a split operation and to prevent contact of organic production and handling operations and products with prohibited substances."[64] The plan must list the substances that will be used in the operation, including a description of the substances' "composition, source, location(s) where it will be used, and documentation of commercial availability, as applicable."[65] Finally, the plan must include any information required by the certifying agent to verify that a producer or handler of a production or handling operation has complied with NOP requirements.[66]

A producer is permitted to submit a substitute organic system plan that is designed to comply with the requirements of another federal, state, or local government's regulatory program.[67] Any substituted plan must comply with all applicable organic production and handling requirements.[68]

[57] *See id.*
[58] *See id.*
[59] *See id.*
[60] 7 C.F.R. § 205.201(a)(4).
[61] 65 Fed. Reg. 80,547, 80,559 (prefatory comments to final rule with request for comments).
[62] *See id.*
[63] *Id.*
[64] 7 C.F.R. § 205.201(a)(5).
[65] *Id.* at § 205.201(a)(2).
[66] *See id.* at § 205.201(a)(6). *See also* 7 C.F.R. § 205.201(b) (stating that "[a] producer may substitute a plan prepared to meet the requirements of another Federal, state, or local government regulatory program for the organic system plan," as long as "the submitted plan meets all the requirements of this subpart.").
[67] 7 C.F.R. § 205.201(b). The final rule does not state that handlers are permitted to substitute an organic system plan designed to comply with the requirements of another federal, state, or local government's regulatory program. Note further that the final rule is unclear as to what meaning, if any, is attached to the phrase "of another federal . . . regulatory program."
[68] *See id.*

3.2.3.2 *Livestock production requirements—7 C.F.R. §§ 205.236-205.240*

The organic production and handling requirements for livestock production establish requirements governing the (1) origins of livestock; (2) composition of livestock feed; (3) health care practices for livestock; and (4) living conditions for livestock.

Origin of livestock—7 C.F.R. § 205.236
As a general rule, livestock products intended to be sold, labeled, or represented as "organic" must be produced from livestock raised "under continuous organic management from the last third of gestation of hatching."[69] There are three exceptions to this general rule.

First, poultry or edible poultry products must be raised under continuous organic management from no later than the second day of life.[70] Second, milk or milk products must be produced from dairy animals "that have been under continuous organic management beginning no later than 1 year prior to the production of the milk or milk products. . . ."[71] However, in the event that a producer converts an "entire, distinct herd" of dairy animals from a nonorganic production system into an organic one, the producer may provide for the first 9 months of the first year in which the herd is converted "a minimum of 80-percent feed that is either organic or raised from land included in the organic system plan and managed in compliance with [the] organic crop requirements."[72] For the final 3 months of the year, producers may provide feed to the herd in accordance with the livestock feed requirements.[73]

Whenever a herd is converted, however, producers must raise all dairy animals from the last third of gestation in accordance with applicable standards set forth in the OFPA and the final rule.[74] The prefatory comments state that "the conversion provision cannot be used routinely to bring nonorganically raised animals into an organic operation.[75] It is a one-time opportunity for producers working with a certifying agent to implement a conversion strategy for an established, discrete dairy herd in conjunction with the land resources that sustain it."[76]

The third exception applies to "breeder livestock."[77] "Breeder livestock" are "[f]emale livestock whose offspring may be incorporated into an organic operation at the time of their birth."[78] Breeder stock can be transferred from a nonorganic operation to an organic one at any time.[79] If breeding stock transferred into the operation is gestating and any offspring of that animal are to be raised as organic, then the breeding stock must be transferred no later than the last third of gestation.[80]

A livestock producer is prohibited from selling, labeling, or representing livestock or edible livestock products as "organic" if the livestock has been transferred from an organic operation to a nonorganic one.[81] A producer is also prohibited from selling, labeling, or representing any breeder stock or dairy stock as "slaughter stock" if the breeder stock has not

[69] *See* 7 C.F.R. § 205.236(a).
[70] *See id.*
[71] *Id.* at § 205.236(a)(2).
[72] *Id.* at § 205.236(a)(2)(i).
[73] *See id.* at § 205.236(a)(2)(ii).
[74] *See id.* at § 205.236(a)(2)(iii).
[75] 65 Fed. Reg. 80,547, 80,570 (prefatory comments to final rule with request for comments).
[76] *Id.*
[77] 7 C.F.R. § 205.236(a)(3).
[78] *Id.* at § 205.2.
[79] *See id.* at § 205.236(a)(3).
[80] *See id.*
[81] *See id.* at § 205.236(b)(1).

been "under continuous organic management since the last third of gestation."[82] "Slaughter stock" is "[a]ny animal that is intended to be slaughtered for consumption by humans or other animals."[83] Finally, a livestock producer is required to "maintain records sufficient to preserve the identity of all organically managed animals and edible and nonedible animal products produced on the operation."[84]

Livestock feed—7 C.F.R. § 205.237

As a general rule, livestock must be given a feed ration that is composed entirely of organically produced agricultural products and, if applicable, organically handled agricultural products.[85] The term "feed" is defined as "edible materials that are consumed by livestock for their nutritional value."[86] "Feed" includes grains, hay, silage, pasture, and fodder.[87] Synthetic and nonsynthetic substances that are listed as allowed on the National List may be used as feed additives and supplements.[88] A "feed additive" is a substance that is added to feed "in micro quantities to fulfill a specific nutritional need," such as "essential nutrients in the form of amino acids, vitamins, and minerals."[89] A "feed supplement" is a

> combination of feed nutrients added to livestock feed to improve the nutrient balance or performance of the total ration and intended to be . . . [d]iluted with other feeds when fed to livestock; . . . [o]ffered free choice with other parts of the ration if separately available; or . . . [f]urther diluted and mixed to produce complete feed.[90]

A livestock producer may not use animal drugs to promote an animal's growth, provide feed supplements or additives in excess of the amount needed for the animal's nutrition and health maintenance, use plastic feed pellets for roughage, or use feed formulas that contain urea or manure.[91] An "animal drug" is "[a]ny drug as defined in section 201 of the Federal Food, Drug, and Cosmetic Act, as amended . . . , that is intended for use in livestock, including such livestock feed."[92] Producers are also prohibited from feeding "mammalian or poultry slaughter by-products to mammals or poultry" and from using "feed, feed additives, and feed supplements in violation of the Federal Food, Drug, and Cosmetic Act."[93] Additionally, producers cannot provide "feed or forage to which any antibiotic including ionophores has been added" and may not "prevent, withhold, restrain, or otherwise restrict ruminant animals from actively obtaining feed grazed from pasture during the grazing season."[94]

New grazing provisions were adopted in February, 2010, to go into effect June 17, 2010. These new regulations require the producer to provide "not more than an average of

[82] *See id.* at § 205.236(b)(2).
[83] *Id.* at § 205.2.
[84] *Id.* at 205.236(c).
[85] *See id.* at § 205.237(a).
[86] *Id.* at § 205.2.
[87] *See id. See also* 7 C.F.R. § 205.237(a).
[88] 7 C.F.R. § 205.237(a).
[89] *See id.* at § 205.2.
[90] 7 C.F.R. § 205.2.
[91] *See id.* at § 205.237(b)(1)-(4).
[92] *Id.* at § 205.2. *See also* 21 U.S.C. §§ 301-395 (Federal Food, Drug, and Cosmetic Act) & 21 U.S.C. § 321(g)(1)-(2) (defining "drug" and "counterfeit drug").
[93] 7 C.F.R. § 205.237(b)(5)-(6).
[94] *Id.* at § 205.237(b)(7)-(8).

70 percent of a ruminant's dry matter demand from dry matter fed."[95] This is calculated as an average over the entire grazing season for each type and class of animal.[96] Producers must graze ruminant animals throughout the entire grazing season for the geographic region, which much be at least 120 days per calendar year.[97] Additionally, producers must provide a pasture of "sufficient quality and quantity to graze throughout the grazing season" and must provide "all ruminants under the organic system plan with an average of not less than 30 percent of their dry matter intake from grazing."[98] There are exceptions to these grazing requirements for breeding bulls and livestock that are temporarily denied pasture access for approved reasons.[99]

The new grazing regulations also include recordkeeping requirements. Producers are required to "describe the total feed ration for each type and class of animal."[100] This must include all feed produced on-farm, all feed purchased from off-farm sources, the percentage of each feed type (including pasture) in the total ration, and a list of all feed supplements and additives.[101] Additional recordkeeping requirements include documenting "the amount of each type of feed actually fed to each type and class of animal," "changes that are made to all rations throughout the year in response to seasonal grazing changes," and "the method for calculating dry matter demand and dry matter intake."[102]

Livestock health care practice standard—7 C.F.R. § 205.238

An organic livestock producer is required to "establish and maintain year-round livestock living conditions which accommodate the health and natural behavior of animals."[103] These practices include selecting livestock suited for site-specific conditions and for resistance to prevalent diseases and parasites, providing a feed ration that satisfies nutritional requirements for particular livestock, and establishing housing, pasture conditions, and sanitation practices that minimize the spreading of diseases and parasites.[104]

A producer must also administer vaccines and other "veterinary biologics" when needed.[105] "Biologics" are defined as "[a]ll viruses, serums, toxins, and analogous products of natural or synthetic origin, such as diagnostics, antitoxins, vaccines, live microorganisms, killed microorganisms, and the antigenic or immunizing components of microorganisms intended for use in the diagnosis, treatment, or prevention of diseases of animals."[106]

In addition, a producer must provide living conditions that allow "for exercise, freedom of movement, and reduction of stress" appropriate for the particular species of livestock.[107] Physical alterations to an animal that are necessary to promote the welfare of the animal, such as debeaking poultry or removing horns from livestock, must be done in a manner that minimizes the animals pain and stress.[108]

95 *Id.* at § 205.237 (c)(1).
96 *Id.*
97 *Id.*
98 *Id.* at § 205.237(c)(2).
99 *Id.* at § 205.237(c)(2)(i)-(ii). *See also* 7 C.F.R. § 205.239(b)(1)-(8) and 7 C.F.R. § 205.239(c)(1)-(3).
100 *Id.* at § 205.237(d)(1).
101 Id.
102 *Id.* at § 205.237(d)(2)-(4).
103 7 C.F.R. § 205.238(a).
104 *See id.* at § 205.238(a)(1)-(3).
105 *See id.* at § 205.238(6). *See supra* text accompanying note 60.
106 *Id.* at § 205.2.
107 *See id.* at § 205.238(a)(4).
108 *See id.* at § 205.238(a)(5). *See also* 65 Fed. Reg. 80,547, 80,572-73 (prefatory comments to final rule with request for comments).

When use of these preventive practices and veterinary biologics do not sufficiently prevent sickness in livestock, a producer is allowed to administer synthetic medications to the livestock, as long as the medication is listed on the National List as a synthetic substance allowed for use in organic livestock production.[109] Parasiticides that are included on the National List of allowed synthetic substances may be used on breeder stock and dairy stock, subject to certain conditions.[110]

For breeder stock, the parasiticide must be used before the last third of gestation, but not during lactation of offspring intended to be sold, labeled, or represented as "organic."[111] The parasiticide must be applied to dairy stock at least 90 days before the production of milk or milk products intended to be sold, labeled, or represented as "organic."[112]

A producer is prohibited from selling, labeling, or representing any animal or edible product derived from an animal that has been treated with antibiotics, any synthetic substance not included on the National List of synthetic substances allowed for organic livestock production, or any nonsynthetic substances allowed for used in organic livestock production.[113] A producer is prohibited from administering any animal drug, other than vaccinations, when there is no presence of illness in the livestock.[114] In addition, a producer is prohibited from administering hormones for the purpose of growth promotion, administering synthetic parasiticides to livestock on a routine basis, and from administering parasiticides to slaughter stock.[115] A producer is also prohibited from administering animal drugs in violation of the Federal Food, Drug, and Cosmetic Act.[116]

Finally, a livestock producer is prohibited from denying medical treatment to a sick animal with the intention of preserving the animal's organic status.[117] A producer must use "all appropriate medications . . . to restore an animal to health when methods acceptable to organic production fail."[118] Livestock that has been treated with a prohibited substance must be clearly identified and cannot be sold, labeled, or represented as organically produced.[119]

Livestock living conditions—7 C.F.R. § 205.239

A livestock producer must create and maintain year-round living conditions for livestock that accommodate the health and natural behavior of animals on that producer's operation.[120] Such living conditions include year-round access to fresh air, the outdoors, exercise, clean and dry bedding, and access to pasture for ruminants.[121] Yards, feeding pads, and feedlots can be used to provide supplemental feeding and access to the outdoors during nongrazing season, however continuous total confinement indoors, in yards, feeding pads, and feedlots.[122] Living conditions also include providing livestock access to shelter that allows for "natural maintenance, comfort behaviors, and opportunity to exercise," establishes a temperature level and air circulation suitable to the particular species of livestock, and reduces the threat of

[109] *See* 7 C.F.R. § 205.238(b). This list is found at 7 C.F.R. § 205.603.
[110] *See id.* at § 205.238(b)(1)-(2).
[111] *See id.* at § 205.238(b)(1).
[112] *See id.* at § 205.238(b)(2).
[113] *See id.* at § 205.238(c)(1).
[114] *See id.* at § 205.238(c)(2).
[115] *See id.* at § 205.238(c)(3)-(5). *See id.* at § 205.2 (defining "routine use of parasiticide").
[116] *See id.* at § 205.238(c)(6).
[117] *See id.* at § 205.238(7).
[118] *See id.*
[119] *See id.*
[120] *See id.* at § 205.239(a).
[121] *See id.* at § 205.239(a)(1)-(3).
[122] *Id.*

livestock injury.[123] A producer is allowed, however, to temporarily confine an animal as a result of inclement weather, the animal's state of production, existing conditions that could jeopardize the animal's health, safety, or well-being, when soil or water quality is threatened, for preventative healthcare procedures, treatment of illness or injury, sorting or shipping or breeding.[124]

It is important to note that the regulations specify that lactation is not a stage of life that would exempt ruminants from any of the living condition requirements and that animal being sorted or shipped must remain under continuous organic management, including feed, throughout their confinement period.[125] There are also special exemptions to the confinement rules for organic livestock that are a part of 4-H, Future Farmers of America and other youth projects.[126] These livestock may be confined for up to a week prior to a fair or other demonstration, through the event and up to 24 hours after the animals have arrived home after the event.[127] The livestock must be managed under continuous organic management, including organic feed, during their confinement.[128]

Other instances where organic livestock may be temporarily confined include "one week at the end of a lactation for dry off, three weeks prior to parturition, during parturition, and up to one week after parturition."[129] Newborn dairy cattle may be confined for up to 6 months and then "must be on pasture during the grazing season and may no longer be individually housed."[130] During the 6 months, the newborn dairy cattle "shall not be confined or tethered in a way that prevents the animal from lying down, standing up, fully extending its limbs, and moving about freely."[131] Livestock may also be confined for short periods for shearing and daily milking, so long as the milking is "scheduled in a manner to ensure sufficient grazing time to provide each animal with an average of at least 30 percent dry matter intake (DMI) from grazing though the grazing season."[132]

Ruminant slaughter stock that is typically grain finished must be maintained "on pasture for each day that the finishing period corresponds with the grazing season," but yards, feeding pads or feedlots may be used to provide finish feeding rations as long as they are large enough to allow all the livestock to feed "simultaneously without crowding and without competition for food."[133] During the finishing period, which cannot exceed the lesser of one-fifth of the animal's life or 120 days, ruminant slaughter stock is exempt from the minimum 30% DMI requirement for grazing.[134]

The manure from the livestock operation must be managed "in a manner that does not contribute to contamination of crops, soil, or water by plant nutrients, heavy metals, or pathogenic organisms and optimizes recycling of nutrients."[135] Under the final rule, "manure" is defined as "[f]eces, urine, other excrement, and bedding produced by livestock that has not

[123] *See id.* at § 205.239(a)(4)(i)-(iii).
[124] *See id.* at § 205.239(b).
[125] 7 C.F.R. § 205.239 (b)(2) and (6).
[126] *Id.* at § 205.239(b)(8).
[127] *Id.*
[128] *Id.*
[129] *Id.* at § 205.239(c)(1).
[130] *Id.* at § 205.239(c)(2).
[131] *Id.*
[132] *Id.* at § 205.239(c)(3)-(4).
[133] 7 C.F.R. § 205.239(d).
[134] *Id.*
[135] *Id.* at § 205.239(e).

been composted."[136] Producers must also manage pastures and other outdoor access areas in a "manner that does not put soil or water quality at risk."[137]

Pasture practice standard—7 C.F.R. § 205.240
Organic livestock producers must maintain a functioning management plan for pasture. The pasture "must be managed as a crop in full compliance" with the organic land requirements, soil fertility and crop nutrient practices and crop pest, weed and disease management practices.[138] The pasture plan must be included as part of the producer's organic system plan and be updated annually and any changes must be approved by the certifying agent.[139] The pasture plan must include a description of the types of pasture provided; "cultural and management practices" to be used to ensure grazing requirements are met; grazing season for the regional location, location and size of pastures (including maps); types of grazing methods, location and types of fences, shade of water; "soil fertility and seeding systems" and "erosion control and protection of natural wetlands and riparian areas practices.[140]

3.2.3.3 *Handling requirements—7 C.F.R. §§ 205.270-205.272*

The handling requirements include general requirements for the handling of "organic" products, standards for managing pests in organic handling facilities, and handling requirements for preventing the commingling and contact of organic products with prohibited substances.

Organic handling requirements—7 C.F.R. § 205.270
A handler may use mechanical or biological methods to process organic products "for the purpose of retarding spoilage or otherwise preparing the agricultural product for market."[141] Such methods include, but are not limited to "cooking, baking, curing, heating, drying, mixing, grinding, churning, separating, distilling, extracting, slaughtering, cutting, fermenting, eviscerating, preserving, dehydrating, freezing, chilling, or otherwise manufacturing, and the packaging, canning, jarring, or otherwise enclosing food in a container."[142]

A handler may use nonagricultural substances and nonorganically produced agricultural products "[i]n or on a processed agricultural product intended to be sold, labeled, or represented as 'organic,'" if the substance or product is not commercially available in organic form.[143] Nonagricultural substances and nonorganically produced agricultural products may also be used if the substance or product is used "[i]n or on a processed agricultural product intended to be sold, labeled, or presented as 'made with organic (specified ingredients or food group(s))'."[144] However, in both instances, the nonagricultural substance or nonorganically produced product that is used must be listed as an allowed substance or product on the National List.[145]

[136] *Id.* at § 205.2.
[137] *Id.* at § 205.239(e).
[138] *Id.* at § 205.240(a).
[139] *Id.* at § 205.240(c).
[140] *Id.* at § 205.240(c)(1)-(8).
[141] *Id.* at § 205.270(a).
[142] *Id. See also* 7 C.F.R. § 205.2 (defining "processing").
[143] 7 C.F.R. § 205.270(b)(1).
[144] 7 C.F.R. § 205.270(b)(2).
[145] 7 C.F.R. § 205.270(b).

A handler is prohibited from using ionizing radiation in or on agricultural products intended to be sold, labeled, or represented as "organic."[146] Ionizing radiation cannot be used in or on any ingredients labeled as "organic."[147] Handlers are also prohibited from using excluded methods, except for vaccines that are listed as approved on the National List.[148]

A handler is also prohibited from using in or on agricultural products intended to be sold, labeled, or represented as "organic," or in or on any ingredients labeled as "organic," any "volatile synthetic solvent or other synthetic processing aid," unless the solvent or aid is listed as allowed on the National List.[149] However, nonorganic ingredients used in products that are labeled as "made with organic (specified ingredients or food group(s)" are not subject to this prohibition.[150]

Facility pest management—7 C.F.R. § 205.271

The first paragraph of the requirements contained in the facility pest management practice standard, § 205.271 refer only to handlers, as do the prefatory comments accompanying this particular requirement.[151] The reader should be aware that the discussion of the pest management standard in this article refers only to handlers and does not refer to producers.

A handler of an organic facility is required to implement practices designed to prevent pests.[152] Permitted practices include, but are not limited to, the removal of habitat, food sources, and breeding areas for pests; prevention of pest access to handling facilities; and manipulation of environmental factors, such as humidity and temperature, designed to prevent pest reproduction.[153] A handler of an organic facility may manage existing pests by using mechanical or physical control such as traps, light, or sound.[154] A handler can also use lures and repellents that contain nonsynthetic or synthetic substances as long as those substances are used in accordance with the standards set forth in the National List.[155] If neither the preventative pest practices nor practices for controlling pests existing in the facility are successful, a nonsynthetic or synthetic substance listed as allowed on the National List may be applied to the organic facility.[156]

In the event that none of the pest management practices described in the previous paragraph effectively prevent or control pests in an organic facility, a synthetic substances not allowed under the National List may be used.[157] In order to take this final measure, however, the handler and the certifying agent must agree on the substance that will be used, the method of its application, and the measures that will be implemented to prevent contact of organic products or ingredients with the substance.[158]

If a synthetic or nonsynthetic substance is used to control pests, the handler must update the handling operation's organic system plan to indicate that the substance was applied.[159]

146 7 C.F.R. § 205.270(c)(1).
147 *See* 7 C.F.R. § 205.270(c)(1).
148 *See id. See also id.* at § 205.105(e)-(f).
149 7 C.F.R. § 205.270(c)(2).
150 *See id.*
151 *See id.* at § 205.271(a).
152 *See id.*
153 7 C.F.R. § 205.271(a)(1)-(3).
154 *See id.* at § 205.271(b)(1).
155 *See id.* at § 205.271(b)(2).
156 *See id.* at § 205.271(c).
157 *See id.* at § 205.271(d).
158 *See id.*
159 *See id.* at § 205.271(e).

The updated plan must also list all of the measures that were implemented to prevent contact with organic products and ingredients in the facility and the substance that was used.[160]

Not withstanding any of these pest management practices, "a handler may otherwise use substances to prevent or control pests as required by Federal, State, or local laws and regulations."[161] In such event, however, the handler must implement measures to prevent contact of organic products and ingredients with the substance used.[162]

Prevention of commingling and contact with prohibited substances—7 C.F.R. § 205.272
The organic production and handling requirements also set forth specific standards for preventing the commingling and contact of organic between nonorganic products and organic products with prohibited substances.[163] This standard prohibits a handler from using "packaging materials, and storage containers, or bins that contain a synthetic fungicide, preservative, or fumigant."[164] It also prohibits a handler from using or reusing

> any bag or container that has been in contact with any substance in such a manner as to compromise the organic integrity of any organically produced product or ingredient placed in those containers, unless such reusable bag or container has been thoroughly cleaned and poses no risk of contact of the organically produced product or ingredient placed in those containers, unless such reusable bad or container has been thoroughly cleaned and poses no risk of contact of the organically produced product or ingredient with the substance used.[165]

3.2.4 Labels, labeling, and market information—7 C.F.R. §§ 205.300-205.311

Numerous NOP standards, requirements, and restrictions govern the labeling and use of marketing information for organically produced products. More specifically, there are (1) guidelines governing the use of the term "organic"; (2) labeling requirements for livestock feed; (3) standards for labeling nonretail containers used for shipping and storage of "organic" products; (4) standards for labeling products in other than packaged form sold, labeled, or represented as "organic" at the point of retail sale; (5) labeling requirements and restrictions for products produced on an exempted or excluded operation; and (6) standards governing the use of the USDA seal on organically produced products. The labels, labeling, and market information requirements are designed to prevent abuses in the marketing of organic products and to assure consumers that organic products and ingredients are labeled in a consistent, reliable, and predictable manner.

The prefatory comments state that the labels, labeling, and market information requirements must be implemented in such a way that they do not conflict with other federal labeling requirements.[166] The comments also state that the implementing regulations for the Federal Meat Inspection Act, Poultry Products Inspection Act, and the Egg Products Inspection Act

[160] *See id.*
[161] *See id.* at § 205.271(f).
[162] *See id.*
[163] *See id.* at § 205.272(a).
[164] *See id.* at § 205.272(b)(1).
[165] 7 C.F.R. § 205.272(b)(1).
[166] 65 Fed. Reg. 80,547, 80, 576 (prefatory comments to final rule with request for comments).

must be followed when labeling meat, poultry, and egg products.[167] In addition, the comments state that the Food and Drug Administration's regulations governing the placement of information on food product packages, the Federal Trade Commission regulations implemented pursuant to the Fair Packaging and Labeling Act, and the Alcohol Tobacco and Firearms regulations implementing the Federal Alcohol Administration Act must also be followed, as applicable to the nature of the particular product.[168]

> The final rule defines a "label" as a display of written, printed, or graphic material on the immediate container of an agricultural product or any such material affixed to any agricultural product or affixed to a bulk container containing an agricultural product, except for package liners or a display of written, printed, or graphic material which contains only information about the weight of the product.[169]

"Labeling" is "[a]ll written, printed, or graphic material accompanying an agricultural product at any time or written, printed, or graphic material about the agricultural product displayed at retail stores about the product."[170]

3.2.4.1 *Use of the term "Organic"—7 C.F.R. § 205.300*

A person may only "sell or label an agricultural product as organically produced... if such product is produced and handled in accordance" with OFPA and the final rule.[171] In addition, "no person may affix a label to, or other [sic] provide market information concerning, an agricultural product if such label or information implies, directly or indirectly, that such product is produced and handles using organic methods, except in accordance" with the OFPA and the final rule.[172]

A product that is produced in the United States for export to a foreign country that is "produced and certified to foreign national organic standards or foreign contract buyer requirements, may be labeled in accordance with the organic labeling requirements of the receiving country or contract buyer."[173] However, the shipping containers and documents must comply with the requirements for labeling nonretail containers that are used only for shipping or storage of "organic" products.[174] A product that is produced in a foreign country and exported into the United States must be produced and handled in accordance with NOP certification requirements and must be labeled in accordance with the applicable labels, labeling, and market information requirements.

[167] *See id. See also* 9 C.F.R. pt. 317 (implementing regulations for the Federal Meat Inspection Act, Poultry Products Inspection Act, and Egg Products Inspection Act); 21 U.S.C. §§ 601-695 (Federal Meat Inspection Act); 21 U.S.C. §§ 451-470 (Poultry Products Inspection Act); and 21 U.S.C. §§ 1031-1056 (Egg Products Inspection Act).

[168] 65 Fed. Reg. 80,547, 80,576 (prefatory comments to final rule with request for comments). *See also* 16 C.F.R. pt. 500 (Federal Trade Commission implementing regulations under the Fair Packaging and Labeling Act); 27 C.F.R. pts, 4,5, and 7 (Alcohol Tobacco and Firearms regulations implementing the Federal Alcohol Administration Act).

[169] 7 C.F.R. § 205.2.

[170] *Id.*

[171] 7 U.S.C. § 6505(a)(1)(A). *See also* 7 C.F.R. § 205.300(a).

[172] 7 U.S.C. § 6505(a)(1)(B).

[173] 7 C.F.R. § 205.300(b).

[174] *See id.*

In the prefatory comments to the final rule, the AMS states that OFPA

> provides the Secretary with the authority to review use of the term, "organic," in agricultural product names and the names of companies that produce agricultural products. While we believe that the term, "organic," in a brand name context does not inherently imply an organic production or handling claim and, thus, does not inherently constitute a false or misleading statement, we intend to monitor the use of the term in the context of the entire label. We will consult with the [Federal Trade Commission] and [the Food and Drug Administration] regarding product and company names that may misrepresent the nature of the product and take action on a case-by-case basis.[175]

"Organic" versus "Natural"
Other labeling terms, such as "natural," can easily be confused with the term "organic." The Food Safety and Inspection Service (FSIS) defines the term "natural" as "a product containing no artificial ingredient or added color and is only minimally processed (a process which does not fundamentally alter the raw product) may be labeled natural (FSIS). The label must explain the use of the term natural (such as – no added coloring or artificial ingredients; minimally processed)."[176] The NOP program requires certification before the organic label can be used, therefore, the terms "organic" and "natural" are different.[177]

3.2.4.2 *Livestock feed—7 C.F.R. § 205.306*

There are only two labeling categories that can be used to indicate that livestock feed has been organically produced: "100 percent organic" and "organic." Raw or processed livestock feed intended to be sold, labeled, or represented as "100 percent organic" must be composed entirely of organically produced ingredients, excluding water and salt.[178] A raw or processed livestock feed to be sold, labeled, or represented as "100 percent organic" or "organic" must be produced in accordance with the organic production and handling requirements for livestock feed.[179]

Livestock feed products labeled as "100 percent organic" and "organic" may display on any package panel "[t]he statement, '100 percent organic' or 'organic,' as applicable, to modify the name of the feed product."[180] The package panel may also display the USDA seal and/or the identifying mark of the certifying agent that certified the production and handling operation producing the raw or processed organic ingredients used in the finished product" as long as the certifying agent's identifying mark is not displayed more prominently than the USDA seal.[181] Finally, the package may identify ingredients that are organically produced with the word "organic" "or an asterisk or other reference mark which is defined on the package" to identify the organic ingredients.[182]

[175] 65 Fed. Reg. 80,576, 80,547 (prefatory comments to final rule with request for comments).
[176] *See id.*
[177] 7 C.F.R. § 205.400.
[178] 7 C.F.R. § 205.301(e)(1).
[179] *See id.* at § 205.305(e)(1)-(2). *See also id.* at § 205.237.
[180] 7 C.F.R. § 205.306(a)(1).
[181] *Id.* at § 205.306(a)(2)-(3).
[182] *Id.* at § 205.306(a)(4).

Livestock feed products must display the name of the certifying agent that certified the handler of the finished product.[183] The agent's name must be preceded by the phrase "Certified organic by..." or another similar phrase, and the package label may include the agent's telephone number, business address, or internet address.[184] Livestock feed products must also be labeled, as applicable, with any other labeling requirements created under federal or state law.[185]

3.2.4.3 *Labeling of nonretail containers used for shipping and storage—7 C.F.R. § 205.307*

The labeling requirements discussed so far have related only to the display of written, printed, or graphic material placed on the container or package of an individual agricultural product as it appears for retail sale. Under NOP, there are also requirements for the labeling of nonretail containers used only for shipping or storage of raw or agricultural products labeled as "100 percent organic," "organic," or "made with organic...."[186] The requirements for labeling of nonretail containers are intended to prevent the commingling of organic and nonorganic products and ingredients and to prevent a product from being handled in a way that would destroy the product's organic status.

Nonretail containers used solely for shipping or storing agricultural products labeled as containing organic ingredients may display (1) the "name and contact information of the certifying agent which certified the handler which assembled the final product"; (2) handling instructions necessary to maintain the product's organic integrity; (3) terms or marks that identify the product as "organic"; (4) the USDA Organic seal; and (5) the identifying mark "of the certifying agent that certified the organic production or handling operation that produced or handled the finished product."[187] The containers are also required to display the "production lot number" of the product, if one is available.[188] The "production lot number" is the "[i]dentification of a product based on the production sequence of the product showing the date, time, and place of production used for quality control purposes."[189]

Nonretail containers used to export domestically produced products labeled as organic to international markets can be labeled in accordance with the labeling requirements of the country of destination or with the specifications of the foreign contract buyer, if certain conditions are satisfied.[190] The containers and shipping documents must be clearly marked "For Export Only," and the handler must maintain verification of such marking and export "in accordance with the recordkeeping requirements for exempt and excluded operations under § 205.101."[191]

This requirement states that there are recordkeeping requirements for exempted *and* excluded operations located at § 205.101 of the final rule. According to this provision of the final rule, the AMS indicates its belief that there are recordkeeping requirements for excluded operations set forth at § 205.101. However, as discussed in Part B of this article, there are

[183] *See id.* at § 205.306(b)(i).
[184] *See id.*
[185] *See id.* at § 205.306(b)(ii).
[186] *See id.* at § 205.307.
[187] *Id.* at § 205.307(a)(1)-(5).
[188] *See id.* at § 205.307(b).
[189] *Id.* at § 205.2.
[190] *See id.* at § 205.307(c).
[191] *See id.*

no recordkeeping requirements specifically applicable to excluded operations located at § 205.101.

3.2.4.4 *Products in other than packaged form at point of sale—7 C.F.R. §§ 205.308-205.309*

Products in other than packaged form at the point of retail sale that are to be sold, labeled, or represented as "100 percent organic" or "organic" "may use the term "100 percent organic" or "organic," as applicable, to modify the name of the product in retail display, labeling, and display containers," as long as the term "organic" "is used to identify the organic ingredients listed in the ingredient statement."[192] This requirement applies to "100 percent organic" and "organic" products that are not packaged before sale and are presented in a way that permits consumers to the select the amount of the product they want to purchase.

Such products, if prepared in a certified facility, may have the USDA seal placed on the retail display, labeling, and display containers.[193] They may place the identifying mark "of the certifying agent which certified the . . . operation producing the finished product and any other certifying agent which certified operations producing raw organic product or organic ingredients used in the finished product" on the retail display, labeling, and display containers.[194] The identifying mark cannot be displayed more prominently than the USDA seal.[195]

Products in other than packaged form at the point of retail sale that are to be sold, labeled, or represented as "made with organic . . ." are allowed to use the term "made with organic . . ." to modify the product name in retail display, labeling, and display containers.[196] The "made with organic . . ." statement cannot list more than three organic ingredients or food groups.[197] The ingredient statement must identify organic ingredients as "organic."[198] As long as the product is prepared in a certified facility, it may display the certifying agent's identifying mark in retail displays, display containers, and market information.[199]

3.2.4.5 *Labeling of products produced on exempted and excluded operations—7 C.F.R. § 205.310*

Excluded or exempted operations must comply with the labeling requirements for exempted and excluded operations.[200] A product that has been produced on an exempted or excluded operation cannot be represented in any way to any buyer that it has been certified as an organically produced product or ingredient.[201] Thus, such products are prohibited from displaying the USDA seal and the identifying mark of a certifying agent that identifies the operation as certified.[202]

[192] 7 C.F.R. § 205.308(a).
[193] *See id.* at § 205.308(b)(1).
[194] *Id.* at § 205.308(b)(2).
[195] *See id.*
[196] *See id.* at § 205.309.
[197] *See id.* at § 205.309(a)(1).
[198] *See id.* at § 205.309(a)(2).
[199] *See id.* at § 205.309(b).
[200] *See id.* at § 205.310.
[201] *See id.* at § 205.310(a)(1)-(2).
[202] *See id.*

A product that has been organically produced or handled on an exempt or excluded operation may be identified as an "organic" product or ingredient in a multi-ingredient product that has been produced by that operation.[203] However, the product or ingredient can only be identified as "organic" if it complies with the applicable labels, labeling, and market information requirements.[204] The product or ingredient cannot be identified as "organic" if it has been processed by other persons or if it is used to modify a nonorganic ingredient in the product.[205]

A product that is produced or handled on an exempt or excluded operation cannot be produced by using excluded methods or sewage sludge.[206] It must also be processed without the use of ionizing radiation.[207] In addition, the product must be processed without the use of processing aids that are not listed as approved or allowed on the National List, except for products labeled as "100 percent organic" which, if processed, must be processed with the use of organically produced processing aids.[208] Finally, products and ingredients produced or handled on an exempt or excluded operation must not (1) "[c]ontain sulfites, nitrates, or nitrites added during the production or handling process", (2) be produced with the use of "nonorganic ingredients when organic ingredients are available"; and (3) "[i]nclude organic and nonorganic forms of the same ingredient."[209]

3.2.4.6 *The organic seal—7 C.F.R. § 205.311*

Only products to be sold, labeled, or represented as "100 percent organic" or "organic," including livestock feed, may contain the USDA Organic seal.[210] The seal indicates to consumers that a product has been certified as organically produced and handled in accordance with all applicable NOP requirements. The USDA seal must "replicate the form and design" of the examples provided by the USDA.[211]

3.2.5 Certification—7 C.F.R. §§ 205.400-205.406

"Certification" is a certifying agent's determination that an operation has complied with all applicable NOP requirements, which is documented by a certificate of organic operation.[212] A "certifying agent" is "[a]ny entity accredited by the Secretary for the purpose of certifying a production or handling operation as a certified production or handling operation."[213]

The final rule establishes general requirements for certification, standards for submitting an application for certification, standards for reviewing the application, and requirements governing the on-site inspections that must be conducted by a certifying agent. It also sets forth standards for the granting, denying, and continuing certification.

[203] *See id.* at § 205.310(b).
[204] *See id.* at § 205.310(c). *See also* 7 C.F.R. § 205.300(a).
[205] 7 C.F.R. § 205.310(c). *See also* 7 C.F.R. § 205.300(a).
[206] 7 C.F.R. § 205.310(c). *See also* 7 C.F.R. § 205.301(f)(1)-(2).
[207] 7 C.F.R. § 205.310(c). *See also* 7 C.F.R. § 205.301(f)(3).
[208] 7 C.F.R. § 205.310(c). *See also* 7 C.F.R. § 205.301(f)(4).
[209] 7 C.F.R. § 205.310(c). *See also* 7 C.F.R. § 205.301(f)(5)-(7).
[210] 7 C.F.R. § 205.311(a).
[211] *Id.* at § 205.311(b).
[212] *See* 7 C.F.R. § 205.2.
[213] *See id.* at § 205.2. *See also* 7 U.S.C. § 6501(3).

3.2.5.1 *General certification requirements—7 C.F.R. § 205.400*

There are six general requirements for certification.[214] First, any person seeking certification must comply with the OFPA and all applicable organic production and handling standards. Second, the producer or handler must also create, implement, and annually update an organic system plan.[215] Third, the producer or handler applying for certification must allow the certifying agent to conduct on-site inspections "with complete access to the production or handling operation, including noncertified production and handling areas, structures, and offices."[216] Fourth, the applicant must also maintain all records pertaining to the organic operation for at least 5 years beyond the creation of the records and allow "authorized representatives of the Secretary, the applicable state organic program's governing State official, and the certifying agent access" to those records during normal business hours.[217] Fifth, the person seeking certification must tender the applicable payments charged by the certifying agent.[218] Sixth, the person must notify the certifying agent when there is an application of a prohibited substance to any part of the operation or when there is a change in the operation that could affect the operation's compliance with NOP.[219]

3.2.5.2 *Application for certification—7 C.F.R. § 205.401*

A producer or handler of a production or handling operation seeking certification must submit an application for certification to a certifying agent.[220] The application must include an organic system plan, the name of the person completing the application, and the applicant's business name, address, and telephone number.[221] If the applicant is a corporate entity, the name, address, and telephone number of the person authorized to act on the entity's behalf must be provided.[222]

The application must also include "[t]he name(s) of any organic certifying agent(s) to which application has previously been made; the year(s) of application; [and] the outcome of the application(s) submission."[223] If an applicant has previously received either a notice of noncompliance or denial of certification from a certifying agent, a copy of the notice must be included, when available, with the application.[224] In the event that an applicant has received a notice of noncompliance or denial of certification, the application should include "a description of the actions taken by the applicant to correct the noncompliances" indicated in the notice, "including evidence of such correction."[225] Finally, the application must contain any additional information needed to demonstrate compliance with all applicable NOP requirements.[226]

214 7 C.F.R. § 205.400.
215 *See id.* at § 205.400(b).
216 *Id.* at § 205.400(c).
217 *See id.* at § 205.400(c)-(d).
218 *See id.* at § 205.400(e).
219 *See id.* at § 205.400(f)(1)-(2).
220 *See id.* at § 205.401.
221 *See id.* at § 205.401(a)-(b).
222 *See id.*
223 *Id.* at § 205.401(c).
224 *See id.*
225 *Id.*
226 *See id.* at § 205.401(d).

3.2.5.3 *Review of application—7 C.F.R. § 205.402*

Once an application for certification is submitted to the USDA, the certifying agent must review the application and determine whether it should be granted, denied, or continued. The final rule sets forth several requirements that certifying agents must follow when making this determination.[227]

The certifying agent must examine the application to determine whether the applicable requirements for submitting an application for certification have been met.[228] Next, the agent must determine whether the applicant "appears to comply or may be able to comply" with the applicable organic production and handling requirements.[229] If the agent determines that the applicant may have satisfied the applicable production and handling requirements, then it must schedule an on-site inspection to determine whether the production and handling requirements have been satisfied.[230] If the applicant has previously received a notice of noncompliance or denial of certification from another certifying agent, the agent must verify that the applicant has submitted documentation demonstrating that the applicant has corrected the noncompliance(s) indicated in the notice.[231]

The certifying agent is required to review the application and inform the applicants of its findings within a reasonable time.[232] It must provide to the applicant a copy of the inspection report developed from an on-site inspection to the applicant and a copy of any test results for samples taken by an inspector to detect the presence of a prohibited substance on the operation.[233]

An applicant is allowed to withdraw its application at any time but must pay costs of services provided up to the time of withdrawal.[234] A notice of noncompliance or denial of certification will not be issued to an applicant if the applicant voluntarily withdrew its application before such notice was issued.[235]

3.2.5.4 *On-Site inspections—7 C.F.R. § 205.403*

Certifying agents are required to "conduct an initial on-site inspection of each production unit, facility, and site that produces or handles organic products and that is included in an operation for which certification is requested."[236] Following the initial on-site inspection, annual inspections must be conducted for certified operations to determine whether a request for certification should be approved or whether certification for the operation should continue.[237]

In addition to the initial and subsequent annual inspections, certifying agents may conduct additional inspections to determine whether an applicant or an already certified operation is in compliance with NOP requirements.[238] The AMS Administrator, the representative

[227] *See id.* at § 205.402.
[228] *See id.* at § 205.402(a)(1).
[229] 7 C.F.R. § 205.402(a)(2).
[230] *See id.* at § 205.402(a)(4).
[231] *See id.* at § 205.402(a)(3).
[232] *See id.* at § 205.402(b)(1).
[233] *See id.* at § 205.402(b)(3).
[234] 7 C.F.R. § 205.402(c).
[235] *See id.*
[236] *See id.* at § 205.403(a)(1).
[237] *See id.*
[238] *See id.* at § 205.403(a)(2)(i). *See also* 65 Fed. Reg. 80,547, 80,590 (prefatory comments to the final rule with request for comments).

delegated responsibility to act on behalf of the Administrator, or the governing official of a State organic program may require a certifying agent to conduct additional inspections "for the purpose of determining compliance" with the OFPA and the final rule.[239] At the discretion of the Administrator, its representative, or the governing official of a state organic program, additional inspections may be announced or unannounced.[240]

The initial inspection must be conducted within a reasonable time after the certifying agent has determined that the applicant "appears or may be able to comply" with the organic production and handling requirements.[241] An initial inspection can be delayed for up to 6 months, however, "to comply with the requirement that the inspection be conducted when the land, facilities, and activities that demonstrate compliance or capacity to comply can be observed."[242]

All inspections, whether initial, annual, or additional must occur when an authorized representative of the operation is present and "at a time when land, facilities, and activities that demonstrate the operation's compliance with or capability to comply" with the applicable organic production and handling requirements can be observed.[243] This requirement does not apply to unannounced inspections.

An inspection must verify an "operation's compliance or capability to comply" with OFPA and the final rule and confirm that the information provided in the application including the organic system plan, "accurately reflects the practices used or to be used by the applicant for certification or by the certified operation."[244] On-site inspections must also verify that no prohibited substances have been applied or are currently being applied to the production or handling operation.[245] Such verification may, at the discretion of the certifying agent, include "the collection and testing of soil; water; waste; seeds; plant tissue; and plant, animal, and processed products samples."[246] If samples are taken, the inspector must provide the operation's authorized representative with a receipt if samples are taken.[247]

An exit interview must be conducted "with an authorized representative of the operation who is knowledgeable about the inspected operation" to verify that observations and information gathered during the inspection are accurate and complete.[248] During the interview, the inspector must inform the authorized representative if additional information is needed and if there are any issues of concern.[249] The prefatory comments to the final rule state that the main purpose of the exit interview "is to present the inspection observations to those in charge of the firm in such a manner so as to ensure they clearly understand the results of the inspection."[250]

3.2.5.5 *Granting certification—7 C.F.R. § 205.404*

Once the initial inspection has been conducted, the certifying agent must review the initial inspection report, the results from any tests conducted to detect the presence of prohibited

[239] *Id.* at § 205.403(a)(2)(ii).

[240] *See id.* at § 205.403(a)(2)(iii).

[241] *Id.* at § 205.403(a)(2)(iii).

[242] *Id.* at § 205.403(b)(1). *See also* 65 Fed. Reg. 80,547, 80,592 (prefatory comments to the final rule with request for comments)

[243] *See* 7 C.F.R. § 205.403(b)(2).

[244] *Id.* at § 205.403(c)(1)-(2).

[245] *See id.* at § 205.403(c)(3).

[246] *Id.*

[247] *See id.* at § 205.403(e).

[248] *Id.* at § 205.403(d).

[249] *See id.*

[250] 65 Fed. Reg. 80,547, 80,589 (prefatory comments to final rule with request for comments).

substances, any information supplied by or requested from the applicant, the organic system plan, and all procedures and activities used in the applicant's operation.[251] Based upon this review, the certifying agent must grant certification if it determines that the applicant's operation complies with the applicable NOP requirements and that the operation will be able to conduct its operation in accordance with its organic system plan.[252] The certifying agent may require the operation to correct minor noncompliances "within a specified time period as a condition of continued certification."[253]

If certification is granted, the certifying agent must issue a certificate of organic operation to the certified operation.[254] The certificate must specify (1) the name and address of the certified operation; (2) the effective date of certification; (3) the name, address, and telephone number of the certifying agent, and (4) the "[c]ategories of organic operation, including crops, wild crops, livestock, or processed products produced by the certified operation."[255] An operation's certification is effective until it is surrendered by the operation, or suspended or revoked by either the certifying agent, the governing official for a State's organic program, or the Administrator.[256]

3.2.5.6 *Denial of certification—7 C.F.R. § 205.405*

If a certifying agent has reason to believe that an applicant "is not able to comply or is not in compliance with the requirements" set forth in the final rule, it must provide a written notice of noncompliance to the applicant.[257] When an operation's noncompliance is such that it would be impossible to correct, the certifying agent may combine a written notification of noncompliance and denial in one notice.[258]

A notice of noncompliance must describe each noncompliance, the factual basis for the noncompliance, and the "date by which the applicant must rebut or correct each noncompliance and submit supporting documentation of each such correction when correction is possible."[259] If a certifying agent determines that an applicant has wilfully provided a false statement or purposefully misrepresented its operation or compliance, the agent may issue a denial of certification without first issuing a notice of noncompliance.[260]

An applicant that has received a notice of noncompliance may exercise one of three options. First, it may correct the noncompliances and explain to the certifying agent in supporting documentation the corrective actions that it implemented.[261] Second, the applicant may also correct the noncompliance(s) and submit a new application for certification to another certifying agent.[262] If an applicant exercises this second option, it must submit a complete application for certification to the new certifying agent that includes a copy of "the notification of noncompliance received from the first certifying agent," a description

[251] *See* 7 C.F.R. § 205.404(a).
[252] *See id.*
[253] *Id.*
[254] *See id.* at § 205.404(b).
[255] *Id.* at § 205.404(b)(1)-(4).
[256] *See id.* at § 205.404(c).
[257] *Id.* at § 205.405(a).
[258] *See id.*
[259] *Id.* at § 205.405(a)(1)-(3).
[260] *See id.* at § 205.405(g).
[261] *See id.* at § 205.405(b)(1).
[262] *See id.* at § 205.405(b)(2).

of the corrective measures that it implemented to remedy the noncompliance, and documentation that evidences the corrective measures that were taken.[263] Finally, the applicant may challenge the certifying agent's determination with written information that rebuts the noncompliance(s) described in the notice of noncompliance.[264]

After issuing a notice of noncompliance, the certifying agent must examine the corrective measures taken by an applicant, the documentation evidencing those corrective actions, or the applicant's written rebuttal.[265] If necessary, the certifying agent must also conduct an on-site inspection.[266] The certifying agent must issue an approval of certification to the applicant if it determines that either the corrective measures implemented by the applicant or the applicant's rebuttal are sufficient to qualify the applicant for certification.[267] The certifying agent must provide a written notice of denial of certification to the applicant if it determines that the corrective measures or the rebuttal are not sufficient to qualify the applicant for certification.[268] A certifying agent must also send a written notice of a denial of certification if the applicant fails to respond to a notice of noncompliance.[269] Notice of either approval or denial must also be sent by the certifying agent to the Administrator of USDA AMS.[270]

Any notice of denial must state the reason(s) for the denial.[271] It must also state that the applicant has a right to reapply for certification, to request mediation, or file an appeal challenging the denial of certification.[272] When a certifying agent receives a new application for certification that includes a notice of noncompliance or notice of denial, it must review the application as if it were a new application and begin an entirely new application process.[273]

3.2.5.7 *Continuation of certification—7 C.F.R. § 205.406*

To maintain its certification status, a certified operation must annually update its organic system plan, submit the updated information to its certifying agent, and pay the necessary fees.[274] The updated organic system plan must contain a summary statement and supporting documentation describing any changes or modifications to the previous year's organic system plan.[275] The updated plan must also describe any deletions or additions to the previous year's organic system plans that are expected to be implemented in the upcoming year.[276]

Any changes in the applicant's business name, address, telephone number, or the name, address, or telephone number of its authorized representative must also be provided to the certifying agent as part of the annual update.[277] If the certifying agent notified the operation that minor noncompliances would need to be corrected when it granted certification, the

[263] *Id.*
[264] *See id.* at § 205.405(b)(3).
[265] *See id.* at § 205.405(c)(1).
[266] *See id.*
[267] *See id.* at § 205.405(c)(1)(i).
[268] *See id.* at § 205.405(c)(1)(ii).
[269] *See id.* at § 205.405(c)(2).
[270] *See id.* at § 205.405(c)(3).
[271] *See id.* at § 205.405(d).
[272] *See id.* at § 205.405(d)(1)-(3).
[273] *See id.* at § 205.405(f).
[274] *See id.* at § 205.406(a).
[275] *See id.* at § 205.406(a)(1)(i).
[276] *See id.* at § 205.406(a)(1)(ii).
[277] *See id.* at § 205.406(a)(2).

operation must provide an update on the corrective measures it implemented.[278] Finally, the operation must provide any information requested by the certifying agent to demonstrate compliance with NOP requirements.[279]

Once the certifying agent has received the required updated information, it must arrange for and conduct an on-site inspection of the operation.[280] When such an inspection is not possible, however, the certifying agent "may allow continuation of certification and issue an updated certificate of organic operation on the basis of the information submitted and the most recent on-site inspection conducted during the previous 12 months."[281] If an operation's certification is continued in this manner, the required annual on-site inspection must be conducted in the first 6 months after the operation's scheduled date of annual update.[282]

The certifying agent must issue a written notice of noncompliance to an operation if the on-site inspection and review of an operation's updated information causes the certifying agent to determine that the operation is not in compliance with NOP requirements.[283] However, if the inspection and review of information demonstrates an operation's compliance with NOP requirements, the certifying agent must issue an updated certificate of organic operation.[284]

3.3 FUTURE DIRECTIONS AND CONCLUSIONS

Congress enacted the OFPA as part of the 1990 Farm Bill, the Food, Agriculture, Conservation, and Trade Act of 1990, and more than a decade later began implementation of the regulatory provisions of the NOP. This was a significant step in that it created uniform standards for members of the organic industry to abide by in order to gain consumers' trust in the marketplace that products were truly "organic."

Since 1990, the organic sector has advanced significantly. For example, it is estimated that since 1990 the organic sector has had a 20% annual growth (Dimitri & Greene, 2002). Overall, the organic sector comprises approximately 3% of total US retail food sales (Johnson, 2008). Only time can tell whether these figures will expand or moderate, but it is unlikely that the organic sector will meaningfully deteriorate in the aggregate so long as consumer demand remains.

However, the organic sector has unique challenges and opportunities on the horizon, all of which arise in the larger context of the expanding debate over what should be the structure and operation of the US food system—organic and nonorganic. For example, the rapid development and investments in developing local food systems directly implicates organic producers. Another issue more specific to organic livestock production and products is the rising tide of animal welfare concerns advanced by organizations such as the Humane Society of the United States (HSUS). Overlapping concerns over the use of antibiotics in animal agriculture factor into the equation as well.

A consequence of this rising tide of animal welfare concerns is that more consumers may look to the organic program as the foundation from which they base their meat-purchasing

[278] *See id.* at § 205.406(a)(3).
[279] *See id.* at § 205.406(a)(4).
[280] *See id.* at § 205.406(b).
[281] *Id.*
[282] *See id.*
[283] *See id.* at § 205.406(c).
[284] *See id.* at § 205.406(d).

considerations. In addition, it could lead to a more invigorated debate over how strict of an interpretation regulatory provisions such as living conditions for livestock should receive.

From a technical perspective, any changes that occur will arise through court decisions, regulatory modifications, or through legislative changes enacted by Congress. Since creation of the NOP, there have been a few matters litigated, though the pace of the litigation seems to have moderated in some respects. However, issues involving coexistence of organic production and biotech crops remain a serious policy and legal issue. Again, the issue of coexistence interplays in the larger context of determining the structure and operation of the overall food system.

On the legislative side, the future is particularly uncertain given concerns over the burgeoning federal budget deficit and appropriations reductions that are likely to occur over the coming years. Further, as most provisions of the 2008 Farm Bill are set to expire in 2012, the elements affecting organic production will be revisited by a new Congress.

In conclusion, the NOP represents arguably one of the most significant national modifications to the food system over the past several years. One can further reasonably conclude that the organic sector is now a permanent component of the nation's food system, so that the issue is not whether it will continue but rather in what form and to what degree it will continue. However, this issue—the future direction of the NOP and specifically its regulatory application to livestock production and handling—is embedded in the larger issue of assessing the future direction of the overall food system.

That said, the US food system is approaching an historic crossroads and is continuously under more policy pressure to undergo fundamental changes than at most any other time in US history. In short, it could be argued that a handful of factors are bringing to bear considerable pressure onto the food system, which has a direct impact on the future direction of the organic sector. These factors principally include: (1) globalization and the desire to liberalize international trade through the World Trade Organization; (2) perceived concerns of many over the environmental impact of agricultural production, including organic production; (3) the interrelationship of addressing energy issues and agricultural production; (4) rising influence of consumers on the food system; and (5) the chronic issue of the federal budget deficit, coupled with states' budget issues.

REFERENCES

Dimitri, C. and Greene, C. 2002. Recent growth patterns in the U.S. organic foods market. *Economic Research Service/USDA/AIB-777* September. Available at: http://www.ers.usda.gov/publications/aib777/.

Greene C. and Kremen A. 2003. U.S. organic farming in 2000-2001: adoption of certified systems. *Economic Research Service/USDA/AIB-780* February. Available at: http://www.ers.usda.gov/publications/aib780/.

Food Safety and Inspection Service (FSIS). Available at: http://www.fsis.usda.gov/Fact_Sheets/Meat_&_Poultry_Labeling_Terms/index.asp.

Johnson R. 2008. Organic agriculture in the United States: program and policy issues. *Congressional Research Service RL31595* November. Available at: http://www.nationalaglawcenter.org/assets/crs/RL31595.pdf.

4 Organic Meat Production in Europe: Market and Regulation

Simona Naspetti and Raffaele Zanoli

Abstract: This chapter provides an overview of the evolution of the European market for organic meat, currently representing 13% of the total European organic market and about 2% of the total meat sales in Western Europe. The main factors affecting this evolution are discussed, and the regulatory framework is described. Future prospects for the development of the organic meat supply and demand are analyzed and implications are discussed.

Keywords: organic meat; production; consumption; willingness to pay; ethical values

4.1 INTRODUCTION

In the early years of the millennium, the organic meat market in Europe has undergone periods of undersupply followed by periods of oversupply (Organic Monitor, 2010). Since the introduction of the European Union (EU) regulations on organic animal farming in 1999, market equilibrium has proven to be a difficult goal for marketers and producers. The reasons for this may be traced back to different degrees of organization of supply chains at the subsector level (beef, sheep, pork, and poultry) and at the country level. The implementation of regulations in various European countries had also been a source of instability in the markets, given that some countries had national standards in place well before the turn of the century. These standards were in some cases more restrictive than the new European regulation, while in other countries they were not (Padel *et al.*, 2004).

In this chapter, we introduce the European regulations on organic livestock and its implications for the development of production and market. We then present a basic description of the organic meat market in Europe followed by focus on production and supply in the various European countries. Finally, some discussion on the implication for the European consumer closes the chapter.

4.2 THE REGULATORY FRAMEWORK

In 1991, when the European Council of Agricultural Ministers adopted Regulation (EC) No. 2092/1991 on organic farming and the related labeling of agricultural products and

Organic Meat Production and Processing, First Edition. Edited by Steven C. Ricke, Ellen J. Van Loo, Michael G. Johnson, and Corliss A. O'Bryan.

foods, and organic agriculture was officially recognized by the 15 EU member states, livestock production was not included. The organic legislation at that time only regulated plant products.

The basic rules of organic livestock farming were introduced later in the Council Regulation (EC) No. 1804/1999 as a supplement to the first regulation on organic production of agricultural products and foodstuffs. These rules relate to animal feed, foodstuffs, disease prevention and veterinary treatments, animal welfare, husbandry practices, livestock breeding in general, and the livestock manure. The import of organic products from third countries, whose production criteria and control systems could be recognized as equivalent to those of the EU, was also regulated.

In June 2007, the European Council of Agricultural Ministers finally agreed to a new Regulation (EC) No. 834/2007 on organic production control and labeling of organic products. The aim of this new regulation was to reduce the complexity of the provisions contained in the process of supplementation and amendment, which followed the Regulation (EC) No. 2092/1991. In addition, the new regulation expanded the scope of previous regulations on organic production, and now included "plant, livestock, and aquaculture production, along with rules for the collection of wild plants and seaweeds."

In addition, in 2008, two "implementing" regulations were issued: (1) (EC) Regulation No. 889/2008, laying down "detailed rules" for the implementation of Council Regulation (EC) No. 834/2007—further amended by (EC) Regulation No. 710/2009—and (2) Council Regulation (EC) No. 1235/2008, which established equally detailed rules with respect to the arrangements for imports of organic products from third countries.

The new goals, principles, and general rules for organic production went into effect on January 1, 2009, even though some of the new provisions on labeling have been gradually introduced after July 1, 2010. Under the new rules, the use of the EU organic logo becomes mandatory for prepackaged organic food produced in the EU but may be accompanied by national or private logos. On March 31, 2010, a new EU organic logo was published as Commission Regulation 271/2010. On July 1, 2010, the use of the logo became mandatory for all prepackaged organic products produced in the EU (with a 2-year transitional period) and optional for products from third countries complying with EU organic standards. In addition to the model logo and the technical reproduction information, the Annex to Commission Regulation 271/2010 also sets out the format of the code number of the control body or authority. This code number together with an indication of the place of farming of the agricultural raw materials must be placed below the EU organic logo.

The new regulation included a new import regime, replacing import certificates released by member states for direct authorization and monitoring of control bodies working in third countries. As mentioned before, an equivalency regime was established for some countries, currently including Argentina, Australia, Costa Rica, India, Israel, New Zealand, Japan, and Switzerland. For other third countries, like the United States, the Commission will compile a list of authorized control bodies and control authorities. To be included in the EU list, US control bodies/authorities must submit a technical dossier before October 31, 2011. To avoid trade disruptions, transitional rules have been established to allow member states to continue to grant authorizations to importers of US organic products on a case-by-case basis. These authorizations will expire at the latest 24 months after the publication of the first EU list of control bodies/authorities. According to IFOAM (2009), allowing for direct equivalency in production and processing standards and inspection encourages the growth of the global organic sector. It also makes it "easier to import the produce required to meet the continued growing demand for organic products."

The Commission, in order to increase the transparency of the regulatory environment, established the Organic Farming Information System (OFIS) as a key instrument for the exchange of agricultural data relating to organic products and for the provision of current information for the public.

The original EU organic regulation created common minimum standards for the entire EU, by strengthening consumers' confidence and letting them purchase organic products from other member states with the certainty that these products fulfilled the same minimum requirements. The current regulation aimed to go beyond, establishing a mandatory common labeling and giving greater emphasis on environmental protection, biodiversity, and high standards of animal protection.

4.3 ORGANIC ANIMAL PRODUCTION: SALIENT FEATURES OF THE NEW EU REGULATION

The implementing Regulation (EC) 889/2008 allows only the following species to be raised organically: "bovine including bubalus and bison, equidae, porcine, ovine, caprine, poultry (species as mentioned in annex iii) and bees." Aquaculture is also included. Wild animals and wild fish are excluded from the regulation. Pet food is covered in the regulations for the first time, but detailed processing rules are still pending; in the transition, private national standards—where existing—are applicable.

The new regulation reaffirms that animals are to be born and raised on organic farms. Where such animals are not available in sufficient numbers, conventional animals may be purchased, under certain conditions, which vary according to the species. Special conversion periods apply to these cases.

Globally, the most relevant change with the previous regulation is that member states can no longer have stricter animal production rules (Schmid, 2009). Stricter private standards may still exist, but these are additional to the basic EU standards laid down by the regulation, which apply to the entire EU.

Other relevant changes are the inclusion of indoor tethering, which was not allowed before: in smallholdings, as defined by member states, indoor tethering is now temporarily permitted on certain conditions that ensure enough animal welfare (e.g., a twice-weekly outdoor exercise).

Indoor finishing for beef is still allowed for a maximum of 3 months. It is, however, no longer allowed for sheep and pigs, after a transition period ending at the end of 2010. In general, herbivores must have access to free range and to grazing when possible. Physical mutilation is now generally banned (although some practices may be authorized for special hygienic and health reasons), while castration is more strictly regulated.

For poultry breeding, free range is now more precisely defined: poultry must be allowed free range for at least one-third of their lifespan. The issue of fast- versus slow-growing breeds has not been fully resolved, since the definition of approved slow-growing strains has been left to member states. However, poultry not coming from a slow strain must be reared until they reach a minimum age, which are generally lower for female gender (e.g., 81 days for chicken and 150 days for capons; 100 days for female turkey; and 140 days for males).

For animal feed, stricter provisions now apply, according to the general principle—set out in the main regulation—that feed should primarily derive for livestock from the holding where the animals are kept or from other organic holdings in the same region.

In the veterinary treatment of animals, as in the old regulation, a preference for natural methods and remedies (e.g., phytotherapy, homeopathy, etc.) is clearly stated. Chemically synthesized allopathic veterinary treatment or antibiotics are allowed as a second-best practice under the responsibility of a veterinarian and are forbidden as preventive measures. Should the latter methods be used, animals, or products thereof, cannot be sold as organic until they have undergone a full conversion period as if they were conventional animals brought in the farm. Vaccinations and antiparasite treatments are an exception to this.

4.4 CHARACTERISTICS OF THE ORGANIC MEAT INDUSTRY

4.4.1 The market for organic meat in Europe

The European organic market is considered the largest and most sophisticated in the world, with a high degree of competition. The European organic market has faced rapid expansion since the mid-1990s until the turn of the century. In the most recent years, growth has continued at rates above 10% until 2009, when the growth rate was "only" 3.5%. The expected growth for the coming year is above 6%, albeit the current economic outlook is still uncertain (Organic Monitor, 2010).

No official figures exist in Europe on the market size, although various sources concur in estimating total sales to be around 18.5 billion euros (2009 figures: Organic Monitor, 2010; Willer & Kilcher, 2011). The largest market is still Germany (5.8 billion), followed by France (3 billion), United Kingdom, Italy, and Switzerland. The largest market share is observed in Denmark (7.2%), followed by Austria (6%) and Switzerland (5.2%). Germany and Italy score at the fourth place, slightly above 3%. The per capita expenditure of organic food is 60.24 euros/year, ranging from 138 euros/year in Denmark to 0.1 euro/year in Turkey (Willer & Kilcher, 2011). Mainstream retailers largely dominate the retail market, although the specialized channel still plays an important role in some countries (Figure 4.1).

Specialist retailers comprise most of the organic food sales in Italy and Spain where they have over 40% of the market share. The market share is above 70% in some Central Eastern European countries. In Nordic countries, mainstream retailers represent over 80% of total sales. Hotels, restaurants, and catering (HoReCa) represents the fastest growing channel during the past few years, while other channels include on-farm direct marketing and other short-supply chain outlets (e.g., box schemes). In Germany, drugstores also play a relevant role, with the leading chains marketing organic foods under their private labels (Organic Monitor, 2010).

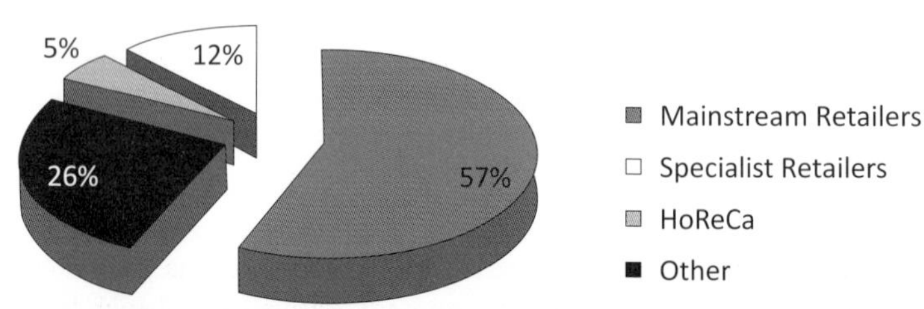

Figure 4.1 Sales breakdown by channels, 2009. HoReCa, hotels, restaurants, and catering.

Valued at 2.4 billion euros, the organic meat market represents 13% of the total European organic market and about 2% of total meat sales in Western Europe (Organic Monitor, 2010: data refer to 2009). It is second only to fruit and vegetables (28%, 5.8 billion euros), and is paired with the dairy market (13%). Beef is the leading segment, representing over half of the organic meat revenues in most European countries. Organic pork (including processed meats such as ham, sausages, etc.) is the second largest product segment, with poultry and lamb following it.

Supply–demand imbalances have become a characteristic of the European organic meat sector. Oversupply was the case in the years just after the approval of the (EC) Regulation No. 2092/91. According to Napolitano *et al.* (2009), a reason for this is that many of the early organic farming pioneers—especially in the Northern and Central Europe—were dairy and beef cattle farmers. Supply fell short of demand during the years of the boom of organic sales, at the beginning of the new millennium. Oversupply was reported in the years 2003 and 2004, while undersupply became prominent again after 2005 until 2009 when demand fell short once again during the negative economic cycle that influenced organic market growth. During periods of oversupply, significant volume of organic beef and pork went into the nonorganic market, while producer prices were declining (Organic Monitor, 2010).

According to Napolitano *et al.* (2009b), foreign exporters are finding it increasingly difficult to meet supply gaps because of the differences in organic standards between regions. Different regional organic standards (such as the US National Organic Program and the EU Regulations) act like nontariff barriers that often hinder the potential of free trade in the global organic meat industry.

Indeed, as we have seen in the previous discussion, organic foods in Europe must meet EU regulations, which is also the basis of national standards of most national markets. The two non-EU member European countries (Norway and Switzerland) have very similar regulations based on the EU rules, which are considered equivalent to EU organic standards. As previously noted, (EC) Regulation No. 834/2007 gives market access to organic products according to national standards which are equivalent to the EU standards.

Other organic products coming into the EU must be inspected and certified that they meet EU regulations, until a list of "authorized" certification bodies and authorities in these countries can be compiled by the Commission.

The European organic meat products market is highly concentrated, with approximately 15 companies leading the scene. Conventional meat companies dominate most national markets entering the organic sector either via directly supplying organic meat products or by acquiring dedicated organic meat companies (Organic Monitor, 2003).

4.4.2 Production and supply of organic meat in Europe

4.4.2.1 *Number and size of organic holdings in Europe*

The period 2002–2008 has been one of slow growth and consolidation for the organic sector, following the very rapid expansion phase in 1999–2000. In 2009, however, the sector has experienced a miniboom in organic agricultural land (Figure 4.2).

At the end of 2009, in Europe, nearly 260,000 farmers manage organically 9.3 million hectares of agricultural land. In Europe, 1.9% of the agricultural area, and in the EU 4.7% of the agricultural area is organic. Non-EU member European countries account for approximately 900,000 hectares. Twenty-five percent of the world's organic land is in Europe. With respect to the previous year, organic land has increased by almost 12%, nearly 1 million

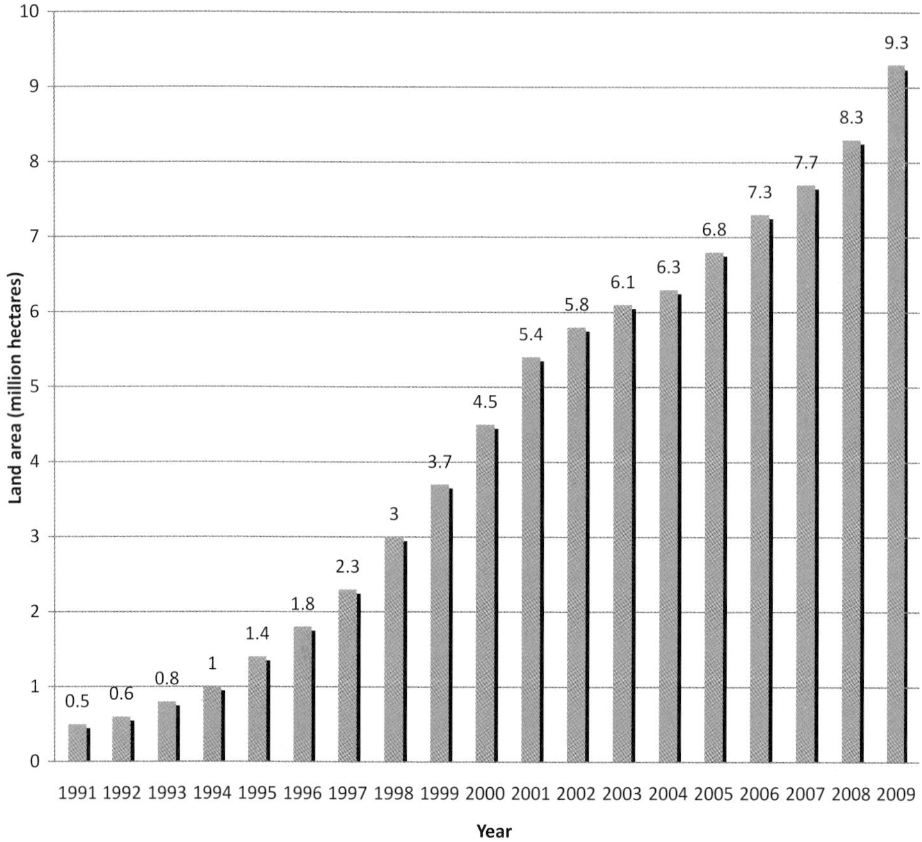

Figure 4.2 Development of organic agricultural land since 1985. (From Willer & Kilcher, 2011.)

hectares. Spain is the country with the largest organic agricultural area (1.3 million hectares) followed by Italy (1.1 million hectares) and Germany (0.95 million hectares) (Willer & Kilcher, 2011).

The share of organic land over total land is above 10% in the following countries: Liechtenstein (26.9%), Austria (18.5%), Sweden (12.6%), Switzerland (10.8%), and Estonia (10.5%). In addition to agricultural land, there are other certified areas such as wild collection, aquaculture, and grazing land outside the agricultural land and forests. Finland has the largest organic wild collection area in the world with 7.8 million hectares, mainly occupied by wild berries (Willer & Kilcher, 2011).

In addition to farmers, there are almost 35,000 processors and 2500 importers in Europe (Willer & Kilcher, 2011).

4.4.2.2 *Number of organic livestock in Europe*

Organic livestock have increased steadily in most parts of Europe between 2007 and 2009, the only exception being goats, which declined in total numbers by 13% to 567,880 in 2009. Eurostat (2011) data show that cattle numbers for meat production have increased by 56% in those 2 years, following the increase in demand. Beef represents the largest meat segment, as has been already mentioned. Other species show different growth rates. Pigs have increased by 15%, sheep are almost steady (growing only by 3.5%), and as we have discussed, goats were reduced in number. Poultry numbers have increased by 21%, with broilers exhibiting the largest increase (38%), while other poultry (ducks, turkey, geese) increased by 22%.

Eurostat (2011) data show marked unexplained variations between years in some countries that may be a reflection of reporting errors in the early stages of establishing the data collection system, especially in the new EU member countries. Most of the reported increases are attributable to these countries, while the rest of the EU exhibits a more steady and slow rate of growth. For this reason, only 2009 detailed data per country are shown here (Table 4.1).

Germany represents 27% of total European stocks of beef cattle, followed by the United Kingdom (19%), Italy (13%), and Sweden (12%). The largest European share of pigs is raised in Denmark (26%), followed by Germany (17%), the Netherlands (12%), and Austria (10%). Sheep are mainly in the United Kingdom (27%) and Italy (20%). Goats are primarily in Greece (55%) and Italy (13%). Italy holds the largest share of almost 14 million poultry (18%), followed by Germany (15%) and Denmark (10%). Eurostat statistics do not have any data for France, but report considerable numbers of broilers (more than 6 million), which—if confirmed—will make France the largest organic poultry producing country in Europe with more than 50% of total broiler production.

4.4.2.3 *Organic meat prices*

Systematically updated information on prices for organic meat and meat products is generally lacking. The most recent systematic survey was performed in 2001 (producer prices) and 2002 (consumer prices) by the OMIaRD (Organic Marketing Initiatives and Rural Development) research project (Hamm & Gronefeld, 2004).

There was considerable variability among the 15 member states in farm gate prices for beef as well as in consumer prices for minced beef and rump steak. Organic price premiums at the farm gate level varied from 17% in Denmark to 190% in Spain and in the EU-15 market were an average of approximately 49%. In general, farm gate and consumer prices in Southern European countries (Spain, Greece, and Italy) were more than 20% above the weighted EU average, reflecting the higher supply imbalance in these countries.

At the consumer level, the price for organic minced beef was highest in France, Greece, and Luxembourg while it was relatively low in Sweden, Spain, and Austria. Organic consumer price premiums ranged from 4% in Portugal to 126% in Luxembourg and reached 57% on EU-15 average. Consumer prices for organic rump steak were particularly high in Germany but low in Spain and Finland. Price premiums for organic steak were, on average, slightly higher than minced beef, reaching 59% on EU-15 average.

Farmer prices for organic pork in 2001 was approximately 2.46 €/kg on EU-15 average. Producer price premiums for organic pork range from 45% in Germany and Austria to 132% in the Netherlands while the EU-15 average was approximately 62% (Hamm & Gronefeld, 2004).

Consumer price for organic pork cutlets differed considerably between the 15 member states: the Danish price was approximately double the price in Spain, Portugal, and Finland. The average EU-15 farmer price premium was 62%. Consumer price premium for organic pork cutlet ranged widely from 0% in Portugal to 165% in Greece, with a weighted EU-15 average of 81%. These prices reflected the relative underdevelopment of pig farming in many countries at that time. Poultry price premiums were the highest in absolute

Table 4.1 Number of organic livestock in Europe (2009).

Country	Bovine animals (total)	Bovine animals for meat production	Other bovine animals	Pigs (total)	Sheep (total)	Goats (total)	Poultry (total)	Other poultry (turkeys, ducks, geese, etc.)
Belgium	53,338.00	1,291.00	40,699.00	10,348.00	9,211.00	3,126.00	1,233,098.00	699.00
Bulgaria	272.00	:	100.00	104.00	5,831.00	2,732.00	:	:
Czech Republic	136,026.00	11,419.00	121,993.00	1,990.00	53,038.00	4,352.00	25,292.00	2,168.00
Denmark	170,155.00	:	108,031.00	185,828.00	10,640.00	3,566.00	1,357,375.00	119,856.00
Estonia	21,074.00	13,698.00	4,322.00	328.00	39,374.00	709.00	8,099.00	287.00
Ireland	32,700.00	7,500.00	23,800.00	840.00	31,400.00	680.00	57,750.00	2,750.00
Germany	250,000.00	130,000.00	120,000.00	122,000.00	132,000.00	:	2,091,000.00	357,000.00
Greece	28,618.00	9,015.00	3,075.00	54,631.00	357,499.00	309,060.00	266,182.00	4,452.00
Spain	128,004.00	:	:	8,052.00	459,364.00	50,488.00	154,137.00	:
France	:	:	:	:	:	:	:	:
Italy	185,513.00	62,949.00	78,254.00	25,961.00	658,709.00	74,500.00	2,399,885.00	42,934.00
Cyprus	0.00	:	:	0.00	612.00	3,333.00	10,760.00	:
Latvia	53,867.00	24,062.00	24,337.00	9,637.00	31,251.00	7,216.00	33,309.00	:
Lithuania	21,927.00	3,941.00	9,678.00	275.00	13,001.00	755.00	1,510.00	240.00
Luxembourg	3,252.00	:	:	830.00	425.00	126.00	22,351.00	1.00
Hungary	25,089.00	20,316.00	1,332.00	6,447.00	11,123.00	2,066.00	113,939.00	13,352.00
Malta	0.00	:	:	0.00	0.00	0.00	0.00	:
Netherlands	59,000.00	18,000.00	17,000.00	85,000.00	23,000.00	29,000.00	:	:
Austria	373,720.00	:	:	69,849.00	94,130.00	35,899.00	1,227,553.00	:
Poland	51,391.00	11,178.00	20,804.00	18,664.00	39,159.00	6,333.00	186,311.00	45,090.00
Portugal	:	:	:	:	:	:	:	:
Romania	8,145.00	0.00	3,842.00	603.00	51,470.00	4,738.00	9,400.00	0.00
Slovenia	18,238.00	870.00	16,196.00	2,149.00	35,751.00	5,569.00	21,904.00	1,474.00
Slovakia	33,486.00	13,524.00	15,551.00	266.00	102,134.00	1,516.00	4,324.00	509.00
Finland	32,354.00	2,444.00	25,018.00	2,607.00	11,935.00	76.00	175,346.00	8.00
Sweden	194,063.00	59,394.00	100,554.00	42,502.00	86,741.00	1,178.00	786,430.00	144.00
United Kingdom	331,156.00	89,735.00	96,320.00	48,151.00	884,810.00	133.00	3,958,669.00	78,182.00
EU27	**2,158,050.00**	**478,045.00**	**790,207.00**	**686,714.00**	**3,133,397.00**	**544,025.00**	**12,911,526.00**	**668,447.00**
Norway	23,863.00	:	16,241.00	2,486.00	45,423.00	1,337.00	244,652.00	26,747.00
Switzerland	150,905.00	:	:	15,406.00	86,997.00	19,392.00	510,749.00	2,387.00
Total Europe	**2,332,818.00**	**478,045.00**	**790,207.00**	**704,606.00**	**3,265,817.00**	**564,754.00**	**13,666,927.00**	**697,581.00**

Source: Eurostat; AMI (Germany only).
: indicates missing data.

terms, reaching an EU-15 average of 266% for farm gate prices. Conversely, consumer price premium on whole chicken were "only" 129% on average.

More recent data are only available for some countries. For example, Cullen *et al.* (2007) reported relatively stable prices for beef, pork, and sheep meat from 2002 to 2007 in the United Kingdom; only poultry farm gate prices showed positive dynamics, increasing approximately 10%/year since 2003. According to Cullen *et al.* (2007), poultry price dynamics reflect producer–processor prices selling to smaller scale specialists or direct sale markets. Prices achieved by large-scale poultry producers supplying multiple retailers would be expected to be lower than this.

In 2007, consumer price premiums in the United Kingdom for organic lamb ranged from −2%–65% for different meat cuts, while the organic premium on beef ranged from 16% for rib eye steak to 47% for roasting meat joint. Higher premiums were scored by poultry meat, again, with the whole chicken scoring 118% but with the thigh fillets reaching 216% (Tesco Pricecheck as reported by Cullen *et al.*, 2007).

In 2010, ISMEA reported increasing farmer price premiums for organic beef in Italy, caused by both increasing organic producer prices and declining conventional prices. A similar dynamic was observed for pork meat. Consumer price premiums were stable or slightly reduced for beef and cooked poultry specialties, while fresh chicken breast and sliced cured pork (e.g., ham and salami) were increasing (ISMEA, 2010).

4.5 CONSUMER ISSUES

Studies carried out in different European countries have found that consumer's motivations for purchasing organic food are mainly related to concerns for health, environment, and animal welfare (Padel *et al.*, 2005; Baker *et al.*, 2004; Gambelli *et al.*, 2003; Harper *et al.* 2002; Zanoli & Naspetti, 2002; Zanoli, 2004). With respect to organic meat and meat products, the main motivations for purchasing them have been reported to be (Zanoli, 2004): health; animal welfare and other ethical and environmental concerns; local and regional origin; better taste (enjoyment of food); and higher trust.

Concerning health, Zanoli (2004) showed that—for meat and meat products—in Europe, the health motivation is strictly linked with the appropriate husbandry. Using means–end chain analysis, Figure 4.3 illustrates how consumers perceive organic meat and its attributes in relation to their personal goals regarding health. Organic meat is appropriately produced by using fewer drugs or hormones, with natural and healthy fodder and higher freedom of movement for animals. This production method—in the eyes of consumers—leads to better quality food and a healthier diet, which enhances well-being and health.

The same study found that for most regular consumers, animal welfare is the main motivation when buying organic meat products. Consumers link free-range animals and lesser chemicals and hormones to appropriate husbandry and animal welfare as a value (Figure 4.4). Knowing that animals are kept with care makes most consumers feel good by soothing their conscience (Zanoli, 2004).

Given the leading role of animal welfare as a purchasing motive for organic meat, other ethical and environmental concerns seem to play a minor role when purchasing organic meat. However, in a recent study, Zanoli *et al.* (2011) concluded that organic beef is attractive to Italian consumers not only because it is associated with higher animal welfare standards but also with other ethical issues, namely lower food miles and preservation of local breeds.

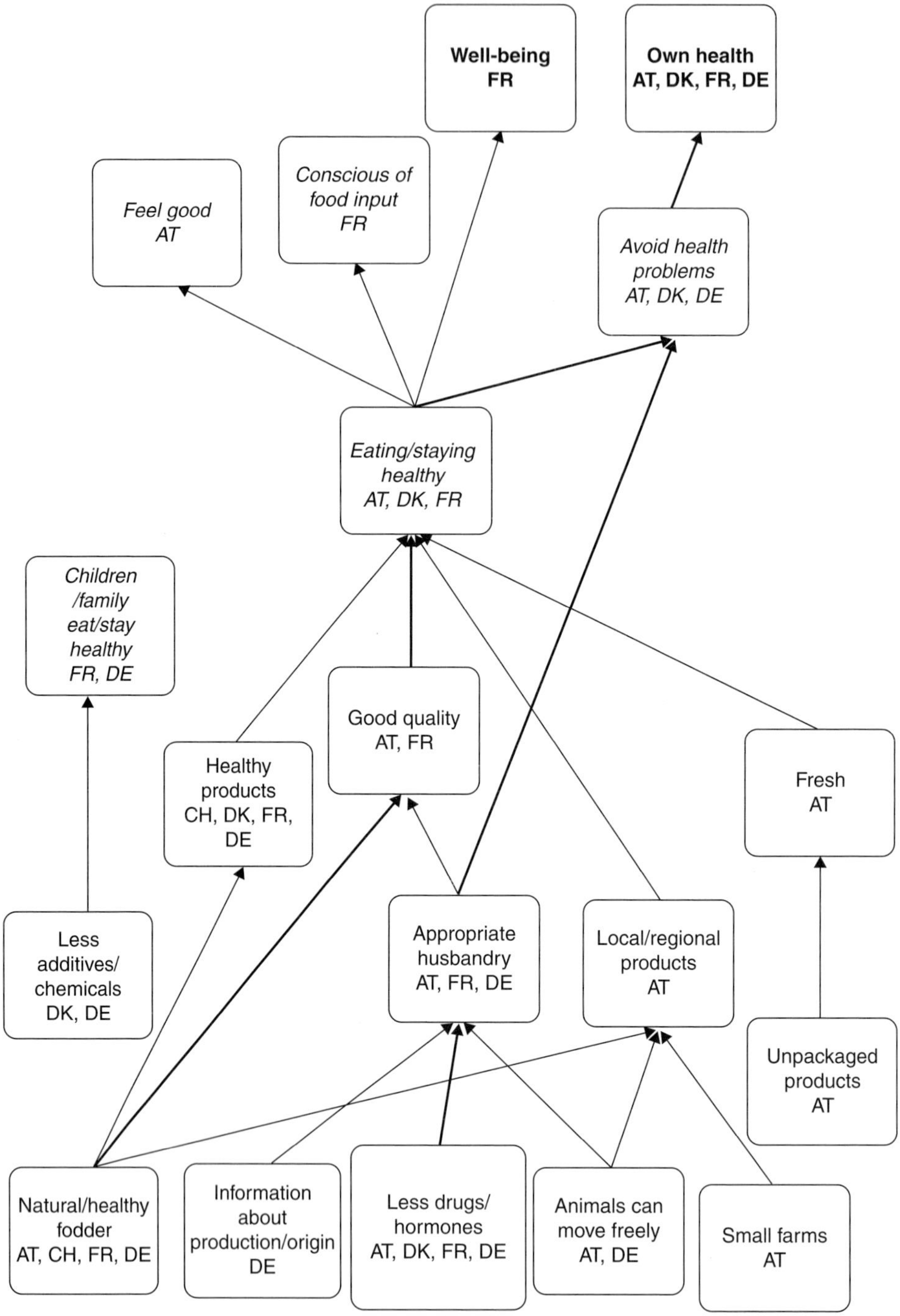

Figure 4.3 Meat and meat products: cognitive structure of the health motivation. (According to Zanoli, 2004.)

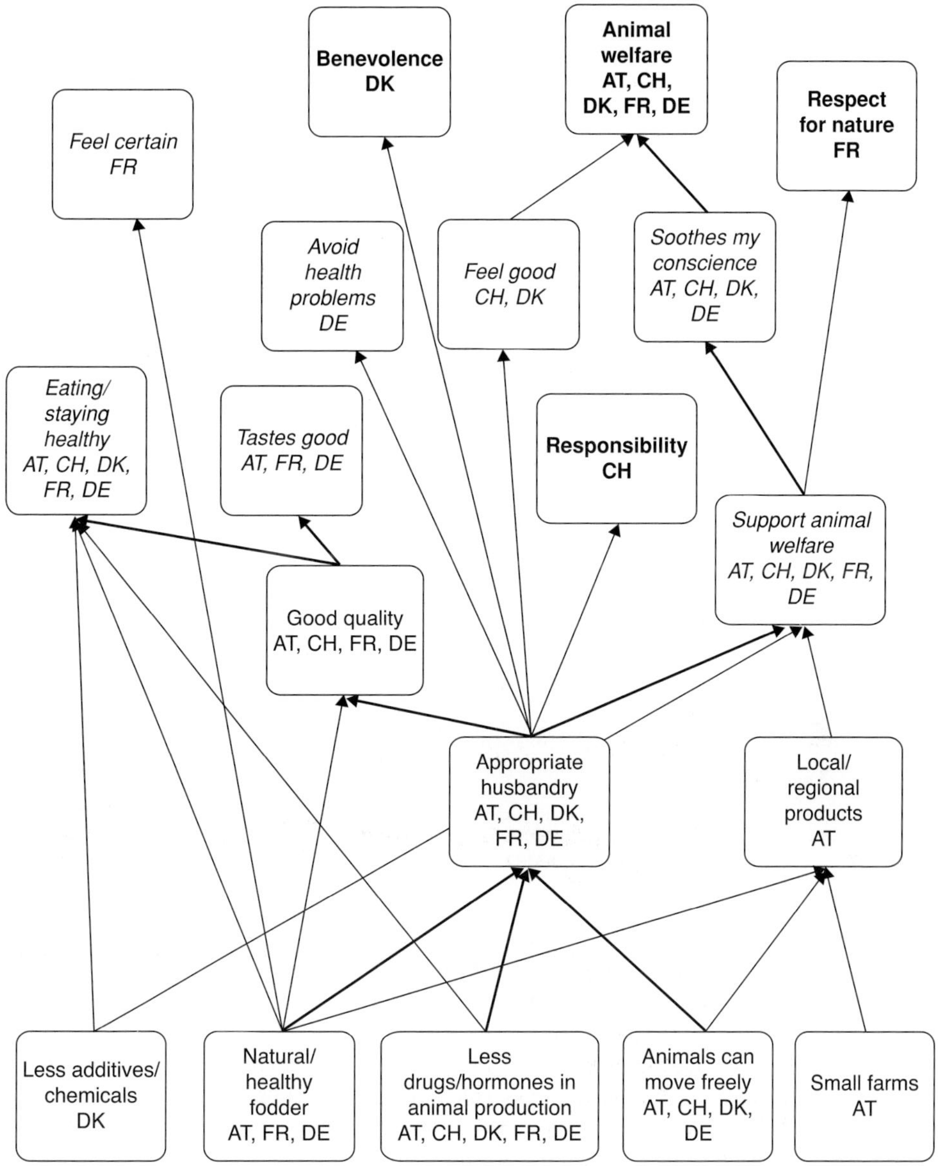

Figure 4.4 Meat and meat products: cognitive structure of the animal welfare motivation. (According to Zanoli, 2004).)

Consumers also care about appropriate husbandry for more selfish reasons such as enjoyment of food. It is associated with higher quality meat (AT, CH, FR, DE) and, therefore, with good taste (AT, FR, DE). The enjoyment of food as a value has a simpler cognitive map (Figure 4.5).

Higher trust in organic meat has been also reported, especially after the bovine spongiform encephalopathy crisis, although in Germany this was severely questioned after a major animal feed scandal in 2002. Germany had the largest organic meat products market in Europe at that

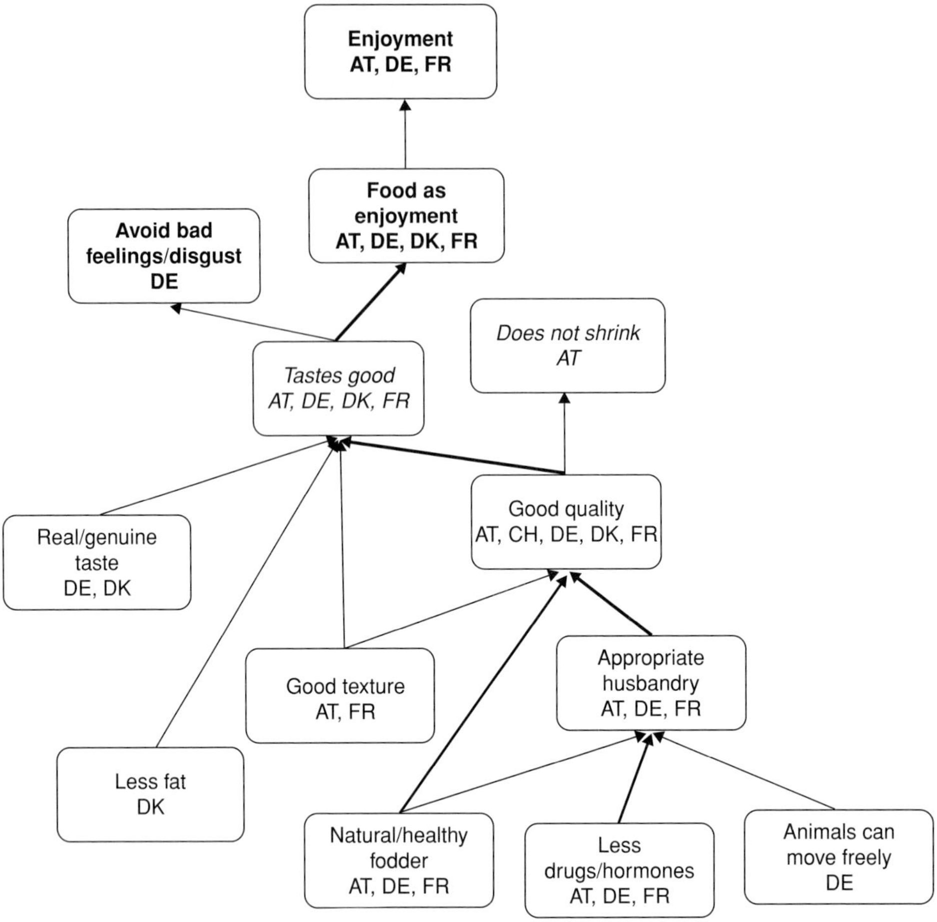

Figure 4.5 Meat and meat products: cognitive structure of the food as enjoyment motivation. (According to Zanoli, 2004).)

time; however, sales drastically fell in 2002 after the banned pesticide Nitrofen was found to be fed to organic poultry on German farms. Hundreds of thousands of organic chickens were slaughtered in June, 2002, as retailers cleared their store shelves of organic poultry meat and organic eggs. Although consumer confidence in organic foods was returning in the latter part of the year, some supermarkets had withdrawn organic meat products indefinitely (Organic Monitor, 2002).

According to Napolitano *et al.* (2009a), the main barrier to purchasing organic meat remains price because of high production costs, which are affected by organic rules (higher space allowance, origin of feedstuffs, etc.) and small-scale production systems. Connected but somehow independent to price is availability, especially in supermarkets or other mainstream marketing channels, which still serves as a barrier to more widespread consumption in some European countries, as has already been pointed out by Zanoli (2004).

However, for European consumers, price is not the only determinant behind animal-food purchases. Consumers do not seek the lowest priced foods but foods with the best value for the price, i.e., the maximum benefit for what they are prepared to spend (Napolitano *et al.*, 2009a). Dransfield *et al.* (2005) found a consistent effect of organic labeling on the price offered by consumers for organic pork. This effect is connected to the trust on organic labels, which—as Naspetti and Zanoli (2005) have shown—are important search cues for organic consumers since "organic" is a credence attribute that cannot be confirmed by experience.

Other barriers may be more perceived barriers than actual ones. As Naspetti (2010) has pointed out in her review on organic livestock and quality, most studies do not report substantial quality and sensory differences between conventional and organic meat, although in some instances organic beef has been reported to be less tender. For the concerns associated with poultry meat, the main barrier for the diffusion of organic poultry meat could be the use of slow-growing breeds, since people are now acquainted to the texture of fast-growing strains (Castellini *et al.*, 2002), which according to the EC Regulation No. 834/2007 are under scrutiny in terms of animal welfare.

Napolitano *et al.* (2009a) examined the effect of information on organic beef preferences. The study has shown that Italian consumers are influenced by information about organic production and shift their actual acceptability in the direction of expected liking, possibly because consumers are aware of the ethical value of organic farming and its effects on product safety. In particular, the hedonic discrepancy was totally assimilated indicating that actual liking of organic beef was markedly affected by information. Conversely, the information concerning conventional production had a detrimental effect on expectancy. Accordingly, consumers were willing to pay more for organic beef as compared to the suggested price, thus indicating that reliable information about the organic farming system may markedly increase consumer willingness to pay.

4.6 CONCLUSIONS

The organic meat sector is a highly dynamic market segment, owing to the specific characteristics of the relevant supply chains. On the supply side, the organic meat market in Europe, despite the new regulations, still has some difficulties in adapting to the demand, not only in terms of quantities but also for aspects related to meat quality, logistic, and distribution. Organic meat producers will be faced with increasing global competition, meaning volatile prices and uncertain returns. This will not favor further conversion, given the magnitude of assets at risk.

On the demand side, the future development of organic meat consumption relies on how the consumer needs will be matched by producers and retailers. Competing conventional meat producers increasingly incorporate animal welfare and other ethical considerations in their production standards, thereby reducing the perceived gap between organic and conventional meat products.

The most promising strategies for the development of the organic meat market rely on both the diversification of the sale channels and the information given to the consumers. The spread of organic meat in large retail chains will increase the stability of the market. This will reduce producer's risk and increase the overall supply, inducing, as a consequence, a price reduction of the consumer organic meat prices. Furthermore, valid and reliable information on product origin as well as on the ethical values connected with the organic production system and its effects on product safety may increase consumers' willingness to buy—and pay—for organic meat.

REFERENCES

Baker, S., K. Thompson, and J. Engelken. 2004. Mapping the values driving organic food choice. *Eur. J. Mark.*, 38:995–1012.

Castellini, C., C. Mugnai, and A. Del Bosco. 2002. Effect of organic production system on broiler carcass and meat quality. *Meat Sci.* 60:219–225.

Cullen, R., N. Lampkin, and S. Moakes. 2007. *Review Of The Market For Welsh Organic Meat.* Organic Centre Wales, Aberystwyth. p. 58.

Dransfield, E., T. M. Ngapo., N. A. Nielsen, L. Bredahl, P. O. Sjödén, M. Magnusson, M. M. Campo, and G. R. Nute. 2005. Consumer choice and suggested price for pork as influenced by its appearance, taste and information concerning country of origin and organic pig production. *Meat Sci.* 69:61–70.

EUROSTAT. 2011. Statistics database. Available at: http://epp.eurostat.ec.europa.eu/portal/page/portal/statistics/search_database (accessed April, 2011).

Gambelli, D., S. Naspetti, and D. Vairo. 2003. Why are consumers buying organic meat and milk? A qualitative study of the Italian market. In: M. Hovi, A. Martini, and S. Padel (eds) *Proc. 1st SAFO Workshop "Socio-Economic Aspects of Animal Health and Food Safety in Organic Farming Systems"*, Florence. pp. 125–142.

Hamm, U. and F. Gronefeld. 2004. The European market for organic food: revised and updated analysis. In: O. Schmid, J. Sanders, and P. Midmore (eds) *Organic Marketing Initiatives and Rural Development.* Vol. 5. University of Wales, Aberystwyth. p. 165.

Harper, G. and A. Makatouni. 2002. Consumer perception of organic food production and farm animal welfare. *Br. Food J.* 104:287–299.

IFOAM EU Group. 2009. The IFOAM EU Group's initial Assessment of the new EU organic regulations. IFOAM EU Group, Brussels. p. 11.

ISMEA. 2010. Osservatorio del mercato dei Prodotti Biologici (August 2010). ISMEA News (Prodotti Biologici). 1/10. p. 14.

Napolitano, F., A. Braghieri, E. Piasentier, S. Favotto, S. Naspetti, and R. Zanoli. 2009a. Effect of information about organic production on beef liking and consumer willingness to pay. *Food Qual. Prefer.* 21:207–212.

Napolitano, F., F. Girolami, and A. Braghieri. 2009b. Organic meat: market development and consumer willingness to pay. In: M. Nelsonand I. Artamova (ed.) *Organic Farming: Methods, Economics and Structure.* Nova Science Publishers, Hauppauge, NY. pp. 113–126.

Naspetti, S. and R. Zanoli. 2005. Consumers' knowledge of organic quality marks. In: U. Köpke, U. Niggli, D. Neuhoff, P. Cornish, W. Lockeretz, and H. Willer (eds) Researching sustainable systems. *Proc. 1st Sci. Conf. Int. Soc.Org. Agric. Res. (ISOFAR)*, Held in Cooperation with the International Federation of Organic Agriculture Movements (IFOAM) and the National Association for Sustainable Agriculture, Australia (NASAA), September 21st–23rd, Adelaide Convention Centre, Adelaide, South Australia FiBL, Frick, CH, and International Society of Organic Agriculture Research (ISOFAR), c/o Institute of Organic Agriculture (IOL), University of Bonn, Bonn, DE. pp. 393–395.

Naspetti, S. 2010. Zootecnia biologica e qualità agroalimentare. *Economia Agro-alimentare* 12:123–143.

Organic Monitor. 2002. *The German Market for Organic Meat Products.* Organic Monitor, London. p. 70.

Organic Monitor. 2003. *The European Market for Organic Meat Products.* Organic Monitor, London. p. 440.

Organic Monitor, 2010. *The Global Market for Organic Food & Drink: Business Opportunities & Future Outlook.* Organic Monitor, London. p. 270.

Padel, S., O. Schmid, and V. Lund. 2004. Organic livestock standards. In: Vaarst, M. (ed.) *Animal Health and Welfare in Organic Agriculture.* CABI, Wallinford. pp. 57–72.

Schmid, O. 2009. Animal production. In: C. Mikkelsen and M. Schluter. *The New EU Regulation for Organic Food and Farming: (EC) No 834/2007. Background, Assessment, Interpretation.* IFOAM EU Group, Brussels, BE. pp. 33–34.

Willer, H. and L. Kilcher. Eds. 2011. The world of organic agriculture. Statistics and emerging trends 2011. FiBL-IFOAM Report. IFOAM, Bonn, DE and FiBL, Frick, CH. p. 288.

Zanoli, R. 2004. *The European Consumer and Organic Food.* Organic marketing initiatives and rural development series: volume 4. University of Wales. Aberystwyth. p. 175.

Zanoli, R. and S. Naspetti. 2002. Consumer motivations in the purchase of organic food: a means-end approach. *Br. Food J.* 104:643–653.

Zanoli, R., R. Scarpa, F. Napolitano, E. Piasentier, S. Naspetti, and V. Bruschi. 2011. Organic label as identifier of environmentally-related quality: a consumer choice experiment on beef in Italy. Paper submitted to *Ren. Agr. and Food Sys.*

5 Organic Meat Marketing

Ellen J. Van Loo, Vincenzina Caputo, Rodolfo M. Nayga, Jr., Maurizio Canavari, and Steven C. Ricke

Abstract: This chapter discusses the marketing aspect of organic meat. Many factors influence the organic food purchase behavior and these factors are described. With organic meat being sold for high premium prices, the consumers' willingness to pay for these products is an important attribute to determine the potential of these products. Sociodemographic variables of the organic food consumers are described to get an overview of the organic consumer profile.

Keywords: marketing; willingness to pay; organic meat; consumer perceptions; natural; local

5.1 INTRODUCTION

In the last decade, the sales of organic foods in the United States have increased every year. In 2000, the total organic food sales were only $6.1 billion and grew to an estimated $26.7 billion in 2010 (Organic Trade Association (OTA), 2011). While the total US food sales grew by less than 1% in 2010, the organic food industry grew by 7.7% (OTA, 2011). Although organic meat is only a small portion of the total meat, the organic meat, poultry, and fish categories have grown remarkably from $23 million in 2000 to $476 million in 2010 (OTA, 2011).

In 2010, one in five consumers (18%) purchased organic or natural meat in the past 3 months before being surveyed (Food Marketing Institute (FMI) & American Meat Institute (AMI), 2010). Organic and natural chicken is the most popular natural or organic meat variety. According to the 2010 study conducted by the FMI and the AMI, 70% of the consumers buying organic meat purchased chicken and 46% purchased beef (Figure 5.1). With an increase in organic meat availability, natural and organic food stores are losing market share (responsible for 19% of the natural and organic meat purchases) and conventional supermarkets are now the top outlets for organic meat (responsible for 50% of the natural and organic meat purchases) (Figure 5.2).

5.2 CONSUMERS' PURCHASING DRIVERS AND DETERRENTS

The organic product purchase decision is influenced by many factors. The consumer's knowledge and awareness are important as well as product-related exogenous factors, economic

Organic Meat Production and Processing, First Edition. Edited by Steven C. Ricke, Ellen J. Van Loo, Michael G. Johnson, and Corliss A. O'Bryan.

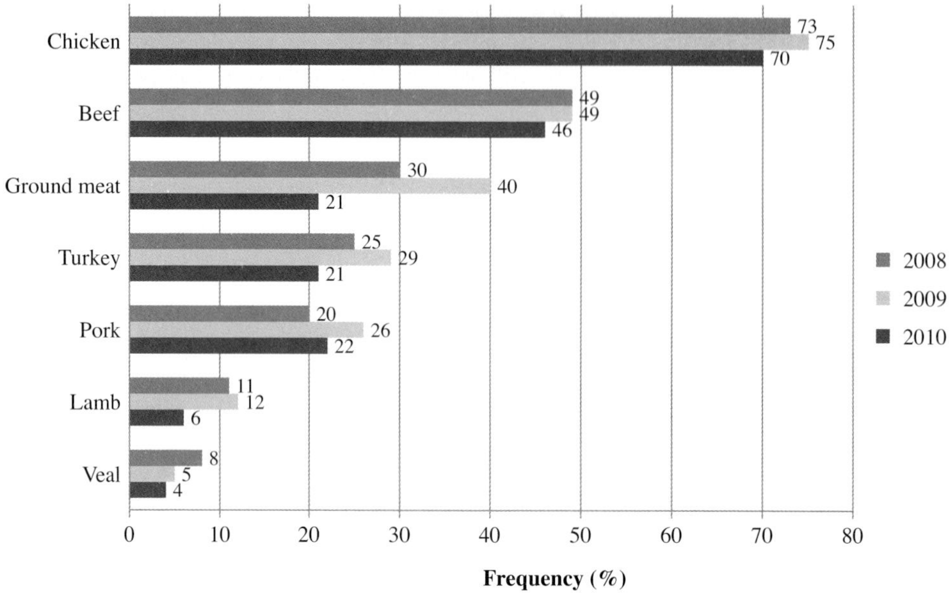

Figure 5.1 Frequency of purchase for natural and organic meat and poultry by those who purchased organic meat in the past 3 months (FMI & AMI, 2010).

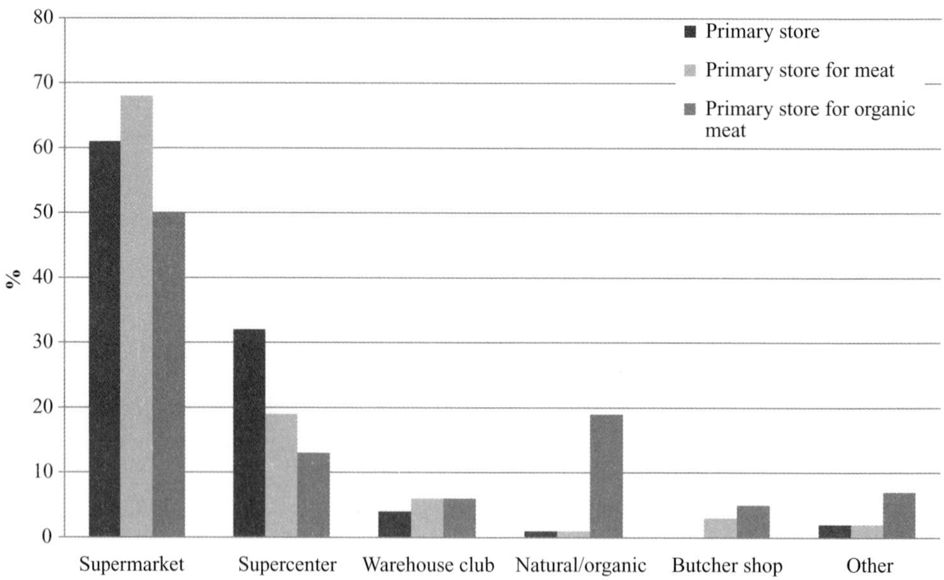

Figure 5.2 Primary store for organic meat purchases versus meat and groceries in general (FMI & AMI, 2010).

issues, and social demographic variables. Whether consumers buy organic food products or not is also influenced by their perceptions and attitudes as well as their ability/willingness to pay premium prices. All these factors will have an effect on the consumer preference and attitude and thus will influence the purchase decision (Harper & Makatouni, 2002; Yiridoe *et al.*, 2005; Aertsens *et al.*, 2009).

Consumer awareness and their respective knowledge have an effect on their attitudes, purchase decisions, and willingness to pay a price premium. The organic food label is an important tool to inform consumers that products are organic. Without a label, the consumer may not be aware that the product is organic because differentiation between conventional and organic food is often difficult. Organic products are produced differently but do not necessary have obvious visual contrasts with conventional products. The knowledge, awareness, and positive attitude toward organic foods do not always translate into purchasing organic foods. Some barriers might prevent the consumer from purchasing it. An overview of the most important motives why consumers purchase or not purchase organic foods are given in Table 5.1.

5.2.1 Purchasing drivers

Multiple studies have explored the rationale of the increased organic food consumption. The most common reported drivers are the belief that organic foods are healthier and safer (Jolly *et al.*, 1989; Harper & Makatouni, 2002; Zanoli & Naspetti, 2002; Kouba, 2003; Magnusson

Table 5.1 Reasons for buyers and nonbuyers of organic food.

Consumers' purchasing drivers	Deterrents
Health and nutritional concern[a–i]	High price premiums[a,c,f,h,i,j]
Food safety, lack of confidence in the conventional food industry[a–c,e,g,i,j] Concern over animal welfare[a–f,h]	Poor organic food availability[a,b,i,j] Skepticism of certification boards and organic labels[b]
Concern for environment[a–h]	Insufficient marketing, poor merchandising[c]
Support of local economy[c]	Satisfaction/content with current food supply[a,c]
Superior taste[a,c,d,f–l]	Disbelieve that organic food is better
No GM and irradiation[k] Freshness[d,l]	Unsatisfactory quality (possible mainly focused on appearance of fresh produce) such as sensory defects[a,c]
Nostalgia[c,d]	Unfamiliar with the term "organic," certification systems, and organic logos[a,l]
Fashionable/curiosity[c]	

[a]Bourn & Prescott (2002).
[b]Harper & Makatouri (2002).
[c]Hughner *et al.* (2007).
[d]Krystallis *et al.* (2006a).
[e]Magnusson *et al.* (2003).
[f]Stobbelaar *et al.* (2007).
[g]Tsakiridou *et al.* (2008).
[h]Wier & Calverley (2002).
[i]Yiridoe *et al.* (2005).
[j]O'Donovan & McCarthy (2002).
[k]Kouba (2003).

et al., 2003; Magkos *et al.*, 2006; Hughner *et al.*, 2007; FMI & AMI, 2010). Although health and food safety concerns are undoubtedly the main motivations for organic food purchases, ethical and moral perceptions such as concerns for the environment and animal welfare are also significant factors in the decision (Harper & Makatouni, 2002). Thus, a mixture of private and public benefits is provided by consuming organic food (Canavari *et al.*, 2005).

According to Wier and Calverley (2002), "the majority of the consumers purchase organic foods because of the product-specific characteristics which directly benefit them rather than the production-process-specific characteristics which indirectly benefit them." Therefore, in most cases, the egoistic motives (health) might better predict organic food purchases than the altruistic motives (environment and animal welfare) (Magnusson *et al.*, 2003). Studies often include four key motivations: (1) health, (2) food safety, (3) impact on the environment, and (4) the well-being of animals (Soil Association UK, 2001; Harper & Makatouni, 2002; Kouba, 2003; Magkos *et al.*, 2006; Van Loo *et al.*, 2010).

Purchasing drivers consumer attitudes, and interest in organic foods may be linked to sociodemographic characteristics of the consumer such as age, gender, level of education, and household size (Davies *et al.*, 1995; Magnusson *et al.*, 2001; Tsakiridou *et al.*, 2008) because of association between purchasing motives and consumer profile. High income and education levels are strongly correlated with food safety and environmental concerns and are drivers to buy organic foods for consumers in this group (Tsakiridou *et al.*, 2008). Women and older people are more willing to pay a premium for organics possibly due to an increased level of concern about food safety and health issues in these consumer groups (Tsakiridou *et al.*, 2008).

5.2.1.1 *Health and food safety*

The purchase of organic foods is strongly related with the perceived benefit for health. There is a widespread belief that organic foods are more nutritious, healthier, and safer than conventionally produced foods. Health concern is often found as the most important driver to purchase organic food, together with food safety (Jolly *et al.*, 1989; Chinnici *et al.*, 2002; Harper & Makatouni, 2002; Zanoli & Naspetti, 2002; Kouba, 2003; Magnusson *et al.*, 2003; Krystallis *et al.*, 2006b; Magkos *et al.*, 2006; Hughner *et al.*, 2007; Mintel, 2008).

There is little information available on the motivation for organic meat purchases specifically. O'Donovan and McCarthy (2002) examined Irish customers' perception of organic meat and concluded that food safety and human health are the key factors for organic meat purchases. Similarly, Van Loo *et al.* (2010) reported health and food safety as the key drivers for organic meat purchases in the United States. The report from FMI and AMI (2010) listed positive long-term personal health effects and better nutritional value as the top reasons for organic and natural meat purchases. According to O'Donovan and McCarthy (2002), consumers rate food safety as the most important meat product attribute and organic meat buyers believe organic meat is superior in quality and food safety compared to conventionally produced meat products.

Even though consumers purchase organic foods mainly because of perceived health and food safety benefits, it remains to be proven that organic food is healthier and/or safer. Most of the research concluded that there is no evidence that organic food is healthier or more nutritious (Williams, 2002; Magkos *et al.*, 2003; Dangour *et al.*, 2010). Consumers are often not aware that the organic standards are only based on the production and processing of the product and not on the final quality of the product (Brennan *et al.*, 2003). This indicates the need to educate the consumer more about the organic certification. The United States

Department of Agriculture (USDA) National Organic Program (NOP) requirements apply to the production process itself rather than the product. It does not directly address food safety or nutrition. It, therefore, does not imply that organic food is safer or healthier than conventional food (Bourn & Prescott, 2002; Winter & Davis, 2006; Gold, 2008; USDA National Organic Program, 2008), but it may have consequences for product quality. With the desire to not use perceived dangerous synthetic chemicals in organic food production, consumers may assume organic food as intrinsically safer, although regulation and public authorities state that food with residual contents below the legal allowance limits is safe.

A majority of the consumers have concerns about chemicals such as hormones, pesticides, herbicides, and antibiotics. Consumers try to avoid the chemicals used in the conventional food industry. The consumers who want to avoid genetically modified organisms (GMOs) or food irradiation might purchase organic food (Kouba, 2003; Magkos *et al.*, 2003; Anderson *et al.*, 2006). Recently, these worries have increased because of the negative publicity about genetically modified (GM) crops, which are sold or fed to livestock, and food irradiation (McEachern & McClean, 2002). Consumers have misconceptions about irradiation techniques such as altering the food quality or making food radioactive (Kouba, 2003). Furthermore, consumer concerns about food safety have been fuelled with food scares such as bovine spongiform encephalopathy, foot and mouth disease, *Salmonella*, and *Escherichia coli* outbreaks. This has influenced the food-buying behavior and has increased the demand for organic foods. The perception of health benefits and food safety are one of the drivers for the growth of the organic market.

5.2.1.2 *Environment*

Although health and food safety concerns are identified as the main reason for organic food purchases, respecting and protecting the environment is for many consumers an important attribute as well (Bourn & Prescott, 2002; Harper & Makatouri, 2002; Wier & Calverley, 2002; Magnusson *et al.*, 2003; Hughner *et al.*, 2007; Van Loo *et al.*, 2010). Organic certification such as the USDA-certified organic food does not directly guarantee that products so labeled are more "environmentally friendly" than nonorganic products. However, organic agricultural practices and organic standards are connected with environmental conservation and ecological sustainability (Gold, 2008). In organic food production, fewer synthetic chemicals are allowed than in the conventional food production; therefore, organic food production can be viewed as more environmentally friendly (Gold, 2008). Ecologically oriented consumers are more likely to consume organic food (Hjelmar, 2011; Honkanen *et al.*, 2006). Other important environmental impacts measurements are food miles and CO_2 emission (life-cycle assessments) related to a food product. However, USDA organic certification does not address food miles or CO_2 emissions (USDA National Organic Program, 2008). In the future, labels specifically developed to inform the environmental impact of a product may become more prevalent. Both health and the environmental impact are important drivers for organic food purchases but health and food safety are the key drivers followed by the concern about the environment (Harper & Makatouni, 2002; Zanoli & Naspetti, 2002; Magnusson *et al.*, 2003; Durham, 2007; Hughner *et al.*, 2007; Van Loo *et al.*, 2010).

5.2.1.3 *Animal welfare*

The expectations of better animal welfare is a driver for organic purchases but is less important than health, food safety, and environment (Bourn & Prescott, 2002; Harper &

Makatouni, 2002; Wier & Calverley, 2002; Magnusson *et al.*, 2003; Krystallis *et al.*, 2006a; Hughner *et al.*, 2007; Stobbelaar *et al.*, 2007; Van Loo *et al.*, 2010). The way that animals live, are fed, and killed is important for the organic consumer. The consumer perceives the conventional food production methods as a threat to animals (e.g., the use of bovine somatotropin to increase milk yields) and perceives organic farming as more animal friendly than the conventional production. Consumers believe that animal welfare indicates food quality, food safety, and humane treatment of livestock (Harper & Makatouni, 2002). For the organic meat production, animal welfare is an important driver for organic meat purchases (Zanoli & Naspetti, 2002; McEachern & Willock, 2004; Van Loo *et al.*, 2010).

5.2.1.4 *Further purchasing drivers: taste, local, fashionable*

As listed in Section 5.1, there are many other consumers' drivers to buy organic foods. Consumers of organic food often believe that the organic products taste better (Bourn & Prescott, 2002; Wier & Calverley, 2002; Yiridoe *et al.*, 2005; Krystallis *et al.*, 2006a; Hughner *et al.*, 2007; Stobbelaar *et al.*, 2007; Tsakiridou *et al.*, 2008; Hjelmar, 2011). Some people believe that buying organic foods supports the local economy because they think the organic food is locally grown (Hughner *et al.*, 2007); however, this may not always be true. Organic food has received significant attention by the media and they have subsequently identified it as a fashionable item. In addition, it is associated with high prices. Therefore, some consumers perceive it as fashionable and trendy (Hughner *et al.*, 2007; Roitner-Schobesberger *et al.*, 2008). Nearly 30% of the respondents were reported as buying organic foods due to either something seen or heard in the news and media (Mintel, 2008).

5.2.2 Deterrents

There is a difference between the consumer attitudes toward organic foods and the actual purchase behavior. Magnusson *et al.* (2001) found that many consumers have positive attitudes toward organic food: from 46% to 67% of the population, depending on the food category. Nevertheless, only a small portion (4% to 10%) actually indicated an intention to purchase it. Similar results were reported by Bellows *et al.* (2008) who found 73% of the consumers that value organic farming do not purchase organic products. The price differentials, lack of knowledge, and product availability represent the major purchasing barriers for organic meat (O'Donovan & McCarthy, 2002; Van Loo *et al.*, 2010). These barriers explain the gap between values and behavior. The different obstacles to the purchase of organic foods are summarized in Table 5.1 and are discussed in the forthcoming sections.

5.2.2.1 *Price*

Price is the most important reason not to purchase organic food, especially the price difference between organic and conventional food products (Magnusson *et al.*, 2001; Bourn & Prescott, 2002; O'Donovan & McCarthy, 2002; Zanoli & Naspetti, 2002; Yiridoe *et al.*, 2005; Krystallis *et al.*, 2006b; Roitner-Schobesberger *et al.*, 2008; Tsakiridou *et al.*, 2008). Nonbuyers have the perception that the high price for organic food is not justified (Mintel, 2008). Nearly 78% of the respondents indicated that price was the major concern when buying organic foods and they would buy more organic food if it was less expensive (Mintel, 2008).

For organic meat in particular, Van Loo *et al.* (2010) reported the high price for organic meat to be the most important factor preventing organic meat purchases. More than half of the respondents strongly agreed with the high price being a discouraging factor (Van Loo *et al.*, 2010). Consumers would be willing to purchase organic meat more often if the price were more in line with the conventional meat prices (FMI & AMI, 2010). The high price is especially a barrier for lower income households. Mintel's US Organic Food Report (2008) suggested that a strong consumer base across a range of income levels would be important for the success of organic food products. As a result, the willingness to pay a price premium is an important characteristic and will be discussed later. According to Brennan *et al.* (2003), the low-income social groups cannot afford those prices and are therefore disadvantaged. Li *et al.* (2007), however, found that income does not significantly influence the decision to buy organic foods. In addition, the cost of organic foods should decrease as they become more widely available. Li *et al.* (2007) describes this as the "Walmart-effect." Siderer *et al.* (2005) believes that the decrease in price will be mainly caused by a cost reduction due to economies of scale in processing and distribution systems. With an increasing production level, there will be gains in the efficiency of production, processing, and distribution. This can keep the prices from deviating too much from conventional poultry products and making the organic meat products a more viable choice for price-sensitive consumers (Holcomb *et al.*, 2008). With an increasing production, organic products are expected to be priced more competitively with the conventional products (FMI, 2008).

5.2.2.2 *Limited organic food availability*

Poor organic food availability is a second obstacle to its purchase (O'Donovan & McCarthy, 2002; Wier & Calverley, 2002; Zanoli & Naspetti, 2002; Hughner *et al.*, 2007; Van Loo *et al.*, 2010). Van Loo *et al.* (2010) identified limited organic meat availability as the second major deterrent for organic meat purchases. Nearly 40% of the consumers currently not buying organic meat as well as 30% of the occasional organic meat consumers perceived it to be not likely or hardly likely that organic meat was available in their supermarket. However, only 14% of the habitual consumers had this perception. The organic food availability is improving since grocery stores increasingly offer organic foods and thus organic meat has become increasingly available in multiple channels (Mintel, 2008). Specialty organic and natural food stores represent 19% of the natural and organic meat purchases (FMI & AMI, 2010) (Figure 5.2). However, nearly half of the organic meat purchases in the United States are made in conventional supermarkets (FMI & AMI, 2010) (Figure 5.2) and this will most likely continue to increase as more conventional stores offer organic meat items.

The expanding availability and types of organic products in the supermarkets will make it possible to reach more consumers potentially interested in organic foods (Padel & Foster, 2005; Wier & Calverley, 2002). However, with an increase in the distribution through the supermarkets, higher volumes of organic foods are necessary. This could cause increased importing of organic foods. Currently, the demand of organic pork, beef, and lamb is greater than the production and thus these products are imported from Latin America, Australia, and Canada (Mintel, 2008). When organic foods are not produced locally or even have to be imported, they will have a greater carbon footprint due to greater transportation distances. According to some consumers, this conflicts with the principles of organic food production, which is perceived to be environmentally friendly (Wier & Calverley, 2002). Organic foods are not that "green" if they are not produced locally and have to be transported between states or between countries. However, as of now, there are no restrictions on food miles for certified organic production.

5.2.2.3 *Skepticism of certification boards and organic labels*

Consumers might distrust the certification systems and consequently question the realness of the organic foods. This is partly due to reported cases of mislabeling and product misrepresentation and partly because of the nonuniform standards and certification systems for organic foods. The lack of confidence in the certification and labeling may hold the consumer back from buying organic products. This may be particularly true with large corporations that include organic lines along with their conventional lines. Consumers may be more doubtful about the trustworthiness of these organic labels. Reliable labels guaranteeing organic food production and unified standards are important to the consumer and are necessary for continued growth of the organic industry (Wier & Calverley, 2002; Conner & Christy, 2004; Padel & Foster, 2005; Yiridoe *et al.*, 2005; Hughner *et al.*, 2007; Li *et al.*, 2007; Canavari *et al.*, 2010).

In order to have stricter rules for organic foods and gain consumer confidence, the USDA initiated the NOP that oversees the USDA-certified organic labeling. Additional information about the regulations of the NOP has been discussed in an earlier chapter.

5.2.2.4 *Insufficient marketing*

Despite the favorable attitude of the media, in some cases, organic food suffers from a lack of commercial promotion, mainly because of the wide prevalence of micro, small, and medium companies in the industry. This causes the consumers to be much less familiar with organic foods and unaware of the organic label (Li *et al.*, 2007; Bellows *et al.*, 2008). Consequently, consumers are not consistent with their interpretation of what organic is. It is important to promote organic foods and make information about organic food more available to increase the organic food knowledge and familiarity. The increase in awareness of the term "organic," the organic label, and a better knowledge of its meaning might increase the consumer's willingness to pay a premium for organic foods (Batte *et al.*, 2007). The lack of consumer knowledge about organic labeling may limit further development of the organic food demand (Zanoli & Naspetti, 2002; Yiridoe *et al.*, 2005; Roitner-Schobesberger *et al.*, 2008).

According to Hill and Lynchechaun (2002), good retail strategies (merchandising and displays) are also important such as the location of the organic food in the store. In their study with organic milk, they found that the consumers would prefer to have the organic milk located in the store next to the conventional milk. This way, they can perform price comparison and keep their habitual shopping behavior.

5.2.2.5 *Other competing alternative products to replace conventional products: natural and local foods*

The organic food industry is competing with similar products such as natural and locally grown foods (Martinez *et al.*, 2010; Mintel, 2008). Consumers are confused about the differences between natural and organic foods causing major challenges to the organic food market (Mintel, 2008, 2010). The fiercest competition between natural and organic is in the meat and egg industries (Mintel, 2008). While natural food is not subject to the strict rules for organic foods, some consumers believe it is a sufficient alternative to replace conventional products. In addition, it is priced less and may offer a more viable alternative than organic foods. Other alternatives competing with the organic meat market include "grass-fed," "pasture raised," and "free-range," which are less regulated than the organic meat production but are priced less.

There is an increasing concern among consumers about sustainability and food miles corresponding to the transportation of the foods from the farm to the consumer. There is a strong "locavore movement" that prefer locally grown food products (Mintel, 2008; Martinez *et al.*, 2010) even if not produced organically (59% of the respondents agreed to buy locally grown food whether it is organic or not) and thus local foods are competing with organic foods (Mintel, 2008).

5.2.2.6 *Minor deterrents*

The consumer satisfaction with the conventional foods is one of the less frequently mentioned reasons for not buying organic foods (Bourn & Prescott, 2002; Hughner *et al.*, 2007). On top of this, some consumers do not believe that organic food is better (Van Loo *et al.*, 2010). Another minor reason is not being satisfied with the quality of organic food, mainly with the appearance of the fresh produce (Bourn & Prescott, 2002; Hughner *et al.*, 2007). Organic food products often do not appear as visually appealing as conventional food products (Hughner *et al.*, 2007).

5.3 ECONOMICS AND PRICE PREMIUM

5.3.1 Price premium

According to the outlook report from the USDA Economic Research Service (Oberholtzer *et al.*, 2006), organic products are sold at a higher price than their conventional counterparts due mainly to three reasons. First, the production, processing, procurement, and distribution costs are higher than for conventional production (Oberholtzer *et al.*, 2006). The higher cost for the production of organic livestock is primarily due to the organic feed costs which are responsible for up to 70% of the cost of raising organic poultry and are approximately 50% to 100% more expensive than conventional feed prices (USDA & AMS, 2003). Claus (2010) reported the higher price for organic beef production to be due to the increased feed costs. Additional extra costs for production include the lower stocking densities, cost of outdoor excess and health costs without antibiotics, higher mortality, and longer production cycles for broilers (Fanatico, 2008). Other factors that contribute to the higher costs of organic livestock production are the possible increased labor costs, increased record keeping as well as extra costs for certification and segregation for organic food (Oberholtzer *et al.*, 2006; Fanatico, 2008). Organic farmland needs to be run under an organic farming management system for 3 years before its produce can be certified organic. This creates another disadvantage for farmers who want to start organic farming since they will have to invest 3 years prior to selling their products as organic. This 3-year conversion requirement also generates a gap between the increase in retail demand and supply (Dimitri & Oberholtzer, 2009).

Secondly, the differences in the relative levels of supply and demand of organic food products can also lead to higher prices. Thirdly, the organic food products are often valued more by the consumer since they perceive it as healthier and more environmentally friendly. Therefore, the consumer is willing to pay a premium price (Oberholtzer *et al.*, 2006).

Sales in organic and natural meat will likely maintain stability in market penetration since (1) more large conventional meat manufacturers are introducing natural and organic meat products, (2) consumers perceive these products as healthier and consumers are less likely to give up on products that they believe improve their health, and (3) organic buyers make in general a higher income than average and have more means to sustain purchasing these products even when the overall economy is not as strong (FMI & AMI, 2010).

The price premiums for organic broilers have increased from 169% in 2004 to 262% in 2006 (Oberholtzer *et al.*, 2006). For organic and natural beef, Claus (2010) reported a price average in retail supermarkets of $5.42/lb in the first quarter of 2010, compared to $3.42/lb for the average price for conventional beef products in the retail supermarkets, indicating that consumers are paying a premium of $2.00/lb or 58% premium for natural or organic beef.

Due to the anticipated growth in consumer demand, the organic meat market should continue its expansion. This could lead to an increasing availability in grocery stores as well as introduction of organic meat products by the conventional firms and may influence the supply and price for those products. It is predicted that for organic poultry, in the near future, the price premium most likely will stay high since production is having difficulties catching up with the fast-growing consumer demand (Oberholtzer *et al.*, 2006; van der Sluis, 2007).

5.3.2 Willingness to pay for organic foods

The yield on an organic livestock farm is generally lower than conventional production farms and it has a higher cost than conventional livestock production. Knowing that the high price premiums are the greatest limiting factor for organic meat purchases (Van Loo *et al.*, 2010), the willingness to pay is an important attribute to determine the profitability of organic farms (Yiridoe *et al.*, 2005) as well as an indicator on how consumers value organic meat which gives insight to their marketability. The proportion of the consumers willing to pay the price premium for organic foods decreases with increase of the premium level. The demand depends more on the price differential with the corresponding conventionally grown foods than on the actual price (Yiridoe *et al.*, 2005).

The price premiums that consumers are willing to pay for organic foods vary and depend on many factors including type of products, country, sociodemographic factors such as gender, age, income, and education (Bonti-Ankomah & Yiridoe, 2006; Ureña *et al.*, 2008; Yiridoe *et al.*, 2005). The sociodemographic factors influencing the willingness to pay are discussed in a later section. Several recent willingness-to-pay studies on organic foods have been conducted and the results vary considerably (Table 5.2). Studies indicated that consumers are willing pay a price premium of 10% to 60% for various organic foods (Table 5.2). Krystallis *et al.* (2006a) and Millock *et al.* (2002) demonstrated that willingness to pay depended on the type of organic food. For Greek organic consumers, the willingness to pay ranged from 19% to 63% (Krystallis *et al.*, 2006a) and in the Danish study, premium varied from 20% to 40% (Millock *et al.*, 2002), depending on the type of organic food.

Most studies discuss the willingness to pay for organic food in general or for organic fruit, vegetables, and milk. Only a few studies reported on willingness to pay for organic meat (Table 5.2). Krystallis *et al.* (2006b) reported that most Greek consumers are willing to pay high premiums for organic meats: 85% to 130% for organic chicken, 103% to 125% for pork, more than 115% for beef, and more than 105% for lamb/goat. O'Donovan and McCarthy (2002) observed significantly lower willingness-to-pay values for organic meat. Most respondents (44%) were willing to pay 1% to 5% extra, 29% of the respondents were

Table 5.2 Willingness to pay for organic foods reported in recent studies.

Author	Country	Main findings
		Organic food
Batte *et al.* (2007)	USA	Specialty/natural grocery shoppers are willing to spend a higher premium for organic foods (50%) compared to conventional grocery store shoppers (10%).
Brugarolas *et al.* (2010)	Spain	Approximately 24% of the respondents would not want to pay a premium for organic wine. However, 38%, 21%, 12%, and 5% of the respondents are willing to pay premiums of <25%, 25% to 49%, 50% to 99%, and more than 100%, respectively.
Gifford *et al.* (2005)	USA	During an auction experiment, half of the respondents offered a premium of 20% to 30% for various organic food products.
Gil *et al.* (2000)	Spain	Organic consumers are willing to spend 15% to 25% extra for organic foods.
Govindasamy *et al.* (2006)	USA	23% of the respondents were willing to pay 10% or more extra for organic produce.
Krystallis & Fotopoulus (2006a)	Greece	Organic consumers are willing to spend 19% to 63% extra for various types of organic foods (olive oil, oranges, bread, wine, and raisins).
Millock *et al.* (2002)	Denmark	Approximately 59% of consumers were willing to pay a 32% premium for organic milk; 48% were willing to spend a 40% premium on organic potatoes; 51% were willing to spend a 23% premium on organic rye bread, and 41% were willing to pay a 19% premium on organic minced beef.
Olesen *et al.* (2010)	Norway	Participants indicated to be willing to pay a price premium of 15% for organic salmon.
Soler *et al.* (2002)	Spain	Around 70% of the participants were willing to pay a premium for organic olive oil.
Ureña *et al.* (2008)	Spain	Organic consumers are willing to pay a 10% to 25% premium for organic foods (dichotomous choice questioning). Women have a more positive attitude toward organic foods; however, men are willing to spend more on organic foods.
Wier & Calverley (2002)	Denmark	Five percent to 20% of consumers are willing to pay price premiums of 30% or higher.
		Organic meat
Krystallis *et al.* (2006b)	Greece	Willingness to pay for organic meat depends on the type of organic meat. Most Greek consumers are willing to pay 85% to 130% extra for organic chicken, 103% to 125% for pork, more than 115% for beef, and more than 105% for lamb/goat.
O'Donovan & McCarthy (2002)	Ireland	Thirteen percent of the respondents were not willing to pay extra for organic meat. However, most respondents (44%) were willing to pay 1% to 5% extra. Approximately 29% of the respondents were willing to pay 6% to 10% premium, 3% were willing to pay up to 50% extra (26% to 50%), and 1% were willing to pay more than 50%.
Ureña *et al.* (2008)	Spain	For regular, occasional, and probable organic food consumers, the willingness to pay for meat and sausages was of 17.6%, 14.8%, and 12.6%, respectively.
Van Loo *et al.* (2011)	USA	Consumers are willing to pay a premium of 34.8% for the general organic label and 103.5% for United States Department of Agriculture (USDA) organic chicken breast. The nonbuyers, occasional, and habitual buyers are willing to pay a premium for USDA organic chicken of 26.2%, 97.3%, and 244.3%, respectively.
Dransfield *et al.* (2005)		Consumers are willing to spend 5% extra for "home country"- and "raised outside"-labeled pork.

willing to pay 6% to 10% premium, only 3% were willing to pay up to 50% extra (26% to 50%), and 1% wanted to pay more than 50%. A third study reported willingness-to-pay values for organic meat. This Spanish study generated willingness-to-pay values for meat and sausages of 17.6% for regular, 14.8% for occasional, and 12.6% for probable organic food consumers (Ureña *et al.*, 2008). Van Loo *et al.* (2011) assessed the willingness to pay for organic chicken in the US with a choice experiment and reported premiums of 35% for a generic organic-labeled and 103% for USDA organic-labeled chicken breast.

Moreover, due to growing anonymity of trade with organic products, organic consumers are becoming increasingly critical of food products that were produced under unknown social conditions. This trend has led organic consumers to question not only how their food is produced but also who benefits from their purchase. Several examples illustrate that consumers of organic food are willing to pay an additional price premium for fair trade products from developing countries (Loureiro & Lodate, 2005; Lusk & Briggeman, 2009; Chang & Lusk, 2009; Briggeman & Lusk, 2011) and for the direct support of small farmers' initiatives in disadvantaged areas (Schmid *et al.*, 2004), thus viewing the organic food system as a way to alleviate inequality of the profit distribution. According to Chang and Lusk (2009), consumers' concerns about the distribution of benefits resulting from food purchase are significant predictors of people's willingness to pay for organic food. Also, Lusk and Briggeman (2009) pointed out that food value of "fairness" was positively correlated with consumers' willingness to pay for organic food. In addition, Briggeman and Lusk (2011), using a model of inequality aversion and a set of real-money experiments, found that about 15% of consumers' willingness to pay for organic food is attributable to altruism and inequality aversion. However, they concluded that fairness premium is significantly influenced by consumers' knowledge about who receives this premium (e.g., small farmers).

5.4 AN ANALYSIS ACROSS ORGANIC BUYER TYPES AND SOCIODEMOGRAPHIC DIMENSIONS

Several findings across organic consumption research studies have identified consumers' purchase behavior as well as aspects relating to consumers' sociodemographic characteristics as factors determining different organic consumers' groups, highlighting how the profile of organic food consumers has become less homogeneous (Krystallis & Chryssochoidis, 2005; Yiridoe, *et al.* 2005; Hughner *et al.*, 2007; Gracia & De Magistris, 2008; Michaelidou & Hassan, 2008; Mintel, 2008; Pellegrini & Farinello, 2009; Dettmann & Dimitri, 2010; Smith & Paladino, 2010). When considering characteristics of consumers buying organic food products, consumers can be classified into different groups based on their declared purchase frequencies (e.g., regular vs. occasional, habitual vs. light organic). Specifically, several empirical studies have profiled different organic consumers' types by relating their buying behavior with other aspects, for example, the deterrent factors of consumers for purchasing organic food products (Gracia & de Magistris, 2008; USDA-FAS, 2010; Van Loo *et al.*, 2010).

However, only a few studies have grouped organic consumer types focusing on an organic meat product. Van Loo *et al.* (2010) assessed the effect of organic poultry purchase frequency on consumers' attitudes toward organic poultry meat in the United States and demonstrated that both habitual and occasional consumers buy organic chicken due to the presence of less residues (e.g., antibiotics, pesticides, and other compounds) as well as for health and safety concerns compared to the conventional products. Other studies on generic organic food consumption reported that health factors tend to lose importance in the case of regular organic buyers (Gracia & de Magistris, 2008), highlighting that the "regular buyers" group, which includes environmentalists, nature lovers, and socially conscious people, shows more emphasis on ethics issues, while organic "light buyers" group buys organic product for various reasons including healthy lifestyle, food safety concerns, quality, taste, animal welfare, and sustainability (USDA-FAS, 2010). As a consequence, when the product under consumers' valuation is represented by meat, safety and health concerns seem to play a more important role in consumers' purchasing process. As mentioned by Naspetti (2010), "eating quality" is considered by organic meat consumers (Brendahl, 2003) as a preface for health quality (Brunsø *et al.*, 2002; Zanoli, 2004). Since the organic production method voluntarily adopted by producers represents a credence quality aspect of a product, it is usually signaled by labels or logos in order to be fully evaluated by the consumer. In many countries, it must be certified by public institutions or by private accredited certification bodies.

Van Loo *et al.* (2011) found a positive relationship between consumers' purchase decision and organic meat labels. Specifically, using a choice experiment approach to assess consumers' willingness to pay for two types of organic labels displayed on chicken breast product (e.g., a general organic label and USDA-certified organic label), the authors reported that consumers prefer to buy chicken breast displaying organic labels rather than purchasing a similar product without any of these labels. Moreover, grouping consumers based on their frequency of organic meat purchases into habitual, occasional, and nonbuyers showed that habitual buyers are willing to pay the highest price premium for organic chicken breast bearing the USDA-certified organic label (244.3%), followed by a generic organic label (146.6%), and occasional buyers are willing to spend 97.3% and 35.7% premium for USDA organic and generic organic chicken breast, respectively. These high willingness-to-pay values suggest a market potential for these products.

With regard to the impact of social contests and demographic characteristics on organic consumers' purchase behavior, several studies have identified mixed consumers' profiles, especially with regard to income and education dimensions (Wilkins & Hillers, 1994; Chinnici *et al.*, 2002; Gracia & de Magistris, 2008; Mintel, 2008). However, some consistent results when examined across studies revealed that organic consumers are typically female (Davies *et al.*, 1995; Yiridoe *et al.*, 2005; Hughner *et al.*, 2007; Bellows *et al.*, 2008; Zander & Hamm, 2010), older (Schifferstein & Oude Ophuis, 1998; Cicia *et al.*, 2002; Hughner *et al.*, 2007), and with children living in the household (Thompson & Kidwell, 1998; Hughner *et al.*, 2007).

Income and gender are one of the most important demographics that have an effect on the willingness to pay. Women tend to purchase organic food more often (Yiridoe *et al.*, 2005; Bellows *et al.*, 2008). This can be explained by the high percentages of grocery shoppers being women and also by women being probably more informed or worried about nutrition and food safety. Nevertheless, Ureña *et al.* (2008) found that men are willing to pay a higher premium than women (11.4% compared to 9.5%). Younger people are also more willing to purchase organic foods (higher amount and higher prices). However, their behavior often does not convert to organic food purchases due to their lower purchasing power (Krystallis *et al.*, 2006a). Income is another factor that influences the buying behavior (Mintel, 2008; Yiridoe *et al.*, 2005). Frequent and occasional organic consumers are often high-income shoppers (Mintel, 2008). According to Yiridoe *et al.* (2005), income and willingness to buy a product are positively correlated up to a certain level of income. Above this level, further increases in income will not cause an increase in organic food purchases. Education has been identified as a major factor associated with organic purchases; the more educated consumers are, the more likely they buy organic foods frequently (Bellows *et al.*, 2008).

However, Krystallis and Chryssohoidis (2005) argue that criteria other than price and socioeconomic factors are better determinants for the consumers' willingness to pay such as food quality and security, trust in certification, and brand name familiarity. Batte *et al.* (2007) reported specialty grocery shoppers to be willing to pay higher price premiums for organic food compared to conventional grocery store shoppers.

With regard to the organic meat consumption, few studies are available that analyze willingness-to-pay differences across demographic groups and their findings are mixed. Zanoli *et al.* (2010) conducted a choice experiment to investigate Italian consumers' preferences and willingness to pay for organic, conventional, and GM-fed beef with respect to intrinsic, extrinsic, and environmental cues. There were no significant differences observed across consumers' demographic characteristics except for gender: females exhibited a higher willingness to pay for animal welfare than males. Similarly, Van Loo *et al.* (2011), using a choice experiment approach to assess consumers' willingness to pay for generic organic labels and USDA organic labels displayed on organic chicken breast, reported that the willingness to pay for females is higher than for males. Conversely, Van Loo *et al.* (2010) and FMI and AMI (2010) reported no differences in the frequency of buying organic chicken between men and women.

Van Loo *et al.* (2011) also found differences across the other demographic variables such as age and income. In particular, they pointed out that consumers' willingness to pay for chicken breast with the general organic label and USDA organic label increases with age, reaching, for example, with the USDA label a maximum of $3.7/lb for age group 35–44 years and fluctuating for higher age groups. Similarly, FMI and AMI (2010) found that consumers aged 25–39 are most likely to purchase organic meats whereas older shoppers are least likely. Finally, their results also suggest that willingness to pay is higher among respondents with higher household income levels but lower among those with children (less than 18 years) in the household.

5.5 CONCLUSIONS

Organic meat is an option for consumers and an opportunity for farmers and food industry operators. The stage of the organic market life cycle as a whole is in its later growth phase and probably approaching maturity (Van Osch *et al.*, 2008). The US organic food market appears to be almost mature since well-developed standards are in place and market regulation is quite strict and well-developed organic market supply chains are available. Growth of the organic food and beverage market has been rapid in the recent past but is now slowing down. Products are highly diversified and although product availability in the main distribution channels is still uneven, many retail chains are selling organic products and in some cases organic private labels with a large product assortment are available.

The organic meat industry, however, appears to be still in its introduction phase, since it is still available mainly in specialized organic stores and thus room for growth exists. The price gap between conventional and organic meat is quite high, resulting in a lack of general consumer acceptance, but the creation of more efficient supply chains will contribute to reducing this gap. Distribution is a key factor for the growth of sales for any product and all organic markets strive to gain access to large retail to achieve economies of scale. However, several relevant critical points for further development of the organic meat market may be identified. This industry may face the problem of "conventionalization of organic farming," since the need to provide supermarkets with a sustained, stable, and reliable supply of organic products requires the adoption of a more market-oriented logic, somewhat deviating from an "idealistic" approach to farming that was at the origin of the organic movement. This may represent a threat for a growing sector such as this, eroding the core value of its credibility. In addition, delayed adoption of organic farming by potentially interested domestic livestock and poultry producers may generate supply bottlenecks, thus creating stock breaking and forcing wholesalers and retailers to import from abroad. This may be questioned for compliance with the organic philosophy in relation to food safety, transparency, and environmental impact due to transportation issues. At the moment, the issue of genetically modified organisms (GMOs) does not directly affect the organic meat sector, although the use of GM-free feed may be problematic in the United States.

Together with the risk of "conventionalization," the organic sector is under constant threat to its credibility, trustworthiness, and reliability. High price gaps, in fact, may attract producers/traders that may try to sell conventional products as organic. Apart from the risk for the consumer to be cheated, scandals may damage the whole sector if they are not appropriately dealt with (Haas *et al.*, 2010).

The growth in market share of the organic market will probably lead to a higher differentiation and the creation of several organic segments. We may expect to find a "basic organic segment," focused on the main drivers for organic purchase (health and environment) and with a reduced price premium, along with one or several "premium organic segments." Consumers belonging to the latter segment(s) may look for organic food that also ensure coherence with specific concepts such as, for instance, (1) a regional/local food, (2) fair trade, (3) superior taste, (4) animal welfare, and (5) ethical business, sustainability, and social responsibility, all commanding a higher premium or the use of a specific market channel. In this scenario, the secondary drivers identified in this chapter may gain more relevance, since the primary drivers are supposed to be equally satisfied by all organic foods.

Considering the current situation of the organic meat market, the "basic" organic segment is likely to be the most important, especially if mainstream consumers increasingly approach the organic choice. The demand for healthy food will continue to grow in a rapidly aging society and media coverage about environmental problems will continue to represent an important hook for marketers to promote environmentally friendly production systems such as organic farming; therefore, a growing demand of organic meat is expected. However, the other segments are not going to be abandoned, since they probably represent both the cultural basis of the organic movement, as well as the target market for innovation in the organic industry.

ACKNOWLEDGMENT

Support is gratefully acknowledged from the National Integrated Food Safety Initiative (NIFSI) (2008-51110-04339) to coauthor Steven C. Ricke.

REFERENCES

Aertsens, J., W. Verbeke, K. Mondelaers, and G. van Huylenbroeck. 2009. Personal determinants of organic food consumption: a review. *Br. Food J.* 111:1140–1167.

Anderson, J. C., C. J. Wachenheim, and W. C. Lesch. 2006. Perceptions of genetically modified and organic foods and processes. *Ag. Bio. Forum.* 9:180–194.

Batte, M. T., N. H. Hooker, T. C. Haab, and J. Beaverson. 2007. Putting their money where their mouths are: consumer willingness to pay for multi-ingredient, processed organic food products. *Food Policy* 32:145–159.

Bellows, A. C., B. Onyango, A. Diamond, and W. K. Hallman. 2008. Understanding consumer interest in organics: production values vs. purchasing behavior. J. Agric. Food Ind. Organ. 6:1–2.

Bonti-Ankomah, S., and E. K. Yiridoe. 2006. Organic and conventional food: a literature review of the economics of consumer perceptions and preferences. Final report. Available at: http://oacc.info/Docs/BONTI&YIRIDOEApril282006Final.pdf (accessed October 13, 2011).

Bourn, D. and J. Prescott. 2002. A comparison of the nutritional value, sensory qualities, and food safety of organically and conventionally produced foods. *Crit. Rev. Food Sci. Nutr.* 42:1–34.

Bredahl, L. 2003. Cue utilization and quality perception with regard to branded beef. *Food Qual. Prefer.* 15:65–75.

Brennan, C., K. Gallagher, and M. McEachern. 2003. A review of the 'consumer interest' in organic meat. *Int. J. Consum. Stud.* 27:381–394.

Briggeman, B. C. and J. L. Lusk. 2011. Preferences for fairness and equity in the food system. *Eur. Rev. Agric. Econ.* 38:1–29.

Brugarolas, M., L. Martinez-Carrasco, R. Bernabeu, R., and A. Martinez-Poveda. 2010. A contingent valuation analysis to determine profitability of establishing local organic wine markets in Spain. *Renew. Agric. Food Syst.* 25:35–44.

Brunsø K., T. A. Fjord, and K. G. Grunert. 2002. *Consumers' Food Choice and Quality Perception.* Mapp, The Aarhus School of Business, Aarhus, Working Paper No. 77.

Canavari, M., N. Cantore, E. Pignatti, and R. Spadoni. 2010. Role of certification bodies in the organic production system. In: R. Haas, M. Canavari, B. Slee, T. Chen, and B. Anurugsa (eds) *Looking East Looking West: Organic and Quality Food Marketing in Asia and Europe.* Wageningen Academic Publishers, Wageningen. pp. 85–99.

Canavari, M., G. Nocella, and R. Scarpa. 2005. Stated willingness-to-pay for organic fruit and pesticide ban: an evaluation using both web-based and face-to-face interviewing. *J. Food Prod. Mark.* 11:107–134.

Chang, J. B. and J. L. Lusk. 2009. Fairness and food. *Food Policy* 34:483–491.

Chinnici, G., M. D'Amico, and B. Pecorino. 2002. A multivariate statistical analysis on the consumers of organic products. *Br. Food J.* 104:187–199.

Cicia, G., T. Del Giudice, and R. Scarpa. 2002. Consumers' perception of quality in organic food: a random utility model under preference heterogeneity and choice correlation from rank-orderings. *Br. Food J.* 104:200–213.

Claus, R. 2010. Organic beef profile. Available at: http://www.agmrc.org/commodities_products/livestock/beef/organic_beef_profile.cfm (accessed May 9, 2011).

Conner, D. and R. Christy. 2004. The organic label: how to reconcile its meaning with consumer preferences. *J. Food Distr. Res.* 35:40–43.

Dangour, A. D., K. Lock, A. Hayter, A. Aikenhead, E. Allen, and R. Uauy. 2010. Nutrition-related health effects of organic foods: a systematic review. *Am. J. Clin. Nutr.* 1:203–210.

Davies, A., A. J. Titterington, and C. Cochrane. 1995. Who buys organic food? A profile of the purchasers of organic food in Northern Ireland. *Br. Food J.* 97:17–23.

Dettmann, R. L. and C. Dimitri. 2010. Who's buying organic vegetables? Demographic characteristics of U.S. consumers. *J. Food Prod. Mark.* 16:79–91.

Dimitri, C. and L. Oberholtzer. 2009. Marketing U.S. organic foods. US Department of Agriculture (USDA), Economic Research Service (ERS). USDA Economic Research Service Economic Information Bulletin No. 58. September 2009. Washington, DC. Available at: http://www.ers.usda.gov/Publications/EIB58/.

Dransfield, E., T. M. Ngapo, N. A. Nielsen, L. Bredahl, P. O. Sjödén, M. Magnusson, M. M. Campo, and G. R. Nute. 2005. Consumer choice and suggested price for pork as influenced by its appearance, taste and information concerning country of origin and organic pig production. *Meat Sci.* 69:61–70.

Durham, C. A. 2007. The impact of environmental and health motivations on the organic share of produce purchases. *Agric. Resour. Econ. Rev.* 36:304–320.

Fanatico, A. 2008. Organic poultry production in the United States (ATTRA). Available at: https://attra.ncat.org/attra-pub/PDF/organicpoultry.pdf (accessed February, 2010).

Food Marketing Institute (FMI) and American Meat Institute (AMI). 2010. *The Power of Meat – An in-Depth Look at Meat through the Shoppers' Eyes.* Joint report AMI/FMI, Arlington, VA. p. 77.

FMI. 2008. Natural and organic foods. Report of FMI. September. p. 6.

Gifford, K., J. C. Bernard, U. C. Toensmeyer, and R. Bacon. 2005. An experimental investigation of willingness to pay for non-GM and organic food products. Agricultural and Applied Economics Association.

Gil, J. M., A. Gracia, and M. Sánchez. 2000. Market segmentation and willingness to pay for organic products in Spain. *Int. Food Agrib. Manag. Rev.* 3:207–226.

Govindasamy, R., M. DeCongelio, and S. Bhuyan. 2006. An evaluation of consumer willingness to pay for organic produce in the Northeastern U.S. *J. Food Products Marketing*. 11:3–20.

Gracia, A. and T. De Magistris. 2008. The demand for organic foods in the South of Italy: a discrete choice model. *Food Policy*. 33:386–396.

Gold, M. 2008. Should I purchase organic foods? Available at: http://www.nal.usda.gov/afsic/pubs/faq/BuyOrganicFoodsIntro.shtml (accessed April, 2009).

Haas, R., M. Canavari., S. Pöchtrager, R. Centonze, and G. Nigro. 2010. Organic food in the European Union: a marketing analysis. In: R. Haas, M. Canavari, B. Slee, C. Tong and B. Anurugsa. *Looking East Looking West*. Wageningen Academic Publishers, Wageningen. pp. 21–46.

Harper, G. C. and A. Makatouni. 2002. Consumer perception of organic food production and farm animal welfare. *Br. Food J.* 104:287–299.

Hill, H. and F. Lynchehaun. 2002. Organic milk: attitudes and consumption patterns. *Br. Food J.* 104:526–542.

Hjelmar, U. 2011. Consumers' purchase of organic food products. A matter of convenience and reflexive practices. *Appetite*. 56:336–344.

Holcomb, R. B., C. Willoughby, E. Early, and K. Reed. 2008. Market research study: organic, free-range and pasture poultry. FACP-153. p. 8.

Honkanen, P., B. Verplanken, and S. O. Olsen. 2006. Ethical values and motives driving organic food choice. *J. Consumer Behav.* 5:420–430.

Hughner, R. S., P. McDonagh, A. Prothero, C. J. Shultz, II, and J. Stanton. 2007. Who are organic food consumers? A compilation and review of why people purchase organic food. *J. Comsumer Behav.* 6:94–110.

Jolly, D. A., H. G. Schutz, K. V. Diaz-Knauf, and J. Johal. 1989. Organic foods: consumer attitudes and uses. *Food Technol.* 43:60–66.

Kouba, M. 2003. Quality of organic animal products. *Livest. Prod. Sci.* 80:33–40.

Krystallis, A. and G. Chryssochoidis. 2005. Consumers' willingness to pay for organic food: factors that affect it and variation per organic product type. *Br. Food J.* 107:320–343.

Krystallis, A., C. Fotopoulos, and Y. Zotos. 2006a. Organic consumers' profile and their willingness to pay (WTP) for selected organic food products in Greece. *J. Int. Consumer Mark.* 19:81–107.

Krystallis, A., I. Arvanitoyannis, and G. Chryssohoidis. 2006b. Is there a real difference between conventional and organic meat? Investigating consumers' attitudes towards both meat types as an indicator of organic meat's market potential. *J. Food Products Mark.* 12:47–78.

Li, J., L. Zepeda, and B. W. Gould. 2007. The demand for organic food in the U.S.: an empirical assessment. *J. Food Distrib. Res.* 38:54–69.

Loureiro, M. L. and J. Lotade. 2005. Do fair trade and eco-labeling in coffee wake up the consumer conscience? *Ecological Economics* 53:129–138.

Lusk, J. L. and B. C. Briggeman. 2009. Food values. *Am. J. Agric. Econ.* 91:184–196.

Magkos, F., F. Arvaniti, and A. Zampelas. 2003. Putting the safety of organic food into perspective. *Nutr. Res. Rev.* 16:211–222.

Magkos, F., F. Arvaniti, and A. Zampelas. 2006. Organic food: Buying more safety or just peace of mind? A critical review of the literature. *Crit. Rev. Food Sci. Nutr.* 46:23–56.

Magnusson, M. K., A. Arvola, U. K. Hursti, L. Aberg, and P. Sjoden. 2001. Attitudes towards organic foods among Swedish consumers. *Br. Food J.* 103:209–227.

Magnusson, M. K., A. Arvola, U. K. Hursti, L. Åberg, and P. Sjödén. 2003. Choice of organic foods is related to perceived consequences for human health and to environmentally friendly behaviour. *Appetite*. 40:109–117.

Martinez, S., M. Hand, M. Da Pra, S. Pollack,K. Ralston, T. Smith, S. Vogel, S. Clark, L. Lohr, S. Low, and C. Newman. 2010. Local food systems – concepts, impacts, issues. ERS/USDA Economic Research Report 97, May 2010. Available at: http://www.ers.usda.gov/Publications/ERR97/ERR97.pdf/.

McEachern, M. G. and P. McClean. 2002. Organic purchasing motivations and attitudes: are they ethical? *Int. J. Consumer Stud.* 26:85–92.

McEachern, M. G. and J. Willock. 2004. Producers and consumers of organic meat: a focus on attitudes and motivations. *Br. Food. J.* 106:534–552.

Michaelidou, N. and L. M. Hassan. 2008. The role of health consciousness, food safety concern and ethical identity on attitudes and intentions towards organic food. *Int. J. Consumer Stud.* 32:163–170.

Millock, K., L. G. Hansen, M. Wier, and L. M. Andersen. 2002. Willingness to pay for organic foods: A comparison between survey data and panel data from Denmark. Available at: http://orgprints.org/1754/ (accessed April, 2010).

Mintel. 2008. Organic food – US - October 2008.
Mintel. 2010. Consumer attitudes toward natural and organic food and beverage - US - March 2010.
Naspetti, S. 2010. Zootecnia biologica e qualita' agroalimentare. *Economia Agroalimentare*. Anno XII (2):123–144.
Oberholtzer, L., C. Greene, and E. Lopez. 2006. Organic poultry and eggs capture high price premiums and growing share of specialty markets. Available at: http://www.ers.usda.gov/Publications/LDP/2006/12Dec/LDPM15001/ (accessed December, 2006).
O'Donovan, P. and M. McCarthy. 2002. Irish consumer preference for organic meat. *Br. Food. J.* 104:353–370.
Olesen, I., F. Alfnes, M. B. Rora and K. Kolstad. 2010. Eliciting consumers' willingness to pay for organic and welfare-labelled salmon in a non-hypothetical choice experiment. *Livest. Sci.* 127:218–226.
Organic Trade Association (OTA). 2011. *Organic Trade Association's 2011 Organic Industry Survey*.
Padel, S. and C. Foster. 2005. Exploring the gap between attitudes and behaviour: understanding why consumers buy or do not buy organic food. *Br. Food J.* 107:606–625.
Pellegrini, G. and F. Farinello. 2009. Organic consumers and new lifestyles: an Italian country survey on consumption patterns. *Br. Food J.* 111:948–974.
Roitner-Schobesberger, B., I. Darnhofer, S. Somsook, and C. R. Vogl. 2008. Consumer perceptions of organic foods in Bangkok, Thailand. *Food Policy*. 33:112–121.
Schifferstein, H. N. J. and P. A. M. Oude Ophuis. 1998. Health-related determinants of organic food consumption in the Netherlands. *Food Qual. Prefer.* 9:119–133.
Schmid, O., U. Hamm, T. Richter, and A. Dahlke. 2004. *A Guide to Successful Marketing Initiatives*. Research Institute of Organic Agriculture, Frick. p. 208.
Siderer, Y., A. Maquet, and E. Anklam. 2005. Need for research to support consumer confidence in the growing organic food market. *Trends Food Sci. Technol.* 16:332–343.
Smith, S. and A. Paladino. 2010. Eating clean and green? Investigating consumer motivations towards the purchase of organic food. *Austr. Mark. J.* 18:93–104.
Soler, F., J. M. Gil and M. Sanchez. 2002. Consumers' acceptability of organic food in Spain: Results from an experimental auction market. *Br. Food J.* 104:670–687.
Soil Association. 2001. *Organic Farming, Food Quality and Human Health: A Review of the Evidence*. Soil Association, Bristol.
Stobbelaar, D. J., G. Casimir, J. Borghuis, I. Marks, L. Meijer, and S. Zebeda. 2007. Adolescents' attitudes towards organic food: a survey of 15- to 16-year old school children. *Int. J. Consumer Stud.* 31:349–356.
Thompson, G. D. and J. Kidwell. 1998. Explaining the choice of organic produce: cosmetic defects prices, and consumer preferences. *Am. J. Agric. Econ.* 80:277–287.
Tsakiridou, E., C. Boutsouki, Y. Zotos, and K. Mattas. 2008. Attitudes and behaviour towards organic products: an exploratory study. *Int. J. Retail Distrib. Manag.* 36:158–175.
Ureña, F., R. Bernabéu, and M. Olmeda. 2008. Women, men and organic food: differences in their attitudes and willingness to pay. A Spanish case study. *Int. J. Consum. Stud.* 32:18–26.
U.S. Department of Agriculture (USDA) and Agricultural Marketing Service (AMS). 2003. Organic feed for poultry and livestock: availability and prices. USDA/AMS.
USDA National Organic Program. 2008. Agricultural Marketing Service 7, Code of Federal Regulations (CFR), Part 205: the National Organic Program.
USDA Foreign Agricultural Service (USDA-FAS). 2010. EU27 organic products market report. GAIN Report No. NL0022, 17 August 2010. Available at: http://gain.fas.usda.gov/Recent%20GAIN%20Publications/EU-27%20Organic%20Products%20Market%20Report_The%20Hague_EU-27_8-17-2010.pdf.
van der Sluis, W. 2007. More organic poultry meat and eggs in the US. *World Poultry*. 23:26–28.
Van Loo, E., V. Caputo, R. M. Nayga, J. F. Meullenet, P. G. Crandall, and S. C. Ricke. 2010. Effect of organic poultry purchase frequency on consumer attitudes toward organic poultry meat. *J. Food Sci.* 75:S384–S397.
Van Loo, E. J., V. Caputo, R. M. Nayga, J.-F. Meullenet, and S. C. Ricke. 2011. Consumers' willingness to pay for organic meat: experimental evidence from chicken breast. *Food Qual. Prefer.* 22:603–613.
van Osch, S., B. Schaer, C. Strauch, and C. Bauer. 2008. *Specialised Organic Retail Report 2008*. Organic Retailers Association, Vienna.
Wier, M. and C. Calverley. 2002. Market potential for organic foods in Europe. *Br. Food J.* 104:45–62.
Williams, C. M. 2002. Nutritional quality of organic food: shades of grey or shades of green? *Proc. Nutr. Soc.* 61:19–24.
Wilkins, J. L. and V. N. Hillers. 1994. Influences of pesticide residue and environmental concerns on organic food preference among food cooperative members and non-members in Washington State. *J. Nutr. Educ.* 26:26–33.

Winter, C. K. and S. F. Davis. 2006. Organic foods. *J. Food Sci.* 71:R117–R124.

Yiridoe, E. K., S. Bonti-Ankomah, and R. C. Martin. 2005. Comparison of consumer perceptions and preference toward organic versus conventionally produced foods: a review and update of the literature. *Renew. Agric. Food Syst.* 20:193–205.

Zander, K. and U. Hamm. 2010. Consumer preferences for additional ethical attributes of organic food. *Food Qual. Prefer.* 21:495–503.

Zanoli, R. and S. Naspetti. 2002. Consumer motivations in the purchase of organic food: a means-end approach. *Br. Food J.* 104:643–653.

Zanoli, R. 2004. The European consumers and organic food. In: R. Zanoli (ed) *Organic Marketing Initiatives and Rural Development.* Vol. 4. Management and Business, University of Wales, Aberystwyth.

Zanoli, R., R. Scarpa, F. Napolitano, E. Piasentier, S. Naspetti, and V. Bruschi. 2010. Organic label as identifier of environmentally-related quality: a consumer choice experiment on beef in Italy. *119th EAAE Semin. 'Sustainability in the Food Sector: Rethinking the Relationship between the Agro-Food System and the Natural, Social, Economic and Institutional Environments'*, Capri, Naples, June 30th–July 2nd.

Section II

Management Issues for Organically Raised and Processed Meat Animals

6 Health and Welfare of Organic Livestock and Its Challenges

Albert Sundrum

Abstract: Animal health and welfare (AHW) have become an important component of consumer motivation to purchase organic products of animal origin. Although organic farms are obliged to follow the same minimum standards, living conditions for farm animals differ markedly between farming systems. Sources of variation with respect to AHW at the herd level are due to the numerous and complex interactions between the individual animals and the various aspects that characterize the living conditions within the farm. Hence, AHW is not related to minimal standards but to the farm management in the first place, and to be seen as an emergent property of the farm system. To strive for a high AHW status requires high management skills combined with the availability of essential resources. Currently, a considerable number of farms cannot cope, in all respects, with the requests for high AHW status. To prevent the loss of credibility, organic farmers and retailers are obliged to take the burden of proof for what they claim toward the consumers. A market that honors evidence-based performances in relation to AHW would provide the best option for an improvement because of monetary incentives. However, retailers currently widely refuse to honor and differentiate diverse performances in relation to AHW, which can be identified as the most relevant constraint. It prevents the farmers from gaining resources and incentives that are desperately needed and boosts the development of unfair competition between organic farmers. The situation appears to be tenuous and irreconcilable without a change in the basic conditions of marketing and changes in the current paradigm of standard-oriented to a new paradigm of a result- and output-oriented approach.

Keywords: animal welfare; animal health; standards; constraints; stakeholders

6.1 INTRODUCTION

Consumers are becoming increasingly sensitive of health and welfare problems in commercial livestock production systems and expect their food to be produced with greater respect for the needs of farm animals. This largely explains the attention given to this issue as a specific object for public policy and market intervention. Indeed, animal health and welfare (AHW) have become an important component of consumer motivation to purchase products that derive from specific brand label programs or from organic farms, claiming to provide animal-friendly living conditions for farm animals (Harper & Makatouni, 2002). For many people, organic farming appears to be a superior alternative to conventional livestock production (Hughner *et al.*, 2007; Yiridoe *et al.*, 2005). Even though organic farming only covers a

Organic Meat Production and Processing, First Edition. Edited by Steven C. Ricke, Ellen J. Van Loo, Michael G. Johnson, and Corliss A. O'Bryan.

small percentage of the food market, the expressed sympathy in the general public in most European Union (EU) countries appears far greater than the market share. The general trend in consumers' awareness of AHW encounters, however, are very complex phenomena, which are not clearly defined. AHW have different meanings to different people, and the interests of one particular group may come into conflict with those of others. Attributes included in the concept of AHW primarily depend on who is making the definition. Typical actors participating in the evaluation of AHW are scientists from the disciplines of veterinary, agricultural, and biological science. Also, stakeholder groups, especially farmers, retailers, government officials, marketing people, and consumer groups are involved.

In the forthcoming sections, it is not the intention to provide a further meaning to the terms "animal health and welfare" and "organic farming," but to contribute to enlightenment on how both of these terms might fit together. In this context, it is of special interest to bridge the gap between the perspectives of the different stakeholders involved and to discuss the processes on different scales. Finally, overall conclusions are drawn with respect to the challenges for organic farmers and for a market-driven label program such as organic livestock farming.

6.2 CHARACTERISTICS OF ORGANIC LIVESTOCK FARMING

Guidelines have been a characteristic feature of organic farming since 1954 and clear criteria to identify organically produced goods have been required since trademark legislation in Germany (Schaumann, 2002). Because the variety of production sites and the resulting product properties did not allow their identification to be linked to products qualitatively in terms that could be described exactly and understood analytically, the production method itself became the identifying criterion. This fundamental principle has been kept to the present day in private standards and has also been adopted by the legislators in the Council Regulation (EEC No. 834/2007) to harmonize the rules of organic livestock production across member states setting a minimum standard for all organic systems across EU member states. Some of the essential purposes of the Regulation are as follows:

- To protect the consumers from unjustified claims.
- To avoid unfair competition between those who label their products as organic.
- To ensure equal conditions for all operators.

Private schemes are still allowed to be applied in Europe; however, this is only when they are stricter than the Community organic production rules.

The organic concept refers to the whole farm as the base of a comprehensive system where the production process is intended to ensure quality production rather than maximizing production. The leading idea is based on the voluntary self-restriction in the use of specific means of production with the objectives to produce food of high quality in an animal-appropriate and environment-friendly manner on the basis of a nearly complete nutrient cycle (Sundrum, 1999). Labeling food as being "organic" identifies the products as deriving from a production method defined by guidelines and minimum standards. In general, these include specifications for the conversion process, housing conditions, animal nutrition, care and breeding, disease prevention, and veterinary treatment, while they create frameworks for organic livestock production and product-labeling.

According to the European Council Regulation (2007), animal health problems shall be controlled mainly by prevention, based on the selection of appropriate breeds or strains of animals, as well as the application of animal husbandry practices appropriate to the requirements of each species, and encouraging strong resistance to disease and the prevention of infections. The use of high-quality feed, together with regular exercise and access to pasture, is expected to encourage the natural immunological defense of the animal. Furthermore, an appropriate density of livestock should be ensured, thus avoiding overstocking and any resulting animal health problems. Concerning veterinary treatment, phytotherapeutic essences and homeopathic products shall be used in preference to chemically synthesized allopathic medicinal products or antibiotics. Furthermore, where animals receive more than three courses of treatment with chemically synthesized allopathic medicinal products or antibiotics within 1 year, the livestock concerned, or product derived from them, may not be sold as being of organic origin.

Detailed information on the differences between national and international organic standards is provided by Schmid *et al.* (2008). According to the authors, the International Federation of Organic Agriculture Movement (IFOAM) Basic Standards and the CODEX Alimentarius Guidelines regulate the area of disease prevention and veterinary treatment in a similar way as the Council Regulation (EEC) 834/2007 with no substantial deviations. At the international level, only the US National Organic Program (NOP) Regulation deviates substantially: animal products cannot be sold as certified organic if antibiotics or other substances not listed in the US NOP positive list have been used just once.

At the international level, the regulation on animal health issues is widely harmonized with the exception of the US NOP regulation, being very restrictive in the use of antibiotics. While the US concept seems to be more consumer driven ("pure" food), the European approach is considering not only the health aspects of the food but also the animal welfare aspects as important issues.

Apart from litigable minimum standards, organic farming commits itself to a number of substantial values, and thereby sets itself apart from conventional farming. The IFOAM states four principles: (1) Health, (2) Ecology, (3) Fairness, and (4) Care. These principles grew out of stakeholder consultations and were agreed upon on a worldwide basis by the members of IFOAM with each principle being accompanied by an explanation (IFOAM, 2006). However, organic agriculture is not organized uniformly, neither with respect to the various objectives nor in the degree of their implementation. There have been and still are different perspectives on organic agriculture with different understandings of what it is and what it develops. Alroe and Noel (2008) identified protest, meaning, and market as significant perspectives on organic agriculture that cannot be merged into one but can be helpful in understanding its past and its future development.

6.3 IMPLICATIONS OF LIVING CONDITIONS ON ANIMAL HEALTH AND WELFARE

In the past, conventional livestock farming has been impressively successful in reference to its ability to increase the performance of farm animals and to decrease production costs. However, a marked decrease in the production costs has a price, and to a great extent that price has been paid by the farmers who were no longer able to stand their ground in the severe competition but also by the farm animals restricted to economically priced living conditions. While efficiency has been achieved by intensification, the issue of AHW has been pushed into the background due to its cost- and labor-intensive demands. Farm animals typically get less space per individual and live in barren environments that do not allow them to exercise their normal range of behavior. On the other hand, in certain respects, the life of most farm animals has also improved, for example, where nutrition and incidence of infectious disease are concerned.

Organic livestock farming, as an integrated part of organic agriculture, claims to be an alternative in comparison to the development in conventional animal production. The minimal standards of the EEC Regulation reach a level that is clearly higher in comparison to conventional production and private branded programs (Sundrum, 2000). In addition, the EEC Regulation prescribes a regular check of the minimum standards by independent certifiers, providing an additional advantage in comparison to conventional ordinances. The crucial question in this context is whether the alternative approach achieves an improved level of AHW. In the forthcoming sections, implications of some relevant prescriptions of the regulation on the AHW issue are briefly discussed.

6.3.1 Space allowance

Concerning the indoor area provided for livestock in organic farming, the EEC Regulation goes far beyond the current space allowance in conventional production with respect to different species. In addition, the indoor area is supplemented by an outdoor area that has to be at least 75% of the indoor area. Furthermore, organic livestock are not allowed to be kept tethered or in individual confinement housing. Close confinement of farm animals has raised welfare concerns into both physical and behavioral issues. Physical concerns arise from the consequences of lack of exercise, such as leg weakness, skin lesions, and lameness (Buchwalder & Huber-Eicher, 2004; Ruis-Heutinck *et al.*, 2000). Major behavioral issues are the restrictions in the execution of locomotion and comfort behavior.

6.3.2 Flooring and provision of bedding

Beef cattle, pigs, and poultry in conventional systems are frequently housed in unbedded systems with partly or fully slatted floors. In contrast, all farm animals in organic farming must be provided with litter while completely slatted floors are forbidden. In addition to its role as a cushioned physical surface, the provision of straw is expected to reduce health and welfare hazards through its role as a behavioral substrate. Enlarging the locomotion area and providing litter bedding have substantial benefits for AHW (Guy *et al.*, 2002; Hindhede *et al.*, 1996). In the case of pigs, the litter bedding can be eaten to provide gut fill, and rooted to permit appropriate expression of foraging behavior. This might prevent the development of stereotyped tail-biting behaviors, and reduces social aggression (Meunier-Salaün *et al.*, 2001).

In contrast, slatted flooring can give improved hygiene conditions, especially if management of bedded systems is poor. It is also well known that poor quality straw bedding can contain high levels of mycotoxins, which causes health and reproductive disorders. Moreover, increased use of straw, especially in deep litter systems, is a significant risk factor for endoparasites (Haugegaard, 2010; Nansen & Roepstorff, 1999).

6.3.3 Outdoor access

Organic standards require animals to have access to outdoor areas, in contrast to conventional systems, which, in most cases, maintain farm animals indoors throughout their life. In some

European countries, organic beef cattle or pigs are kept outdoors throughout long periods of the year, whereas in other countries, farmed animals in organic farming are primarily housed indoors, and may have access to an outdoor run or pasture. Keeping farm animals in outdoor systems potentially has both advantages and disadvantages for the AHW issue, frequently described in the literature (Edwards, 2005; Redbo *et al.*, 1996; Van de Weerd *et al.*, 2009).

Animals in outdoor production face greater challenges from climatic extremes and social competition, but greater space and environmental diversity permit expression of a wider range of behaviors. Health challenges may be reduced by the lower animal density and better air quality, but there are also negative influences of reduced biosecurity, contact with wildlife disease reservoirs, and increased numbers of endoparasites, some of which can only be transmitted outdoors.

6.3.4 Nutrition/feeding

Possible welfare problems associated with nutrition include hunger and thirst, relating to the availability and quality of food and water, and nutritional imbalances. In organic farming, beef cattle are primarily kept on a grass-based system, often consisting of a period of summer pasture feeding. The regulation also states that at least 60% of the dry matter in daily rations has to consist of roughage, fresh or dried fodder, or silage. Due to a lack in the availability of energy-rich forage such as maize silage and fodder beets as well as restrictions in the use of concentrate, farmers are challenged to provide an appropriate energy supply (Sundrum *et al.*, 2008).

In the case of monogastric animals, the first limitation in exploiting the genetic potential of monogastric animals in organic farming is the supply of limited amino acids (Høøk Presto *et al.*, 2007; Sundrum *et al.*, 2006a). As soybean meal from conventional processing and synthetic amino acids are banned as feed ingredients, grain legumes represent the main protein source. These are characterized by comparable low concentrations of limited amino acids. Further limitations in the availability of limited amino acids are due to the total ban of conventionally cultivated bought-in feedstuffs from the beginning of 2012. Correspondingly, the performance capacity under organic conditions is, in general, diminished compared to conventional production (Sundrum, 2010).

The presence of detrimental levels of antinutritive factors in grain legumes and toxins in feed can arise through inappropriate selection of ingredients, or poor storage of feed. Growth or storage of raw materials in warm moist conditions can promote fungal growth and production of mycotoxins, which impair health and reproductive performance (Osweiler, 2006).

However, the requirement to provide roughage in the diet of pigs can confer significant health and welfare benefits (Carlson *et al.*, 1999; Meunier-Salaün *et al.*, 2000; Stewart & Boyle, 2008). Roughage can reduce risk of constipation and gastric ulcers, and increase satiety from increased feeding time and greater physical bulk in the gut.

A significant risk to AHW arises if animals are not genetically suited to the production systems in which they are placed. Pigs in organic systems receive diets of poorer nutrient quality, with fewer high-energy and high-protein digestibility ingredients (Dietze *et al.*, 2007; Sundrum *et al.*, 2010). The preferred use of homegrown versus bought-in feedstuffs can cause a high variation in the nutrient content of feeding rations and this has consequences for the growth processes within the different stages of life, for the performance level and the occurrence of imbalances and metabolic disturbances. The restriction in the availability of ingredients does not necessarily give rise to specific nutrient deficiencies, provided that careful ingredient control, ration formulation, and mixing of feed are carried out.

6.3.5 Human care

The number of animals per stockman has significantly increased in recent decades and consequently the time spent by humans with each animal has decreased. The skill and care of the stockpersons and the way in which they interact with the animals has a considerable influence on the behavior and welfare of the animals. The extent of qualification, training, and the use of advice from veterinarians and other qualified professionals are relevant aspects of development of improved quality of care. They assist in the correct diagnosis of problems and treatment of sick animals. However, there is little information on the extent to which such supporting mechanisms are exploited in organic herds.

Most production diseases have a multifactorial etiology, meaning that various factors contribute to the occurrence of disease. In addition to pathogens and animal-related conditions, other contributing factors may be considered as stressors in the environment, disturbing homeostasis in the animal (Fisher *et al.*, 1996). Thus, most diseases appear to be multivariate responses to a complex set of interrelated causal factors and are often due to mistakes of the farmer, inadequate handling, and inappropriate housing conditions (Enevoldsen & Gröhn, 1996). Usually, there is a dynamic balance between the pathogen burden in the environment and the disease resistance of the animal. Genetic selection over the past decades has focused on the improvement of productivity. At the same time, this selection was accompanied by a negative selection in physiological adaptability and immunological responsiveness, which could render the animal more susceptible to stress impact and pathogens (Noordhuizen, 1999).

Cleaning and disinfection are regarded as one of the most important disease control strategies. Contaminated buildings can harbor infectious agents over long periods of time, continuing to spread infection to new animals in a disease cycle. Effective disinfection breaks this cycle, removing, or at the very least, reducing the exposure of new batches to the pathogens of their predecessors. Options for disinfection are often hindered by the fact that many farms are not able to implement an all-in all-out concept as they do not possess partitioned buildings that can be cleaned and disinfected separately without the risk to contaminate animals in the same building with water and solvent. This is of special importance in the case of pigs and poultry.

In general, organic pig herds and poultry flocks are smaller in size with less ability to operate batch systems with all-in all-out use of housing. A survey of Austrian pig herds indicated that all farms used a continuous flow production system (Baumgartner *et al.*, 2003). A survey of German pig herds indicated that only 25% operated as an all-in all-out system, and only 25% of herds used disinfection (Dietze *et al.*, 2007). This means that vertical transmission of infections from older to younger pigs can occur unchecked, which is a significant disease risk. Where endemic disease is known to be present, it can be controlled by a sound vaccination strategy. However, this is often not adopted in organic herds. In Austrian pig herds, only a few farms vaccinated sows against erysipelas and parvovirus infection (Baumgartner *et al.*, 2003), while in a German survey, 85% of the organic pig herds used vaccination protocols (Dietze *et al.*, 2007). On the other hand, organic units generally import fewer animals than conventional units, which might reduce pressure of pathogens; however, they may also be less likely to have rigorous quarantine facilities.

The results of slaughter animal and carcass inspection were evaluated by Machold *et al.* (2005) for pigs and cattle, reared in organic or conventional production systems. The most common pathological findings of the pigs were liver diseases, condemned carcasses, and lesions of the tail tip. The most common pathological findings of the cattle were claw and leg disorders, diseases of liver, kidney and pleura, and parasitic infections. From cattle reared in organic systems, less carcasses, livers, and kidneys were condemned in comparison to conventional cattle. Pigs from organic systems showed no higher incidence of "milk spots" in the liver caused by *Ascaris suum*; and there was no pronounced difference concerning the number of condemned carcasses and the incidence of tail tip lesions between pigs from organic and conventional farms. In a study from Austria, the prevalence of pneumonic lesions at slaughter was 74% in conventional pigs and approximately 25% in organic pigs (Leeb & Baumgartner, 2000). A comparison at a German abattoir showed that in a large proportion of the organic pigs (47%), no abnormalities were detected in the lungs versus 41.4% lung lesions among conventional pigs Sundrum & Ebke, 2004); in contrast, only 36% of the livers from organic pigs were free from milk spots, while 57% of the conventional livers were not affected.

6.3.6 Mutilation

In intensive livestock farming, several surgical mutilations are carried out as a matter of routine in farm animals, including castration, dehorning, tail docking, and beak trimming, and have been shown to cause fear, pain, and distress (Gentle *et al.*, 1997; Mellor & Stafford, 1999). The amount of pain and distress caused by a given mutilation will depend, among others, on the method used, the skill of the person who carries it out, and the methods to alleviate the pain and distress.

In contrast to conventional livestock production, mutilations are widely banned in organic livestock farming, and are restricted to waivers that require a separate permission by the certifying bodies. An exception is the castration of male pigs, which is also practiced to a large extent in organic pig farming. However, from the beginning of the year 2012, castration of piglets has to be carried out with the application of anesthesia and/or analgesia (Commission Regulation, 2008), which differs from the current practices in conventional production.

6.3.7 Treatment

Observation of the animals is particularly important as problems are likely to be expressed through animal behavior. Regular, patient, and careful observations of the animals are required in order to detect signs of disease at an early stage. Due to a generally short time spent on the farm with fattening animals mainly restricted to the time around feeding, diagnosis of diseases and disorders can be quite difficult to carry out. Farm animals in outdoor systems may be difficult to observe, while health control and treatment in the case of diseases are hindered (Turner & Dwyer, 2007). In order to perform clinical examinations, animals showing disease signs have to be separated from the group and restrained before examination for reasons of thoroughness as well as human safety.

With regard to the therapeutic use of veterinary products, misuse occurs, especially when veterinary products are used in the absence of a specific diagnosis. The prophylactic use of veterinary products usually occurs at stages of the production where an increased risk of various diseases is present, in particular at the beginning of the fattening period. While the prophylactic use of veterinary products may be seen, especially from the farmers' point of view, as a contribution to animal health, it may also serve to conceal deficits in housing conditions and farm management.

In organic farming, phytotherapeutic essences and homeopathic products shall be used in preference to chemically synthesized allopathic medicinal products or antibiotics, provided that their therapeutic effect is effective for the species of animal, and the condition for which the treatment is intended. It is emphasized that if the use of the authorized substances and measures is not effective and if the treatment is essential to avoid suffering or distress, chemically synthesized products may be used with some limitations. Where animals receive more than two courses of treatment with chemically synthesized allopathic medicinal products or antibiotics within 1 year, the livestock concerned, or products derived from them, may not be sold as being of organic origin.

Information about medication in organic livestock production is scarce. An investigation of the use of antibiotics to fattening pigs in Denmark showed that the conventional herds consumed three times as much as the organic herds (Hegelund *et al.*, 2006). There was no significant difference in mortality rate between conventional and organic herds and clinical examinations in the herds did not reveal more pigs in need of treatment in the organic herds.

6.4 HETEROGENEITY OF LIVING CONDITIONS BETWEEN ORGANIC FARMS

As all organic farms are obliged to follow the same minimum standards, one might expect a higher level of uniformity in the framework of livestock production in comparison to conventional production with less requirements, bans, and rules. Several aspects contradict this expectation.

Living conditions for farm animals differ markedly between organic production systems within and between countries (Vaarst *et al.*, 2006). The production conditions vary from outdoor production to indoor production and encompass huge differences between regions, not only in relation to environmental conditions and the genotype used, but also in relation to the nutrient supply and the management measures (Sundrum, 2010). Although most organic farmers make use of conventional breeds and genotypes, other farmers also keep indigenous breeds providing a large variation in the performance level and in the nutrient resources used. Feed should be based to a high degree on homegrown feedstuffs, which are expected to provide a higher variability in ingredients and composition than feed from the feed millers. In addition, housing conditions include not only indoor but also outdoor conditions with varying options with respect to space allowance, climate and air conditions, etc., depending on the farm-specific and local conditions.

Recently, an epidemiological study was performed on 101 pig herds in six European countries (Denmark, Germany, Austria, Sweden, Italy, and France) (Sundrum *et al.*, 2010). Farmers differed widely in their feeding regimes, in the cleaning management, and in the implementation of disinfection measures. The study also revealed large differences in animal health management with respect to the use of quarantine, clinical examination of the individual pigs, separation of diseased sows, piglets, or weaners from healthy animals, presence of the farmer during farrowing, and regarding routine measures such as castration, grinding teeth, and iron supply of piglets.

In contrast, intensive systems have closely controlled environments to maximize aspects of animal productivity. Living conditions have typically been adopted across regions and countries, thereby increasing environmental homogeneity between farms and leading to the predominance of a small number of breeds and genotypes that are particularly productive under these specific conditions. Currently, the number of breeds used has declined rapidly to the point of very few breeding companies remaining in the world. Parallel to the increasing performance of livestock, nutrient demands have increased the call for the use of highly concentrated concentrate to meet the nutrient requirements of the animals at their various life stages. The equipment of indoor housing and the feeding ration is offered by specialized enterprises, becoming more and more standardized all over the world. The trend toward a high uniformity in input resources and tools is accompanied by a comparable high level of uniformity in the output with respect to product quality traits.

While conventional livestock production has a strong shift toward specialization, the basic concept of organic farming is focused on mixed farms, although the degree of mixture can vary widely (Hermansen & Kristensen, 1998). It can be supposed that farmers on highly specialized livestock farms have a more specific management qualification and are more aware of the relevant health-related factors than farmers on mixed farms. Because time capacity and competence of the farmers can be limited, excessive demands in several fields at the same time provoke conflicts within the farm management and, in consequence, lead to deficits on one or more of the various agricultural fields. Thus, there are reasons for the assumption that on organic mixed farms, handling and management of the farm animals are in a far greater competition with various other farm activities compared to highly specialized conventional livestock farms.

While common ideas and guidelines exist only on a meta-level, the concrete implementation in farm practice results in a large variety of stocking densities, performance levels, nutrient availabilities, housing conditions, and hygiene measures. Thus, low input and highly intensified organic livestock farms mark the corner stones of different objectives, different priorities, and different living conditions for farm animals, and thus providing divergent implications for the living conditions and for the implementation of a high level of AHW.

6.5 STATUS OF ANIMAL HEALTH AND WELFARE IN ORGANIC FARMING

The current situation in organic livestock farming in relation to AHW has been described in various reviews covering different species (Cabaret, 2003; Hovi *et al.*, 2003), pigs (Baumgartner *et al.*, 2003; Sundrum *et al.*, 2010), or poultry (Tuyttens *et al.*, 2008; Van de Weerd *et al.*, 2009). In this context, animal health is related to the occurrence of disease patterns while animal welfare is primarily used in relation to the possibility to execute species-appropriate behavioral patterns. It is assumed that both these issues contribute to the well-being of the animals.

In general, indicators of AHW are characterized by a large variation between the animals of a herd and between the preconditions of farm systems, while at the same time the degree and the source of the variation differs widely. Sources of variation in relation to AHW attributes are due to the numerous interactions between genotype, age, sex, nutrient supply, housing conditions, care, and their varying impacts on the dynamic nature of AHW status. Striving for a high status of AHW requires high management skills; one must be capable to gain an overall picture of the complex interactions within a farming system, to reflect on the most relevant factors, to implement feedback mechanisms, and guide the production process. Thus, it primarily depends on the management whether the potentials for a high level of AHW are fully realized.

The studies mentioned previously indicate that a considerable number of farms cannot cope in all respects with the requests for high AHW status. As differences between farms appear to be greater than those between production methods, organic livestock farming defined by minimum standards does not provide a homogenous outcome with respect to the AHW status.

The reviews clearly indicate that organic standards do not automatically lead to a high status of animal health that exceeds the level in conventional production. Comparable high rates of mortality and morbidity in livestock production interfere with the well-being of farm animals and indicate that the animals are not able to cope appropriately with their environment.

Obviously, the issue of AHW often is not the first priority in organic livestock farming. Differences in management practices, restrictions in the availability of resources (such as labor time, financial budget), and a lack of feedback and control mechanism within the farm system appear to be primary reasons for the substantial variation.

6.6 DIFFERENT PERSPECTIVES

Discussing a complex phenomenon such as AHW necessarily requires a reduction of complexity by using general and abstract terms that will generate a certain level of understanding between dialogue partners. As a consequence, any statement about AHW is more or less influenced by specific perspectives and world views involved. Also, organic farming is value driven and as such represents a certain perspective. These different perspectives must be taken into account when describing AHW in organic livestock farming and providing some general results and conclusions.

6.6.1 Farmers

Personal values play an important role in decision-making and also in the perceptions of symptoms related to AHW (Kristensen & Enevoldsen, 2008; Whay *et al.*, 2002). In general, producers anchor their idea of a good life for animals largely in their own daily practices and link welfare to the technical conditions necessary to ensure profitability and good welfare—a discourse that is also influenced by the official standards (Skarstad *et al.*, 2007). However, farmers are not a homogenous group, neither those who have recently converted to organic agriculture nor those being organic farmers for decades (Flaten *et al.*, 2006). According to Darnhofer *et al.* (2005), farmers can be categorized into different types (e.g., "pragmatic" or "committed") according to a cluster of values governing their actual practice. Wilkie (2005) distinguished four types of relationship of varying levels of attachment and detachment. This suggests that the frequency, intensity, and intimacy of farmers' contact with the animals define the level of attachment or detachment that farmers feel for their livestock. On the basis of this concept, Bock *et al.* (2007) confirmed with their analysis that animal species matters as does the purpose for which they are kept, the number of animals, the length of their stay on the farm, and the frequency, intensity, and intimacy of the contact between the farmer and the individual animal.

Farmers are not driven in the first place by concerns about AHW status but by the need to increase productivity to maintain competitiveness in an increasingly globalized and price-driven market. Profits from increased efficiency are generally short term, as they are regularly pared away by competition to reduce selling prices. As resulting costs of production for most organic farm types are higher than for conventional systems, price premiums are urgently needed to achieve an appropriate income (Offermann & Nieberg, 2000). Consequently, each costly and time-consuming requirement with respect to the issue of AHW is more or less met with an obstacle or at least with reservation.

6.6.2 Retailers

Consumers cannot directly discern attributes of process quality, such as AHW, in a product. Only additional information will identify the nature of the origin, or the production process of these foods. Correspondingly, perception of the consumers is to a high degree influenced by information through advertising. However, advertising campaigns do not define their view on AHW or by which criteria the status is assessed.

While the production processes at the farm level are characterized in the first place by heterogeneity, marketing strategies reduce the heterogeneity of reality to very simple messages. The less that is known by the consumer about a product's production process and traits, the easier it is to construct an image, and this enables the consumer to project and connect individual and favorable ideas with the product while leaving plenty of room for speculation. Therefore, food marketing ignores existing and production-related variability but nonetheless tries to sell a more or less uniform image of the product. The market is full of oversimplifications and inaccurate images and this makes it difficult to employ more meaningful quality assessments into practical shopping/consuming situations. Since the consumer is seldom able to evaluate whether the food product has delivered promised product qualities, process-related qualities of food products are almost exclusively based on trust.

While a specific food item's image becomes more and more decisive in the purchase situation, positive interferences do not necessarily lead to a purchase if consumers do not think that the trade-off between give and take components is sufficiently favorable (Grunert *et al.*, 2004). Labeling offers the options for marketing schemes using food labeling communicate animal welfare as a differentiating factor (Bock & Van Huik, 2007).

Marketing in organic products is committed to outperforming conventional farming in a number of areas, albeit with somewhat higher costs. The promotion of products from organic production systems is based more on simple and anecdotal information rather than on real facts. Introducing AHW as a marketing attribute can be interpreted as a strategy of retailers seeking niches when facing greater competition (Skarstad *et al.*, 2007). In the first place, retailers are interested to increase turnover rates, while the sold commodities are to a large extent exchangeable as long as they are not linked with a negative image.

6.6.3 Consumers

Interest in organic food has grown remarkably as consumers react to popular media about health effects and have then gradually evolved attitudes toward the origins and to the production process of food. For example, a change from confinement to more free-range systems has been one of the tools to evoke positive associations with the product (Andersen *et al.*, 2005). Thus, AHW has been turned into a quality attribute of food.

Introduction of the wholesomeness concept in livestock production, most often represented by organic production, is mainly due to a wish for reestablishing a positive image of food safety and animal welfare aspects (Verbeke & Viaene, 2000). Largely, the widespread sympathy for organic agriculture seems to stem from its value-based approach. Many consumers directly associate organic farming with enhanced animal welfare and conflate organic and animal-friendly products (Harper & Makatouni, 2002; McEachern & Willock, 2004). Nevertheless, the literature clearly indicates that the word "organic" has many meanings to consumers and that consumers of organic foods are neither homogenous in demographics nor in beliefs (Aertsens *et al.*, 2009; Hughner *et al.*, 2007; Martelli, 2009). According to Yiridoe *et al.* (2005), many consumers tend not to understand the complexities and niceties of organic farming practice and organic food quality attributes. They hold a huge variety of motivations, perceptions, and attitudes regarding organic foods and their consumption. All of these factors drive the decision-making process to buy those products (Harper & Makatouni, 2002).

Some consumers appear to delegate responsibility for ethical issues in food production to the retailer or the government as many consumers do not like to be reminded about issues connected with the animal when choosing meat (Bernués *et al.*, 2003; Skarstad *et al.*, 2007). Moreover, knowledge of production systems often appears of little importance for food market potentials as consumer groups often freely remark that there is no link between the negative images of production methods and consumers' purchase behavior (Ngapo *et al.*, 2003). According to the authors, consumer groups are often confused and mistrust the limited information available at the point of purchase, whereas price is an extremely visible attribute of products related to quality by the notion of value (McEachern & Schröder, 2002). Demand tends to depend more on the price differential with respect to conventionally grown products than on actual price. In contrast to sensitivity of demand to changes in price, income elasticity of demand for organic foods is generally small (Martelli, 2009). Hence, consumers are neither a uniform group covering common interests, nor are they experts who can decide on how to evaluate AHW and what is needed to provide housing conditions and management that are appropriate to farm animals in relation to AHW.

6.6.4 Politicians and lobbyists

Even though organic farming covers only a small percentage of the food market, the expressed sympathy in the general public in most European countries appears far greater than the market share. It is probably for this reason that organic farming in Europe has become the subject of increased attention from the political sphere. Thus, public bodies promoting organic products are motivated by a perceived consumer demand for such products. Governments intend to support initiatives that can be expected to improve common goods such as AHW. In general, they are not interested in the details but in satisfying the expectations of possible voters. In particular, they seem not interested in regulating or controlling the process as any initiative that is expected to increase bureaucracy will be parried.

Over the last decades, a sizeable number of initiatives promoting legislation that defines minimum standards of animal care in farm animal production have appeared. However, minimum standards are restricted to very few measurable and auditable properties of the production. They are the outcome of a political process involving numerous stakeholder groups, which, therefore, results in compromises between various concerns (Sundrum, 1999).

More important is that all initiatives by politicians and farmers' organizations to improve AHW are limited by the conditional acceptance that they do not undermine the competitiveness of European agriculture. Due to a strong competition between the different European countries and toward countries outside of Europe, no relevant legislative initiatives with respect to AHW are expected in the near future, neither in organic nor in conventional agriculture.

6.6.5 Scientific perspective

The recent development and growth of organic livestock farming and the related development of national and international regulations has fueled discussions among scientists concerning the proper conceptualization of AHW. The emerging discipline of animal welfare science grew out of veterinary science but has more or less been taken over by applied ethology. Besides ethology, stress physiology, pathology, and veterinary epidemiology play a significant role in the field. In the literature, a variety of definitions of AHW exists, thoroughly discussed by Rushen and de Passillé (1992) and Stafleu *et al.* (1996). Hence, there is no generally accepted definition of AHW within the scientific community. Three types of definitions are often distinguished, depending on what is considered important for the well-being of the animal (Fraser *et al.*, 1997):

(i) **The natural living approach:** The welfare of an animal depends on it being allowed to perform its natural behavior and live a life as natural as possible.
(ii) **The biological functioning approach:** Animal welfare is related to the normal functioning of physiological and behavioral processes.
(iii) **The subjective experience approach:** The feelings of the animal (suffering, pain, and pleasure) determine the welfare of the animal.

It seems comprehensible that AHW is a concept based on values and that animal welfare science cannot be made independent of questions of values and ethics (Alroe *et al.*, 2001; Fraser, 2003). The authors investigated whether those values that underpin organic farming in particular, also affect the interpretation of AHW. While some of the issues raised in connection with organic farming are relatively uncontroversial, others are not. The introduction of organic farming values seems to introduce new criteria for what counts as good animal welfare, as well as a different ethical basis for making moral decisions on welfare. Organic farming embodies distinctive systemic or communitarian ethical ideas and the organic values are connected to a systemic conception of nature, agriculture, the farm, and the animal (Verhoog *et al.*, 2004). While the organic values overlap with those involved in the conventional discussion of animal welfare, some of them suggest a need to set new priorities and to reconceptualize animal welfare, for example, with respect to "naturalness," in relation to the possibilities for expression of natural behavior and in relation to animal integrity as a concept for organismic harmony. According to Alroe *et al.* (2001) and Lund (2006), the organic values may also call for sacrifices of individual welfare in a conventional sense in order to advance welfare from the perspective of organic farming.

The assessment of AHW raises several methodological problems while the results are often hard to interpret (Webster & Main, 2003). Due to differences in genetic origin, age, sex, or in the experiences during ontogenesis, farm animals vary widely in their requirements in relation to the housing condition. Appropriate living conditions in relation to AHW can not only be assessed with respect to the species but also has to take the individual and heterogeneous demands of farm animals into account (Sundrum, 2000).

In general, legislators and brand label programs are using technical indicators, which refer to single aspects of housing conditions (e.g., space allowance, laying surface, feeding regime), to mark a difference to lower levels of minimum standards. The prescriptions of the EU Regulation on organic livestock production clearly exceed the legal standards of conventional production (Schmid *et al.*, 2008). However, the meaningfulness of technical indicators used in basic standards with regard to AHW is limited (Sundrum & Rubelowski, 2001). Technical indicators represent only a small section of the complex interrelationship among farm animals and their living conditions, thus lacking validation when the responses of the animals are not assessed directly in the specific situation.

According to the conclusions from the comprehensive scientific EU project "Welfare Quality," assessments of animal welfare should be based on housing conditions, management, animal health, and animal behavior (Botreau *et al.*, 2009). Within the European Welfare Quality Project (Welfare Quality, 2009), protocols have been elaborated to assess the situation of AHW with respect to different species. The scientists from different disciplines and other stakeholders involved agreed upon the list of indicators as a pragmatic approach to provide orientation, being well aware that it is not possible to cover all aspects of AHW with these indicators.

From a systemic point of view, AHW is the outcome of a very complex process that emerges on the level of the single organism and on the farm level. Correspondingly, status of AHW can be described as the emergent property of the individual animal or of the specific farm (Sundrum, 2007a). Numerous factors such as genotype, age, sex, growth, and development interact within each organism and react with the various factors outside of the organism, such as feeding, housing, and management conditions. Accordingly, the large variation within a herd in relation to AHW is due to the potential of numerous interactions. There is a growing understanding within the scientific community that it is necessary to develop more comprehensive concepts, which simultaneously consider a larger number of causal relationships. The isolated view under *ceteris paribus* assumptions appears guilty of selective attention and overgeneralization, and is beginning to be replaced by a systemic approach (DFG, 2005). This requires interdisciplinary collaboration since this problem's definition cuts across conventional commodity and disciplinary lines. The key feature of a systemic approach is that it captures the dynamics and interactions between the various elements within each system. Thus, there is a need for a higher level of plausibility and coherence in relation to general statements and conclusions regarding marketing, labeling, and branding. Furthermore, efficient feedback mechanisms, such as monitoring concepts and quality assurance schemes, have to be promoted to increase the predictability of AHW status.

6.7 INCONSISTENCIES AND COGNITIVE DISSONANCES

Organic livestock production is generally understood as being based on ethical values. Investigations that try to assess the relevance of additional ethical attributes of organic food for consumers' purchase decisions are based on the assumption that organic farming, in general, provides optimal living conditions that ensure a high level of AHW (Zander & Hamm, 2010). Taking into account the large variation between organic farms, the assumption that organic is offering an "ethical product" is questionable and so are the results of questionnaires when they do not correspond to reality.

Producers and consumers have different understandings of what naturally produced food is and also of what optimal animal life is (Te Velde *et al.*, 2002). Moreover, if consumers care about AHW, they often have an outlook that is quite different from that of animal scientists (Lassen *et al.*, 2006). Within a quantitative study (Vanhonacker *et al.*, 2008), a similar interpretation of farm animal welfare in terms of animal welfare-related aspect's ranking was found between consumers and farmers, whereas differences were observed related to aspects dealing with the ability to engage in natural behavior on the one hand and with production process-related aspects on the other hand. Consumers evaluated the current state of animal welfare as rather problematic, while farmers reported a more satisfactory evaluation of the present condition of farm welfare. When imagining a good farm life, consumers often refer to a romantic view of nature, where animals could live freely (Murdoch & Miele, 1999). At the same time, they emphasize a close and caring relationship between farmers and their animals as important for animal welfare.

The fact that both "good" and "bad" food products are commercially available in the market causes considerable ambivalence. Many consumers refuse to blame the individual farmer, but instead point to structural factors, mostly of an economic nature, which together force producers to produce farm animals in the way they do (Lassen *et al.*, 2006). Some consumers recognize that they themselves are part of the problem, because they sustain the market for conventionally produced foods by purchasing cheap products rather than paying more for alternative products.

The pressure on the producers to reduce production costs consecutively has become market driven, with competition between producers and between retailers to sell food as cheaply as possible, thereby acquiring its own momentum. In contrast, implementation of measures capable of improving the level of AHW on the farms is expected to clearly increase the production costs (Sundrum *et al.*, 2007). However, organic farmers will not adequately benefit from those efforts as an increased health status is not honored directly through the market by extra prices. Often, prices for organic animal products do not even cover the previous additional expenditures of organic livestock farming. As a consequence, those producers who aim for a high level of AHW by increasing their labor and management efforts compete with their products on the same markets as those who widely ignore the issue of AHW (Sundrum, 2007b).

The current market conditions widely ignore the large variability in quality traits and in the impacts on common goods, promoting unfair competition when enabling equal prices for very different performances. Farmers who gain economical benefits by selling organic products but only providing a low level of AHW undermine efforts of other farmers attempting to maintain a high level of AHW. It is difficult to discuss ethical attributes of organic food when simultaneously fairness between farmers and credibility between farmers and consumers is eroding.

There is reason for the concern that the current handling of the issue of AHW in organic livestock production by different stakeholder groups provides huge discrepancies between the claim and reality and causes various conflicting areas on different levels:

- Retailers and/or producers claim to offer products that are derived from healthy animals, without providing transparency and evidence of AHW status of farm animals.
- Retailers want to increase the turnover by offering organic food with comparable low prices and at the expense of the possibilities of the farmer to investigate substantial improvements of AHW.
- Producers who strive for a high status of AHW by using appropriate management concepts and encountering higher production costs are confronted with unfair competition when

competing with their products on the same markets as those who widely make use of minimum standards and produce on a low-cost base.

- A high percentage of consumers announce their special interest in the issue of AHW and their willingness to pay premium prices, but hesitate to do so when corresponding food is offered and instead prefer purchasing cheaper food.
- Many consumers prefer to delegate responsibility for ethical issues when choosing animal products to the retailer or the government and are by their ignorance jointly responsible for the severe deficits in AHW within livestock production.
- Consumers partly wish to buy animal products related to the issue of AHW but are not willing to pay premium prices that cover the higher expenditures.

While farmers as owners are initially responsible for the well-being of their farm animals, they are very limited in their options of decision-making as they, in general, possess little financial flexibility that can be used for improvements. In contrast, consumers are able to make a choice between a large range of products. Expenditures for food in relation to the total budget of a household have dramatically decreased during the last decades. Hence, consumers, in general, can afford more expensive food products if their priorities were inclined in this fashion.

According to Vaarst *et al.* (2007), establishing farmer groups ("stable schools") seems to be a beneficial way of stimulating a dynamic development on the farms toward continuous improvement in the case of AHW planning. However, farmers start from different levels of AHW and are generally facing limited availability of resources, especially with respect to labor time and investments. They often act on the basis of best practice assumptions without being able to prove if the assumptions are valid, and thus implement measures and make decisions by following their own appreciation and world views. Baars (2010) believes in the intuitive and experimental knowledge held by organic farmers and propagates an "experimental science" in a dialogue-based culture of equality and mutual exchange focusing on best practices and case studies.

However, neither the approach of "stable schools" nor "experimental science" explains how case studies derived from farm-specific situations might provide results that can be generalized and transferred to different farm systems with different dynamics and interactions between the various elements. Furthermore, they do not provide options on how to prevent the development of primarily self-centered and self-referential individual perspectives.

Reasons for the various inconsistencies on and between different process levels can partially be traced back to the selective and single-minded perceptions that members of the different stakeholder groups may have on this issue, causing cognitive dissonances. The terms "animal health and welfare" and "organic farming" seem to be a cover without content. They serve as a projection surface for partly contrasting interests within and between the different stakeholders. However, it does not make sense to blame single stakeholder groups for being responsible for the discrepancies between the claim and reality, as all stakeholder groups are a part of the complex field of interests.

It is obvious that profound knowledge alone is not sufficient to change the current unsatisfactory situation. Indeed, no progress can be expected with regard to AHW if the farmer will not gain any benefit and profit from the market. Clearly, higher production costs in organic compared to conventional livestock production and ongoing competition between organic farmers has created a situation where the on-farm availability of resources (labor time, high-quality feed, investments, knowledge, etc.) essential for any improvements in relation to AHW are limited to a high degree. In this context, it is often argued that farmers will benefit from a high status of AHW by higher performances and less veterinary costs and thus should have an inherent incentive. This assumption, however, ignores the monetary efforts that have to be investigated before reliable effects can be obtained and congruously blames the majority of farmers for not being able to follow economical rules.

Gaining a high status of AHW in organic farming is not the responsibility of the farmer alone. Retailers of organic products so far have failed to provide a clear profile about AHW traits and have made no significant efforts to explain any differences to consumers' groups interested in the AHW issue. Thus, there is a need for more transparency and guidance in the complex and confusing issue both for the farmers and the consumers. Retailers have to make sure that a high level of AHW will be honored by adequate prices to cover the additional costs and efforts needed to ensure a high level of AHW. Retailers and organic farmers' associations have led many consumers to expect organic farming to take greater care for AHW than conventional farming. A structure, however, that only attempts to support price premium without providing sound evidence for a premium quality contributes to the erosion of organic practices.

6.8 CHALLENGES

Organic farming has committed itself to outperforming conventional farming in a number of areas, including the AHW issue, albeit with somewhat higher costs. Consumers' interests and expectations are very important as they are closely linked to their willingness to pay premium prices. Premiums are an essential precondition to cover at least partly the higher production costs in comparison to conventional production. Hence, it is of high relevance for organic farming to clarify how to cover consumers' interests and confidence without making use of a misleading labeling. The organic movement is challenged to ensure that its credibility and the consumers' confidence does not get lost due to discrepancies between different expectations.

The large variation in the living conditions and risk factors between organic farms, and the large differences in AHW status emerging from them, contradict the expectations of consumers with respect to a nearly invariable process quality advertised with food labeling. The variability found in practice has definitely been overextended to justify the label of organic in general as representative for a high level of AHW. Nevertheless, clear benefits can be expected from the enhanced levels of minimum standards in organic farming with respect to the possibilities for the execution of animal-specific behavioral patterns. In contrast, animal health status does not differ significantly from conventional production. That does not mean that the situation cannot be changed. It is easier to deal with health problems in alternative systems through good management than to change conventional housing systems to meet animals' behavioral needs (Millet *et al.*, 2005).

In the earlier development of organic agriculture, uniform regulations were put into legislation to harmonize organic standards throughout Europe and to prevent unfair competition between organic and other labels acting as copycats. However, regulations based on minimum standards are of restricted relevance for the AHW issue and not only prevent but are expected to even fuel the development of unfair competition between organic farms.

Each stakeholder group involved in the AHW issue is acting within certain world views, enabling only a cutout of the complexity of the processes involved and as such is inevitably limited. Coherency within world views is obtained primarily by blocking out other relevant areas and side effects. From an overriding perspective, statements from single stakeholder groups appear as a patchwork of separated positions not interweaved into a coherent overall picture. Thus, establishing a higher level of consistency between different process levels and stakeholders appears to be the real challenge, especially with regard to the credibility of the whole production chain.

Communication between stakeholders about world views and perspectives via labeling is doomed to fail. When attempting to apply a value-based approach to very down-to-earth choices in practice (top–down approach), misunderstanding is unavoidable. Striving for abstract objectives, such as "natural living," "integrity," or "ethical values," often named as attributes of organic livestock farming to distinguish organic from conventional production, do not add to clarification. They are even less defined than the terms "animal health and welfare" or "organic" and as such cannot provide a better understanding but are fueling those prejudices that blame organic to be ideologically exaggerated.

As always in confusing and unclear areas, concepts based on reproducible facts might be able to provide orientation. To prevent the loss of credibility, organic farmers and retailers are obliged to take the burden of proof for what they claim toward consumers and politicians. Retailers and producers of organic food seem to be the victim of their own announcements, which, in turn, has increased consumer expectations in relation to process quality but is not accompanied by the claim for the actual price needed to cover all costs for this enhanced level of process qualities (Sundrum, 2008). Currently, retailers widely refuse to honor and differentiate diverse performances in relation to AHW. Following the previous line of reasoning, the refusal of the retailers in relation to AHW can be identified as the most relevant constraint. It prevents the farmers from gaining resources and incentives that are desperately needed and boosts the development of unfair competition between organic farmers, therewith losing credibility toward the expectation of consumers. The situation appears to be tenuous and irreconcilable without a change in the basic conditions of marketing and changes in previous paradigms.

A market that honors evidence-based performances in relation to AHW would provide an actual opportunity for an improvement because of monetary incentives. Reliable monitoring systems for assessing AHW are already available and partly implemented, such as records of all incidences of treatment, mortality and morbidity rates, and pathological findings gained from slaughterhouse data of fattening animals. These data do not cover the whole issue of AHW. Improvements in specific parameters would, however, contribute to a general improvement of AHW of farm animals. Producers failing to meet certain health standards in the longer term should face consequences, such as imposition of improvement schemes and loss of product certification (Sundrum *et al.*, 2006b).

However, reference values and thresholds cannot be defined by natural science but require an agreement between different stakeholders. An example is given by Sundrum and Löser (2008), describing the process of a common agreement obtained between scientists, veterinarians, and advisory service when defining an acceptable health status on the farm level recommended for use in organic pig production. Reference values are essential not only to differentiate between on-farm situations but especially to provide orientation for the farmer on what to strive for. There is consensus in the scientific community that farm management is the most relevant factor at the farm level for the implementation of a high AHW level. To strive for any improvements needs a clear goal and an assessment on how far the current state differs from the goal. Assessment of the on-farm AHW level in relation to reference values enables a diagnostic procedure with respect to the possible reasons for the farm-specific situation and inaugurates the search for solutions. Identification of farm-specific weak points in the management is an essential precondition for any advisory service to estimate the most effective and most efficient measures that might be appropriate to improve the very specific situation on the specific farm.

There seems to be no alternative than to increase transparency and to provide credible information about AHW, which consumers are able to understand, and which can serve as a prerequisite for launching such products into the market. The organic movement still has the option to make use of the head start by defining a clear goal or improved categories of AHW levels that provide an orientation to the farmer and enables him to decide which level he can or should strive for. However, this requires a change in the paradigm from a standard-oriented to a result- and output-oriented approach. Without the implementation of a monitoring system and fixed thresholds that should not be exceeded without severe consequences, many producers will fail to meet a high AHW status needed to justify premium prices.

Simultaneously, retailers have to make sure that a high level of AWH will be honored by adequate premium prices to cover the additional costs that are needed to produce food that is derived from farm animals with a high status of AHW. According to the hierarchy of the different process levels, there is reason to assume that those stakeholders that do not have direct contact to farm animals nevertheless have a high impact on AHW by influencing the business environment of organic farming, and thus the living conditions of farm animals and of farmers.

6.9 NEW APPROACH

Livestock production is an integrative part of an agro-ecosystem, encompassing various subsystems within a nested hierarchy of scales. In contrast to natural ecosystems, livestock farms are not self-organizing systems, but are managed by an operator. Farm managers are not only agents of control but are also components of the system itself (DFG, 2005). They follow internal and external perceptions, rituals, doctrines, world views, economical incentives, or individual objectives. Constraints in the previous approaches have emerged from a merely functional and deterministic understanding of the production process, focusing primarily on input criteria, while ignoring the structural and hierarchical characteristics of agro-ecosystems, and the topological and dynamical coupling of the processes involved. Interactions within and between subsystems and scales are deemphasized as are the potentials and synergistic effects that emerge from an improved interoperability between scales. Facing the complexity of agro-ecosystems, the farmer is largely left alone when confronted with various conflicting and complex aims and often makes decisions on a daily basis.

AHW emerges from very complex interactions between the individual animals and the environment within the farm system. Emergent effects on the farm level capture the dynamics and interactions between the various elements of the system, with particular emphasis on the nonlinear relationships, which are typical for biological systems and which result in extraordinarily complex interactions. Thus, AHW is not primarily related to minimal standards but to the farm management in the first place. From a systemic approach, AHW is to be seen as an emergent property of the living organism and the farm system. It largely depends on the capability of the farm management to understand the complexity of the processes on different process levels and to organize a well-balanced farm system and optimal living conditions for farm animals while facing severe limitations in the availability of relevant resources (labor, time, investments, knowledge, etc.). Adaptive management relies on local feedback and on the progressive accumulation of knowledge while supporting the agro-ecosystem to develop along its natural trajectory. As such, it is the opposite of an engineering and deterministic type of management.

Successful agro-ecosystem management requires detailed knowledge, the capacity to overlook the complexity of the production processes on various scales and the ability to detect disturbances based on both bottom–up and top–down approaches. It is inherent within the systemic approach that the same objectives and reference values are obtained from different starting points and different living conditions. While in any closed system, the final state is unequivocally determined by the initial conditions, the final state in open systems, such as agro-ecosystems, may be reached from different initial conditions and in different ways. This is what Bertalanffy (1968) in his general system theory defined as "equifinality." However, equifinality can be accomplished only by management factors, which govern the process in foresight of the goal. The approach of organic farming in the context of labeling is, at least to some degree, based on the assumption that a defined production method, although used in very different situations and under different local conditions, can be expected to produce an output (products) that includes common features and that the quality of the products and the production processes emerges from the processes within the farm system.

In view of the large heterogeneity between organic farms in relation to the status of AHW, it appears a considerable carelessness at the beginning of the organic agriculture movement when the pioneers neglected to define minimum standards with respect to the qualitative outcome of the production process, particularly the status of AHW.

REFERENCES

Aertsens, J., W. Verbeke, K. Mondelaers, and G. Van Huylenbroeck. 2009. Personal determinants of organic food consumption: a review. *Br. Food J.* 111:1140–1167.

Alroe, H. F., and E. Noel. 2008. What makes organic agriculture move - protest, meaning or market? A polyocular approach to the dynamics and governance of organic agriculture. *Int. J. Agric. Res. Governance and Ecology* 7:5–22.

Alroe, H. F., M. Vaarst, and E. Kristensen. 2001. Does organic farming face distinctive livestock welfare issues? A conceptual analysis. *J. Agric. Environ. Ethics* 14:275–299.

Andersen, H. J., N. Oksbjerg, and M. Therkildsen. 2005. Potential quality control tools in the production of fresh pork, beef and lamb demanded by the European society. *Livest. Prod. Sci.* 94:105–124.

Baars, T. 2010. Experiential science; towards an integration of implicit and reflected practitioner-expert knowledge in the scientific development of organic farming. *Journal of Agriculture and Environmental Ethics*. Available at: http://www.springerlink.com/content/j473211484056175/ Published online: July 22, 2010.

Baumgartner, J., T. Leeb, T. Gruber, and R. Tiefenbacher. 2003. Husbandry and animal health on organic pig farms in Austria. *Anim. Welf.* 12:631–635.

Bernués, A., A. Olaizola, and K. Corcoran. 2003. Labelling information demanded by European consumers and relationships with purchasing motives, quality and safety of meat. *Meat Sci.* 65:1095–1106.

Bertalanffy, von L. 1968. *General System Theory - Foundations, Development, Application*. George Braziller, New York. p. 295.

Bock, B. B., M. M. Van Huik. 2007. Animal welfare: attitudes and behavior of European pig farmers. *Br. Food J.* 109:931–944.

Bock, B. B., M. M. Van Huik, M. Prutzer, F. Kling Eveillard, and A. Dockes. 2007. Farmers' relationship with different animals: the importance of getting close to the animals. Case studies of French, Swedish and Dutch cattle, pig and poultry farmers. *Int. J. Sociol. Food Agric.* 15:108–125.

Botreau, R., I. Veissier, and P. Perny. 2009. Overall assessment of animal welfare: strategy adopted in welfare quality. *Anim. Welf.* 18:363–370.

Buchwalder, T. and B. Huber-Eicher. 2004. Effect of increased floor space on aggressive behaviour in male turkeys. *Appl. Anim. Behav. Sci.* 89:207–214.

Cabaret, J. 2003. Animal health problems in organic farming: subjective and objective assessments and farmers actions. *Livest. Prod. Sci.* 80:99–108.

Carlson, D., H. N. Lærke, H. D. Poulsen, and H. Jørgensen. 1999. Roughages for growing pigs, with emphasis on chemical composition, ingestion and faecal digestibility. *Acta Agric. Scand. A Anim. Sci.* 49: 129–136.

Commission Regulation (EC). 2008. No. 889/2008 of 5 September 2008 laying down detailed rules for the implementation of Council Regulation (EC) No. 834/2007 on organic production and labeling of organic products with regard to organic production, labeling and control.

Council Regulation (EC). 2007. (EC-No. 834/2007) of 28 June 2007 on organic production and labeling of organic products and repealing Regulation (EEC) No. 2092/91.

Darnhofer, I., W. Schneeberger, and B. Freyer. 2005. Converting or not converting to organic farming in Austria: farmer types and their rationale. *Agric. Human Values* 22:39–52.

DFG (German Research Foundation). 2005. *Future perspectives of agricultural science and research.* Wiley-VCH, Bonn. p. 80.

Dietze, K., C. Werner, and A. Sundrum. 2007. Status quo of animal health of sows and piglets in organic farms. In: U. Niggli, U. C. Leiffert, T. Alföldi, L. Lück, H. Willer (eds) *Improving Sustainability in Organic and Low Input Food Production Systems. Proc. 3rd QLIF Congr.*, Hohenheim. pp. 366–369.

Edwards, S. A. 2005. Product quality attributes associated with outdoor pig production. *Livest. Prod. Sci.* 94:5–14.

Fisher, A. D., M. A. Crowe, M. E. Alonso de la Varga, and W. J. Enright. 1996. Effect of castration method and the provision of local anesthesia on plasma cortisol, scrotal circumference, growth, and feed intake of bull calves. *J. Anim. Sci.* 74:2336–2341.

Flaten, O., G. Lien, M. Ebbesviek, M. Koesling, and P. S. Valle. 2006. Do the new organic producers differ from the 'old guard'? Empirical results from Norwegian dairy farming. *Renew. Agric. and Food Syst.* 21:174–182.

Fraser, D., D. M. Weary, E. A. Pajor, and B. N. Milligan. 1997. A scientific conception of animal welfare that reflects ethical concerns. *Anim. Welf.* 6:187–205.

Fraser, D. 2003. Assessing animal welfare at the farm and group level: the interplay of science and values. *Anim. Welf.* 12:433–443.

Enevoldsen, C. and Y. T. Gröhn. 1996. A methodology for assessment of the health-production complex in dairy herds to promote welfare. *Acta Agric. Scand. A Anim. Sci.* 27:86–90.

Gentle, M. J., N. Hunter, and S. A. Corr. 1997. Effects of caudolateral neostriatal ablations on pain-related behaviour in the chicken. *Physiol. Behav.* 61:493–498.

Grunert, K. G., L. Bredahl, and K. Brunso. 2004. Consumer perception of meat quality and implications for product development in the meat sector – a review. *Meat. Sci.* 66:227–259.

Guy, J. H., P. Rowlinson, J. P. Chadwick, and M. Ellis. 2002. Health conditions of two genotypes of growing/finishing pig in three different housing systems: implications for welfare. *Livest. Prod. Sci.* 75:233–243.

Harper, G. C. and A. Makatouni. 2002. Consumer perception of organic food production and farm animal welfare. *Br. Food J.* 104:287–299.

Haugegaard, J. 2010. Prevalence of nematodes in Danish industrialized sow farms with loose housed sows in dynamic groups. *Vet. Parasitol.* 168:156–159.

Hegelund, L., M. Bonde, and J. T. Sørensen. 2006. Medicinforbrug og dødelighed i økologisk og konventionel slagtesvineproduktion. Intern rapport. Sundhed og medicinforbrug hos økologiske og konventionelle slagtesvin. Husdyrbrug Nr1, Januar 2006. Ministeriet for Fødevarer, Landbrug og Fiskeri. Danmarks JordbrugsForskning.

Hermansen, J. E. and T. Kristensen. 1998. Research and evaluation of mixed farming systems for ecological animal production in Denmark. *Workshop Proc.: Mixed Farming Syst Eur.*, Dronten. *APMinderhoudhoeve-reeks* 2:97–101.

Hindhede, J., J. Sorensen, M. Bak Jensen, and C. Krohn. 1996. Effect of space allowance, access to bedding and flock size in slatted floor systems on the production and health of dairy heifers. *Acta Agric. Scand. A Anim. Sci.* 46:46–56.

Høøk Presto, M., H. K. Andersson, P. Wallgren, and J. E. Lindberg. 2007. Influence of dietary amino acid level on performance, carcass quality and health of organic pigs reared indoors and outdoors. *Acta Agric. Scand. A Anim. Sci.* 57:61–72.

Hovi, M., A. Sundrum, and S. M. Thamsborg. 2003. Animal health and welfare in organic livestock production in Europe – current state and future challenges. *Livest. Prod. Sci.* 80:41–53.

Hughner, R. S., P. McDonagh, A. Prothero, C. J. Shultz II, and J. Stanton. 2007. Who are organic food consumers? A compilation and review of why people purchase organic food. *J. Cons. Behav.* 6:94–111.

International Federation of Organic Agricultural Movement (IFOAM). 2006. *Principles of Organic Agriculture*. IFOAM, Bonn.

Kristensen, E., Enevoldsen, C. 2008. A mixed methods inquiry: how dairy farmers perceive the value(s) of their involvement in an intensive dairy herd health management program. *Acta Vet. Scand.* 50:50.

Lassen, J., P. Sandøe, and B. Forkman. 2006. Happy pigs are dirty! – conflicting perspectives on animal welfare. *Livest. Sci.* 103:221–230.

Leeb, T. and J. Baumgartner. 2000. Present status of pig fattening on selected organic farms in Austria. Animals in organic farming. *Proc. 13th Int. IFOAM Sci. Conf.*, Basel. p. 365.

Lund, V. 2006. Natural living – a precondition for animal welfare in organic farming. *Livest. Sci.* 100:71–83.

Machold, U., K. Troeger, and M. Moje. 2005. Evaluation of health status of pigs and cattle in organic production compared to those in conventional production based on the results of the official slaughter animal and meat inspection. *Mitteilungsblatt der Fleischforschung Kulmbach* 44:1–8.

Martelli, G. 2009. Consumers' perception of farm animal welfare: an Italian and European perspective. *Ital. J. Anim. Sci.* 8:31–41.

McEachern, M. G. and M. J. Schröder. 2002. The role of livestock production ethics in consumer values towards meat. *J. Agric. Environ. Ethics.* 15:221–237.

McEachern, M. G. and J. Willock. 2004. Producers and consumers of organic meat: a focus on attitudes and motivations. *Brit. Food J.* 106:534–552.

Mellor, D. and K. Stafford. 1999. Assessing and minimising the distress caused by painful husbandry procedures in rumiants. *Pract.* 21:436–446.

Meunier-Salaun M. C., S. A. Edwards, and S. Robert. 2001. Effect of dietary fibre on the behaviour and health of the restricted-fed sow. *Anim. Feed Sci. Technol.* 90:53–69.

Millet, S., C. P. Moons, M. J. Van Oeckel, and G. P. Janssens. 2005. Welfare, performance and meet quality of fattening pigs in alternative housing and management systems: a review. *J. Sci. Food Agric.* 85:709–719.

Murdoch, J. and M. Miele. 1999. Back to nature: Changing worlds of production in the food sector. *Sociol. Ruralis.* 39:465–483.

Nansen, P. and A. Roepstorff. 1999. Parasitic helminths of the pig: factors influencing transmission and infection levels. *Int. J. Parasitol.* 29:877–891.

Ngapo, T. M., E. Dransfield, J. F. Martin, M. Magnusson, L. Bredahl, and G. R. Nute. 2003. Consumer perceptions: pork and pig production. Insights from France, England, Sweden and Denmark. *Meat Sci.* 66:125–134.

Noordhuizen, J. P. 1999. Production diseases in dairy cattle. The veterinarian's role in disease control in the new millennium. In: T. Wensing (ed.) *Production Diseases in Farm Animals.* International Conference on Production Diseases, Utrecht, Nl. p 1–9.

Offermann, F. and H. Nieberg. 2000. Economic performance of organic farms in Europe. In: F. Offerman and H. Nieberg (eds) *Organic Farming in Europe: Economics and Policy.* Vol. 5. University of Hohenheim, Stuttgart.

Osweiler, G. D. 2006. Occurrence of mycotoxins in grains and feeds. In: B. E. Straw, J. J. Zimmerman, S. D'Allaire, D. J. Taylor (eds) *Diseases of Swine.* 9th edn. Blackwall Publishing Ltd., Ames, IA. p. 915–929.

Redbo, I., I. Mossberg, A. Ehrmark, and M. S. Högber. 1996. Keeping growing cattle outside during winter: behaviour, production and climatic demand. *Anim. Sci.* 62:35–41.

Ruis-Heutinck, L., M. Smits, A. C. Smits, and J. J. Heeres. 2000. Effect of floor type and floor area on behaviour and carpal joint lesions in beef bulls. *EAAP-Publ.* 102:29–36.

Rushen, J. and A. M. de Passillé. 1992. The scientific assessment of the impact of housing on animal welfare: a critical review. *Can. J. Sci.* 72:721–743.

Schaumann W. 2002. Der wissenschaftliche und praktische Entwicklungsweg des ökologischen Landbaus und seine Zukunftsperspektive. In: W. Schaumann, G. Siebeneicher, I. Lünzer (eds) *Geschichte des ökologischen Landbaus.* SÖL-Sonderausgabe. 65:11–58.

Schmid O., B. Huber, K. Ziegler, L. M. Jespersen, and G. Plakolm. 2008. Analysis of differences between EU regulation 2092/91 in relation to other standards. In: *Proc. 16th IFOAM Org. World Congr. "Culitvate the future" and the 2nd Sci. Conf. "Cultivating the future based on science" of ISOFAR (Int. Soc. Org. Agric. Res.).* Vol. 2. June 18–20th, pp. 382–385.

Skarstad, F. A., L. Terragni, and H. Torjusen. 2007. Animal welfare according to Norwegian consumers and producers: definitions and implications. *Int. J. Sociol. Food Agric.* 15:74–90.

Stafleu, F. R., F. J. Grommers, and J. Vorstenbosch. 1996. Animal welfare, evolution and erosion of a concept. *Anim. Welf.* 5:225–234.

Stewart, C. O. and L. Boyle. 2008. Influence of access to straw provided in racks on the welfare of sows in large dynamic groups. *App. Anim. Behav. Sci.* 112:235–247.

Sundrum, A. 1999. EEC-Regulation on organic livestock production and their contribution to the animal welfare issue. In: *Regulation of Animal Production in Europe*. 270:93–97, KTBL-Schrift, Darmstadt.

Sundrum, A. 2000. Preconditions of organic livestock farming to improve animal health and welfare. *5th Int. Livest. Farming Syst. Symp.* European Association for Animal Production. 97:81–88.

Sundrum, A. and I. Rubelowski. 2001. Meaningfulness of design criteria in relation to animal health. *Acta Agric. Scand. A Anim. Sci.* 30:48–52.

Sundrum, A. and M. Ebke. 2004. Problems and challenges with the certification of organic pigs. In: M. Hovi, A. Sundrum, S. Padel (eds) *Proceedings of the 2nd SAFO-Workshop*. University of Kassel, Kassel. p. 193–198.

Sundrum, A., K. Schneider, and U. Richter. 2006a. Possibilities and limitations of protein supply in organic poultry and pig production. In: A. Sundrum, K. Schneider, U. Richter (eds) *Proc. Eur. Jt. Cong. Org. Farming Eur. Rural Dev.*. Odense. p. 528–529.

Sundrum, A., M. Vaarst, G. Arsenos, A. Kuzniar, B. I. F. Henriksen, M. Walkenhorst, and S. Padel. 2006b. Recommendations to the formulation of the EU regulation 2092/91 on livestock production. In: A. Sundrum, M. Vaarst, G. Arsenos, A. Kuzniar, B. I. F. Henriksen, M. Walkenhorst, S. Padel (eds) *Proc. 1st IFOAM Int. Conf. Anim. Org. Prod.*. St. Paul. p. 121–127.

Sundrum, A. 2007a. Achievements of research in the field of livestock systems and quality production. In: A. Rosati, A. Tewolde, C. Mosconi (eds) *Animal Production and Animal Science Worldwide. WAAP Book of the Year 2006*. Wageningen Academic Publishers, Wageningen. pp. 95–106.

Sundrum, A. 2007b. Conflicting areas in the ethical debate on animal health and welfare. In: W. Zollitsch, C. Winckler, S. Waiblinger, A. Haslberger (eds) *Sustainable Food Production and Ethics*. Wageningen Academic Publishers, Wageningen. p. 257–262.

Sundrum, A., K. Dietze, and C., Werner. 2007. System approach to improve animal health. In: W. Zollitsch, C. Winckler, S. Waiblinger, A. Halsberger (eds) *Sustainable Food Production and Ethics*. Wageningen Academic Publishers, Wageningen. pp. 360–364.

Sundrum, A. 2008. Organic livestock production – trapped between aroused consumer expectations and limited resources. *Proceed. 2nd Sci. Conf. Int. Soc. Org. Agric. Res. (ISOFAR)*, Modena. pp. 208–211.

Sundrum, A. and R. Löser. 2008. Zielvorgaben für die Tiergesundheit. *Ökologie & Landbau.* 145:39–42.

Sundrum, A., P. Nicholas, and S. Padel. 2008. Organic farming: challenges for farmers and feed suppliers. In: P. S. Garnsworthy, and J. Wiseman (eds) *Recent Advances in Animal Nutrition 2007*. Nottingham University Press, Nottingham. pp. 239–260.

Sundrum, A. 2010. Assessing impacts of organic production on beef and pork quality. CAB Review: Perspectives in Agriculture, Veterinary Science, Nutrition and Natural Resources, 5, No. 004, p. 13. Available at: http://www.cabi.org/cabreviews.

Sundrum, A., A. Goebel, D. Bochicchio, M. Bonde, A. Bourgoin, K. Dietze, S. Dippel, L. Hegelund, T. Leeb, K. Lindgren, A. Prunier, and S. Wiberg. 2010. Health status in organic pig herds in Europe. *Proceed. 21st Int. Pig Vet. Soc. (IPVS) Cong.*, Vancouver. p. 227.

Te Velde, H., N. Aarts, and C. van Woerkum. 2002. Dealing with ambivalence: farmers' and consumers' perception of animal welfare in livestock breeding. *J. Agr. Environ. Ethics* 15:203–219.

Turner, S. P. and D. M. Dwyer. 2007. Welfare assessment in extensive animal production systems: challenges and opportunities. *Anim. Welf.* 16:189–192.

Tuyttens, F., M. Heyndrickx, M. De Boeck, A. Moreels, A. Van Nuffel, E. Van Poucke, E. Van Coillie, S. Van Dongen, and L. Lens. 2008. Broiler chicken health, welfare and fluctuating asymmetry in organic versus conventional production systems. *Livest. Sci.* 113:123–132.

Vaarst, M., S. Padel, G. Arsenos, A. Sundrum, A. Kuzniar, M. Walkenhorst, L. Grova, and I. B. Henriksen. 2006. Challenges for animal health and welfare in the implementation of the EU legislation on organic livestock production: analysis of questionnaire survey among SAFO participants. In: C. Rymer, M. Vaarst, S. Padel (eds) *Future Perspectives for Animal Health on Organic Farms: Main Findings, Conclusions and Recommendations from the SAFO Network. 5th SAFO-Workshop*, Odense. pp. 43–74.

Vaarst, M., T. B. Nissen, S. Östergaard, I. C. Klaas, T. W. Bennedsgaard, and J. Christensen. 2007. Danish stable schools for experiential common learning in groups of organic dairy farmers. *J. Dairy Sci.* 90:2543–2554.

Van de Weerd, H. A., R. Keating, and S. Roderick. 2009. A review of key health-related welfare issues in organic poultry production. *Worlds Poult. Sci. J.* 65:649–683.

Vanhonacker, F., W. Verbeke, E. Van Poucke, and F. A. Tuyttens. 2008. Do citizens and farmers interpret the concept of animal welfare differently? *Livest. Sci.* 116:126–136.

Verbeke, W. A. and J. Viaene. 2000. Ethical challenges for livestock production: meeting consumer concerns about meat safety and animal welfare. *J. Agric. Environ. Ethics.* 12:141–151.

Verhoog, H., V. Lund, and H. Alrøe. 2004. Animal welfare, ethics and organic farming. In: M. Vaarst, R. Roderick, V. Lund, W. Lockeretz (eds) *Animal Health and Welfare in Organic Agriculture.* CABI Publishing, Wallingford, Oxon. pp. 73–94.

Welfare Quality. 2009. *Welfare Quality Assessment Protocol for Cattle.* Welfare Quality Consortium, Lelystad.

Whay, H., A. Watermann, A. Webster, and J. O'Brien. 2002. Farmer perception of lameness prevalence. In: H. Whay, A. Watermann, A. Webster, J. O'Brien (eds) *12th Int. Symp. Lameness Ruminants*, Orlando, FL. pp. 355–358.

Webster, A. J., and D. C. Main. 2003. Proceedings of the 2nd International Workshop on the assessment of animal welfare at farm and group level. *Anim. Welf.* 12:429–708.

Wilkie, R. 2005. Sentient commodities and productive paradoxes: the ambiguous nature of human livestock relations in Northeast Scotland. *J. Rural Stud.* 21:213–230.

Yiridoe, K. Y., S. Bonti-Ankomah, and R. C. Martin. 2005. Comparison of consumer perceptions and preference toward organic versus conventionally produced foods: a review and update of the literature. *Renew. Agric. Food Syst.* 20:193–205.

Zander, K. and U. Hamm. 2010. Consumer preferences for additional ethical attributes of organic food. *Food Qual. Prefer.* 21:495–503.

7 Environmental Impacts and Life Cycle Analysis of Organic Meat Production and Processing

Cesare Castellini, Antonio Boggia, Luisa Paolotti, Greg J. Thoma, and Dae-soo Kim

Abstract: The main objective of organic agriculture is the attainment of an ecological production to enhance biodiversity, environmental sustainability, and food safety. The organic farming system, in contrast with the conventional one, can greatly vary due to a wide range of breed types, feed ingredients, and rearing systems used by producers. Accordingly, the effects of operating as an organic production (OP) system on several key points (performance, animal welfare, impact, characteristics of product) are different and often only comparisons between case studies can be performed. In this chapter, a case study is presented in which the environmental impacts caused by three different poultry production systems—(1) conventional, (2) organic, and (3) organic-plus—are compared, by means of the life cycle assessment (LCA) method. The same method is then used in the context of another case study, concerning a national scan-level carbon footprint for the US swine production.

Keywords: organic; conventional; sustainability; life cycle assessment; environmental impacts; poultry production; swine production; carbon footprint

7.1 ORGANIC MEAT AND ENVIRONMENTAL IMPACTS

The main objective of organic agriculture is the attainment of an ecological production to enhance biodiversity, environmental sustainability, and food safety. Organic production (OP) is established according to strict regulations (Council Regulation 2092/91 and Council Regulation 1804/99; US Organic Foods Production Act (OFPA), 1990) outlining specific productive protocols for agriculture and animal husbandry. European Union (EU) Regulation comprises compulsory rules that define the feeding protocols (no chemical additives added to the feed, no genetically modified organisms (GMO) feed ingredients), prophylaxis, and therapy standards as well as rearing system characteristics (origin of animals, age at slaughtering, and dimension of external area), and recommendations (use of autochthonous strains and pasture availability). The recommendations—especially those related to the use of autochthonous breeds—seem to concern animal welfare, quality, and environmental impact more than the compulsory rules do.

Thus, the organic farming system, in contrast with the conventional one, can greatly vary due to a wide range of breed types, feed ingredients, and rearing systems used by

Organic Meat Production and Processing, First Edition. Edited by Steven C. Ricke, Ellen J. Van Loo, Michael G. Johnson, and Corliss A. O'Bryan.

producers. Accordingly, the effects of operating as an OP system on several key points (performance, animal welfare, impact, and characteristics of product) are different and often only comparisons between case studies can be performed.

Several authors (e.g., Foster *et al.*, 2006; de Vries & de Boer, 2010) have placed meat production among the top contributors to environmental impact. Compared to other food-related products, meat production and processing is most significant in term of eutrophication, for which a contribution of approximately 18% of the total EU impact is estimated.

Within the animal chain, the production of feedstuffs is the main source of impact; in terms of energy requirements, beef needs 28 MJ/kg of body weight (BW), sheep 23 MJ/kg BW, pork 17 MJ/kg BW, and poultry 12 MJ/kg BW (Foster *et al.*, 2006).

Organic feed production is generally associated with lower energy demands, even if discrepancies are present for some species. Most organic animal production reduces primary energy use by 15%–40%, but when the productive performance of animals is significantly affected (poultry and pig sector), the benefits of the lower energy needs for production of organic feeds are sometimes overridden because more total feed is needed per kg meat produced.

Direct emission of methane from the enteric processes of ruminant and nitrous oxide emission from soil are much more relevant than energy use as a source of greenhouse gases (GHGs) in meat production systems; as a result, OP does not necessarily have lower global warming potential (GWP) than conventional production. For sheep and pig meat, OP seems to reduce overall GHG emission, while for beef and poultry, OP increases emission (Foster *et al.*, 2006).

However, such comparisons theoretically assume the homology of organic products, whereas there are sufficient references stating the presence of a different animal welfare (Meluzzi *et al.*, 2009; Mugnai *et al.*, 2009; Dal Bosco *et al.*, 2010), different characteristics of products (Castellini *et al.*, 2008; Branciari *et al.*, 2009; Sirri *et al.*, 2010), and landscape esthetics due to the farming system.

This lack of homology points to the necessity of a multitrait evaluation that can simultaneously show how the main factors of animal production (welfare, quality, and environmental impact) vary, because what is appropriate for organic animal production sites (e.g., use of autochthonous breed, etc.) does not always reduce the environmental impact.

7.2 THE LIFE CYCLE ASSESSMENT METHOD

7.2.1 LCA methodology

Life cycle assessment (LCA) is a method to evaluate environmental impacts associated with all the stages of a product or process throughout the entire life cycle, which for agricultural products begins with production of fertilizers, soil preparation, crop planting, cultivation, harvesting, and animal husbandry, through processing, use, and disposal of wastes after the product's final end use (Cederberg & Mattsson, 2000; Hogass, 2002; Keoleian *et al.*, 2004; Thomassen *et al.*, 2008). This includes identifying and quantifying an inventory of relevant energy and materials used and wastes released to the environment, calculating the potential environmental impacts associated with identified inputs and outputs, interpreting the results to make informed decisions, and evaluating improvement opportunities. By using LCA, we can choose the least burdensome alternative by comparing the full range of environmental and social damages attributable to products and services.

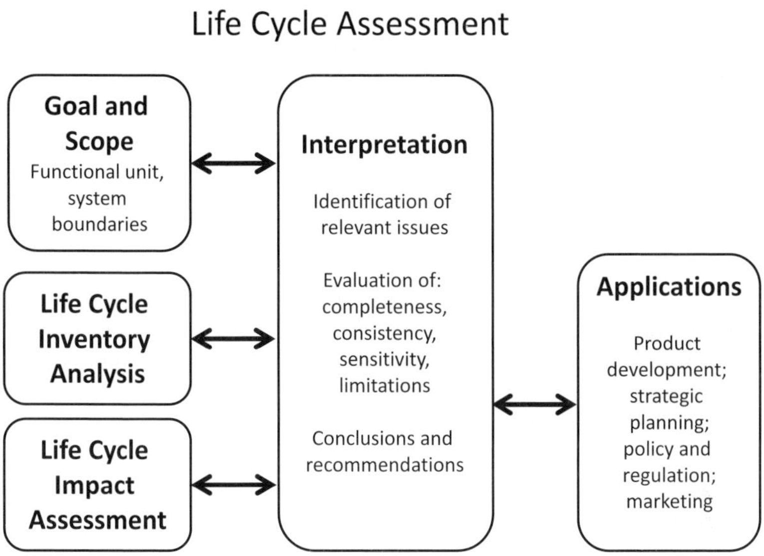

Figure 7.1 Stages of a life cycle assessment.

Figure 7.1 depicts the core LCA phases and highlights the iterative nature of the process. The goal and scope definition phase is a planning process that involves defining and describing the product, process, or activity to be studied, establishing the aims and context in which the LCA is to be performed, and identifying the life cycle stages and environmental impact categories to be reviewed for the assessment.

The object of study is described in terms of a functional unit that characterizes the specific use of the product and must be carefully defined, especially for comparative purposes. Apart from describing the functional unit, the goal and scope stage should address the overall approach used to establish the system boundaries. The system boundary determines which unit processes are included in the LCA and must reflect the goal of the study.

The life cycle inventory (LCI) analysis phase (LCI phase is the second phase of LCA) is an inventory of input–output material and energy flows with regard to the system being studied; it involves identifying and quantifying energy, water, materials, and environmental releases (e.g., air emissions, solid wastes, wastewater discharge). This encompasses all data related to technical and environmental quantities for relevant unit processes within the system boundaries. Other types of interventions and exchanges such as radiation or land use can also be included. Usually, LCIs and modeling are conducted using a dedicated software package, such as SimaPro (http://www.pre.nl/simapro/) or GaBi (http://www.gabi-software.com/). LCA software analyzes each stage of the product's life cycle based on data input by the analyst.

Life cycle impact assessment (LCIA) phase is the third phase of the LCA. This step calculates human and ecological effects of material consumption and environmental releases identified during the inventory analysis. There are several common LCIA midpoint impact categories, such as climate change (global warming), eutrophication, acidification, ozone depletion, land use, etc. The first step in the LCIA is termed characterization. Here, impact potentials are calculated based on the LCI results. The next steps are normalization and weighting; these are both voluntary according to the International Organization for Standardization (ISO) standard. Normalization provides a basis for comparing different types of environmental impact categories; in short, a normalization flow is defined (e.g., all GHGs emitted in the EU), and the emissions from the life cycle under study are normalized to this flow. By comparing the normalized contribution, the life cycle of different impacts can be ranked. It is possible that a particular product contributes to 0.01% of GWP, but a much smaller relative contribution to total eutrophication. Weighting is used to combine environmental midpoint impact categories into a single score or small number of categorical indices (such as human health impact, which is affected by several midpoint categories) and requires assigning a weighting factor to each midpoint impact category depending on perceived relative importance. Readers should be cautioned that interdependencies may exist between impact categories and poor decisions can be made when only a single impact metric is used as the basis.

Life cycle interpretation is the final phase of the LCA procedure, in which the results are summarized and discussed. Its goal is to identify the most significant environmental impact categories and the associated life cycle stage, and highlight opportunities for potential change or innovation. Another purpose of life cycle interpretation is to determine the level of confidence in the final results and communicate them in an accurate, fair, and complete manner. This is accomplished by identifying the data elements that contribute most significantly to each impact category, evaluating the sensitivity of these significant data elements, assessing the consistency and completeness of the study, and drawing conclusions and recommendations based on a clear understanding of how the LCA was carried out and the results were developed.

7.2.2 The LCA method applied to animal husbandry

The LCA method has been applied to numerous research fields, varying from the industrial to the agri-food sector. In the field of animal husbandry, several studies have been conducted, especially concerning cattle and pig production. The majority of the works involved the study of OP systems, analyzed separately, or in comparison with conventional or other systems. This demonstrates the relevance assigned to organic animal production in the scientific research context.

Regarding milk production, an ecological advantage of organic compared to conventional milk production was shown by several studies (e.g., Cederberg, 1998; Haas *et al.*, 2001; Stonehouse *et al.*, 2001; de Boer, 2003). Thomassen *et al.* (2008) compared the environmental impact of two Dutch milk production systems, i.e. a conventional and an organic system. Hotspots were identified in both milk production chains. Results showed better environmental performance (per kg milk) concerning energy use and eutrophication potential for organic farms than for conventional farms. Total acidification potential and GWP did not differ between the selected conventional and organic farms. In addition, results showed lower land use of conventional compared with the organic dairy farm. The study of Müller-Lindenlauf *et al.* (2010) focused attention on a systemic comparison of different dairy farm types within the organic dairy sector, considering intrafarm interactions and all relevant environmental impact categories. The study showed how different farm types have significant and relevant differences in environmental impact. This strengthens the concept that organic products are not all homologous, but their performance strictly depends on the type of system production.

The effectiveness of the environmental indicators of three methods widely used in the animal husbandry field—(1) input–output accounting, (2) ecological footprint analysis, and (3) LCA—was also evaluated (Thomassen & de Boer, 2005). The data concerning organic dairy farms in the Netherlands were used to evaluate the effectiveness of the environmental indicators, in terms of relevance, quality, and availability of data (Mitchell *et al.*, 1995; Bell & Morse, 1999; Cornelissen, 2003). An indicator is relevant if it is significant, and if it is understandable to all the stakeholders involved, whereas indicator quality is related to its reliability, sensitivity, and the feasibility of determining a target value. According to the results, the indicators from input–output accounting are effective, but do not include all of the environmental impact categories; the indicators from the ecological footprint analysis are not effective due to their low quality and relevance, whereas those from LCA, although difficult to collect, exhibit a high quality and relevance.

Cederberg and Mattsson (2000) applied LCA to organic and conventional milk production in Sweden. The particular elements investigated were the flows of ingredients in feed production and the flows of farm nutrients. The results showed that the different feeding strategies in both production methods greatly affect several impact categories.

Halberg *et al.* (2005) made an analysis on European livestock production systems, comparing organic and conventional production of milk and pig meat. They used different environmental assessment tools, differing in terms of which environmental objectives were included, and how indicators were developed and interpreted. Among them, LCA and the ecological footprint analysis were considered. The authors showed that the comparison between organic and conventional production of meat gives different results depending on the choice of functional unit (i.e., per hectare or per kg product). Some recommendations about the choice of the agricultural and environmental indicators were included, taking into account the geographical scale, the system boundaries, and methods of interpretation.

Also, Haas *et al.* (2001) applied the LCA method to estimate the impacts caused by three different types of pasture production systems: (1) intensive, (2) extensive, and (3) organic. The results showed that the impact categories, climate change, acidification–eutrophication, and fossil fuels, had lower values in both extensive and organic systems, compared to conventional systems.

Several LCA studies were also performed in the field of pig production. Basset-Mens and van der Werf (2005), for example, compared three production systems: (1) conventional, (2) organic, and (3) the French extensive system "label rouge," following its own regulations. According to the study, the impact category acidification–eutrophication had lower values in the organic and in the label rouge systems, whereas the category global warming exhibited similar values in all of the systems. The influence, in terms of impacts, of feed choice in pig production was also investigated (Eriksson *et al.*, 2005), by comparing three alternative scenarios of protein supply. The LCA results demonstrated that feed production contributes more than animal husbandry to the environmental impact of the system for the following impact categories: energy use, global warming, and eutrophication. The opposite situation was observed for acidification.

Pelletier *et al.* (2010) used LCA to compare the environmental performance of high and low profitability commodity, and deep-bedded niche swine production systems, in the Upper Midwestern United States. They found that commodity systems generally outperformed deep-bedded niche systems for the environmental criteria considered in the study. Drivers of impacts differed between commodity and deep-bedded niche systems. In particular, feed production was the key consideration in both, but proportionally more important in niche production due to lower feed use efficiencies.

Only a few LCA studies have been performed for the poultry sector. Ellingsen and Aanondsen (2006) assessed the environmental impacts of Norwegian cod fishing and salmon farming compared to those produced by poultry farming. The main aim of the study was to find reference levels for environmental performance, and to identify critical areas and potential improvements. The results showed that the fishing phase for the cod and the feeding phase for both salmon and poultry had higher environmental impacts. Poultry was the most energy effective, followed by salmon and cod, which showed almost the same value as each other in terms of energy consumption. The area of sea floor affected by bottom trawling was approximately 100 times larger than the land area needed to produce the chicken feed.

Pelletier (2008) investigated the environmental performance along the US broiler poultry chain by means of LCA. The study showed that the feed production phase accounted for 80% of the energy consumption, 82% of GHG emissions, 98% of ozone depleting emissions, 96% of acidifying emissions, and 97% of eutrophying emissions associated with the cradle-to-farm gate production of broiler poultry. Whereas on farm inputs and emissions, which are largely related to heating and ventilation, these contributed on average to only 9% of these impacts. Hatchery chick production phase contributed negligibly to the total amount.

Boggia *et al.* (2010) used LCA to compare the environmental impact of three different poultry production systems: (1) conventional, (2) organic, and (3) organic-plus. Organic-plus has more restrictive requirements than the organic system for improving animal welfare (i.e., use of slow-growing strains and 10 m^2 pasture per bird) and results in higher quality of the product. The study provided useful information for reducing the environmental impacts of poultry production. An important recommendation was drawn: great attention has to be paid to the feed production phase as it contributes more than animal rearing to the environmental impact of the overall system. With reference to the comparison among the systems, the results of the study showed that the organic system had the best environmental performance because it not only had the lowest impact values for two of the most important impact categories (i.e., respiratory inorganics and fossil fuels), but it also had the lowest values for most of the remaining categories. LCA, by means of its approach and its specific indicators, proved to be a useful tool to be applied in a wider process for the analysis of sustainability and to adapt and improve production systems.

Finally, some studies were performed analyzing different livestock products simultaneously for direct comparison. De Vries and de Boer (2010) analyzed 16 peer-reviewed studies, compared the environmental impacts of production of beef, pork, chicken, milk, and eggs using LCA. According to their review analysis, production of 1 kg of beef used the most land and energy, and had highest GWP, followed by production of 1 kg of pork, chicken, eggs, and milk. Differences in environmental impact among pork, chicken, and beef could be explained mainly by three factors: (1) differences in feed efficiency, (2) differences in enteric CH_4 emission between monogastric animals and ruminants, and (3) differences in reproduction rates. Production of 1 kg of beef protein also had the highest impact, followed by pork protein, whereas chicken protein had the lowest impact.

Williams *et al.* (2006) conducted an LCA analysis of 10 major commodities in England and Wales, including beef, pig, sheep and poultry meats, eggs, and milk. In general, OP was less energy consuming, except for poultry meat and eggs. Environmental burdens, such as GWP or eutrophication, were often greater per unit of production in organic than nonorganic systems.

7.3 CASE STUDY—ENVIRONMENTAL IMPACT EVALUATION OF POULTRY PRODUCTION SYSTEMS, BY MEANS OF LCA: COMPARISON AMONG CONVENTIONAL, ORGANIC, AND ORGANIC-PLUS

7.3.1 The three production systems

In this section, three different poultry production systems—(1) conventional, (2) organic, and (3) organic-plus—are compared by means of LCA. The three systems are located in the same region of Italy. Although the analysis can offer useful information about the environmental impacts of the three systems, the results cannot be generalized or extrapolated to other countries, as they come from a case study and not from analysis of a representative sample of farms.

The conventional rearing system is located in a rural zone of Central Italy. The animals' buildings cover a surface of 2585 m^2. The shelters are air-conditioned to maintain a correct temperature and humidity level for maximizing the chickens' performance. The density of birds per m^2 is 17.5. The number of animals in a year is approximately 261,120. Each cycle lasts 50 days (six production cycles per year). Broilers (Ross 308) of both sexes are used; the weight of the animals on arrival at the farm is approximately 40 g, whereas at the end of the cycle, their weight is, on average, 2.6 kg (feed to gain ratio 2.0) with a dressing percentage of 83%. The conventional diets (starter: 1–14 days, grower: 15–28 days, and finisher diets: 29–50 days) were formulated with common ingredients according to the standard recommendations (Aviagen Technical Team, 1999). The relative consumption of the different diets—starter (8%), grower (47%), and finisher (45%)—was considered for LCA calculations (Table 7.1).

The organic farm has a surface of 8.0 hectares; 0.2 hectares are designated as animal buildings, 0.2 hectares for infrastructure, and 7.6 hectares are used as pasture. Each building consists of two houses (regulated by the organic rules), each containing 4800 animals, for a total of 9600 animals per building and per cycle, with a density of 9.6 birds per m^2. Therefore, the density is approximately half that of the intensive system. The density outside is 0.25 birds per m^2. The number of animals produced in a year is approximately 67,200. Each cycle lasts 81 days (3.5 cycles per year). The strain used is the same as in the conventional system (Ross 308), but only female birds are reared to avoid excessive final weights. The weight of the animals at the end of the cycle is, on average, 3.5 kg (feed to gain ratio 2.8) with a slaughter yield of 83%. Whereas typical fast-growing diets start at 22% crude protein and finish at 17%–18% protein, organic rations start at approximately 20%–21% protein and finish at 15%–16% (Fanatico, 2006). Such low-protein ration is also used to slow down the rapid growth of meat type broiler. Starter, growing, and finisher diets were intermediate between National Research Council (1994) recommendations (Table 7.1) and label rouge diets (Lewis *et al.*, 1997) with ingredients entirely coming from organic crops. The starter, grower, and finisher diets represented, respectively, the 12% (1–21 days), 52% (22–63 days), and 36% (64–81 days) of the total bird consumption.

Chemical compounds are not present since they are not permitted in organic feed. The main differences in the organic feed formulation, compared to the feed used in conventional rearing, are the absence of GMO ingredients, synthetic amino acids, and coccidiostats.

The organic-plus farm has a total surface of 70 hectares, of which 0.3 hectares are designated as animal "rural" buildings, 3.5 hectares are for roads network, and 61 hectares are

Table 7.1 Mean ingredients (% of diets) and calculated composition (%) of starter, grower, and finisher diets used in conventional, organic, and organic-plus systems.

(a) Conventional	**Starter (1–14 days)**	**Grower (15–28 days)**	**Finisher (29–50 days)**
Maize	38.0	54.0	51.5
Solvent-extracted soybean	39.0	33.75	28.0
Sorghum	15.64	7.0	13.0
Alfalfa meal	0.4	–	2.96
Soybean oil	3.0	2.0	2.5
Vitamin–mineral premix	1.2	1.0	0.9
Calcium carbonate	1.3	1.1	1.0
Dicalcium phosphate	0.9	0.6	0.6
Sodium bicarbonate	0.3	0.3	0.3
Salt	0.2	0.2	0.2
DL-methionine	0.02	0.01	0.01
L-lysine HCl	0.02	0.01	0.01
Coccidiostat	0.03	0.03	0.02
Crude protein	22.1	20.1	18.2
AME MJ/kg	12.2	12.5	12.6
(b) Organic/organic-plus	**Starter (1–21 days)**	**Grower (22–63 days)**	**Finisher (63 days–slaughtering)**
Maize	42.2	49.8	42.8
Extruded soybean flakes	29.0	23.0	18.0
Maize gluten feed	24.0	13.0	7.0
Barley	–	3.0	12.0
Wheat	–	5.0	12.0
Alfalfa meal	1.0	3.0	5.0
Vitamin mineral premix[a]	1.0	1.0	1.0
Calcium carbonate	1.4	1.0	1.0
Dicalcium phosphate	1.0	0.8	0.8
Sodium bicarbonate	0.3	0.2	0.2
Salt	0.2	0.2	0.2
Crude protein	19.7	16.8	15.2
AME MJ/kg	13.1	13.0	12.7

Source: Aviagen Technical Team, 1999.
[a]Vitamins in the organic diet were provided by cod liver oil and malt yeast.
AME, apparent metabolizable energy.

used as rotated pasture (the remaining 5.2 hectares are forest and unused surface). The density of birds per m^2 is 0.10. The number of animals produced in a year is approximately 217,140. Each cycle lasts 110 days (3.3 production cycles per year). Slow-growing strains (less than 25 g per day) are used (Castellini *et al.*, 2008): at the beginning of the cycle they weigh 38 g, whereas at the end of the cycle their weight is 2.5 kg (feed index 3.5) on average. A slaughter yield of 80% was assumed. The feed formulation is the same as in the organic system.

Table 7.2 shows the main characteristics of the three animal rearing systems. Table 7.2 shows the composition of starter, grower, and finisher diets. Figures 7.2 and 7.3 show the conventional and the organic systems. The organic-plus system appears similar to the organic one, but with more surface available.

Table 7.2 Main characteristics of the three rearing systems.

Building and space distribution	**Conventional**	**Organic**	**Organic-plus**
Surface of the farm (ha)	1.5	8.0	70.0
Total birds per cycle (*n*)	45,334	9600	65,800
Buildings area (m^2)	2585	2000	3000[a]
Density indoor (birds/m^2)	17.5	9.6	16.6[a]
Surface for pasture (ha)	–	7.6	61.0
Density outdoor (birds/m^2)	–	0.25	0.10
Animal characteristics[b]	**Conventional**	**Organic**	**Organic-plus**
Final weight (kg)	2.6	3.5	2.5
Age at slaughtering (days)	50	81	110
Cycles of production (*n*/year)	6.0	3.5	3.3
Feed to gain ratio	2.0	2.8	3.5
Mortality rate (%)	4.0	4.0	6.0
Dressing percentage at slaughtering (%)	83	83	80

Source: Direct surveys of the three systems.
[a]Buildings only in case of bad weather and during the night.
[b]For conventional and organic-plus: mean performance considering a female/male ratio = 1. For organic: only female.
ha, hectare.

7.3.2 The LCA analysis for the three systems

An LCA analysis for three different systems was performed, according to ISO guidelines. The main goal of the LCA was a comparative evaluation of the environmental impacts of each of the three production systems. The functional unit considered in the LCA is 1 kg of poultry meat. In regards to the definition of the scope, an LCA study "from cradle to gate"

Figure 7.2 The conventional system.

Figure 7.3 The organic system.

was performed; it means that only the production of raw materials and poultry meat were considered. Factors that were not considered include: the transport to the slaughterhouse, slaughtering, processing of carcasses, and distribution. This choice was made in order to focus attention on production methods. The specific data necessary for the study (foreground data), describing the production systems, were collected by means of direct surveys of local production scenarios, in particular for the cultivation of maize, grain, barley, sorghum, and soy (which represent the main components of feed); for the transformation of maize (gluten feed) and soy (extrusion and solvent extraction of lipids) in feed; as well as for poultry rearing. Background data, taken from databases and literature, were mainly used for constructing the remaining cultivation processes, and for the general data of transport, fuel, and electricity consumption, with reference to databases into SimaPro software (Product Ecology Consultants, 1990). The Ecoinvent database was primarily used (Nemecek *et al.*, 2004). The impact assessment phase was developed using the "Eco-Indicator 99" method (Goedkoop & Spriensma, 2001) that measures various environmental impacts and is based on a damage function approach. The damage function represents the relation between the impact and the damage to human health or to the ecosystem. Impacts can be computed according to 11 midpoint impact categories, which can also be aggregated into the three endpoint categories: (1) human health, (2) ecosystem quality, and (3) resources consumption.

Table 7.3 presents a brief explanation of the categories' meaning. The results of the impact assessment are presented as normalized scores; therefore, the final results are expressed in points: the higher the score, the more important the impact. Even though optional, normalization is an important step, because it enables the evaluation of the extent an impact category can contribute to the overall environmental problem, allowing comparison of the different impact categories having different reference units.

The LCA analysis is referred to as the production from a year in each system. Each system analysis is divided into three main phases: (i) production of raw materials (crops and other

Table 7.3 Meaning of the impact categories.

Impact categories	Meaning
Human health	
(i) Carcinogens	Arsenic, benzene, cadmium, heavy metals, etc.
(ii) Respiratory organics	All kinds of organic emissions to air
(iii) Respiratory inorganics	Emissions to air, mainly of SO_2 and NO_x
(iv) Climate change	Emissions to air of hydrocarbons, carbon dioxide, methane, etc.
(v) Radiation	All kinds of radioactive materials
(vi) Ozone layer	All kinds of ozone-depleting substances (CFC, etc.)
Ecosystem quality	
(vii) Ecotoxicity	Emission to water, air, and soil, which cause toxic stress for the ecosystem
(viii) Acidification/eutrophication	Emissions to air, mainly of nitrogen and ammonia
(ix) Land use	Occupation and transformation of land
Resources consumption	
(x) Minerals	Aluminum, iron, lead, copper, etc.
(xi) Fossil fuels	Consumption of nonrenewable resources

Source: Nemecek *et al.*, 2004.

minor components); (ii) transformation of crops into feed; and (iii) animal rearing. Every phase includes different subprocesses:

(i) **Production of raw materials:** All of the processes related to crop cultivation that constitute the main feed ingredients were considered in the analysis. For each crop, the flows related to basic field operations and to the use of seed, fertilizers, and crop protection compounds were computed. Regarding the emissions derived from the use of fertilizers, a national manual of emissions was considered (Bini & Magistro, 2002). Obviously, the organic and organic-plus systems do not use chemical fertilizers or other chemical products, rather they use organic manure. The emissions to water and air coming from manure were considered 0. With reference to the eventual emissions to water, they are not present because of the consistency of poultry manure (solid) and due to the characteristics of the ground (primarily clay with a high ions retention capacity). With reference to emissions to air, according to the best available techniques of EU for poultry rearing (which the three farms follow), poultry manure must be immediately incorporated into the ground; therefore, emissions to air are so undetectable that they can be considered null (BREF European Commission, 2003).

Other minor components were considered, such as the wheat straw necessary for the rearing, and the amino acids and coccidiostats, which were produced for use only in the conventional feed. Finally, all of the transport data were included, together with the emissions associated with fuel production.

(ii) **Transformation of crops into feed:** The transformation of the crops into feed ingredients, together with the assembly[1] of the ingredients, is the second series of processes

[1] The final feed is a mix of the transformed crops added with natural (in the organic and organic-plus) or synthetic (in the conventional) compounds (see table 2).

considered in the analysis. For each crop, the flows related to the transformation from crop to feed were computed, including the phases of milling, extrusion, and solvent extraction of oil. The main elements involved in the transformation flows were the consumption of electricity and water, and emissions to air. It is important to evaluate the requirement of energy by the system, and it is advisable to distinguish the energy used for the cultivation from the energy used for feed processing (Mourad *et al.*, 2007).

(iii) **Animal rearing:** In this step, together with the main input (feed), data concerning consumption (water, electrical energy, and fuel), infrastructure (all the materials used), and emissions caused by animals, which consist mainly of ammonia, methane, nitrogen, and dust (BREF European Commission, 2003), were taken into account.

The main differences detected in the three farming systems, which consequently influenced the LCA results, concern are as follows:

- The area involved, and consequently, the density of birds per m^2.
- The cultivation phase, with or without the use of chemical products.
- In regards to the transformation phase, no great differences were detected except for the feed formulation (i.e., the use of synthetic versus natural compounds).

Figures 7.4a, 7.4b, and 7.4c show a simplified life cycle for the three systems, and they report the most important activity flows as well as their relative percentages of environmental impact.

The production of feed, considering both crops' cultivation, and their transformation into feed, is the segment that contributes most to the overall environmental impact of the three systems, (78.8% in conventional, 87.3% in organic, and 87.8% in organic-plus). In particular, cultivation and transformation of soy[2] (in the conventional) and maize (in all the three systems) are the processes with the greatest impact. The impact assessment shows that land use is the most important category, followed by fossil fuels and respiratory inorganics (Figures 7.5a, 7.5b, and 7.5c). The analysis of impact assessment made for the single phases shows that the production of feed is what carries the most weight on the previously mentioned categories, whereas animal rearing especially influences acidification and eutrophication, respiratory inorganics, and to a lesser degree climate change. The category of fossil fuels is also clearly influenced by transport flows.

The LCA shows that the relative importance of the different impact categories is the same in all the systems analyzed. As in aforementioned studies (Cederberg & Mattsson, 2000; Eriksson *et al.*, 2005; Ellingsen & Aanondsen, 2006; Pelletier, 2008; Pelletier *et al.*, 2010), our results show that feed production contributes the greatest impact associated with poultry production systems, influencing more than the animal-rearing phase. Comparing the systems in terms of values, the organic-plus system shows the highest land use (8.96 E-3 normalized Pt) and respiratory inorganics (1.38 E-3 Pt) values. The first value is due to the large use

[2] According to Eriksson *et al.* (1985), who compared imported soybean meal, locally grown peas, and rapeseed cake, simple or complemented with synthetic amino acids in pigs, the optimal environmental solution resulted in avoiding, almost entirely, soybean meal because of its poor environmental record. Standard diets for chicken are based on maize and soybean meal, and the complete substitution of soybean with other protein sources is quite impossible. However, our case study confirms that more emphasis should be placed on alternative crop proteins (faba bean, peas; Farrel *et al.*, 1999), with significant implications for the protein and amino acid composition of the diet.

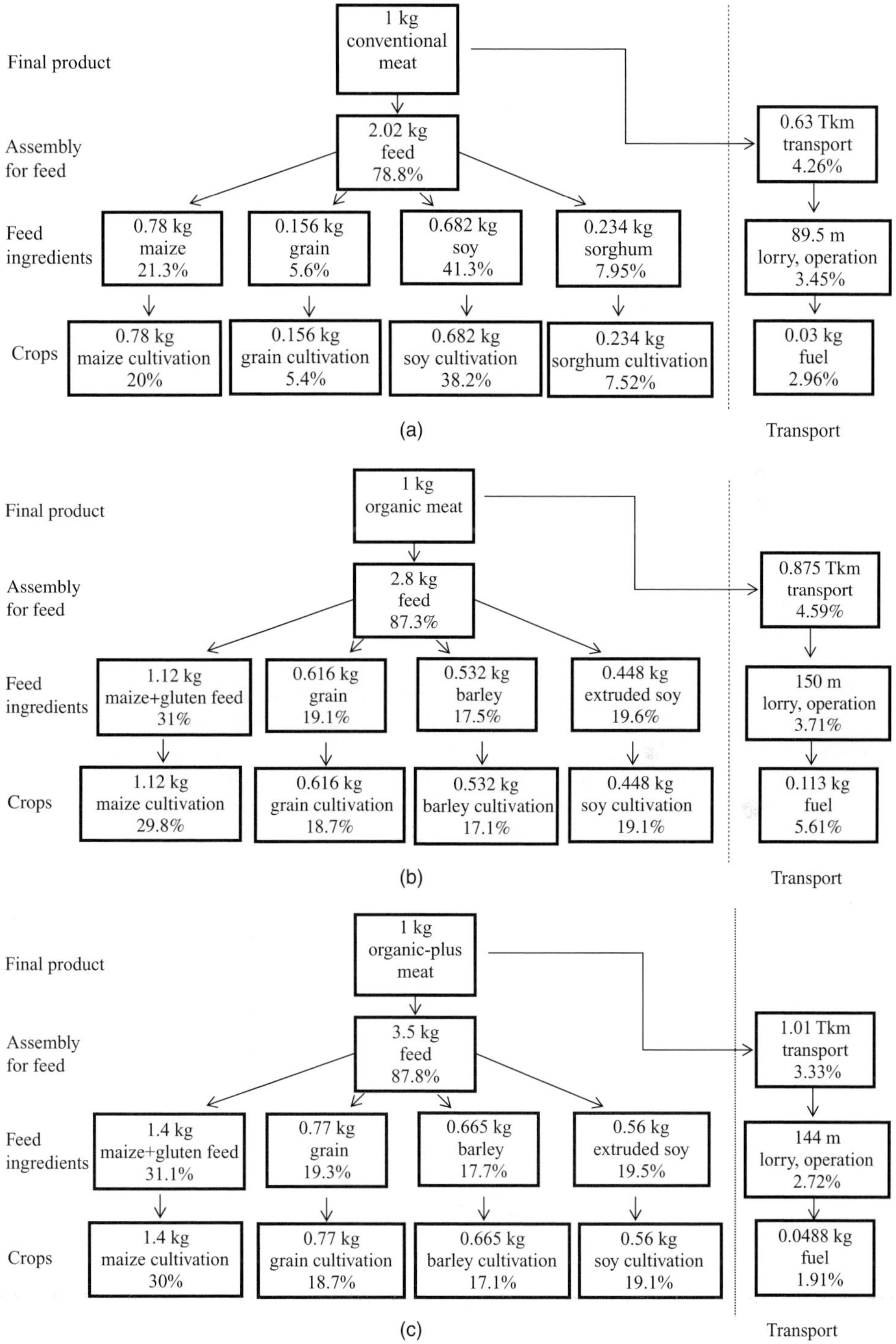

Figure 7.4 Life cycles of the three poultry systems: conventional, organic, and organic-plus. (a) Conventional poultry system. (b) Organic poultry system. (c) Organic-plus poultry system.

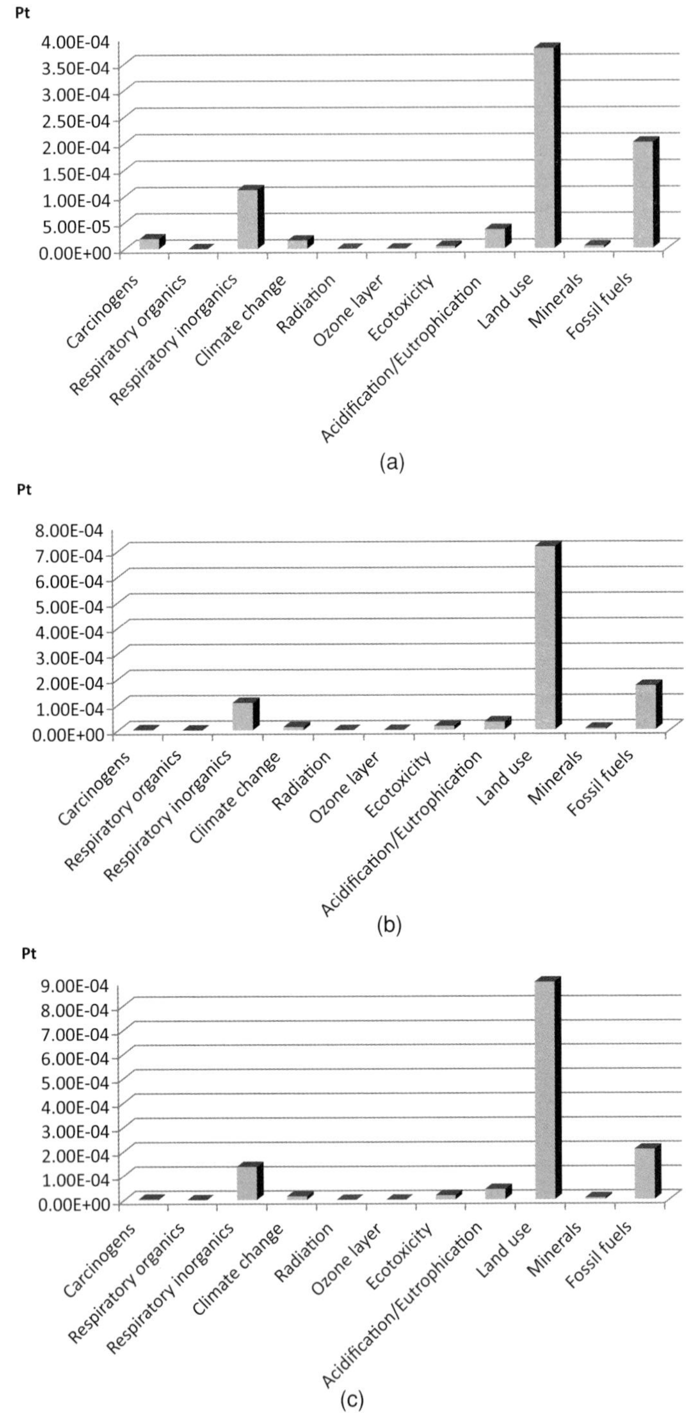

Figure 7.5 Life cycle assessment (LCA) of the three systems: conventional, organic, and organic-plus poultry system (Method: Eco-indicator 99). (a) LCA of conventional poultry system. (b) LCA of organic poultry system. (c) LCA of organic-plus poultry system.

of land, whereas the latter is mainly caused by feed production and by the rearing phase. However, regarding fossil fuels, the conventional and organic-plus systems have similar values (2.01 E-3 and 2.07 E-3, respectively), which are higher than in the organic system. The highest values for carcinogens and climate change are present in the conventional system, and the lowest values are in the organic system. However, the conventional system has the lowest values for respiratory organics, ozone layer, and minerals. The organic-plus system has the lowest value only for the category of radiation, which has the highest value in the conventional system.

7.3.3 Discussion

To sum up, we can say that the organic system has a better environmental performance, as it not only has the lowest values for two of the most important impact categories (respiratory inorganics and fossil fuels), but it also has the lowest values for most of the remaining categories (carcinogens, climate change, acidification–eutrophication) (Table 7.4).

In relation to impact categories, land use was the category most involved, followed by respiratory inorganics and fossil fuels. Considerable attention must be paid to the feed production phase, since, for all of the systems, it contributes the greatest to the environmental impact of the overall system than animal rearing.

LCA was a fundamental instrument to apply a wider process of sustainability analysis. When attempting to measure the sustainability of human activities, one of the most vital gaps is the lack of information on environmental impacts. An environmental impact analysis that refers only to a single phase or a limited set of indicators does not allow the appropriate information to be guaranteed and to be integrated with other information. Therefore, LCA can be seen as a tool to provide environmental information in a broader context (Miettinen & Hämäläinen, 1997), taking into account the whole life cycle of a product. Often, the results of LCA can change the perspective of an environmental analysis due to the consideration of the entire life cycle. Information from LCA can be integrated using additional methodologies, such as Ecological Footprint and Energy Analysis, to broaden the environmental information (Bastianoni *et al.*, 2010).

Table 7.4 Normalized values[a] of the impact categories for the three systems.

Impact category	Unit	Conventional (C)	Organic (O)	Organic-plus (OP)	Highest value	Lowest value
Carcinogens	Pt	1.93 E-05	2.30 E-06	3.05 E-06	C	O
Respiratory organics	Pt	7.01 E-08	9.06 E-08	1.08 E-07	OP	C
Respiratory inorganics	Pt	1.12 E-04	1.07 E-04	1.38 E-04	OP	O
Climate change	Pt	1.60 E-05	1.22 E-05	1.46 E-05	C	O
Radiation	Pt	1.65 E-07	1.24 E-07	1.03 E-07	C	OP
Ozone layer	Pt	5.80 E-09	6.37 E-09	6.78 E-09	OP	C
Ecotoxicity	Pt	4.83 E-06	1.49 E-05	1.79 E-05	OP	C
Acidification/eutrophication	Pt	3.64 E-05	3.17 E-05	4.24 E-05	OP	O
Land use	Pt	3.79 E-04	7.17 E-04	8.96 E-04	OP	C
Minerals	Pt	4.71 E-06	5.88 E-06	7.06 E-06	OP	C
Fossil fuels	Pt	2.01 E-04	1.73 E-04	2.07 E-04	OP	O

[a]Normalization consists of dividing the impact category indicators by a normal value. The most common procedure to determine the normal value is to determine the impact category indicators for a region during an entire year, and divide this result by the number of inhabitants in that area.

Naturally, in a global evaluation of farming systems, many other factors, exogenous to environmental impacts should be taken into account (mainly animal welfare and product characteristics, and social and economic factors).

Although Council Regulation (EC) 1804/99 recommends using local strains, organic systems often use the same genetic strains as in conventional systems. Even if this option is not consistent with the aim of organic agriculture, it leads to a higher productive efficiency of highly selected birds, and consequently to a lower environmental impact. Indeed, the organic-plus system, which uses slow-growing strains and higher pasture availability, surely improves animal welfare and meat quality (Castellini *et al.*, 2008), but the lower productive efficiency has a negative repercussion on the environmental impact. Therefore, to reach equilibrium among all of these factors, namely environment protection, animal welfare, and meat quality, it would be necessary to find a production system that coordinates them into one coherent scheme.

The application of a "multifunctional analysis," permitting identification of a competitive production system in terms of animal welfare, health, and environmental impacts would be the optimal final achievement of a complete sustainability evaluation process.

Since the productive efficiency of an autochthonous breed is lower, possible strategies to overcome the problems include: the substitution of "expensive" crops (e.g., soybean) with alternative feed ingredients (Perella *et al.*, 2009), and a genetic improvement of the low productive performance of local strains while maintaining their adaptability to free range systems.

7.4 CASE STUDY—NATIONAL SCAN-LEVEL CARBON FOOTPRINT FOR US SWINE PRODUCTION

7.4.1 Introduction

This section reports an LCA of US pork supply chain that focused on defining GHG emissions. The pork supply chain is broadly divided into eight stages, each receiving a separate analysis that was combined to provide the entire life cycle carbon footprint. These stages are: (1) feed production, (2) live swine production, (3) delivery to processor, (4) processing, (5) packaging, (6) distribution, (7) retail, and (8) consumption/disposal. This study has been conducted in compliance with ISO guidance.

7.4.2 LCA Methodology

7.4.2.1 *Functional unit and system boundaries*

The functional unit was one serving (4 ounces) of boneless pork prepared for consumption by a US consumer. The system boundaries are shown schematically in Figure 7.6.

Figure 7.6 encompassed effects beginning with the energy and GHG emissions associated with production of fertilizer, and with estimates of GHG emissions from either landfill or municipal waste incineration of the packaging. The impacts of distribution and refrigeration, as well as product loss through the supply chain, were included.

7.4.2.2 *Allocation*

There are three stages in the supply chain where allocation occurs: first, for by-products of feed processing (e.g., distiller's grains and soy meal), second, at the processing gate where

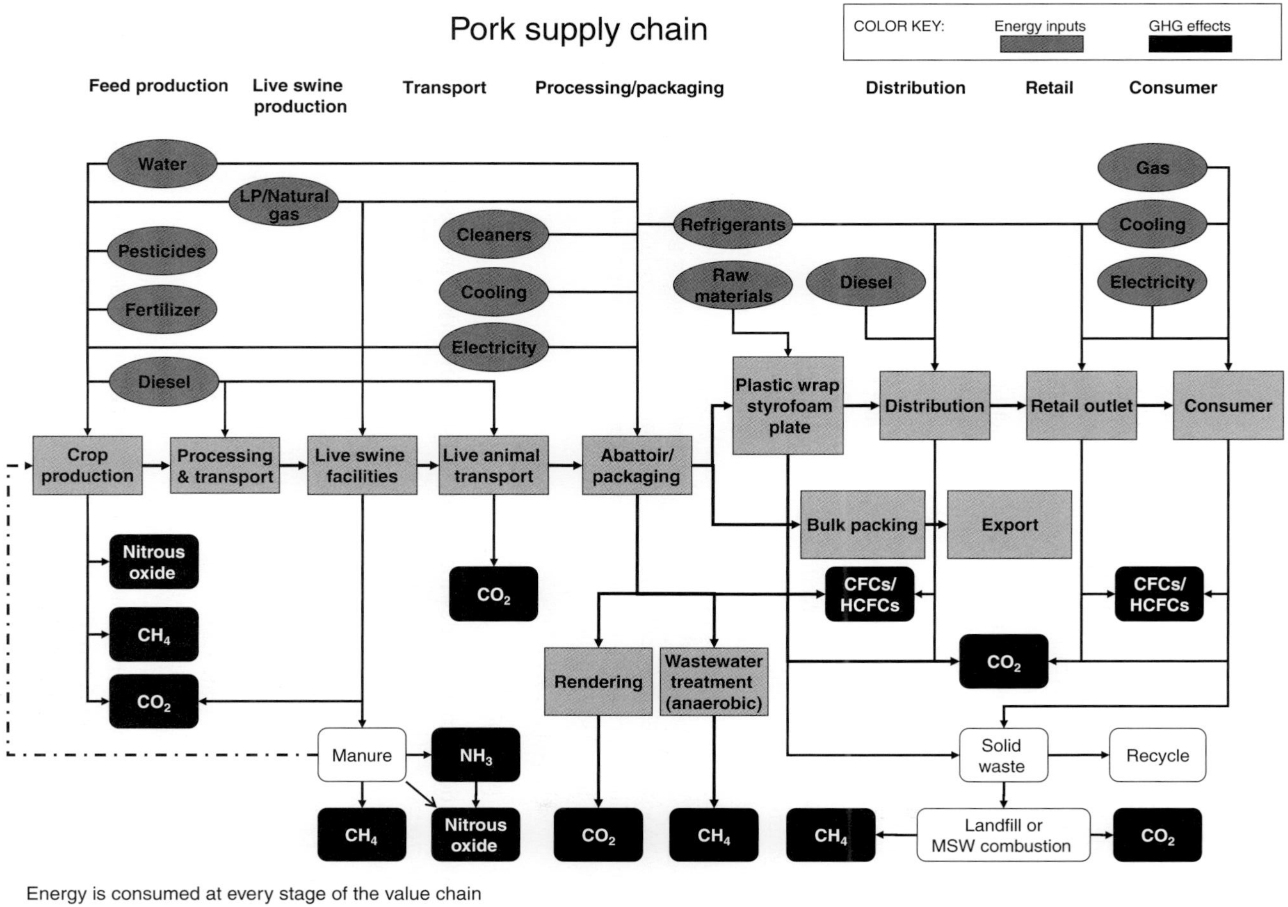

Figure 7.6 Schematic of pork production supply chain showing major inputs and outputs relevant to greenhouse gas emissions.

allocation between dressed carcass and rendering products occurs, and finally, at retail and consumption where an allocation of refrigeration burdens is necessary. We have adopted, as a base case approach, an economic value allocation, but consider a mass allocation for feed by-products for comparison.

7.4.2.3 *Life cycle inventory data*

A literature review was used as the basis for much of the LCI data, and additional discussions with industry representatives and other experts helped fill in the data gaps. Regionally specific data for feed crops and agricultural chemical usage statistics on a state-by-state basis were taken from farm extension and the United States Department of Agriculture National Agricultural Statistical Service (USDA NASS, 2009). Additional input data for fuels and electricity consumption for crop production were obtained from the technical literature (Landis *et al.*, 2007), state agricultural extension services, the USDA (USDA NASS, 2007), the US Department of Energy and academic institutions. The Ecoinvent database was used to provide upstream (background) information and the Intergovernmental Panel on Climate Change (IPCC) GWP 100a impact assessment methodology (Forster *et al.*, 2007) was used to summarize the GWP ($CO_2 = 1$, $CH_4 = 25$, and $N_2O = 298$). The farm model is based on a two-barn system: breeding/gestation/lactation barn (or sow barn) followed by a nursery/finish (N/F) barn. The IPCC provides guidance on estimating the quantities of GHGs that are emitted as a function of the specific on-farm manure management systems (Dong *et al.*, 2006). We have used the American Society of Agriculture Engineers (ASAE) recommendations (ASAE, 2005) to predict the quantity of manure generated, as well as to estimate the amount of nitrogen excreted in the manure. Cradle-to-grave contributions from packaging included production of raw materials (polystyrene, shrink wrap, paper, etc.) and ultimate disposal of the materials.

7.4.2.4 *Transportation*

We used an estimate of 500 miles round trip transportation distance between the farm and the pork processor. These calculations are based on an estimate of 160 head with an average weight of 268 pounds per truck for delivery of finished hogs. Transport emissions from producer to processor and from processor to distributor were calculated from Ecoinvent unit processes based on information provided from industrial reported fuel consumption. A 30-mile round trip distance was assumed for delivery of feed to farms because of the structure of the US industry where feed mills are built near animal production facilities.

7.4.2.5 *Pork processing and packaging*

We received data from several meat processing facilities and aggregated the data to preserve anonymity. The information included the quantity of processed meat and the amount of electricity, natural gas, and other fuels consumed for the entire facility. Estimates of GHG emissions from on-site wastewater treatment facilities and loss of refrigerants were also reported. An economic allocation among the coproducts (meat and rendering products) was applied. The allocation ratio in this method assigned 89% of the GHG burden to the meat processing and 11% to coproducts from slaughter operations. For retail distribution, most meats, including pork, are packed on a polystyrene plate with an absorbent pad and wrapped with shrink wrap. We have estimated approximately 8 g of polystyrene and 0.5 g of shrink wrap material per pound of packaged pork. We used information in a patent (Elves & Ley, 1985) to characterize a typical absorbent pad made from viscose and expanded viscose as the basis for the calculations. For plant electricity consumption, this study used emission factors (in kg equivalent carbon dioxide (CO_2e)/kWh) for the three US regional interconnection grids (Eastern, Western, and the Electric Reliability Council of Texas (ERCOT) Interconnection) (Deru & Torcellini, 2007) and each emission factor was constructed using Ecoinvent unit processes based on regional primary fuel mixes.

7.4.2.6 *Retail, consumer, and postconsumer solid waste*

During this phase, there are four distinct emissions sources: (1) refrigerant leakage, (2) refrigeration electricity, (3) overhead (needed for facility operation) electricity, and (4) fuel. Estimates of the sales volume, space occupancy, and energy demands of pork were used to determine the burden of this supply chain stage. Impacts accounted in the consumer phase include transport from retail to home, refrigeration energy, and cooking energy. Cooking energy requirements were estimated using information from the US EPA Energy Star Program. After consumption, there is a relatively small quantity of postconsumer waste generated, and it was modeled using an Ecoinvent process for landfill disposal.

7.4.3 Results and discussion

7.4.3.1 *Cradle-to-farm gate analysis*

Figure 7.7 presents a comparison of five manure management practices (US EPA, 2011) effect on the cradle-to-farm gate GHG emissions. In these comparisons, the only parameter changed was the type of manure management used for each of the barns modeled. The "piglet" contribution is different between the scenarios as a result of the manure management. It is

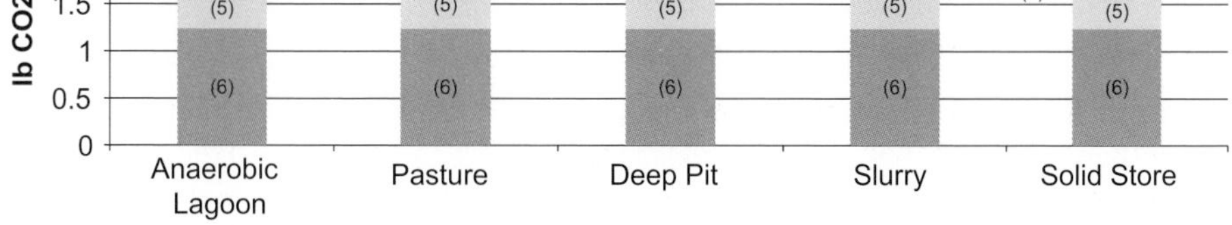

Figure 7.7 Comparison of manure management system choice on farm gate cumulative greenhouse gas (GHG) emissions. Note that the bar for piglets is different between the scenarios because of differences in manure management at the sow barn. The same manure management was selected for both the nursery/finish and sow barns. These scenarios all include distillers dried grains (DDGs) in the ration.

Table 7.5 Summary of European Union pork production global warming potential data.

	Global warming potential, CO_2e	Reference
US pork	3.8	This Chapter
Pork produced in Denmark	3.6	Dalgaard *et al.*, 2007
Pork produced in United Kingdom	3.3	Dalgaard *et al.*, 2007
Organic pork (Denmark)	3.8–4.3	Halberg *et al.*, 2007
Pork produced in Sweden	2.6	Cited in Dalgaard *et al.*, 2007
Pork produced in France	3.0[a]	Basset-Mens and van der Werf, 2005 for GAP production
Pork produced in United Kingdom	5.6	Williams *et al.*, 2006
Pork produced in Canada	3.1[a]	Verge *et al.*, 2008

[a]Value corrected from live weight to carcass.
Functional unit = 1 kg carcass at farm gate.
CO_2e, equivalent carbon dioxide.

not surprising that anaerobic lagoons make a larger contribution to GHG emissions than other management options. It is interesting to note that the IPCC approach does not show a significant difference between pasture-based manure management and deep pit systems.

Based on USDA production statistics, the regional production-weighted US national average carbon footprint is 2.8 kg CO_2e per kg live weight (equivalent to 3.8 kg/kg dressed carcass or 5.8 kg CO_2e/kg boneless meat), which places this cradle-to-farm gate GHG emissions estimate within the same range as similar studies performed on European swine production systems that range from 3–5 kg CO_2e per kg of dressed carcass (Table 7.5). As can be seen from Figure 7.7, for the deep pit manure management system, the burden per kg of live weight at the farm gate is approximately 2.27 kg CO_2e, which on a dressed carcass basis is equivalent to 3.03 kg CO_2e per kg dressed carcass (divide by the factor of 0.75, the dressed fraction of live weight)—this is at the lower end of the range from Table 7.5. Using the results from Figure 7.7 for anaerobic lagoon systems, the farm gate footprint is significantly higher at 3.8/0.75 = 5.1 kg CO_2e/kg dressed carcass. One interesting point to note in this analysis is that these results show that feed production is approximately 50% of the farm gate GHG emissions burden, in contrast to some previous work (cited earlier) where the feed contribution was approximately 80%. The source of this difference may result from differing assumptions regarding the emissions from manure management and application.

7.4.3.2 *Cumulative cradle-to-grave GHG emissions from the US Swine Sector*

Figure 7.8 presents the breakdown of the domestic pork supply chain GHG emissions. Each major supply chain stage has been subdivided into the primary contributing activities. For each barn phase shown, manure management is the largest single contributor. Much of the remaining contribution is associated with production of the animal's rations. The primary contribution from retail and home consumption is associated with electricity used for refrigeration, with some loss of refrigerants at retail. Because it is a legal requirement that refrigerants be captured from disposed household refrigerators, we have assumed that there is no significant GHG emissions associated with end-of-life appliance disposal. One somewhat

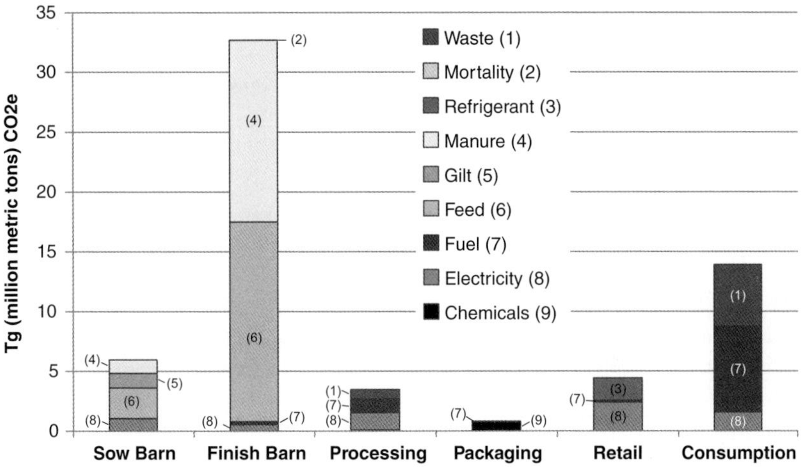

Figure 7.8 Cumulative greenhouse gas (GHG) emissions associated with consumption of pork in the United States. Field emissions are associated with crop production; the waste category primarily captures end-of-life disposal of primary packaging materials. The estimates in this analysis do not include GHG emissions from pork destined for overseas consumption or food service industry emissions.

surprising result of this analysis is that the contribution at processing and packaging is relatively low, and that retail refrigeration and in-home refrigeration are rather significant contributors to the overall carbon footprint. It is not surprising that transportation makes a relatively small contribution to the overall carbon footprint, despite its importance in the economics of the swine industry.

7.4.4 Conclusions from the case study

The factor that changes the most for cradle-to-grave carbon footprint of US swine production is the manure management system. The results also show that both retail and in-home electricity use for refrigeration are important contributions to the overall footprint, while processing and packaging contribute to a relatively smaller amount. The overall cumulative GHG emissions for consumption of one 4-oz serving of US domestic pork based on reported manure management practices and accounting for an estimated 10% waste of product by consumers was found to be 1.02 kg CO_2e with a 95% confidence band from 0.88 kg CO_2e to 1.18 kg CO_2e. Overall, the contribution of emission burden for each stage was found to be:

- 10.2%: sow barn (including feed and manure handling);
- 54.7%: nursery to finish (including feed and manure handling);
- 7.4%: processing (6.3%) and packaging (1.1%);
- 12%: retail (electricity and refrigerants); and
- 15.8%: the consumer (refrigeration and cooking).

This model assumes gas oven cooking in the home; if in-home cooking is assumed to be an electric oven, the overall impact increases to 1.05 kg CO_2e per 4-oz serving.

7.5 CONCLUSIONS

The presentation of the case studies illustrated that LCA can be considered a useful instrument in the analysis of all types of production systems, finalized to perform a process of sustainability assessment. The monitoring of environmental impacts of human activities, in particular those caused by farming or rearing systems, is fundamental. The lack of information on environmental impacts often influences the process of sustainability analysis negatively. An environmental impact analysis that refers only to a single phase or a limited set of indicators does not provide a complete vision of the burdens of the investigated system on environment. Therefore, LCA can be seen as a tool to provide environmental information in a broader context (Miettinen & Hämäläinen, 1997), taking into account the whole life cycle of a product. Often, the results of LCA can change the perspective of an environmental analysis due to the consideration of the entire life cycle.

Considering other factors besides the environmental impacts, such as animal welfare and meat quality in the case of rearing systems, product characteristics, and social and economic factors, completes the process of sustainability assessment of the systems. Therefore, the application of a multifunctional analysis would be the optimal final achievement of a complete sustainability evaluation process.

REFERENCES

American Society of Agricultural Engineers. 2005. *ASAE D384.2. Manure Production and Characteristics.* The Society for Engineering in Agricultural, Food, and Biological systems, St. Joseph, MI. p. 20.

Aviagen Technical Team. 1999. *Ross Breeders Broiler Management Manual.* Aviagen Ltd., Newbridge, Midlothian. p. 114.

Basset-Mens, C. and H. M. G. van der Werf. 2005. Scenario-based environmental assessment of farming systems: the case of pig production in France Agriculture. *Agric. Ecosyst. Environ.* 105:127–144.

Bastianoni, S., A. Boggia, C. Castellini, C. Di Stefano, V. Niccolucci, E. Novelli, L. Paolotti, and A. Pizzigallo. 2010. Measuring environmental sustainability of intensive poultry-rearing system. In: E. Lichtfouse, M. Navarrete, P. Debaeke. (eds) *Genetic Engineering, Biofertilisation, Soil Quality and Organic Farming. Sustainable Agriculture Reviews.* Vol. 4. Springer, New York. pp. 277–309.

Bell, S. and S. Morse. 1999. Sustainability indicators. In: S. Bell and S. Morse (eds) *Measuring the Immeasurable.* Earthscan Publications Ltd., London. p. 175.

Bini, G. and S. Magistro. 2002. Manuale dei fattori di emissione nazionali. Centro Tematico Nazionale Atmosfera Clima ed Emissioni in Aria, Rapporto n. 01.

Boggia, A., L. Paolotti, and C. Castellini. 2010. Environmental impact evaluation of conventional, organic and organic-plus poultry production systems using life cycle assessment. *Worlds Poult. Sci. J.* 66: 95–114.

Branciari, R., C. Mugnai, R. Mammoli, D. Miraglia, D. Ranucci, A. Dal Bosco, and C. Castellini. 2009. Effect of genotype and rearing system on chicken behavior and muscle fiber characteristics. *J. Anim. Sci.* 87:4109–4117.

Castellini, C., C. Berri, E. Le Bihan-duval, and G. Martino. 2008. Qualitative attributes and consumer perception of organic and free-range poultry meat. *Worlds Poult. Sci. J.* 65:120–135.

Cederberg, C. 1998. Life cycle assessment of milk production—a comparison of conventional and organic farming. In: C. Cederberg. (ed) *SIK, Report No. 643.* The Swedish Institute for Food and Biotechnology, Gothenburg. pp. 1–86.

Cederberg, C. and B. Mattsson. 2000. Life cycle assessment of milk production—a comparison of conventional and organic farming. *J. Clean. Prod.* 8:49–60.

Cornelissen, A. M. G. 2003. Common ground for sustainable development, and ground covered by selected sustainability indicators. The Two Faces of sustainability. Ph.D. Thesis. Animal Production Systems Group. Wageningen University, Wageningen. p. 202.

Council Regulation (EEC). 1991. No 2092/91 of 24 June 1991, on organic production of agricultural products and indications referring thereto on agricultural products and foodstuffs. Off J., L 198. p. 1.

Council Regulation (EC). 1999. No 1804/99 of July 1999, supplementing Regulation (EEC) No 2092/91 on organic production of agricultural products. Off. J., L 222, 24/08/1999, pp. 1–28.

Dal Bosco, A., C. Mugnai, F. Sirri, C. Zamparini, and C. Castellini. 2010. Assessment of a global positioning system to evaluate activity of organic chickens at pasture. *J. Appl. Poult. Res.* 19:213–218.

De Boer, I. 2003. Environmental impact assessment of conventional and organic milk production. *Livest. Prod. Sci.* 80:69–77.

Dalgaard, R., N. Halberg, and J. E. Hermansen. 2007. Danish pork production. An environmental assessment. *DJF Anim. Sci.* 82:1–34.

Deru, M. and P. Torcellini. 2007. Source energy and emissions factors for energy use in buildings. Technical Report, NREL/TP-550-38617. National Renewable Energy Laboratory.

De Vries, M. and I. J. M. de Boer. 2010. Comparing environmental impacts for livestock products: a review of life cycle assessments. *Livest. Sci.* 128:1–11.

Dong, H., J. Mangino, T. Mcallister, J. L. Hatfield, D. Johnson, K. R. Lassey, 2006. Emissions from livestock and manure management. In: Dong, H., J. Mangino, T. Mcallister, J. L. Hatfield, D. Johnson, K. R. Lassey. (eds) *2006 IPCC Guidelines for National Greenhouse Gas Inventories.* Institute for Global Environmental Strategies (IGES), Hayama.

Ellingsen, H. and A. Aanondsen. 2006. Environmental impacts of wild caught cod and farmed salmon—a comparison with chicken. *Int. J. Life Cycle Assess.* 11:60–65.

Elves, J. and S. J. Ley. 1985. Absorbent pads United States Chicopee (New Brunswick, NJ) 4551377. Available at: http://www.freepatentsonline.com/4551377.html.

Eriksson, I. S., H. Elmquist, S. Stern and T. Nybrant. 2005. Enviromental systems analysis of pig production, the impact of feed choice. *Int. J. Life Cycle Assess.* 10:143–154.

European Commission. 2003. Integrated pollution prevention and control (IPPC). Reference Document on Best Available Techniques for Intensive Rearing of Poultry and Pigs (BREF).

Fanatico, F. 2006. Alternative poultry production systems and outdoor access. ATTRA Publication IP300. Available at: http://attra.ncat.org/attra-pub/poultryoverview.html. p. 24

Forster, P., V. Ramaswamy, P. Artaxo, T. Berntsen, R. Betts, D. W. Fahey, J. Haywood, J. Lean, D. C. Lowe, G. Myhre, J. Nganga, R. Prinn, G. Raga, M. Schulz, and R. Van Dorland. 2007. Changes in atmospheric constituents and in radiative forcing. In: P. Forster, V. Ramaswamy, P. Artaxo, T. Berntsen, R. Betts, D. W. Fahey, J. Haywood, J. Lean, D. C. Lowe, G. Myhre, J. Nganga, R. Prinn, G. Raga, M. Schulz, and R. Van Dorland (eds) *Climate Change 2007: The Physical Science Basis. Contribution of Working Group I to the Fourth Assessment Report of the Intergovernmental Panel on Climate Change.* Cambridge University Press, Cambridge and New York. p. 996.

Foster, C., K. Green, M. Bleda, P. Dewick, B. Evans, A. Flynn, and J. Mylan. 2006. *Environmental Impact of Food Production and Consumption: A Report to the Department for Environment, Food and Rural Affair.* Manchester Business School, DEFRA, London. p. 199.

Goedkoop, M. and R. Spriensma. 2001. *The Eco-Indicator 99—A Damage Oriented Method for Life Cycle Impact Assessment. Methodology Report.* 3rd edn. Product Ecology Consultants, Plotterweg. p. 49.

Haas, G., F. Wetterich, and U. Köpke. 2001. Comparing intensive, extensified and organic grassland farming in southern Germany by process life cycle assessment. *Agric. Ecosyst. Environ.* 83:43–53.

Halberg, N., J. Hermansen, I. S. Kristensen, J. Eriksen, and N. Tvedegaard. 2007. Comparative environmental assessment of three systems for organic production in Denmark. *Proc. Org. Agric. Asia 2008,* 249–261.

Halberg, N., H. M. G. van der Werf, C. Basset-Mens, R. Dalgaard, and I. J. M. de Boer. 2005. Environmental assessment tools for the evaluation and improvement of European livestock production systems. *Livest. Prod. Sci.* 96:33–50.

Hogass, E. M. 2002. Life cycle assessment (LCA) of industrial milk production. *Int. J. Life Cycle Assess.* 7:115–126.

Keoleian, G. A., A. Phipps, T. Dritz, and D. Brachfeld. 2004. Life cycle environmental performance and improvement of a yogurt product delivery system. *Packag. Technol. Sci.* 17:85–103.

Landis, A. E., S. A. Miller, and T. L. Theis. 2007. Life cycle of the corn-soybean agroecosystem for biobased production. *Environ. Sci. Technol.* 41:1457–1464.

Lewis, P. D., G. C. Perry, L. J. Farmer, and R. L. S. Patterson. 1997. Responses of two genotypes of chicken to the diets and stocking densities typical of UK and 'Label Rouge' production systems: I. performance, behaviour and carcass composition. *Meat. Sci.* 45:501–516.

Meluzzi, A., F. Sirri, C. Mugnai, and A. Dal Bosco. 2009. Effect of genotype on welfare conditions of broilers reared under organic conditions. In: *Proc. viii Eur. Symp. Poult. Welfare*. World Poultry Science Association, Department of Food Science, University of Bologna, Bologna. p. 117.

Miettinen, P., and R. P. Hämäläinen. 1997. How to benefit from decision analysis in environmental life cycle assessment (LCA). *Eur. J. Oper. Res.* 102:279–294.

Mitchell G., A. May, and A. McDonald. 1995. PICABUE: a methodological framework for the development of indicators of sustainable development. *Int. J. Sustain. Dev. World Ecol.* 2:104–123.

Mourad, A. L., L. Coltro, P. A. P. L. V. Oliveira, R. M. Kletecke, and J. P. O. A. Baddini. 2007. A simple methodology for elaborating the life cycle inventory of agricultural products. *Int. J. Life Cycle Assess.* 12:408–413.

Mugnai, C., A. Dal Bosco, and C. Castellini. 2009. Effect of rearing system and season on the performance and egg characteristics of ancona laying hens. *Ital. J. Anim. Sci.* 8:175–188.

Müller-Lindenlauf, M., C. Deittert, and U. Köpke. 2010. Assessment of environmental effects, animal welfare and milk quality among organic dairy farms. *Livest. Sci.* 128:140–148.

Nemecek, T., A. Heil, O. Huguenin, S. Meier, S. Erzinger, S. Blaser, D. Dux, and A. Zimmermann. 2004. Life cycle inventories of agricultural production systems. Final report ecoinvent 2000, No. 15, Agroscope FAL Reckenholz and FAT Taenikon, Swiss Centre for Life Cycle Inventories, Dübendorf, Switzerland. Available at: www.ecoinvent.ch.

Organic foods production act (OFPA). 1990. Available at: http://www.ams.usda.gov/amsv1.0/getfile?ddocname=stelprdc5060370&acct=nopgeninfo.

Pelletier, N. 2008. Environmental performance in the US broiler poultry sector: life cycle energy use and greenhouse gas, ozone depleting, acidifying and eutrophying emissions. *Agric. Syst.* 98:67–73.

Pelletier, N., P. Lammers, D. Stender, and R. Pirog. 2010. Life cycle assessment of high- and low-profitability commodity and deep-bedded niche swine production systems in the Upper Midwestern United States. *Agric. Syst.* 103: 599–608.

Perella, F., C. Mugnai, A. Dal Bosco, F. Sirri, E. Cestola, and C. Castellini. 2009. Faba bean (vicia faba var. minor) as a protein source for organic chickens: performance and carcass characteristics. *Ital. J. Anim. Sci.* 8:575–584.

Product Ecology Consultants. 1990. *SimaPro LCA Software*. Product Ecology Consultants, Plotterweg.

Sirri, F., C. Castellini, A. Roncarati, and A. Meluzzi. 2010. Effect of feeding and genotype on lipid profile of organic broiler chickens. *Eur. J. Lipid Sci. Technol.* 112: 994–1002.

Stonehouse, D. P., E. A. Clark, and Y. A. Ogini. 2001. Organic and conventional dairy farm comparisons in Ontario, Canada. *Biol. Agric. Hortic.* 19:115–125.

Thomassen, M. A., and I. J. M. de Boer. 2005. Evaluation of indicators to assess the environmental impact of dairy production systems. *Agric. Ecosyst. Environ.* 111:185–199.

Thomassen, M. A., K. J. van Calker, M. C. J. Smits, G. L. Iepema, and I. J. M. de Boer. 2008. Life cycle assessment of conventional and organic milk production in the Netherlands. *Agric. Syst.* 96:95–107.

USDA NASS. 2007. *Crop Production 2007 Annual Summary*. USDA NASS, Washington, DC.

USDA NASS. 2009. *Agricultural Chemical Usage Summary, Crop Production Summary*. Available at: http://www.nass.usda.gov/.

US EPA. 2011. Inventory of U.S. greenhouse gas emissions and sinks: 1990–2009 report #430-R-11-005 ANNEX 3 methodological descriptions for additional source or sink categories. Available at: http://epa.gov/climatechange/emissions/usinventoryreport.html (accessed October 17, 2011).

Verge, X. P. C., J. A. Dyer, R. L. Desjardins, and D. Worth. 2008. Greenhouse gas emissions from the Canadian pork industry. *Livest. Sci.* 121:92–101.

Williams, A. G, E. Audsley, D. L. Sandars. 2006. Final report to Defra on project IS0205: determining the environmental burdens and resource use in the production of agricultural and horticultural commodities. Available at: www.defra.gov.uk/science/default.htm.

8 Genetics of Poultry Meat Production in Organic Systems

Poul Sørensen

Abstract: The breeding material used for conventional broiler meat production has a number of disadvantages when used for meat production in organic systems. Growth capacity of the breeding stock should be limited to about 40–50 g per day and should be genetically adapted to perform and behave well in outdoor facilities. As organically grown feed may not always be well balanced regarding amino acids, the chickens should be genetically adapted to such unbalanced feed. In a world with competition for food between humans and farm animals, the organic philosophy tells that food suitable for humans should not be used for farm animals.

Breeding stock is available having a few of these genetically adapted characteristics. Instead of hunting for unknown breeding materials, it may be more profitable to go for direct breeding for genetic adaptation to the organic environment preferably in a cooperative effort between commercial breeders and the organic organisations.

Keywords: broiler; organic; genetically adapted; growth capacity; breeding material; pastured poultry; Label Rouge; dual purpose

8.1 INTRODUCTION

The challenge for the organic poultry meat producers in choosing genetic stock is the question if the present hybrids are suitable for production in the organic environment. The present hybrids have been selected over a long period for a breeding goal that emphasised fast growth based on feed ensuring maximal growth. Also, these chickens are housed under conditions that are optimal in respect to temperature, lighting, ventilation, climate and litter and numerous precautions are taken to avoid outbreaks of disease. If diseases appear, there is hardly any hesitation to use medication.

The dilemma of using breeding stock in organic production with such a breeding history was considered by Price *et al.* (2004) in Chapter 16 of a book developed on the basis of the activity of a European Union (EU)-founded network 'Network on Animal Health and Welfare in Organic Agriculture' consisting of 17 researchers working in 17 universities and research institutes in 13 European countries during 1999–2001. They stated that:

1. The high productivity of chicks from these hybrids depends on high quality of feed, health treatment and other inputs. If these are not available, or are not allowed under the organic rearing system, the chickens may be prone to ill-thrift and disease.

Organic Meat Production and Processing, First Edition. Edited by Steven C. Ricke, Ellen J. Van Loo, Michael G. Johnson, and Corliss A. O'Bryan.

2. Most hybrids are developed for the processing and ready-meal market, or the roasting marked of 2 to 3 kg chicks. These birds will not meet the current EU requirement for organic production of being at least 12 week at slaughter corresponding to an average daily gain of 30 g on an organic regimen, that is far below the daily gain of more than 80 g typical for these hybrids.

Thorp and Luiting (2000) stated that for conventional poultry breeding

> Disease resistance will not be selected for if the cost in a loss of genetic improvement in production traits is too great, and there are other effective methods of disease control.

Later, they slightly modified it by stating, 'although decisions in the poultry industry are largely driven by economic considerations, the psychological impact of flock morbidity on the farmer and society cannot be ignored'. Therefore, it is likely that the breeding goals these days may reflect values that are not purely economic.

The conventional breeding has long ago specialised the hybrids into lines that have egg laying as the major goal and lines used for broiler production. They are specialised to a degree that none can be used as dual purpose which is an ideal seen in the light of the organic concept. It is particularly a problem for the layer-type hybrids as the male chickens grow so poorly and have such poor body composition that they have no economic value to rear them and therefore they are killed immediately after hatch. As to broilers, the female may be able to lay about 60% of the eggs compared to the laying-type hens, but that would add cost such that the egg price becomes more than double the egg price of conventional laying-type hens.

Feed quality and in particular genetic adaptation to special diets have been subject for a number of studies regarding concentration of amino acids/protein. Nesheim and Hutt's (1962) studies of arginine and Sørensen's (1986) studies of methionine, lysine, arginine and threonine discovered a considerable genetic-based effect on utilisation of the low level diets. Regarding digestibility, Mignon-Grasteau *et al.* (2004) observed variable responses to selection for high or low apparent metabolic energy (AME) when feeding the bird a wheat variety with low digestibility. The AME exhibited a considerable genetic variation that after five generations of divergent selection resulted in a difference of 10% between the two lines.

Outdoor access of broilers was studied by Nielsen *et al.* (2003) in comparing a slow-growing line and female of a fast-growing broiler line (Ross208). It was found that the fast-growing birds were not suitable to be used in a system in which they had to use free-range management from 6 to 12 weeks. This is the age required by the EU rules for slaughtering organically grown chickens. During this prolonged period, 3%–9% of the female birds of Ross208 developed deep pectoral myopathy that made them unsuitable for human consumption. In United States, Fanatico *et al.* (2005a, 2005b, 2007a, 2007b) published a series of results on growth, feed efficiency, slaughter yield and meat quality of fast- and slow-growing female hybrids slaughtered at the same weight and kept inside as well as in range from 3 weeks of age. The feed was produced such that it fulfilled requirements of the National Research Council (1994) and therefore the usual organic principles were not followed. The major results were a better feed efficiency of the fast-growing bird, breast meat from the slow-growing birds was more tender, outdoor access resulted in leaner meat in the case of slow-growing birds, slaughter yield and breast meat were better for the fast-growing chickens.

There are an increasing numbers of reports from the United States dealing with pastured poultry chicks raised on the Label Rouge principle from France but most of them represent an organic background (Fanatico & Born, 2002). The US Organization Sustainable Agriculture Research and Education (SARE) offers grants to practical farmers for initiating research. There are 10-year-old reports that indicate the problems with raising conventional broiler hybrids on pasture during 3–11 weeks of age fed on an organic diet (SARE Project FNE01-372, 2002). This is later supplemented by grants to farmers to investigate alternative broiler hybrids/breeds, including the Label Rouge type. The main conclusions are that the farmers prefer the slower-growing breeds kept at pasture during 3–11 weeks of age that obtain a dressed weight of about 2 kg, while hybrids for conventional production obtained a dressed weight of 3–4 kg and often have a considerably better feed efficiency. The farmers disliked these fast-growing chickens due to the risk of heart attack, weak legs, temperature sensitivity and their inactivity and lethargy, and sometimes unacceptably high mortality.

From the previous mentioned reports, a number of issues have to be taken into account when discussing an optimal breeding goal for breeds that follow a selection programme enabling them to produce poultry meat under organic rules. The key issues for such a selection programme are as follows:

(i) The growth capacity of the birds should be much less than the modern broiler.
(ii) The birds should have access to outdoor facilities during the test period, to be genetically adapted to the organic environment.
(iii) Concentration and/or quality of nutrient fed to organically grown chicken.
(iv) Perhaps the parent stock should be kept organic?
(v) Where to buy such genetic material or is it necessary to breed especially for organically grown chicken?
(vi) The choice between dual purpose and specialised breeds.

The rest of the chapter discusses these matters as seen from the poultry point of view, but many of the arguments would also be applicable for organically grown pigs.

8.2 THE GROWTH

The EU Regulation on organic livestock production states that in the choice of breed or hybrids, several factors must be taken into account such as the capacity of the animals to local conditions, and the rules that chickens should be at least 81 days old before slaughter unless slow-growing breeds are used (European Community, 1999).

Modern broiler chickens produced from the two dominating breeding companies – Aviagen Group (Huntsville, AL, USA) and Cobb-Vantress (Siloam Springs, AR, USA) – have a daily gain at maximum growth of approximately 90 g resulting in a body weight of 2.5 kg at 6 weeks and probably more than 5 kg at 81 days when grown under indoor facility conditions. Nielsen *et al.* (2003) reported that females of the Aviagen hybrid (Ross208) obtained a live body weight of 3.9 kg at 12 weeks of age in an outdoor system. It was stated that these female birds were unsuitable for a 12-week growth period in free-range production partly due to the occurrence of deep pectoral myopathy in 3% to 9% of the chickens resulting in a large proportion of the pectoralis minor muscle being dry and green and having an almost cooked appearance and partly due to an unacceptable high frequency of birds that exhibited poor walking ability.

Cobb-Vantress (2011) has recently marketed a hybrid termed Cobb*Sasso150* that may be suitable for organic production displaying a growth capacity of 40–41 g per day to 56 and 70 days corresponding to a body weight of 2.25 kg and 2.85 kg and a maximum daily gain of 45 g. In their commercial brochure, they claim that it has been grown successfully in an indoor certified system, in free range for more than 56 days and in organic systems for over 70 days. A former US breeding company Hubbard operating in France (Hubbard, Walpole, NH, USA) marketed a series of hybrids with a growth capacity of 2.0 kg at 70 days, 2.3 kg at 77 days, 2.4 kg at 63 days and 2.4 kg at 56 days. The most slowly growing hybrid is the one used for the French Label Rouge production termed I 657, which is produced by mating a JA 57 parent stock female with the male I66. The JA 57 hen is a dwarf-type hen reaching 2.2 kg as adult weight and may be fed without restriction. Pedersen *et* al. (2003) described a farm study on nine farms in Denmark in which I 657 was used as parent stock. In total, 35,616 chickens were grown up to 81 days and obtained a live weight of 2.1 kg having a food conversion rate of 2.9.

Of the hybrids on the market that are suitable for organic production, the Hubbard company has probably the most promising hybrids, but they have, to the author's knowledge, not been tested thoroughly under organic conditions.

8.3 ADAPTATION TO OUTDOOR FACILITIES

One of the basic rules in organic production is that animals should have access to a free-range area well covered with grass from the time they do not need heat corresponding to 4–6 weeks of age. There are a number of practical observations of how to operate such areas regarding density of animals, change of fields and other management issues. The chickens must learn to use such areas, which means that the chickens should have the opportunity to explore for learning to find what is of value for the birds in terms of seeds, roots, worms and other grazing sources. However, being exposed to the free-range area may also cause the chickens to be infected/infested with antigens that may challenge the immune system. Infestations with gastrointestinal helminths are known to inevitably invade the intestinal track of birds when they have access to fields with grass (Permin *et al.*, 1999). Studies have thus far been on laying stock where induced infection studies have revealed that 5 weeks after infection the birds began to shed parasite eggs when the manure was examined (Schou *et al.*, 2003). *Ascaridia galli* is the most common helminth found in chickens, and there seem to be clear differences among breeds as White Leghorns were demonstrated to have higher levels of parasite excretion (Gauly *et al.*, 2002; Schou *et al.*, 2003). Gauly *et al.* (2002) found a heritability of 0.05–0.20 regarding faecal *Ascaridia* egg counts, which means it is possible to select for *A. galli* resistance in chickens. Thus, if organically produced broilers are to be kept on pasture for at least 5 weeks, there are risks that they will be infected and some of the chicks will reach the stage that they shed parasite eggs in the grass. The influence of a helminth infection on the birds has been shown to be immune suppressive to the point that they are more easily infected with other bacterial organisms, for example *Salmonella* spp. (Eigaard *et al.*, 2006) and the growth of the chicks can sometimes also be reduced (Gauly *et al.*, 2002; Schou *et al.*, 2003).

The severity to which breeds react to infections appears to depend on the breeds. Therefore, a breeding company should be aware of their lines with respect to the resistance level to the infective agents during the period they are expected to be on the free-range and subsequently act either by exchanging the susceptible lines or including the resistance as a breeding goal.

Similar considerations should be given to other infectious diseases that may occur, especially during free-range occupation by the birds.

8.4 CONCENTRATION AND/OR QUALITY OF NUTRIENTS FED TO THE ORGANICALLY GROWN CHICKEN

Growth based on meat and bone requires protein in the diet, and the higher growth compared to body weight, the higher concentration of protein in the diet is required. This physiologically based principle is the rule used by nutritionists when formulating diets for broilers. In organic production, there are other principles that eclipse the physiologically based principles. These are as follows:

(i) The plant source used in a diet should be organically grown.
(ii) Addition of synthetically produced feed components to the diet is not allowed.
(iii) The plant ingredients should be raised near the location where the animals are grown.

In the Nordic Hemisphere, for example Denmark (Pedersen *et al.*, 2003), this means that protein sources are derived from peas, rape seed and some fish meal, which makes it difficult to achieve a sufficient protein concentration in diet and a well-balanced amino acid composition that is high enough for fast-growing broilers during the starter period of 1–3 weeks.

The consequences of using diets with low concentrations of protein/amino acids cause the chickens to grow poorly, exhibit higher mortalities and ultimately contain 2%–4% more fat in the carcasses (Sørensen, 1980). During a six-generation selection experiment in which one line was selected for high body weight when fed a low-protein diet, the birds of this line grew fairly well on the low-protein diet. This phenomenon is sometimes termed as genetic adaptation to a given environment. A more genetically based term is the concept of genotype (G) × environment (E) interaction, which may result in a genetic correlation between growth response based on two environments. Sørensen (1977) found a genetic correlation of approximately 0.4 between growth on a low-protein diet and a normal protein level. Therefore, performing a breeding programme in which improvement was 50 g per year in each of the respective environments A and B will lead to an improvement of only 20 g per year when growing the chickens in the other environment.

Yet another example of a G × E interaction that may play a role in organic meat production is the digestibility in grains fed to chickens. Mignon-Grasteau *et al.* (2004 studied the effect of feeding wheat varieties with different qualities regarding digestibility. They found a considerable heritability of the ability of chickens to digest wheat varieties with poor digestibility ($h^2 = 0.2$), and they also found considerable G × E interactions. According to an organic-based philosophy, there have been suggestions that food suitable for humans should not be used for animals, which leaves more animal feed sources such as poorly digestible wheat, for example for birds.

The conventional broiler breeding has so far been operating on the concept that various sources of feed were in principle unlimited, except the quantity. This means that genes of significance for better growth, when the quality of nutrition elements is not optimal, have not been promoted. This serves as background for the statement that chicken breeds suitable for organic production should have undergone a number of generations in which the selection for the relevant traits have taken place on feed mixtures that are prevalent under the respective organic production conditions.

8.5 THE PARENT STOCK SHOULD BE ORGANICALLY KEPT – PERHAPS?

Ideally, an organic production system would take place in the historical context with hens roaming around the homestead and the hens themselves organising the reproduction, as they still do in quite a number of developing countries. In the developed world, this concept is almost impossible for the following reasons: the product coming out of such a system would be so costly that there would hardly be any consumers to purchase the products, the hygienic standard would not be acceptable and there would not be sufficient numbers of eggs or broilers to cover the market.

Organic production has evolved from conventional production and adapted to the rules set up to satisfy the organic ideals. There has not been any constrains against the habit of the 'all in all out' principle which contributes considerably to the improvement of the hygienic standard. This management approach makes it possible to initiate a flock at day one without any contamination, since the parent stock used are derived from the breeding stock through a multiplying chain that has followed the general rules for poultry breeding and ensures that the day-old chickens delivered at the farmer are specific pathogen free (SPF) with respect to the common microbial agents in the poultry industry. During discussion of the organic rules, it has been claimed that the parent stock should be raised organically to satisfy the organic philosophy. The consequence of this would be that the day-old chickens would not be SPF, as the biosecurity programme of the parent stock would not be able to carry through, first of all, because the access to free range will risk a contamination of the parent stock with some of the microbial pathogens to such an extent that the biosecurity will break down.

8.6 WHERE TO BUY GENETIC MATERIAL, OR IS IT NECESSARY TO BREED FOR ORGANICALLY GROWN CHICKENS?

Previously, during the discussion on growth, the current review noted that two international hatchery companies offered breeding materials that fulfilled the requirement of slow growth (Cobb-Vantress, 2011; Hubbard Color, 2011). However, it is not clear to what extent the lines behind these hybrids have been genetically adapted to the respective organic environments that have been discussed previously. It should be recognised that the I 657 from the Hubbard Color appears to have a high frequency of breast blister, reported to be as high as 6% by Pedersen *et al.* (2003). In the Danish investigations at the farm level, a mortality of 5.7% for the first 12-week period was observed which was slightly higher than expected, but an additional 5% were rejected at the slaughter plant mainly due to emaciation. Diseases such as coccidiosis did not dominate at any of the farms; therefore, it may be assumed that the hybrid I 657, although suited for the production of Label Rouge, lacks genetic adaptation to the Danish organic environment which results in a little less than 5% that do not thrive well. From a screening of local breeds in developing countries, FAO (2010) received reports from Vietnam and India on birds which may have both the growth potential and also the genetic potential to grow on free range under sub-optimal conditions regarding nutrients and infection risk.

If the conclusion is that no hybrids/breeds on the market are applicable for the organic production environment, or are so far from the ideal, there are three possibilities:

(i) Either to convince the existing breeding companies to develop their hybrids to fit this niche and grow well under the existing sets of rules for organic production of broilers;
(ii) Or, alternatively to initiate breeding programmes via the direction of the appropriate worldwide organic organisations;
(iii) Or, to simply generate a partnership between the two organisations.

The major tools in such a genetic approach would be to select for the traits of interest with the test bird reared under the organic environment with the selection taking place among siblings or parents of the tested bird reared and kept in the SPF environment to retain the hygienic standards required for producing the parent stock. The driving principle would be to keep the elite individuals of breeding birds in a confined environment, but conducting the selection based on conditions associated with extensive environmental issues that could harm the more vulnerable conventional chickens. This is similar in concept to the principle applied by Hutt and Cole (1947) in selection for disease resistance.

8.7 DUAL PURPOSE OR SPECIALISED BREEDS

The organic ideal calls for the principle of developing dual purpose birds because it appears to be a better utilisation of the chickens. Conventional production uses specialised breeds specifically for either egg production or for broiler meat production. In particular, egg production represents a somewhat wasteful practice since the male birds are of no use at all and therefore are killed as soon they are recognised as males after hatching. As opposed to mammals, most species of birds are female as the heterogametic sex (Zw); therefore, the sex determining the reproductive cell is the egg. While sorting of sperm according to its sex determination is in use for mammalian species such as cows, such a technique is far from being possible for eggs of hens.

Among the breeds used for agricultural purpose none appear to be suitable for both egg production and meat production; therefore, one has to go back to the traditional breeds to find some which are suitable, and among these there may be some which may be potential candidates. Among these, the Wyandotte and the Plymouth Rock were actually used in the European system in the 1950s, but they are rarely in use these days because they produce too few eggs and grow too slowly. In Southeast Asia, there may be suitable breeds that produce a daily gain of 15–20 g that yields a body weight of 1.3–1.7 kg at 84 days, and lay 150–200 eggs per year (FAO, 2010) However, it remains to be examined whether they fit into the organic environment with a reasonable feed conversion and whether the meat quality would be acceptable for consumers of organic products.

8.8 CONCLUSION

A number of rules in organic production often cause problems if chickens of hybrids developed for conventional broiler production are used under organic regimes, and therefore other hybrids are to be recommended.

The rules may differ from continent to continent or from region to region due to political opinions on the organic philosophy, climate, consumers' perception and acceptation of the organic ideas and what price difference they accept and which types of farmers are going to produce the product.

Some fundamental rules apply, which are as follows:

(i) Restriction to the feed which makes it impossible to be sure to prepare diet that is balanced completely according to the recommendations from the nutritionist.
(ii) The birds should have access to free-range areas from about 3 weeks of age.
(iii) Restrictive administration of prophylactic agents.

The question of age at slaughter is an issue in European countries as the EU has stated that they should be 81 days at slaughter unless the breed has a moderate growth capacity (<40 g/day). Such rules do not seem to exist in the United States.

As welfare of the animals is a major issue in organic production, the use of chicken hybrids developed to grow in environments that are optimised to produce in conventional systems will often suffer when one or more factors are sub-optimal and this will be reflected in the farmer's choice of hybrid/breed.

Thus, a number of reasons support hybrids or breeds that grow slower, have a better genetic adaptation to the nutrition that is offered under the organic ideas, have a better genetic disease resistance to meet the diseases in a production system in which biosecurity is not optimal.

The chapter gives a discussion of hybrids/breeds that partly have some of the qualities asked for, but none of them have been developed for the organic production system. Instead of searching for unknown hybrids/breeds, it may be more profitable to go to direct breeding for genetic adaptation to the organic environment.

REFERENCES

Cobb-Vantress. 2011. CobbSasso150, the natural choice. www.cobb-vantress.com/products/cobbsasso150.aspx (accessed February, 2011).

Eigaard, N. M., T. W. Schou, A. Permin, J. P. Christensen, C. T. Ekstrøm, F. Ambrosini, D. Cianci, and M. Bisgaard. 2006. Infection and excretion of *Salmonella* Enteritidis in two different lines with concurrent *Ascaridia galli* infection. *Avian Pathol.* 35:487–493.

European Community. 1999. Council Regulation (EC) No. 1804/1999 of 19 July 1999 Supplementing Regulation (EEC) No.2092/91 on organic production of agricultural products and indications referring thereto on agricultural products and foodstuffs to include livestock production. *Off. J. Eur. Communities* L222; 1–28.

Fanatico, A. C. and H. Born. 2002. Label Rouge: pasture-based poultry production in France. *Publ. Nat. Center Appropr. Technol.*, Fayetteville, AR.

Fanatico, A. C., P. B. Pillai, L. C. Cavitt, C. M. Owens, and J. L. Emmert. 2005a. Evaluation of slower-growing broiler genotypes grown with and without outdoor access: growth performance and carcass yield. *Poult. Sci.* 84:1321–1327.

Fanatico, A. C., L. C. Cavitt, P. B. Pillai, J. L. Emmert, and C. M. Owens. 2005b. Evaluation of slower-growing broiler genotypes grown with and without outdoor access: meat quality. *Poult. Sci.* 84:1785–1790.

Fanatico, A. C., P. B. Pillai, J. L. Emmert, and C. M. Owens. 2007a. Meat quality of slow- and fast-growing chicken genotypes fed low-nutrient or standard diets and raised indoors or with outdoor access. *Poult. Sci.* 86:2245–2255.

Fanatico, A. C., P. B. Pillai, J. L. Emmert, E. E. Gbur, J. F. Meullenet, and C. M. Owens. 2007b. Sensory attributes of slow- and fast-growing chicken genotypes raised indoors or with outdoor access. *Poult. Sci.* 86:2441–2449.

FAO. 2010. Chicken genetic resources in smallholder production system and opportunities for their development, by P. Sørensen. FAO Smallholder Poultry Production Paper No. 5 Rome. Available at www.fao.org/ag/againfo/themes/documents/poultry/SPP5.pdf.

Gauly, M., C. Bauer, R. Preisinger, and G. Erhardt. 2002. Genetic differences of *Ascaridia galli* egg output in laying hens following a single dose infection. *Vet. Parasitol.* 103:99–107.

Hubbard. 2011. Hubbard color. http://www.hubbardbreeders.com/products.php?id=11 (accessed February, 2011).

Hutt, F. B. and R. K. Cole. 1947. Genetic control of lymphomatosis in the fowl. *Science* 106:379–384.

Mignon-Grasteau, S., N. Muley, D. Bastianelli, J. Gomez, A. Péron, N. Sellier, N. Millet, J. Besnard, J.-M. Hallouis, and B. Carré. 2004. Heritability of digestibilities and divergent selection for digestion ability in growing chicks fed a wheat diet. *Poult. Sci.* 83:860–867.

National Research Council. 1994. Nutrient requirements of poultry. 9th edn, National Academy Press, Washington, DC. p. 176.

Nesheim, M. C. and F. B. Hutt. 1962. Genetic difference among White Leghorn chicks in requirements of arginine. *Science* 137:691–692.

Nielsen, B. L., M. G. Thomsen, P. Sørensen, and J. F. Young. 2003. Feed and strain effects on the use of outdoor areas by broilers. *Br. Poult. Sci.* 44:161–169.

Pedersen, M. A., S. M. Thamsborg, C. Fisker, H. Ranvig, and J. P. Christensen. 2003. New production systems: evaluation of organic broiler production in Denmark. *J. Appl. Poult. Res.* 12:493–508.

Permin, A., M. Bisgaard, F. Frandsen, M. Peardman, J. Kold, and P. Nansen. 1999. Prevalence of gastrointestinal helminths in different poultry production systems. *Br. Poult. Sci.* 40:439–443.

Price, J. E., J. Conington, P. Sørensen, H. R. C. Kelly, and L. Rydhmer. 2004. Breeding strategies for Organic Livestock. In: M. Vaarst, S. Roderick, V. Lund and W. Lockeretz (eds) *Animal Health and Welfare in Organic Agriculture.* CABI Publishing, Wallingford, Oxon. pp. 357–388.

SARE Project FNE01-372, 2002 Comparison of standard broiler chicken breeds on pasture. Sustainable Agriculture Research and Education. http://mysare.sare.org/mySARE/ProjectReport.aspx?do=viewRept&pn=FNE01-372&y=2010&t=0.

Schou, T., A. Permin, A. Roepstorff, P. Sørensen, and J. Kjær. 2003. Comparative genetic resistance to *Ascaridia galli* infections of 4 different commercial layer lines. *Br. Poult. Sci.* 44:182–185.

Sørensen, P. 1977. Genotype-level of protein interaction for growth rate in broiler chickens. *Br. Poult. Sci.* 18:625–632.

Sørensen, P. 1980. Selection for growth rate in broilers fed on diets with different protein level. VI. *European Poultry Conference* 2:64–71. Published by World's Poultry Science Association.

Sørensen, P. 1986. Study of the effect of selection for growth in meat type chickens. Thesis defended for the Dr. Agro. Degree at the Royal Veterinarian and Agricultural University, Copenhagen. Published as Report No. 612 from National Institute for Animal Sciences, Denmark. p. 314.

Thorp, B. H. and E. Luiting. 2000. Breeding for resistance to production disease in poultry. In: R. F. E. Axford, S. C. Bishop, F. W. Nicholas and J. B. Owen (eds) *Breeding for disease resistance in farm animals.* CABI Publishing, Wallingford. pp. 357–378.

9 Organic Meat By-Products for Affiliated Food Industries

Claudia S. Dunkley, Dave Carter, and Kingsley Dunkley

Abstract: The ability to *market the whole animal* is important for all livestock producers, but can be particularly challenging for organic farmers and ranchers. Every animal brought to slaughter brings with it the costs that the producer has invested throughout its life cycle. Capturing that cost of production, covering the harvest expense, and clearing a net profit are the keys to success. Covering these costs are particularly challenging for producers of organic meat and poultry because of the added expense of raising animals without using growth hormones, antibiotics, or intensive confinement systems. Emerging consumer demand for organic pet food is helping to pave the way for capturing some value for organic by-products. Budding interest in items such as organic leather and other by-products may also create a new path for profitability. All of these paths, however, are filled with significant potholes and roadblocks.

Keywords: American Association of Feed Control Officials (AAFCO); by-products; clothing; cosmetics; fats; leather; pet food; pharmaceuticals; rendering; slaughter; tanning

9.1 INTRODUCTION

Organic foods can no longer be considered a niche market, as this is one of the fastest growing sectors in the food agriculture market around the world as consumers show they are willing to pay higher prices for these specialty foods. Before meat can be labeled "certified organic," not only should the animals be grown organically but they should also be handled organically throughout processing and packaging. In 2002, the United States Department of Agriculture (USDA) National Organic Program (NOP) put in effect the national organic rule (7 CFR Part 205) (USDA-AMS, 2000). The NOP established consistent national standards for organic production, facilities, and interstate and international commerce. The NOP also assumes that organic foods meet a consistent standard and protect consumers from organic claims that are fraudulent. In the United States, on-farm processing is allowed in limited numbers in some states if their processing operation is included in their farm plan (MOSES, 2007). Plants that process organic meat must complete organic certification (including organic inspection) for the processing facility (USDA-AMS, 2000). In some cases, organic and nonorganic animals are processed in the same facility. In this case, certified organic animals and meat products must be separated from the nonorganic. Refrigeration of the organic and nonorganic meats

Organic Meat Production and Processing, First Edition. Edited by Steven C. Ricke, Ellen J. Van Loo, Michael G. Johnson, and Corliss A. O'Bryan.

must also be segregated and all equipment must be thoroughly washed with products that comply with the NOP standards before it comes in contact with the organic carcass.

9.2 MEAT BY-PRODUCTS

Meat by-products are generated from the slaughter and processing of livestock and poultry by slaughterhouses, meat processors, wholesalers, and rendering plants. Due to low prices and health concerns, the traditional markets for edible by-products have been disappearing. As a result, processors have redirected their marketing strategies toward nonfood uses such as the pet food industry, pharmaceuticals, cosmetics, and animal feed (Deng-Cheng, 2002). The USDA Economic Research Service has stated that 11.4% of the gross income from beef and 7.5% from pork come from the by-products (USDA-AMS, 2000). This is a good source of profit for meat processors for a by-product that would typically be regarded as waste. Even though the organic livestock and poultry industry is still in its developmental stages, processors of organic meat and poultry can adopt the strategies of the conventional meat industry in seeking alternative markets for the by-products that are generated from processing. Organic meat by-products can be utilized in the same way and should be prepared with the same food safety concerns as would be used for preparing by-products from conventional meat by-products. By-products from the processing of organic meats can be used for human consumption and also in the pet food industry. However, there is the potential for expansion in other areas such as clothing, pharmaceutical, and cosmetic industries as these industries represents relatively untapped sources for organic by-products.

In the meat industry, everything except the meat that is produced by the animal is considered to be a by-product and in the United States by-products are divided into two classes, edible and inedible.

9.2.1 Edible by-product

As with the conventional meat industry, the edible by-products of organic meat are subjected to the USDA regulations as they relate to processing and labeling. Organic meats are processed and regulated by an arm of the USDA, the NOP (USDA-AMS, 2000). The edible by-products (Table 9.1) are the clean parts of slaughtered animals or poultry, not including the meat. Variety meats are the wholesale edible by-products and include meats such as brains, liver, heart, kidneys, spleen, sweetbreads, tongue, and tail. These can be sold as fresh or frozen items or can be used to make other processed items. They are processed under sanitary conditions and inspected by the US Meat Inspection Service on the basis of the Federal Meat Inspection Act (1906). Other edible by-products include sausage ingredients and components such as cheek and head trimmings, blood, pork and beef stomach, esophagus, and small and large intestines. Gelatin used for making confectionaries, ice cream, and jellied food products is obtained from bones, pork skin, and calfskin trimmings. Edible fats are obtained during slaughter, for example, the caul fat that surrounds the stomach and the rumen and cutting fat that can be obtained from back fat, rumen fat, and pork leaf fat. Shortening, chewing gum, and confectionaries are made from fats. Some edible lard is used in emulsified products and sausages. Edible by-products are more perishable than the carcass and therefore must be chilled quickly after slaughter and be processed or moved into retail trade.

Table 9.1 Potential uses of edible meat by-products.

Raw by-product	By-product use
Brain	Variety meat
Liver	Variety meat
Heart	Variety meat
Gizzard	Variety meat
Kidney	Variety meat
Spleen	Variety meat
Sweetbread	Variety meat
Tongue	Variety meat
Oxtail	Variety meat
Cheek and Head trimmings	Sausage ingredient
Beef Extract	Soup and Bouillon
Blood	Sausage component Medicine and research
Stomach	
(i) Suckling calf	Rennet for cheese making
(ii) Pork	Sausage container and ingredient
(iii) Beef	Sausage ingredient and tripe
Bones	Gelatin for confectionaries, ice cream, and jellied food product
Fats	
(i) Cattle, calves, lambs, and sheep	Candies, shortening, and chewing gum
(ii) Pork	Shortening/lard
Small and large intestine	Sausage casing
Large intestines (pork)	Chitterlings
Esophagus	Sausage ingredients
Pork skins	Gelatin for confectionaries, ice cream, jellied foods, and fried pork skins
Calf skin trimmings	Gelatin for confectionaries, ice cream, jellied foods

Source: Reproduced from Hedrick *et al.* (1994), copyright 2001, 1994 by Kendall Hunt Publishing Co. with permission.

9.2.1.1 *Inedible by-products*

Inedible meat by-products (Table 9.2) are considered unsuitable for human consumption. These are usually salvaged for animal feed, medicinal purposes or, are by their nature not edible. These include the hides of cattle, calves, pork skins, and pelts, which are used to make leather products such as shoes, bags, belts, athletic equipment, rawhide, sausage skins, glue, and edible gelatin. Animal hides and skins have historically been used by humans for shelter, clothing, and as containers. This is the most valuable by-product originating from animals. Fats from all animals can be used to make lubricants, industrial oils, and are included as feed ingredients in livestock and poultry feeds. The major animal fats are lard from pigs and tallow from cattle and sheep. Lard and tallow have traditionally been used for deep-frying but due to health concerns this practice is declining fast. Extracts from glands have traditionally been used as medicine in many countries. They have also been used for enzyme preparations for industrial uses. The most commonly used glands are pituitary, thyroid, parathyroid, adrenal, and ovaries. Glands should only be collected from healthy animals and the function of the harvested glands will depend on the species, sex, and age of the animal (Hedrick *et al.*, 1994). After preparation, glands are milled into a powder and made into capsules or used in a liquid form (Deng-Cheng, 2002). Blood can be used for therapeutic purposes in both human and animal medicine. It is used in laboratory research as a culture medium supplement to grow

Table 9.2 Potential uses of inedible meat by-products.

By-product	Processed by-product	Use of by-product
Hide—cattle and calf	Leather and glue	Leather goods, paper boxes, sand paper, and plywood
	Hair	Felts and plaster binder upholstery
Pork skins	Tanned skin	Leather goods
	Hair	Brush bristles
Pelts	Wool	Textiles
	Skin	Leather good
	Lanolin	Ointments
Fats	Inedible tallow	Industrial oils, lubricants, glycerin
	Tankage	
	Cracklings	Livestock and poultry feeds
	Stick	
	Grease	Industrial oils and animal feeds
Bones	Dry bone	Glue, hardening steel
	Bone meal	Animal feed and fertilizer
	Blood albumin	Leather preparations, textile sizing
Cattle Feet	Neatsfoot stock	Fine lubricants
	Neatsfoot oil	Leather preparations
Glands	Pharmaceuticals	Medicines
	Enzyme preparations	Industrial uses
Blood	Pharmaceuticals	Medicine
	Paste	Industrial uses
Feathers		Pillows, animal feed
Lungs		Pet food

Source: Reproduced from Hedrick *et al.* (1994), copyright 2001, 1994 by Kendall Hunt Publishing Co. with permission.

mammalian cells *in vitro*. Purified bovine albumin is used to help replenish blood or fluid loss in animal medicine. It is also used for Rh factor testing in humans and as a stabilizer in vaccines (Deng-Cheng, 2002). Blood is used in industry as an adhesive and also in the manufacturing of plywood, paper fiber, plastics, and glue. Blood is also used in the cosmetic industry and as a foaming agent in fire extinguishers. Feathers are used in pillows and also as a feed ingredient while hair can be utilized in the manufacturing of brush bristles.

9.3 MARKETING ORGANIC BY-PRODUCTS

The meat and poultry industry is often referred to as "a margin business," because the profit margin earned by producers equates to pennies per pound (Carter, 2010a). This is economically viable as long as considerable product volume is produced. Every animal brought to slaughter brings with it the costs that the producer has invested throughout its life cycle. Capturing that cost of production, covering the harvest expense, and clearing a net profit are the keys to success. Covering those costs are particularly challenging for producers of organic meat and poultry because of the added expense of raising animals without using growth hormones, antibiotics, or intensive confinement systems. In his book, *The Jungle*, Upton Sinclair famously wrote that meat-packing companies "used everything but the squeal." That phrase holds as true today as when Sinclair penned those words in

1906. The conventional meat and poultry industries have developed markets not only for the products eaten by humans, but also for the nonedible parts like bone, fat, and offal.

Over the past decade, organic meat and poultry producers have struggled with the challenge to encourage their customers to purchase meat products beyond the prime cuts of tenderloin, pork chops, rib eyes, and strip steaks (or, breasts and thighs, when it comes to chicken). The growing popularity for items such as organic short ribs and pot roast are helping producers recapture some of those additional costs. However, capturing the organic value for organic by-products still remains a work in progress.

Emerging consumer demand for organic pet food is helping to pave the way for capturing some value for organic by-products. Budding interest in items such as organic leather may also create a new path for profitability. All of these paths, however, are filled with significant potholes and roadblocks. The organic pet food industry is working to clear those roadblocks, but is hindered by several challenges. Understanding those challenges requires some background in the organic pet food business. From the demand side, there is no challenge. According to the Pet Food Institute, sales of pet food certified as organic—or made with at least 70% organic ingredients—have grown to a $262 million segment of the $18 billion pet food industry (Pet Food Institute, 2010).

Producing pet food products, though, is not as simple as cutting up by-products and putting them into a can. The pet food industry is governed by a myriad of complex state and federal regulations. Those ground rules become even more complex when moving into the realm of organic pet food.

9.4 CURRENT REGULATIONS REGARDING THE PET-FOOD INDUSTRY

The baseline standard for any pet food product is to be allowed to be labeled as "complete and balanced," according to the nutritional requirements established by the American Association of Feed Control Officials (AAFCO). Technically, AAFCO has no formal regulatory powers. However, the nutritional standards contained in AAFCO's "model regulations" serve as the basis for nearly every state-level regulation (AAFCO, 2007).

It is very important that pet food provides complete and balanced nutrition because companion animals generally rely on that commercially produced food as their only source of nutrition each day. The AAFCO model regulations specify minimum (and in some cases, maximum) levels of the components of protein, fat, vitamins, minerals, and other nutrients required for healthy growth, or for adult maintenance (AAFCO, 2010).

Meeting the AAFCO standards is only the beginning for organic pet food manufacturers. Those manufacturers must navigate through the complex and not always obvious instructions of the USDA organic regulations. Technically, pet food products do not have to be certified to the USDA standards to carry organic claims. However, AAFCO standards specify that the term "organic" in pet food products must be limited to

> a formula feed or a specific ingredient within a formula feed that has been produced and handled in compliance with the requirements of the USDA National Organic Program (7 CFR Part 205)
>
> AAFCO, 2010

Because state regulators defer to the AAFCO standards, those regulators are increasingly requiring that pet food products containing any organic claim do so in compliance with the

USDA rules. Currently, USDA organic regulations do not have any specific provisions for pet food. Following a major controversy over this topic several years ago, the USDA NOP established a special task force to develop proposed regulations for pet food (USDA-AMS, 2008). The National Organic Standards Board, which advises the NOP on the development of new regulations, unanimously endorsed the task force's proposal for new regulations in 2008. Implementing those regulations will require formal rulemaking procedures, and the NOP has not yet started that process (National Organic Program USDA-AMS, 2010).

In the meantime, the NOP has ruled that pet food products can be certified under the USDA standards *if* those products are produced and manufactured in compliance with the organic regulations for human food (7 CFR § 205.605 and § 205.606 USDA-AMS, 2000). As of late 2010, the NOP was embroiled in discussion regarding whether certain micronutrients required by cats and dogs could be included in pet food products that were manufactured under the organic human food regulations. The Pet Food Institute is working on behalf of the organic pet food manufacturers to address this issue with USDA, and to push for formal rulemaking on the proposed pet food regulations (Carter, 2010b).

9.5 ORGANIC PRODUCT AND BY-PRODUCT USE IN THE PET-FOOD INDUSTRY

Other barriers to the growth of the organic pet food market include the available supply, and the available infrastructure for processing and manufacturing. Pet food manufacturers need to be able to procure a steady supply of ingredients to manufacture and sell organic products in the commercial marketplace. Those companies are not interested in purchasing small lots of ingredients only on a seasonal basis.

To date, most organic pet food products are formulated with chicken ingredients because of the available supply of organic chicken frames, necks, and backs. According to USDA 2008 organic survey, as many as 10.8 million organic broilers and laying hens were in the US flock at any one time during that year (USDA-NASS, 2008a). A small number of companies handle the processing and distribution of those chickens to the national marketplace (USDA-NASS, 2008b).

The available supply of organic beef has been extremely limited, but that is beginning to change. Organic beef production in the United States today is decentralized, with the animals processed in smaller packing plants located across the country. One specific provision in the organic regulations allowed for the initial herds of dairy cattle to be brought into organic production after a 1-year conversion process, whereas any organic slaughter stock must be reared under organic protocols from the last third of gestation. Those regulations, however, require organic dairy replacement cows to be reared under the same standards as slaughter stock (7 CFR § 205.236(a)(2) USDA-AMS, 2000).

As more and more dairies complete the initial "whole herd" transition and begin to move to the replacement stock requirements, a growing number of organic cull dairy cows will be eligible to be processed and marketed into the organic meat market.

The desire among pet food manufacturers to have a consistent supply of organic ingredients does not mean that they have to work exclusively with large slaughter facilities. One idea that has been discussed over the past few years is to work with smaller slaughter facilities to establish a virtual "bank" of organic by-products (Carter, 2010c). Under this concept, the participating plants would adhere to uniform standards regarding how they handle and package the organic materials. They would supply the virtual bank with daily information regarding the available supply of those materials. Manufacturers would subsequently contact the bank to purchase the material. The actual supply could originate from multiple facilities, but would arrive at the manufacturing facility in a uniform manner.

Although this concept has been widely discussed, no formal effort has yet been moved forward to establish this type of virtual by-product bank. However, this concept for an increased supply would conceptually help boost interest among pet food manufacturers in developing or expanding lines of certified organic products.

For poultry, the primary products sought for organic pet food are the frames, backs, and necks. Those parts can be processed through a mechanical sieve (often called a "Beehive") under high pressure to remove the meat from most of the bone and cartilage. Under the AAFCO model standards, poultry ingredients are defined simply as

> the clean combination of flesh and skin with or without accompanying bone, derived from the parts or whole carcasses of poultry or a combination thereof, exclusive of feathers, heads, feed and entrails

Under this definition, chicken processed through a mechanical sieve can still be labeled as chicken, even if it contains the bone or cartilage (AAFCO, 2010). The inclusion of bone and cartilage can actually enhance the level of glucosamine and chondroitin in the finished products (Aldrich, 2007).

The other two poultry ingredients widely utilized in organic pet food include the hearts and livers. Hearts comprise muscle, and under the AAFCO regulations, any muscle meat can simply be labeled as meat. Thus, products containing chicken hearts will likely just list the ingredient as chicken. Livers, however, are organs, and organ ingredients must be listed individually on an ingredient panel, according to the AAFCO standards (AAFCO, 2010).

For beef, hearts, livers, and 50–50 trim (50% protein and 50% fat) comprise the primary ingredients for certified organic pet food products. As with poultry, the 50–50 trim and the heart meat can both be labeled as beef on an ingredient panel, while any liver must be identified as "beef liver." Both chicken and beef livers are popular ingredients in pet food because they enhance the palatability. However, formulators have to be careful not to include too much of these ingredients because the high levels of vitamin A that they contain can be detrimental to pets' health (NRC, 2006).

Pork and lamb producers have not yet been able to capitalize on the growing popularity of organic pet food. Pork is not generally considered a popular pet food ingredient, so the market has not developed for conventional or organic products and by-products (Carter, 2010c).

The US lamb producers have been stifled in marketing their ingredients to pet food manufacturers because of the concern among those manufacturers about the incidence of "scrapies" in domestic flocks (Carter, 2010c). In fact, most pet food companies that manufacture any lamb-based products generally source their ingredients from New Zealand. As the US lamb industry continues to expand certified scrapies-free flocks, those manufacturers are beginning to consider sourcing domestic ingredients. This would potentially be a market for slaughterhouse by-products from the lamb industry.

Unfortunately, most of the offal from any of these species has still not been able to be utilized for pet food production. That is because most offal is processed through rendering facilities and converted in a meal before being utilized as a pet food ingredient. Rendering is generally conducted by large manufacturers that utilize a continuous flow system that has no capability for segregating organic ingredients, or for preventing cross-contamination from other material.

9.6 WHERE DO WE GO FROM HERE?

Some companies are beginning to explore the feasibility of developing a stand-alone rendering system for organic by-products (Carter, 2010c). Development of this system would represent a major breakthrough for the organic pet food sector, and for organic meat and poultry producers. In the meantime, most manufacturers wanting to offer their customers any type of organic product are opting to manufacture under the USDA's "Made with Organic" standards, which require that the product contain at least 70% organic ingredients.

Traditional pet food manufacturing facilities face similar challenges in segregating organic ingredients, and in preventing cross-contamination. During the past 15 years, specialized manufacturers have built new plants specifically designed to produce organic and specialized pet food products. Those plants are designed to segregate all organic ingredients, and the processing lines are built so that they can be completely cleaned out between batches of organic and conventional product lines (Carter, 2010c).

While pet food manufacturing is extremely complex, many organic producers are capitalizing on the growing market for organic pet treat products. Because treats do not make any nutritional claims, they are less highly regulated under the AAFCO standards. However, any treat carrying the organic seal must be processed in accordance with the USDA regulations.

Bones, pigs' ears, liver chews, pizzels, and other by-products can be manufactured into pet treats relatively easily. There are numerous certified organic processing kitchens across the country that can process these products.

9.7 OTHER USES OF ORGANIC BY-PRODUCTS

As the organic marketplace grows, customers are beginning to expand their horizons beyond the food and pet food segment. The growing popularity of clothing made from organic wool and cotton is beginning to create a spillover demand for organic leather.

Just as the rendering system limits the ability to convert organic offal into pet food, the conventional tanning process prevents the hides for organic animals from being converted into organic leather. The harsh chemicals used in that process could never qualify for organic handling. Tanning, however, does not have to involve those harsh chemicals. Traditionally, Native Americans used a paste made of the animals' brains to soften and condition hides (Richards, 2004). Traditional vegetable tanning processes use tree bark and other natural compounds, rather than chromium. Brain tanning and vegetable tanning have not traditionally produced the soft leather that is widely sought by the apparel and accessory industry. Here, too, new developments may soon change that perspective. In recent years, a small group of innovators, and a handful of conventional tanning companies, have begun to investigate how these traditional practices can be adapted to produce soft, pliable leather. With the prospect of more cull organic dairy cows being eligible to be slaughtered as organic beef, the supply of organic hides will increase. This boost in available inputs may intensify the search to develop tanning processes that can be certified as organic under the USDA standards.

9.8 CONCLUSIONS

The organic meat industry is relatively new and has been growing steadily as more and more consumers are demanding products in its most natural form. With the increasing amount of animals that are organically reared being slaughtered, there is the potential for additional income with the increased amount of by-products that are generated during the slaughtering and rendering processes. The difficulty in marketing "the whole animal" has served as a major impediment in the growth of the organic meat and poultry sectors. As consumers continue to expand their interest in organic products, the demand for pet food, pet treats, and even organic leather goods may spur development of new processing systems that will enable farmers and ranchers to capture the added value that will help them grow their herds and flocks.

The organic meat industry has not yet realized its full potential in utilizing by-products from animal slaughter and processing. New and recent demands for organic pet food is paving the way for capturing the organic value for organic by-products but the progress is hindered by some challenges. Presently, some by-products from the processing of organically grown animals are being utilized for human consumption and in the pet food industry. The pet food industry is governed by a system of complex state and federal regulations. These rules become more complex when it comes to organic pet food. For example, although pet food products do not have to be certified to the USDA standards to carry organic claims, the AAFCO standards specify that the term "organic" must be limited to a "formula feed" or a specific ingredient within a formula feed that has been produced and handled on the basis of the requirements of the USDA NOP. The NOP has ruled that pet food products can be certified under USDA standards if those products are produced and manufactured in compliance with the organic regulations for human food.

The by-products can be edible or inedible. The markets for the inedible by-products are untapped. There is a growing market for organic clothing such as leather products and the potential exists for expansion into the pharmaceutical and cosmetic industries. There is also a small market for specialty fertilizers for home gardening that is made from animal by-products. This can also be an outlet for organic animal by-products as consumers seek to use products that are natural without the addition of nonorganic chemicals and products that are not biodegradable. By-products from organic meat processing are the same by-products (in terms of the part of the animal) as those produced in the conventional meat industry and as such can be utilized in the same manner. While we have not gotten to the point where everything is used but the "squeal", the potential exists in the organic meat industry to utilize as much of the slaughtered animal as possible and reduce waste.

REFERENCES

AAFCO. 2007. *How Pet Food is Regulated.* American Association of Feed Control Officials, West Lafayette, IN.

AAFCO. 2010. *Official Publication.* Association of American Feed Control Officials, West Lafayette, IN.

Aldrich, G. 2007. Glucosamine gains popularity. Available at: Petfoodindustry.com (accessed May 16, 2007).

Carter, D. 2010a. Oral testimony on behalf of the Pet Food Institute to the National Organic Standards Board. Madison, WI, October 25.

Carter, D. 2010b. *The True Cost of Unhealthy Food.* Natural Food Merchandiser, Boulder, CO. April, 2010.

Carter, D. 2010c. Personal communications.

Deng-Cheng, L. 2002. Better utilization of byproducts from the meat industry. Available at: www.agnet.org/library.

Federal Meat Inspection Act. 1906. Title 21. Chapter 12, subchapter 1; § 605.
Hedrick, H. B., E. D. Aberle, J. C. Forrest, M. D. Judge, and R. A. Merkel. 1994. By-products of the meat industry (Chapter 16). In: H. B. Hedrick (ed.) *Principles of Meat Science*. 3rd edn. Kendall/Hunt Publishing Company, Dubuque, IA. pp. 333–344.
MOSES. 2007. *Organic Poultry Production: Meat. MOSES Organic Fact Sheet*. Midwest Organic and Sustainable Education Service, Spring Valley, WI.
National Organic Program, USDA-AMS. 2010. *Organic Integrity from Farm to Table, Consumers Trust the Organic Label. Strategic Plan 2010–2012*. USDA publications, Washington, DC.
National Research Council. 2006. *Nutrient Requirements of Dogs and Cats. Animal Nutrition Series*. NRC of the National Academies, Washington, DC.
Pet Food Institute. 2010. *Letter to USDA National Organic Program*. Pet Food Institute, Washington, DC, April 22.
Richards, M. 2004. *The History of Brain Tan, Deerskins into Buckskins*. Backcountry Publishing, Cave Junction, OR.
USDA–AMS. 2008. Organic pet food task force. Available at: http://www.ams.usda.gov/AMSv1.0/ams.fetchTemplateData.do?template=TemplateC&navID=NationalOrganicProgram&leftNav=NationalOrganicProgram&page=NOPTaskForces&description=Task%20Forces.
USDA-AMS. 2000. 7 CFR Part 205. National Organic Program; Final Rule.
USDA-NASS. 2008a. Organic Survey, 2007 Census Publications. USDA, Washington, DC Available at: http://www.agcensus.usda.gov/Publications/2007/Online_Highlights/Organics/index.asp.
USDA-NASS. 2008b. Organic livestock and poultry products sold on certified and exempt organic farms, 2008 Organic Survey. USDA Publications, Washington, DC.

10 Organic Animal Nutrition and Feed Supplementations

Vesela I. Chalova and Steven C. Ricke

Abstract: Organic meat production is an alternative to conventional one and is rapidly developing in response to increasing consumers' demand for better meat quality and enhanced food safety. It is certified organic only if animals are raised under organic management in compliance with specific standards regarding stock origin, housing, nutrition, and animal health care. Formulating animal diets under current policies for organic production that completely satisfy animal nutritive needs is a challenge. This chapter is focused on feed ingredients and supplementations as main sources of proteins, minerals, and vitamins that are suitable for use in organically raised animal practice while providing optimal animal performance.

Keywords: organic animal nutrition; organic standards; feed ingredients

10.1 INTRODUCTION

Organically produced food including meat is currently a small but rapidly developing segment of the worldwide market. In 1997, Western Europe was the major consumer of organic food with a retail sales value of $5.3 billion followed by United States and Japan with $4.2 and $1 billion retail sales, respectively (Kortbech-Olesen *et al.*, 1999). The growth rate of the organic market in these regions was projected to increase by 25% in future years as a response to changed consumption patterns and complex agricultural activities toward development of enhanced biological diversity and soil activity, efficient utilization of plant and animal wastes, and reusable energy sources while still producing high-quality foods.

Changing agricultural production patterns from conventional to sustainable is subject to strict regulations, monitoring, and control that may require considerable time and cost inputs (Lampkin, 1997; Department of Economic and Social Affairs (DESA), 2000; Lamine & Bellon, 2009). Organic livestock is a vital part of the sustainable agriculture and, although not considered to become dominant over conventional forms in the near future, is growing and providing products with perceived better quality and health benefits. Hansson *et al.* (2000) reported fewer postmortem lesions in organically reared pigs (17%) than in conventionally grown ones (28%) and a higher classification rate for carcasses (EUROP system) obtained from cattle from organic than from conventional herds. *Ad libitum* feeding pigs on pasture in the organic production system gave superior meat quality compared to a restrictive feeding

Organic Meat Production and Processing, First Edition. Edited by Steven C. Ricke, Ellen J. Van Loo, Michael G. Johnson, and Corliss A. O'Bryan.

strategy (Oksbjerg *et al.*, 2005). Significantly higher levels of β-tocopherol, cholesterol, and potassium were established in organically produced eggs when compared to conventional system eggs (Matt *et al.*, 2009). In addition to the expectations for better quality of organically produced meat and eggs, the ban for feed supplementation with antibiotics in organic meat production was proposed to decrease incidences of antibiotic resistance observed in patients consuming meat contaminated with food-borne pathogens (Smith *et al.*, 1999; Hooper, 2001; McDermott *et al.*, 2002).

Animal production is certified organic only if animals are raised under organic management and specific standards regarding stock origin, housing, nutrition, and animal health care are met (Lampkin *et al.*, 1999; Dimitri & Greene, 2007). Feed ingredients are of particular interest since they should fulfill both the nutrient requirements of animals and the established standards of organic farming. This chapter focuses on feed ingredients and supplementations that are suitable for use in organically raised animals and provide optimal animal performance.

10.2 ORGANIC ANIMAL NUTRITION: GENERAL CONSIDERATIONS

Raising animals organically is meant to be a part of a self-supportive and well-balanced agroecosystem (Thamsborg, 2001). A basic principle in the system is nutrient recycling, which should be applied at least at a regional level if an individual farm is not suitable (Thamsborg, 2001). In this system, for bringing organically raised animals and feed production together, no synthetic chemical application is allowed at any step in the production cycle. According to the DESA (2000), agricultural products that are harvested in a period shorter than 3 years after chemical treatments are stopped are no longer considered organic. A 3-year transition period is required by the USDA National Organic Program (NOP) (2010) for land used in organic production unless records prove that no prohibited substances were used in or near the production area during the previous 3 years.

Formulating diets under one of the current policies for organic production while providing optimal performance of animals is a challenge. In general, diets allowed in organic animal nutrition should be formulated from ingredients that are produced and handled organically. Supplementations of substances such as enzymes, probiotics, and vitamins should have a natural origin. Ideally, 100% of the diets should be organic. However, due to difficulties faced by feed producers in diet optimization, the relatively high costs of organic feed ingredients, and their limited availability, there is potential to generate certain variations among the currently existing standards. For example, the International Federation of Organic Agriculture Movements (IFOAM, 2005) allows a limited percentage of nonorganic feed when there is inadequate quantity or quality of organic feed or in areas where organic agriculture is in early stages of development. The maximum amount of nonorganic feed for ruminants is 10% dry matter on annual basis; for nonruminants, it is 15% (IFOAM, 2005). At least 50% of the feed needs to be produced at the farm itself or at other nearby organic farms. Until the end of 2011, the European Commission allowed only 5% nonorganic ingredients in organic feed where farmers are unable to obtain feed exclusively from organic production (EC, 2007, 2008), while the USDA NOP requires 100% organic feed (USDA NOP, 2008) with the exception of a few synthetic substances that are allowed for use in the organic feed such as vaccines and certain feed additives including synthetic vitamins and trace minerals approved by the

US Food and Drug Administration (FDA) as well a synthetic methionine, which is allowed for organic poultry only until October 2012.

However, all three standards are agreed on for the full exception of growth promoters in poultry formulations. Another example of controversy is whether to allow synthetic feed additives. Synthetic vitamin and trace elements are allowed in all three standards (EC, 2007, 2008; IFOAM, 2005; USDA NOP 2008). However, the EC requires the vitamins to be identical to the natural vitamins and the IFOAM (2005) only allows synthetic vitamins when natural sources are not available in sufficient quantity and quality. The USDA NOP (2008) permits the use of synthetic vitamins and minerals when FDA approved. The supplementation of animal diets with synthetic amino acids or amino acids produced with genetically modified microorganisms is not allowed for all standards. In Europe, 5% nonorganic food ingredients are allowed until the end of 2011, which can include synthetic amino acids (EC, 2007, 2008), while the USDA NOP (2008) bans all synthetic amino acids from use in all organic livestock feed, with one exception, i.e., the use of synthetic methionine for organic poultry feed only until October 2012 (USDA, 2010). Although organically raised animals are not produced to achieve maximal body gain, suboptimal levels of essential amino acids including lysine, methionine, and tryptophan may affect animal protein metabolism and make the adaptation of the animals to stress conditions less efficient (Tesseraud *et al.*, 2001; Srinongkote *et al.*, 2004). Diet formulation, however, based only on plant-derived proteins may lead to unbalanced protein composition, inefficient protein utilization, and increased nitrogen release to the environment (Blair, 2007). In a similar manner, vitamins necessary for animal development should be provided by feed ingredients. However, some essential vitamins such as B_{12} are deficient in grains and plant materials and are mainly produced via fermentation processes conducted with genetically modified microbial strains (Wilson, 2003).

Antibiotics as well as hormones as feed additives are not allowed for organic production (USDA, 2008). In Europe, these substances are not even allowed for conventional production for growth promotion purposes. The European Union (EU) banned the feeding of all antibiotics and related drugs to livestock for growth promotion purposes, starting on January 1, 2006 (EC, 2003). Similarly, hormones are not allowed for any conventional livestock production in Europe, but are allowed for conventional beef production in the United States.

Organically certified feed ingredients are either available to farmers as single components or are used in the development of complete meals. While preparing the feed ration, except for the standards for organic origin of feed ingredients, precautions should be taken with regard to any physical or chemical treatments that may be in conflict with the regulations of the organic animal nutrition. Grinding or any other form of particle size reduction is the most common feed ingredient treatment. Since it is accomplished by mechanical forcing, no limitations to this specific feed processing under organic management exist. However, feed particle size and physical properties of feed were demonstrated to influence pH, volatile fatty acids, and microbial populations in the digestive tract of broilers (Engberg *et al.*, 2002). Higher incidence of gastrointestinal pathogens including *Salmonella* may also be particle size- and structure-related (Funk & Gebreyes, 2004). Mikkelsen *et al.* (2004) reported lower survival of *Salmonella* in the stomach of pigs fed coarsely ground feed than in pigs fed finely ground feed. Similarly, a higher *in vitro Salmonella* death rate in the gizzard of growing broilers given the coarse mash diet compared to those given the fine mash diet was observed by Huang *et al.* (2006). Therefore, since no antibiotic supplementation is allowed in organic feed, physical properties of the animal meals including structure, size, and forms should be considered as important factors and chosen carefully to maintain animals in good health.

While mechanical treatment of feed ingredients when preparing the complete diet ration does not interfere with organic nutrition standards, intervention with chemicals not allowed for use in organic animal nutrition in the corresponding country is not permitted. Flakes and pellets are forms of solid feeds that have the advantages to provide increased feed density and decreased storage cost (Behnke, 1996). However, conventional processes of feed pellet production may include chemicals such as lignin sulfonate to prevent the crumbling of pellets (Huntington *et al.*, 1981). Alternative pellet-binding agents in the production of pellets designated for organically fed animals are molasses and colloidal clay that improve pellet durability during transportation when added at the levels of 25–50 g/kg and 5–12 g/kg, respectively (Blair, 2007).

Regardless of the species, optimal performance and physiological status of animals require that feed formulation contain all ingredients that provide the necessary quantities of macro- and micronutrients. The main function of macronutrients is to provide energy but they also possess some developmental, structural, and regulatory functions (Groff & Gropper, 2000; Swennen *et al.*, 2005). Proteins are macronutrients, and if compared to micronutrients, are required in much greater amounts. Section 10.3 is focused on protein nutrition in animal production, general sources, and availability.

10.3 PROTEINS

10.3.1 Protein supply

Proteins represent a large group of organic compounds that play an important role in animal physiology. They have diverse functions in living cells, which may involve energetic, structural, or regulatory functions (Denhardt & Guo, 1993; Groff & Gropper, 2000; Razani *et al.*, 2002). In the skin, eyes, and muscles, proteins hold together and build structural elements of cells. Enzymes, hormones, antibodies, and globulins catalyze and regulate biochemical processes (Changeux, 2006). Functionally critical biomolecules such as hemoglobin, myoglobin, and various lipoproteins that carry oxygen and other substances within the body are also proteins (Nelson & Cox, 2004).

Proteins that are necessary for the growth and maintenance of animals are provided by feed ingredients. Once ingested, they are digested to small peptides and amino acids, which are further reused as building blocks in intermediate metabolism or in the body's own protein biosynthesis (Sundrum *et al.*, 2005; Chalova *et al.*, 2009). The amino acids are the building monomers of protein molecules and determine their primary structure. In contrast to prokaryotes, mammals are not capable of *de novo* biosynthesis of all amino acids and rely on their preformed availability in food ingredients. These amino acids, namely methionine, lysine, isoleucine, leucine, threonine, phenylalanine, tryptophan, and valine, are referred to as essential since they are required for growth, viability, and health maintenance. In some studies (Heger *et al.*, 1998; Lenis *et al.*, 1999), arginine is considered an essential amino acid in pig diets, while in other reports (Gotterbarm *et al.*, 1998) this amino acid is classified as nonessential. Glycine and serine are indispensable only for young birds (Taylor *et al.*, 1994; Waguespack *et al.*, 2009). Cysteine, histidine, proline, and tyrosine are considered conditionally essential and are not normally required in the diet but exogenous supplementation is needed when sufficient amounts of these compounds cannot be synthesized by a specific animal population (Reeds, 2000; Fürst & Stehle, 2004).

Inadequate amino acid ratio or the deficiency of even one of the essential amino acids arrests protein synthesis and restricts animal growth and performances. According to Lenis *et al.* (1999) and Blair (2007), if an essential amino acid is present at 50% of the animal requirement, the efficiency of the utilization of the other amino acids would decrease to 50%. The most distinctive effect of an essential amino acid deficiency is weight loss or diminished weight gain (D'Mello, 1994). In poultry, if lysine is the limiting amino acid in the diet, suboptimal body gain, proportional to lysine bioavailability, is achieved (Baker *et al.*, 1996; Tesseraud *et al.*, 2001). The complete deprivation in chicken diets of lysine and histidine limited viability of these birds to no longer than 53–60 days after which mortality was observed (D'Mello, 2003). In addition to suboptimal body gain, which may not be a problem when raising animals organically, lysine deficiency affects muscle and liver protein turnover in growing chicks (Tesseraud *et al.*, 1996), which contradicts the holistic concept of organic management. Methionine is considered the first and second most important amino acid in poultry and pig diets, respectively (Sundrum *et al.*, 2005; Kim *et al.*, 2006). Barnes *et al.* (1995) reported that chicks fed diets deficient in methionine exhibited lower growth rates and decreased efficiency of feed conversion and protein synthesis in the gastrocnemius and pectoralis muscles. Reductions in weight gain and bone growth, concomitant with elevations in plasma growth hormone in tryptophan-deficient chicks, have been reported (Carew *et al.*, 1983). Threonine is a limiting amino acid in most cereals including wheat, sorghum, and barley and their use in diet formulation may cause threonine to be the pressure point when linear programming of rations is used (Kidd & Kerr, 1996).

While in conventional animal production the amino acid balance in rations is commonly achieved by addition of crystalline amino acids, synthetic compound supplementation including amino acids in organic farming such as livestock is not permitted (EC 1991, 1999; UKROFS, 2003). In the United States, to facilitate the transition to complete exclusion of pure amino acids as additives in animal diets, synthetic methionine was made an exception for organic poultry production until October 2012 (Fanatico *et al.*, 2009; USDA, 2010). Utilization of methionine produced from fermentation may appear to be an alternative but wild-type microorganisms do not biosynthesize this amino acid in large quantities as genetically modified microorganisms do. Microbial biosynthesis of methionine occurs via a highly branched pathway with complicated regulation, which may be the reason for the lack of commercialized methionine fermentation so far (Chalova *et al.*, 2010). However, according to Kumar and Gomes (2005), there is potential for isolation of naturally synthesizing microorganisms with more relaxed regulatory mechanisms. Defining appropriate media composition and culture conditions may increase biosynthetic capacity of microorganisms and optimize commercial organic methionine production. Similarly, the utilization of lysine, tryptophan, and threonine produced by microbial fermentation for use in organic animal nutrition is not allowed if these amino acids originate from genetically modified microorganism (Blair, 2008). However, production of these amino acids from naturally non-GMO (genetically modified organisms) occurring isolates would be acceptable.

Preparation of diets optimally fulfilling the amino acid needs of animals without the use of synthetic amino acids is a difficult task and very often leads to protein overload. Feeding animals with excessive protein may be harmful to both the animals and the environment (Nahm, 2007; Sundrum *et al.*, 2005). Klasing and Austic (2003) postulated that excess of individual amino acids may be toxic to animals, which is evidenced by reduced weight gain and feed intake and deviations of amino acid plasma levels from those observed in animals fed amino acid-balanced diets. Langer *et al.* (2000) reported significant reductions in the plasma concentrations of isoleucine and valine in both the postprandial and postabsorptive states in growing pigs fed diets containing 50% excessive amount of leucine. Data from Edmonds and Baker (1987a,1987b,1987c), summarized by Baker (2004), indicated stronger repression of the body gain and feed intake of pigs fed excessive methionine for all levels of supplementation (from 5 to 40 g/kg) compared to pigs fed excessive amounts of the other three amino acids, threonine, leucine, and lysine, at the same levels. The influence of excessive methionine in the diets of chicks was even more pronounced, leading to reduced chick body gain from 118 to 11 over 8 days. The excess of phenylalanine and tryptophan in chick rations also resulted in approximately three- and two-fold reduction of the chick weight gain, respectively. Imbalances between lysine and arginine may also lead to antagonism that can be manifested as anorexia, induction of kidney arginase, inhibition of hepatic glycine transamidinase, and urinary spillage of both lysine and arginine in chicks fed excess dietary lysine (Austic & Scott, 1975).

Excessive protein nitrogen is not accumulated in birds but is excreted as uric acid, which is further converted to ammonia via hydrolysis, mineralization, and volatilization (Oenema *et al.*, 2001; Ritz *et al.*, 2004). Ammonia is a colorless gas, which upon accumulation in poultry barns represents a serious concern (Carlile, 1984; Kristensen & Wathes, 2000). According to Chambers and Smith (1998), 19% of estimated UK ammonia emissions in 1996 were due to the combined contribution of housing, storage, and land application of poultry manure. In poultry barns, the concentration of ammonia increased especially in winter when the ventilation rate was decreased to prevent the heat loss. Ammonia is considered toxic for humans when the concentration is over 25 ppm (Kristensen & Wathes, 2000). The symptomatic level of atmospheric ammonia for birds is even lower than this range (Kristensen & Wathes, 2000). According to Anderson *et al.* (1964), prolonged exposure of birds to ammonia concentrations at the level of 20 ppm throughout the production period can be detrimental to their health and performance.

10.3.2 Protein sources

After the completion of transition periods determined by respective country's organic standards, the required amino acid levels in animal diets has been proposed to be achieved via combining various protein sources that may either be materials of plant or animal origin (Blair, 2008). However, animal by-products are forbidden for use in organic feed (USDA, 2008). The use of the feed ingredients and their combination is subjected to specific regulations to comply with organic management. The EU's Council Regulation (EC No. 834/2007) postulates that feed should be obtained from the livestock holding or from other organic holdings in the same region. A part of the ration may contain feed from holdings that are in transition to organic farming.

Cereal grains, such as maize, barley, wheat, sorghum, and oats, are common ingredients of pig and poultry diets and usually provide 30%–60% of the total amino acid requirements (Sundrum *et al.*, 2005). They also appear to be the primary source of energy with maize (3525 kcal/kg digestible energy) contributing the most in the calculation of the energy balance of diet formulation followed by sorghum (3380 kcal/kg), rye (3270 kcal/kg), wheat (3210 kcal/kg), barley (3050 kcal/kg), and oats (2770 kcal/kg) (Blair, 2007).

Maize is a main ingredient of animal diets and approximately 50% of maize produced in the United States is used in feed formulation (Johnson, 2007). While excessive yellow corn is not desired in pig diets to avoid fat coloration, in layer rations, it may be used as the main source of carotenoids, thus determining the intensity of yolk pigmentation (Na *et al.*, 2004; Vasanthakumar *et al.*, 2009). To avoid supplementation of synthetic methionine in animal nutrition, high-methionine maize varieties containing up to 50% higher levels of methionine than the incurrent parent were developed (Phillips *et al.*, 2008). They were generated under sustainable agricultural practices, which permitted their utilization in organic meat production. After examining the efficiency of high-methionine corn inclusion in pullet diets, Jacob *et al.* (2008) concluded that it was a suitable replacement of conventional maize varieties used in organic practices. However, the expectations for lower yield and relatively high cost are substantial limiting factors that may reduce farmers' interests to high-methionine varieties application in animal production (Methionine Task Force, 2008).

Barley, wheat, sorghum, and oats are alternative protein sources used in organic animal nutrition. Currently, no genetically modified wheat, sorghum, barley, or oats are allowed for field cultivation as indicated in the GMO database (GMO Compass, 2010). The field cultivation of bioengineered varieties of a respective crop may compromise the purity of the nontransgenic variety crop due to field crosspollination. This phenomenon was observed, for example, in nontransgenic corn crop designated for organic nutrition, which was found to be contaminated with an engineered variety (Devos *et al.*, 2005; Messeguer *et al.*, 2006; Akiyama *et al.*, 2008). The uncertainty about the organic origin of a crop due to crosspollination with an engineered variety under field condition may reduce farmers' interest in using this particular grain. Therefore, grains such as barley, wheat, and sorghum with no transgenic counterparts, although with lower yields, may be of higher interest to farmers compared to grains with transgenic varieties.

Other protein sources such as soybean, canola, and sunflower meals may be used to complement the diet to ensure adequate amounts of the essential amino acids. They have high protein content with a relatively balanced and highly digestible essential amino acid composition that make them useful supplements in the diets of monogastric animals (Meng & Slominski, 2005; Gandhi *et al.*, 2008). This is especially valid for soybean meal where digestibility of lysine, methionine, and threonine may reach 84%, 86%, and 75%, respectively (Sundrum *et al.*, 2005). Both canola and sunflower meals are deficient in lysine but rich in sulfur-containing amino acids (Bell, 1984; Senkoylua & Dale, 1999). The three meal products are obtained after defattening the respective beans, which, in contrast to conventional processing, must be conducted by mechanical treatment (expeller processing) without using organic solvents to meet the standards for organic feed formulation (Blair, 2008). Soy, sunflower, and canola beans are rich in oil that contains relatively high amounts of polyunsaturated fatty acids including oleic, linoleic, and linolenic acids that are valuable in human nutrition. However, their abundance in pig diets may result in carcass fat softening and therefore are undesirable (Nishioka & Irie, 2006). Like maize, genetically modified soy and canola varieties are currently field cultivated and potentially represent a risk to crosspollinate and contaminate the crop gathered from the genetically nonmodified counterparts (James *et al.*, 2003). To the best of our knowledge, no genetically modified sunflower varieties are field cultivated yet (Cantamutto & Poverene, 2007).

Fish meal, which is rich in methionine and lysine, is already allowed as an additive in organic practices (EC No. 834/2007) but only if not contaminated by commonly used preservatives, such as ethoxyquin, which almost completely precludes its utilization (Fanatico *et al.*, 2009). By replacing fish meal in broiler diets with dried skim milk at levels up to 160 g/kg, Stevenson and Jackson (1981) observed increased feed intake and no significant alteration of live-weight gain. Growing-finishing pigs fed diets containing 10% dried skim milk exhibited growth performance, carcass traits, and nitrogen digestibility similar to those fed typical corn–soybean meal diets (Yen *et al.*, 2004). Milk and milk by-products are allowed in organic nutrition and may be used not only as protein sources but also for prevention or reduction of gastrointestinal pathogen populations. Reconstituted dry milk supplemented in the drinking water (5%) of broiler chickens reduced the mean $\log_{10}$ number of *Salmonella* Typhimurium from 5.68 to 2.11 (DeLoach *et al.*, 1990).

Grain legumes differ from sunflower and canola meals with respect to their essential amino acid composition by being rich in lysine and low in sulfur-containing amino acids (Dryden, 2008; Wang *et al.*, 2003). The highest levels of lysine are estimated in grass pea (19.4 g/kg), followed by faba beans (16.5 g/kg) and lupins (14.1 g/kg) (Mihailović *et al.*, 2007; Viveros *et al.*, 2007). Faba beans (*Vicia faba L.*), peas (*Pisum sativum L.*), and lupins (*Lupinus spec.*) are widely grown in countries with temperate climates, including most European countries, and are well accepted as alternative protein sources in organic animal nutrition (Sundrum *et al.*, 2005; Písaříková & Zralý, 2009). Perez-Maldonado *et al.* (1999) reported that inclusion of field peas in layer diets at 250 g/kg provided optimal egg production. Diaz *et al.* (2006) demonstrated that peas and faba beans supplemented at 350 g/kg and 500 g/kg, respectively, did not exhibit negative effects on bodyweight gain of Ross male chicks and their feed intake compared to the control. At 300 g/kg, row and extruded lupins slightly reduced the bodyweight gain of broilers from 55.1 g/day (control) to 51.9 g/day and 44.7 g/day, respectively. However, regardless of the numerous studies and reports on the inclusion of grain legumes in animal diets, their application in animal nutrition is still limited due to various antinutritional factors, which will be further discussed in more detail in Section 10.3.3.

10.3.3 Antinutritional factors and availability of amino acids in organic protein sources

10.3.3.1 *Antinutrients in organic protein sources*

Antinutrients are organic compounds with different chemical structures, which lower nutritive values of feed ingredients. When fed in higher amounts, they may cause health problems or even be lethal to animals (Mikić *et al.*, 2009). Therefore, their levels in feedstuffs and potential interaction with the other nutrients should be considered when formulating animal diets. Potential antinutrients include but are not limited to tannins, phytate, pyrimids-glycosids, lectins, and protease-inhibitors. Tannins are phenol molecules that complex predominantly not only with proteins but also with carbohydrates, thus affecting their digestibility and nutritional values (Mangan, 1988). Tannins are also known to inhibit intestinal enzymes in pigs and poultry, and reduce the *in vitro* activity of ruminal cellulase (Kumar & D'Mello, 1995; Norton, 2000). Considerable quantities of tannins are present in some varieties of sorghum (3.7 g/kg), canola (8.8 g/kg), faba beans (16 g/kg), and field peas (20.2 g/kg) (Farrell & Perez-Maldonado, 2000; Ebadi *et al.*, 2005). The inclusion of sorghum and canola meal in cockerels and pig diets at levels 25 g/kg and 85 g/kg, respectively, resulted in reduced growth rates (Connor *et al.*, 1969; Aherne & Baidoo, 1991). When diets of breeding sows were supplemented with faba beans at 170 g/kg, decreases in the number of pigs born alive and milk yield were observed (Thacker, 1990). However, low-tannin sorghum containing less than 2.6 g/kg tannins exhibited nutritive characteristics similar to those of maize and can be used as the main or only grain source in pig growth diets (Kemm & Brand, 1996; Nyachoti *et al.*, 1997). Tannin-free cultivars of sorghum, field peas, and faba beans are currently commercially available to farmers in Europe, Canada, and Australia and therefore could be used to better and more efficiently formulate animal diets (Farrell & Perez-Maldonado, 2000; Zijlstra *et al.*, 2004; Blair, 2007).

Phytic acid (phytate) is a naturally occurring compound, which is known to complex with proteins thus decreasing their digestibility and susceptibility to enzymes such as pepsin, trypsin, and α-amylase. The major problem caused by phytic acid is due to its chelating capabilities. The total content of phosphorus in cereals, which are used as major feedstuffs in monogastric animal nutrition, ranges from 0.35% to 0.45% (Eeckhout & De Paepe, 1994) and up to 80% of it may be bound by phytate (Kirby & Nelson, 1988). Phytate-bound phosphorus is poorly available to animals because of the lack of sufficient endogenous phytase in the feedstuffs. Phytase is an enzyme that liberates orthophosphate from the phytate molecule in the gastrointestinal tract into the phosphate form available to animals. Steiner *et al.* (2007) estimated that the highest phytase activities occurred in cereal by-products (9241–9945 U/kg DM), intermediate in cereals (except oats) (2323–6016 U/kg DM), and lowest in legume seeds and oats (262–496 U/kg dry matter). After analyzing a total of 183 samples representing 24 feedstuffs for total and phytate phosphorus content, and phytase activity, Viveros *et al.* (2000) concluded that from the cereals and cereal byproducts analyzed, only rye (5147 U/kg; 21 955 U/g), wheat (1637 U/kg; 10 252 U/g), rye bran (7339 U/kg; 56 722 U/g), and wheat bran (4624 U/kg; 14 106 U/g) exhibited high phytase activities. Their observations regarding legume seeds agreed with those of Steiner *et al.* (2007) by reporting negligible activity of the enzyme.

Phosphorus availability in phosphorus-deficient feed ingredients may be increased by supplementation of feeds with phytase, which is among the few enzymes allowed for use in organic animal nutrition (Blair, 2007; EU No. 277/2010). Sebastian *et al.* (1998) reported increased digestibility and availability of phytate-bound phosphorus in cereal-based poultry diets after supplementation with microbial phytase, which consequently resulted in improved ileal digestibility of crude protein and amino acids in female broiler chickens and in female turkeys. The addition of microbial phytase to corn–soybean meal diets fed to young pigs improved phosphorus utilization but had only a slight effect on ileal amino acid digestibilities (Omogbenigun *et al.*, 2003).

Pyrimids-glycosids, including vicin and convicin, are limited only to faba beans. Their microbial digestion in animal intestines releases products that may cause hemolisis, decrease egg quality and egg size, and limit the inclusion of faba beans in poultry formulation to 7% (Lacassagne, 1988). However, vicin and convicin in French faba beans exhibited little effect on protein and energy digestibility (Grosjean *et al.*, 2001).

Lectins and protease-inhibitors are substances of proteinaceous nature and are widely distributed in the plant kingdom. Lectins are very reactive to carbohydrates and form glycoproteids, which restrict the absorption capacity of villi in the intestine. They are stable proteins that have been identified at high concentrations in intestinal digesta of pea-fed piglets (30% of the total dietary protein) at all times as reported by le Gall *et al.* (2007) who studied the biochemistry of the digestion of field pea (*Pisum sativum* L.) in weaned piglets. Protease-inhibitors can lower the activity of trypsin and chymotrypsin and therefore impair the protein digestion as well as amino acid absorption. Both groups are present in significant amounts in soybeans as the protease inhibitors may reach up to 6% of the total protein content (Armour *et al.*, 1998; Mikić *et al.*, 2009). Palacios *et al.* (2004) reported that chicks and pigs fed diets free of lectins and Kunitz trypsin inhibitor performed better than the animals grown on raw soybeans. Compared to raw soybean, raw field peas included in poultry diets resulted in lower mortality rates and relative pancreatic weights, which indicated the presence of lower levels of antinutrients including lectins and protease inhibitors (Ravindran *et al.*, 2010). Lectin, trypsin, and chymotrypsin protease inhibitory activities can be easily abolished by aqueous heat-treatment of the seeds at 100°C for 10 minutes. (Armour *et al.*, 1998). However, such procedures could greatly affect availability of essential amino acids and thus diminish the quality of plant proteins (Finley, 1989). Lupins contain negligible amounts of lectins and trypsin inhibitor and therefore could supplement monogastric animal feeds without thermal treatment (Písaříková & Zralý, 2009).

10.3.3.2 *Amino acid availability in organic protein sources*

Nutrient composition of feed ingredients can be quite variable, depending on differences in crop varieties, fertilizers used, harvesting and storage conditions (van Keulen & Stol, 1991). Variability in nutrient composition in organically produced feeds is considered higher than in conventional feeds because of the quantity, type, and application time of organic amendments (Liu *et al.*, 2009). By estimating the nutrient content of feedstuffs, organically grown by producers throughout the US Midwest, Jacob (2007) observed significant variations in protein contents ranging from 9.9% to 15.7% for barley and from 30.4% to 49.4% for soybean samples. Lysine levels were not consistent throughout the samples as well, as the most pronounced variations reaching almost twofold differences were observed in barley. Regardless of recent efforts of industrial farmers to take advantage of global positioning systems to minimize the differences within and among the farms, the variability in nutrient composition of organically grown feedstuffs is still significant and impedes the achievement of consistency in diet formulations.

In organic feed formulation, an additional concern is amino acid availability. The concept of amino acid availability has been introduced and largely accepted in animal nutrition to recognize the incomplete digestibility and utilization of proteins by animals (Johnson, 1992). The susceptibility of peptide bonds to hydrolytic cleavage by digestive enzymes depends on the protein's capacity to complex with other nutrients present in feeds and the reactivity of the radical forms of the amino acids present in the protein molecule. This is especially true for proteins rich in lysine. This essential amino acid contains an ε-amino group, which is very reactive and easily interacts with carbonyl group of the nutritive compounds in feed ingredients (van Boekel, 1998). Feed ingredients with a plant origin are rich in reducing carbohydrates, which can form a complex with the ε-amino group of lysine (Rutherfurd & Moughan, 2007). This process is known as "browning" or "Maillard" reaction and is favored by elevated temperatures (Thomsen *et al.*, 2005). The final products, melanoidines, are nonutilizable by animals and humans, thus reducing lysine availability.

Anderson-Hafermann *et al.* (1993) conducted a 10-day chick bioassay to determine the effect of steam heating and commercial processing on protein quality of canola meal. Lysine quantities as well as that of other amino acids were determined by caectomization. They found marked reduction of bioavailable lysine with the increase in autoclaving time. By using a force-fed rooster technique, Undi *et al.* (1996) established that a 15-minute steam heating of canola meal reduced the true amino acid availabilities of methionine, lysine, and threonine only, while increasing the heating time to 90 minutes affected the true amino acid availability of all essential amino acids. The bioavailability of lysine in distiller's dried grains with solubles after being exposed to 315°C during preparation was estimated to be 80% of total lysine (Lumpkins & Batal, 2005). Soybean and cottonseeds, which otherwise are rich in lysine, undergo oil extraction and/or heat treatment for destruction of inhibitors (National Research Council (NRC) 1994). Depending on the manufacturing procedure, lysine bioavailability can be impacted less or more thus ending up with high variability of protein quality. In the study of lysine bioavailability in 14 meat and bone meals produced by using different procedures, Parsons *et al.* (1997) found its bioavailability level to vary from 43% to 89%.

10.4 MINERAL AND VITAMIN SUPPLEMENTATIONS

10.4.1 Minerals

Minerals such as calcium, phosphorus, sodium, chloride, potassium, and magnesium are required in large quantities and are known as macrominerals (O'Dell & Sunde, 1997). Minerals that are needed in amounts less than 12 mg/day are microminerals and include iron, zinc, copper, manganese, iodine, and selenium. Mineral deficiencies in animal diets interfere with proper animal growth and physiology, reproduction, and behavior. As high as 70% of the mineral content of an animal body consists of calcium and phosphorus, and as expected, inadequate supplementation of these compounds can result in poor overall growth, bone mineralization, milk production, and lactation (Wu *et al.*, 2001; Heinola *et al.*, 2006; Blair, 2007; Alexander *et al.*, 2010). Sodium, potassium, and chloride are involved in maintaining electrolytic balance and homeostasis (Sun *et al.*, 1996). They were also found to participate in the development of nerve membrane potential and active transport across cell membranes (Webb, 1990; Matthews, 2003). In more recent years, a coupled cotransport of Na^+, K^+, and Cl^- by cell membranes has been reported in nearly every animal cell type with the potential of mediating the net influx of osmotically active ions confirming the importance of maintaining optimal levels of these elements (Russell, 2000). Magnesium is also an essential mineral but deficiency of this compound in animal diets is not considered common.

Trace elements, although in small amounts, are mandatory for proper function of animal organisms. Iron takes part in the biosynthesis of hemoglobin in red blood cells and myoglobin in muscle, which participate in oxygen transport. Iodine is necessary for proper functioning of the thyroid gland. Zinc, copper, manganese, and selenium have catalytic or structural functions in numerous enzyme systems. Therefore, inadequate amounts of these elements in animal body can lead to metabolic disorders evidenced by suppressed appetite and growth, abnormal skeletal growth, and dystrophy of the skeletal muscles (Mertz, 1981).

Feed ingredients of plant origin are a major source for macro- and micronutrients. In general, cereals are deficient in calcium (maize 0.02% to 0.06%), manganese (0.13% to 0.23%), and sodium but are rich in iron (60–210 ppm) (Drinceanu *et al.*, 2009). These same authors reported low zinc content (20 ppm) in corn, sorghum, barley, and wheat and over 70 ppm in sunflower and soybean meals. Cereal grains exhibit low Ca/P (0.08) and Ca/Mg (0.19) ratios, which are partially due to the low Ca availability determined by the high phytate content (Cordain, 1999). Low iron and zinc availability in cereal grains due to various antinutrients is also well established (Salunkhe *et al.*, 1982; McKenzie-Parnell & Guthrie, 1986; Torre *et al.*, 1991).

Grain legumes such as soybean and sweet lupins contain more calcium and phosphorus than cereals (Drinceanu *et al.*, 2009). After evaluation of selected grain legumes grown in New Zealand, it was estimated that soybean and lupins (sweet and white) were richer in potassium, phosphorus, magnesium, and manganese than peas and chickpeas (Nalle, 2009). Although not nutritionally well balanced, grain legumes were also reported (chickpea, lentil, cowpea, and green pea) to be optimal sources of potassium, phosphorus, calcium, iron, and zinc (Iqbal *et al.*, 2006).

However, the mineral content of feed ingredients produced in different regions may greatly vary. The margins of variations depend on plant genetics, stage of vegetation, plant section analyzed, soil composition, and climate, and therefore could cause suboptimal concentrations of some of the elements in the feed ration (McKenzie-Parnell & Guthrie, 1986; NRC 1994; Underwood & Suttle, 2001). Based on the mineral level characterization of various seeds by the *Pisum* Core Collection, a part of the US Department of Agriculture's germplasm holdings, (http://www.ars-grin.gov/cgi-bin/npgs/html/crop.pl?177), Wang *et al.* (2003) reported differences approximating 3.5-fold variations for iron, 1.6-fold for magnesium, and 8.6-fold for calcium. Selenium analysis in Yugoslavia revealed a correlation between selenium soil deficiencies and selenium deficiency in the cereal crops and garlic grown on these soils (Maksimović *et al.*, 1992). In Asia, pigs fed US-produced maize and soybean grown on selenium-deficient soils performed more poorly compared to animals fed locally grown feed (Blair, 2007). In a similar manner, a direct linkage between soil contents of iodine and zinc and the respective levels of these elements in plants grown on these soils have been established as well (Nubé & Voortman, 2006).

To compensate and prevent the potential deficiency of trace and macroelements in feed rations, some sources with nonagricultural origin are allowed to supplement feed rations of animals under organic management. In the United States, forms of minerals approved by the FDA are allowed in organic diets of animals raised in the United States even though they may not be considered natural substances or appear on the national list of synthetic substances allowed for use in organic production (Blair, 2007). A summary of mineral substances allowed in EU based on EU regulation 889/2008 is presented by Drinceanu *et al.* (2009).

10.4.2 Vitamins

Vitamins are organic substances, which, although required in small amounts (micrograms to milligrams per day), have essential functions to maintain normal growth and reproduction of animals (Coelho, 2002). Feed ingredients with plant origin are major sources of vitamins except for B_{12}, which, if present in plant materials, may only be due to microbial contamination. It should be noted that poultry, swine, and other monogastric animals are more dependent on their diets for vitamins than ruminants, which are capable of efficient intestinal biosynthesis of most of the B vitamins (Coelho, 2002). Vitamins allowed in organic animal nutrition should be naturally occurring substances in feedstuffs (EC No. 834/2007). Plant materials, however, may not be reliable sources that completely satisfy animal needs for vitamins. Some natural forms of fat-soluble vitamins are unstable and lose potency very readily. Certain natural water-soluble vitamins may have reduced biological availabilities for the animals (Blair, 2007). Tavčar-Kalcher and Vengušt (2007) studied the stability of vitamins A, E, and K3 during storage under controlled conditions over a 1-year period. They established that the concentrations of the vitamins A, E, and K3 in the samples containing no choline chloride decreased to 53%, 59%, and 80% of their initial values, respectively. The natural form of vitamin E, α-tocopherol found in plant oils and seeds, is an antioxidant itself and is easily oxidized in feed (Malayoğlu *et al.*, 2009).

A particular problem in organic animal nutrition is the supplementation with vitamins B_2 and B_{12} that are commercially produced via bacterial fermentation, which may include genetically modified microorganisms as well (Burgess *et al.*, 2009). Insufficient or adequate amounts of vitamins in animal diets lead to specific syndromes or diseases, which are avoided by inclusion of vitamin premixes in diet formulation (McDowell, 2000).

However, synthetic forms are allowed for monogastric animals only if identical to natural vitamins. No GMO-produced vitamins may serve as additives in organically raised animals (EC No. 834/2007).

10.5 CONCLUSIONS AND PERSPECTIVES

Organic meat production is an alternative to conventional one and is rapidly developing in response to increasing consumers' demand for better meat quality and enhanced food safety. Raising organic livestock is in close conjunction with organic agronomy, which is the main and mandatory source of the animal diets. Both, although appearing as independent systems, are in fact parts of a complete holistic agricultural management practice, which is considered to promote healthy and ecological use of natural resources such as soil, plants, and animals. However, fulfilling animal needs while following governments and/or international standards for organic nutrition is a difficult task, which raises the cost of organic meat production. An additional challenge is the rapid evolvement of agricultural biotechnologies that involve the development and application of genetically modified plants and microorganisms. Regardless of the difficulties faced by producers, the organic meat production is projected to grow to meet requirements of contemporary and future societies.

ACKNOWLEDGMENT

This review was supported in part by the USDA/CSREES NIFSI grant no. 406-2008 51110 to Steven C. Ricke.

REFERENCES

Aherne, F. X. and S. Baidoo. 1991. An evaluation of the nutritive value of canola meal for young pigs. In: D. J. Farrell (ed.) *Recent Advances in Animal Nutrition in Australia*. University of New England, Armidale. p. 24A.

Akiyama, H., K. Sakata, K. Kondo, A. Tanaka, M. S. Lui, T. Oguchi, S. Furui, K. Kitta, A. Hino, and R. Teshima. 2008. Individual detection of genetically modified maize varieties in non-identity-preserved maize sample. *J. Agric. Food Chem.* 56:1977–1983.

Alexander, L. S., A. Mahajan, J. Odle, K. L. Flann, R. P. Rhoads, and C. H. Stahl. 2010. Dietary phosphate restriction decreases stem cell proliferation and subsequent growth potential in neonatal pigs. *J. Nutr.* 140:477–482.

Anderson, D. P., C. W. Beard, and R. P. Hanson. 1964. Adverse effects of ammonia on chickens including resistance to infection with Newcastle disease virus. *Avian Dis.* 8:369–378.

Anderson-Hafermann, J. C., Y. Zhang, and C. M. Parsons. 1993. Effects of processing on the nutritional quality of canola meal. *Poult. Sci.* 72:326–333.

Armour, J. C., R. L. C. Perera, W. C. Buchan, and G. Grant. 1998. Protease inhibitors and lectins in soya beans and effects of aqueous heat-treatment. J. Sci. Food Agric. 78:225–231.

Austic, R. E. and R. L. Scott 1975. Involvement of food intake in the lysine-arginine antagonism in chicks. *J. Nutr.* 105:1122–1131.

Baker, D. H. 2004. Animal models of human amino acid responses. *J. Nutr.* 134:1646S–1650S.

Baker, D. H., S. R. Fernandez, H. M. Edwards, 3rd, and C. M. Parsons. 1996. Efficacy of a lysine-tryptophan blend for growth of chicks. *J. Anim. Sci.* 74:1063–1066.

Barnes, D. M., C. C. Calvert, and K. C. Klasing. 1995. Methionine deficiency decreases protein accretion and synthesis but not tRNA acylation in muscles of chicks. *J. Nutr.* 125:2623–2630.

Behnke, K. C. 1996. Feed manufacturing technology: current issues and challenges. *Anim. Feed Sci. Technol.* 62:49–57.

Bell, J. M. 1984. Nutrients and toxicants in rapeseed meal: a review. *J. Anim. Sci.* 58:996–1010.

Blair, R. 2007. *Nutrition and Feeding of Organic pigs*. 1st edn. CABI Publishing, Ames, IA. p. 322.

Blair, R. 2008. *Nutrition and Feeding of Organic poultry*. CABI Publishing, Cambridge, MA. p. 314.

Burgess, C. M., E. J. Smid, and D. van Sinderen. 2009. Bacterial vitamin B2, B11 and B12 overproduction: an overview. *Int. J. Food Microbiol.* 133:1–7.

Cantamutto, M. and M. Poverene. 2007. Genetically modified sunflower release: opportunities and risks. *Field Crops Res.* 101:133–144.

Carew, L. B., Jr., F. A. Alster, D. C. Foss, and C. G. Scanes. 1983. Effect of a tryptophan deficiency on thyroid gland, growth hormone and testicular functions in chickens. *J. Nutr.* 113:1756–1765.

Carlile, F. S. 1984. Ammonia in poultry houses: a literature review. *Worlds Poultry Sci. J.* 40:99–113.

Chalova, V. I., C. A. Froelich, Jr., and S. C. Ricke. 2010. Potential for development of an *Escherichia coli*-based biosensor for assessing bioavailable methionine: a review. *Sensors* 10:3562–3584.

Chalova, V. I., S. A. Sirsat, C. A. O'Bryan, P. G. Crandall, and S. C. Ricke. 2009. *Escherichia coli*, an intestinal microorganism, as a biosensor for quantification of amino acid bioavailability. *Sensors* 9:7038–7057.

Chambers, B. J. and K. Smith. 1998. Nitrogen: some practical solutions for the poultry industry. *Worlds Poult. Sci. J.* 54:353–357.

Changeux, J. 2006. Allosteric proteins: from regulatory enzymes to receptors. *Rend. Lincei Sci. Fis. Nat.* 17:11–29.

Coelho, M. 2002. Vitamin stability in premixes and feeds: a practical approach in ruminant diets. Proc. 13th Annu. Fla. Ruminant Nutr. Symp., Gainesville, FL. P. 127–145.

Connor, J. K., I. S. Hurwood, H. W. Burlin, and D. E. Fuelling. 1969. Some nutritional aspects of feeding sorghum grain of high tannin content to growing chickens. *Aust. J. Exp. Agric.* 9:497–501.

Cordain, L. 1999. Cereal grains: humanity's double-edged sword. *World Rev. Nutr. Diet.* 84:19–73.

D'Mello, J. P. F. 1994. Responses of growing poultry to amino acids. In: J. P. F. D'Mello (ed.) *Amino Acids in Farm Animal nutrition*. CAB International, Wallingford, Oxfordshire. pp. 205–243.

D'Mello, J. P. F. 2003. Amino acids as multifunctional molecules. In: J. P. F. D'Mello (ed.) *Amino Acids in Animal Nutrition*. 2nd edn. CAB International, Cambridge. pp. 1–13.

DeLoach, J. R., B. A. Oyofo, D. E. Corrier, L. F. Kubena, R. L. Ziprin, and J. O. Norman. 1990. Reduction of Salmonella typhimurium concentration in broiler chickens by milk or whey. *Avian Dis.* 34:389–392.

Denhardt, D. T. and X. Guo. 1993. Osteopontin: a protein with diverse functions. *FASEB J.* 7:1475–1482.

Department of Economic and Social Affairs (DESA), Division for Sustainable Development. 2000. Changing consumption and production patterns: organic agriculture. Background Paper No. 4, New York.

Devos, Y., D. Reheul, and A. De Schrijver. 2005. The co-existence between transgenic and non-transgenic maize in the European Union: a focus on pollen flow and cross-fertilization. *Environ. Biosafety Res.* 4:71–87.

Diaz, D., M. Morlacchini, F. Masoero, M. Moschini, G. Fusconi, and G. Piva. 2006. Pea seeds (*Pisum sativum*), faba beans (*Vicia faba* var. *minor*) and lupin seeds (*Lupinus albus* var. *multitalia*) as protein sources in broiler diets: effect of extrusion on growth performance. *Ital. J. Anim. Sci.* 5:43–53.

Dimitri, C. and C. Greene. 2007. Recent growth patterns in the U.S. organic foods market. In: A. J. Wellson (ed.) *Organic Agriculture in the U.S.* Nova Science Publishers, Inc., New York. pp. 129–190.

Drinceanu, D., I. Luca, L. Ştef, E. Simiz, C. Julean, and D. Ştef. 2009. The mineral supplementation of poultry feed in organic farms (Review). *Lucrări ştiinţifice Zootehnie şi Biotehnologii* 42:351–358.

Dryden, G. M. 2008. *Animal Nutrition Science*. CABI, Cambridge, MA. p. 320.

Ebadi, M. R., J. Pourreza, J. Jamalian, M. A. Edriss, A. H. Samie, and S. A. Mirhadi. 2005. Amino acid content and availability in low, medium and high tannin sorghum grain for poultry. *Int. J. Poult. Sci.* 4:27–31.

EC. 1991. Council Regulation (EEC) No. 2092/91 of 24 June 1991 on organic production of agricultural products and indications referring thereto on agricultural products and foodstuffs. *Off. J. Eur. Communities* L198 (22/7/91):1–15.

EC. 1999. Council Regulation No 1804/1999 of July 1999 supplementing Regulation (EEC) No 2092/91 on organic production of agricultural products. *Off. J. Eur. Communities* 222(24/08/1999):1–28.

EC. 2003. Regulation No 1831/2003 of the European Parliament and of the Council (September 22, 2003) on additives for use in animal nutrition. *Off. J. Eur. Communities* L268:29–43.

EC. 2007. Council Regulation (EC) No 834/2007 (June 28, 2007) on organic production and labeling of organic products. *Off. J. Eur. Communities* L189:1–23.

EC. 2008. Commission Regulation Regulation (EC) No 889/2008 (September 5, 2008) laying down detailed rules for the implementation of Council Regulation (EC) No 834/2007 on organic production and labelling of organic products with regard to organic production, labelling and control. *Off. J. Eur. Communities* L205:1–83.

Edmonds, M. S. and D. H. Baker. 1987a. Comparative effects of individual amino acid excesses when added to a corn-soybean meal diet: effects on growth and dietary choice in the chick. *J. Anim. Sci.* 65:699–705.

Edmonds, M. S. and D. H. Baker. 1987b. Amino acid excesses for young pigs: Effects of excess methionine, tryptophan, threonine or leucine. *J. Anim. Sci.* 64:1664–1671.

Edmonds, M. S. and D. H. Baker. 1987c. Failure of excess dietary lysine to antagonize arginine in young pigs. J. Nutr. 117:1396–1401.

Eeckhout, W. and M. De Paepe. 1994. Total phosphorus, phytate-phosphorus and phytase activity in plant feedstuffs. *Anim. Feed Sci. Technol.* 47:19–29.

Engberg, R. M., M. S. Hedemann, and B. B. Jensen. 2002. The influence of grinding and pelleting of feed on the microbial composition and activity in the digestive tract of broiler chickens. *Br. Poult. Sci.* 43:569–579.

EU. 2010. Commission Regulation (EU) No 277/2010 (31/03/2010) concerning the authorisation of 6-phytase as a feed additive for poultry for fattening and breeding other than turkeys for fattening, for poultry for laying and for pigs other than sows (holder of authorisation Roal Oy). *Off. J. Eur. Communities* L86:13–14.

Fanatico, A. C., C. M. Owens, and J. L. Emmert. 2009. Organic poultry production in the United States: broilers. *J. Appl. Poult. Res.* 18:355–366.

Farrell, D. J. and R. A. Perez-Maldonado. 2000. Tannins in feedstuffs used in the diets of pigs and poultry in Australia. In: J. D. Brooker (ed.) *Tannins in Livestock and Human Nutrition.* Australian Centre for International Agricultural Research, Canberra. pp. 24–29.

Finley, J. 1989. Effects of processing on proteins: an overview. In: R. D. Phillips and J. W. Finley (eds) *Protein Quality and the Effects of Processing.* Marcel Dekker, Inc., New York. p. 8.

Funk, J. and W. A. Gebreyes. 2004. Risk factors associated with *Salmonella* prevalence on swine farms. *J. Swine Health Prod.* 12:246–251.

Fürst, P. and P. Stehle. 2004. What are the essential elements needed for the determination of amino acid requirements in humans? *J. Nutr.* 134:1558S–1565S.

Gandhi, A. P., K. Jha, and V. Gupta. 2008. Studies on the production of defatted sunflower meal with low polyphenol and phytate contents and its nutritional profile. *ASEAN Food J.* 15:97–100.

GMO Compass. 2010. http://www.gmo-compass.org/eng/home (accessed February, 2010).

Gotterbarm, G. G., F. X. Roth, and M. Kirchgessner. 1998. Influence of the ratio of indispensable: dispensable amino acids on whole-body protein turnover in growing pigs. *J. Anim. Physiol. Anim. Nutr.* 79:174–183.

Groff, J. L. and S. S. Gropper. 2000. Protein. In: J. L. Groff and S. S. Gropper (eds) *Advanced Nutrition and Human Metabolism.* 3rd edn. Wadsworth Publishing, Florence, KY. pp. 163–219.

Grosjean, F., P. Cerneau, A. Bourdillon, D. Bastianelli, C. Peyronnet, and G. Duc. 2001. Valeur alimentaire, pour le porc, de féveroles presque isogéniques contenant ou non des tanins et à forte ou faible teneur en vicine et convicine. *Journ. Rech. Porcine Fr.* 33:205–210.

Hansson, I., C. Hamilton, T. Ekman, and K. Forslund. 2000. Carcass quality in certified organic production compared with conventional livestock production. *J. Vet. Med. B Infect. Dis. Vet. Public Health* 47:111–120.

Heger, J., S. Mengesha, and D. Vodehnal. 1998. Effect of essential: total nitrogen ratio on protein utilization in the growing pig. *Br. J. Nutr.* 80:537–544.

Heinola, T., E. Jukola, P. Näkki, and A. Sukura. 2006. Consequences of hazardous dietary calcium deficiency for fattening bulls. *Acta Vet. Scand.* 48:25–31.

Hooper, D. C. 2001. Emerging mechanisms of fluoroquinolone resistance. *Emerging Infect. Dis.* 7:337–341.

Huang, D. S., D. F. Li, J. J. Xing, X. Y. Ma, Z. J. Li, and S. Q. Lv. 2006. Effects of feed particle size and feed form on survival of *Salmonella typhimurium* in the alimentary tract and cecal *S. typhimurium* reduction in growing broilers. *Poult. Sci.* 85:831–836.

Huntington, G. B., R. A. Britton, and R. L. Prior. 1981. Feed intake, rumen fluid volume and turnover, nitrogen and mineral balance and acid-base status of wethers changed from low to high concentrate diets. *J. Anim. Sci.* 52:1376–1387.

International Federation of Organic Agriculture Movements (IFOAM). 2005. *The IFOAM Norms for Organic Production and Processing, Version 2005.* IFOAM, Bonn. p. 130.

Iqbal, A., I. A. Khalil, N. Ateeq, and M. Sayyar Khan. 2006. Nutritional quality of important food legumes. *Food Chem.* 97:331–335.

Jacob, J. P. 2007. Nutrient content of organically grown feedstuffs. *J. Appl. Poult. Res.* 16:642–651.

Jacob, J. P., N. Levendoski, and W. Goldstein. 2008. Inclusion of high methionine corn in pullet diets. *J. Appl. Poult. Res.* 17:440–445.

James, D., A. Schmidt, E. Wall, M. Green, and S. Masri. 2003. Reliable detection and identification of genetically modified maize, soybean, and canola by multiplex PCR analysis. *J. Agric. Food Chem.* 51:5829–5834.

Johnson, J. 2007. Ethanol-is it worth it? *Chem. Eng. News* 85:19–21.

Johnson, R. J. 1992. Principles, problems and application of amino acid digestibility in poultry. *Worlds Poult. Sci. J.* 48:232–246.

Kemm, E. H. and T. S. Brand. 1996. Grain sorghum as an energy source for growing pigs. *Pig News Info.* 3:87N–89N.

Kidd, M. T. and B. J. Kerr. 1996. L-threonine for poultry: a review. *J. Appl. Poult. Res.* 5:358–367.

Kim, B. G., M. D. Lindemann, M. Rademacher, J. J. Brennan, and G. L. Cromwell 2006. Efficacy of DL-methionine hydroxy analog free acid and DL-methionine as methionine sources for pigs. *J. Anim. Sci.* 84:104–111.

Kirby, L. and T. Nelson 1988. Total and phytate phosphorus content of some feed ingredients derived from grains. *Nutr. Report Intl.* 37:277–280.

Klasing, K. C. and R. E. Austic. 2003. Nutritional diseases. In: Y. M. Saif (ed.) *Diseases of Poultry*. 11th edn. Iowa State Press, Ames, IA. pp. 1027–1053.

Kortbech-Olesen, R. 1999. Organic food and beverages: world supply and major European markets. *6th IFOAM Trade Conf*, Florence.

Kristensen, H. H. and C. M. Wathes. 2000. Ammonia and poultry welfare: a review. *Worlds Poult. Sci. J.* 56:235–245.

Kumar, D. and J. Gomes. 2005. Methionine production by fermentation. *Biotechnol. Adv.* 23:41–61.

Kumar, R. and J. P. F. D'Mello. 1995. Anti-nutritional factors in forage legumes. In: J. P. F. D'Mello and C. Devendra (eds) *Tropical Legumes in Animal Nutrition*. CAB International, Wallingford. pp. 95–134.

Lacassagne, L. 1988. Leguminous seeds on substitutes of soybean meal in poultry diets. *INRA Prod. Anim.* 1:47–57.

Lamine, C. and S. Bellon. 2009. Conversion to organic farming: a multidimensional research object at the crossroads of agricultural and social sciences. *A review. Agron. Sustain. Dev.* 29: 97–112.

Lampkin, N. 1997. *Organic Poultry Production*. The University of Wales, Aberystwyth, Cardiff. p. 90.

Lampkin, N., C. Foster, S. Padel, and P. Midmore. 1999. *The Policy and Regulatory Environment for Organic Farming in Europe*. Hohenheim, Stuttgart. p. 147.

Langer, S., P. W. D. Scislowski, D. S. Brown, P. Dewey, and M. F. Fuller. 2000. Interactions among the branched-chain amino acids and their effects on methionine utilization in growing pigs: effects on plasma amino- and keto-acid concentrations and branched-chain keto-acid dehydrogenase activity. *Br. J. Nutr.* 83:49–58.

Le Gall, M., L. Quillien, B. Seve, J. Gueguen, and J. P. Lalles. 2007. Weaned piglets display low gastrointestinal digestion of pea (*Pisum sativum L.*) lectin and pea albumin 2. *J. Anim. Sci.* 85:2972–2981.

Lenis, N. P., H. T. van Diepen, P. Bikker, A. W. Jongbloed, and J. van der Meulen. 1999. Effect of the ratio between essential and nonessential amino acids in the diet on utilization of nitrogen and amino acids by growing pigs. *J. Anim. Sci.* 77:1777–1787.

Liu, M., F. Hu, X. Chen, Q. Huang, J. Jiao, B. Zhang, and H. Li. 2009. Organic amendments with reduced chemical fertilizer promote soil microbial development and nutrient availability in a subtropical paddy field: the influence of quantity, type and application time of organic amendments. *Agric., Ecosyst. Environ., Appl. Soil Ecol.* 42:166–175.

Lumpkins, B. S. and A. B. Batal. 2005. The bioavailability of lysine and phosphorus in distillers dried grains with solubles. *Poult. Sci.* 84:581–586.

Maksimović, Z. J., I. Djujić, V. Jović, and M. Ršumović. 1992. Selenium deficiency in Yugoslavia. *Biol. Trace Elem. Res.* 33:187–196.

Malayoğlu, H. B., S. Özkan, S. Koçtürk, G. Oktay, and M. Ergül. 2009. Dietary vitamin E (α-tocopheryl acetate) and organic selenium supplementation: performance and antioxidant status of broilers fed n-3 PUFA-enriched feeds. *S. Afr. J. Anim. Sci.* 39:274–285.

Mangan, J. L. 1988. Nutritional effects of tannins in animal feeds. *Nutr. Res. Rev.* 1:209–231.

Matt, D., E. Veromann, and A. Luik. 2009. Effect of housing systems on biochemical composition of chicken eggs. *Agron. Res.* 7(Special issue II):662–667.

Matthews, G. G. 2003. Membrane potential: ionic steady state. In: G. G. Matthews (ed.) *Cellular Physiology of Nerve and Muscle*. 4th edn. Blackwell Publishing Ltd., Malden, MA. pp. 40–54.

McDermott, P. F., S. M. Bodeis, L. L. English, D. G. White, R. D. Walker, S. Zhao, S. Simjee, and D. D. Wagner. 2002. Ciprofloxacin resistance in *Campylobacter jejuni* evolves rapidly in chickens treated with fluoroquinolones. *J. Infect. Dis.* 185:837–840.

McDowell, L. R. 2000. *Vitamins in Animal and Human Nutrition.* 2nd edn. Iowa State University Press, Ames, IA. p. 809.

McKenzie-Parnell, J. M. and B. E. Guthrie. 1986. The phytate and mineral content of some cereals, cereal products, legumes, legume products, snack bars, and nuts available in New Zealand. *Biol. Trace Elem. Res.* 10:107–121.

Meng, X. and B. A. Slominski. 2005. Nutritive values of corn, soybean meal, canola meal, and peas for broiler chickens as affected by a multicarbohydrase preparation of cell wall degrading enzymes. *Poult. Sci.* 84:1242–1251.

Mertz, W. 1981. The essential trace elements. *Science* 213:1332–1338.

Messeguer, J., G. Peñas, J. Ballester, M. Bas, J. Serra, J. Salvia, M. Palaudelmàs, and E. Melé. 2006. Pollen-mediated gene flow in maize in real situations of coexistence. *Plant Biotechnol. J.* 4: 1–13.

Methionine Task Force. 2008. Transcripts of May 21, 2009, National Organic Standards Board Meeting, Baltimore, MD. Available at: http://www.ams.usda.gov/AMSv1.0/getfile?dDocName=STELPRDC506 9980 (accessed February, 2010).

Mihailović, V., A. Mikić, and B. Ćupina. 2007. Potential of annual legumes for utilisation in animal feeding. *Biotechnol. Anim. Husb.* 23:573–581.

Mikić, A., V. Perić, V. Đorđević, M. Srebrić, and V. Mihailović. 2009. Anti-nutritional factors in some grain legumes. *Biotechnol. Anim. Husb.* 25:1181–1188.

Mikkelsen, L. L., P. J. Naughton, M. S. Hedemann, and B. B. Jensen. 2004. Effects of physical properties of feed on microbial ecology and survival of *Salmonella* enterica serovar Typhimurium in the pig gastrointestinal tract. *Appl. Environ. Microbiol.* 70:3485–3492.

Na, J., J. Song, B. Lee, S. Lee, C. Lee, and G. An. 2004. Effect of polarity on absorption and accumulation of carotenoids by laying hens. *Anim. Feed Sci. Technol.* 117:305–315.

Nahm, K. H. 2007. Feed formulations to reduce N excretion and ammonia emission from poultry manure. *Bioresour. Technol.* 98:2282–2300.

Nalle, C. L. 2009. *Nutritional Evaluation of Grain Legumes for Poultry.* Diss. Massey University, Palmerston North.

National Research Council (NRC). 1994. *Nutrient Requirements of Poultry.* 9th edn. National Academy Press, Washington, DC. p. 176.

Nelson, D. L. and Cox, M. M. 2004. *Lehninger Principles of Biochemistry.* 4th edn. W.H. Freeman, New York. p. 1100.

Nishioka, T. and M. Irie. 2006. Fluctuation and criteria of porcine fat firmness. *Anim. Sci.* 82:929–935.

Norton, B. W. 2000. The significance of tannins in tropical animal production. In: J. D. Brooker (ed.) *Tannins in Livestock and Human Nutrition.* Australian Centre for International Agricultural Research, Canberra. pp. 14–23.

Nubé, M. and R. L. Voortman. 2006. Simultaneously addressing micronutrient deficiencies in soils, crops, animal and human nutrition: opportunities for higher yields and better health. Working Paper-06-02, Centre for World Food Studies, Amsterdam.

Nyachoti, C. M., J. L. Atkinson, and S. Leeson. 1997. Sorghum tannins: a review. *Worlds Poult. Sci. J.* 53:5–21.

O'Dell, B. L. and R. A. Sunde. 1997. Introduction. In: B. O'Dell, B. L., R. A. Sunde (eds) *Handbook of Nutritionally Essential Mineral Elements.* Marcel Dekker, Inc., New York. pp. 1–12.

Oenema, O., A. Bannink, S. G. Sommer, and L. Velthof. 2001. Gaseous nitrogen emissions from livestock farming systems. In: R. F. Follet and J. L. Hatfield (eds) *Nitrogen in the Environment: Sources, Problems, and Management.* Elsevier, New York. pp. 255–289.

Oksbjerg, N., K. Strudsholm, G. Lindahl, and J. E. Hermansen. 2005. Meat quality of fully or partly outdoor reared pigs in organic production. *Acta Agric. Scand. A Anim. Sci.* 55:106–112.

Omogbenigun, F. O., C. M. Nyachoti, and B. A. Slominski. 2003. The effect of supplementing microbial phytase and organic acids to a corn-soybean based diet fed to early-weaned pigs. *J. Anim. Sci.* 81:1806–1813.

Palacios, M. F., R. A. Easter, K. T. Soltwedel, C. M. Parsons, M. W. Douglas, T. Hymowitz, and J. E. Pettigrew. 2004. Effect of soybean variety and processing on growth performance of young chicks and pigs. *J. Anim. Sci.* 82:1108–1114.

Parsons, C. M., F. Castanon, and Y. Han. 1997. Protein and amino acid quality of meat and bone meal. *Poult. Sci.* 76:361–368.

Perez-Maldonado, R. A., P. F. Mannion, and D. J. Farrell. 1999. Optimum inclusion of field peas, faba beans, chick peas and sweet lupins in poultry diets. I. Chemical composition and layer experiments. *Br. Poult. Sci.* 40:667–673.

Phillips, R. L., J. Suresh, M. Olsen, and T. Krone. 2008. Registration of high-methionine versions of maize inbreds A632, B73, and Mo17. *J. Plant Reg.* 2:243–245.

Písaříková, B. and Z. Zraly. 2009. Nutritional value of lupine in the diets for pigs. *Acta Vet. Brno.* 78:399–409.

Ravindran, G., C. L. Nalle, A. Molan, and V. Ravindran. 2010. Nutritional and biochemical assessment of field peas (*Pisum sativum* L.) as a protein source in poultry diets. *J. Poult. Sci.* 47:48–52.

Razani, B., S. E. Woodman, and M. P. Lisanti. 2002. Caveolae: from cell biology to animal physiology. *Pharmacol. Rev.* 54:431–467.

Reeds, P. J. 2000. Dispensable and indispensable amino acids for humans. *J. Nutr.* 130:1835S–1840S.

Ritz, C. W., B. D. Fairchild, and M. P. Lacy. 2004. Implications of ammonia production and emissions from commercial poultry facilities: a review. *J. Appl. Poult. Res.* 13:684–692.

Russell, J. M. 2000. Sodium-potassium-chloride cotransport. *Physiol. Rev.* 80:211–276.

Rutherfurd, S. M. and P. J. Moughan. 2007. Development of a novel bioassay for determining the available lysine contents of foods and feedstuffs. *Nutr. Res. Rev.* 20:3–16.

Salunkhe, D. K., S. J. Jadhav, S. S. Kadam, and J. K. Chavan. 1982. Chemical, biochemical and biological significance of polyphenols in cereals and legumes. *Crit. Rev. Food Sci. Nutr.* 17:277–305.

Sebastian, S., S. P. Touchburna, and E. R. Chavez. 1998. Implications of phytic acid and supplemental microbial phytase in poultry nutrition: a review. *Worlds Poult. Sci. J.* 54:27–47.

Senkoylua, N. and N. Dale. 1999. Sunflower meal in poultry diets: a review. *Worlds Poult. Sci. J.* 55:153–174.

Smith, K. E., J. M. Besser, C. W. Hedberg, F. T. Leano, J. B. Bender, J. H. Wicklund, B. P. Johnson, K. A. Moore, and M. T. Osterholm, The Investigation Team. 1999. Quinolone-resistant *Campylobacter jejuni* infections in Minnesota, 1992–1998. *N. Engl. J. Med.* 340:1525–1532.

Srinongkote, S., M. Smriga, and Y. Toride. 2004. Diet supplied with L-lysine and L-arginine during chronic stress of high stock density normalizes growth of broilers. *Anim. Sci. J.* 75: 339–343.

Steiner, T., R. Mosenthin, B. Zimmermann, R. Greiner, and S. Roth. 2007. Distribution of phytase activity, total phosphorus and phytate phosphorus in legume seeds, cereals and cereal by-products as influenced by harvest year and cultivar. *Anim. Feed Sci. Technol.* 133:320–334.

Stevenson, M. H. and N. Jackson. 1981. The nutritional value of dried skim milk in broiler diets. *J. Sci. Food Agric.* 32:79–86.

Sun, B., C. H. Leem, and R. D. Vaughan-Jones. 1996. Novel chloride-dependent acid loader in the guinea-pig ventricular myocyte: part of a dual acid-loading mechanism. *J. Physiol.* 495:65–82.

Sundrum, A., K. Schneider, and U. Richter. 2005. Possibilities and limitations of protein supply in organic poultry and pig production. Research to support revision of the EU Regulation on organic agriculture SSPE-CT-2004-502397. Available at: http://www.organic-revision.org/pub/Final_Report_EC_Revision.pdf (accessed February, 2010).

Swennen, Q., G. Janssens, S. Millet, G. Vansant, E. Decuypere, and J. Buyse. 2005. Effects of substitution between fat and protein on feed intake and its regulatory mechanisms in broiler chickens: endocrine functioning and intermediary metabolism. *Poult. Sci.* 84:1051–1057.

Tavčar-Kalcher, G. and A. Vengušt. 2007. Stability of vitamins in premixes. *Anim. Feed Sci. Technol.* 132: 148–154.

Taylor, E. J., H. M. R. Nott, and K. E. Earle. 1994. Dietary glycine: its importance in growth and development of the budgerigar (*Melopsittacus undulatus*). *J. Nutr.* 124:2555S–2558S.

Tesseraud, S., R. Peresson, J. Lopes, and A. M. Chagneau. 1996. Dietary lysine deficiency greatly affects muscle and liver protein turnover in growing chickens. *Br. J. Nutr.* 75:853–865.

Tesseraud, S., S. Temim, E. Le Bihan-Duval, and A. M. Chagneau. 2001. Increased responsiveness to dietary lysine deficiency of pectoralis major muscle protein turnover in broilers selected on breast development. *J. Anim. Sci.* 79:927–933.

Thacker, P. A. 1990. Faba beans. In: P. A. Thacker and R. N. Kirkwood (eds) *Non-traditional Feed Sources for Use in Swine Production*. Butterworths, Stoneham, MA. pp. 175–184.

Thamsborg, S. M. 2001. Organic farming in the Nordic countries – animal health and production. *Acta Vet. Scand. Suppl.* 95:7–15.

Thomsen, M. K., L. Lauridsen, L. H. Skibsted, and J. Risbo. 2005. Temperature effect on lactose crystallization, Maillard reactions, and lipid oxidation in whole milk powder. *J. Agric. Food Chem.* 53:7082–7090.

Torre, M., A. R. Rodriguez, and F. Saura-Calixto. 1991. Effects of dietary fiber and phytic acid on mineral availability. *Crit. Rev. Food Sci. Nutr.* 1:1–22.

UKROFS. 2003. United Kingdom register of organic food standards. Available at: http://archive.defra.gov.uk/foodfarm/growing/organic/standards/acos/ukrofs/ (accessed September 03, 2010).

Underwood, E. J. and N. F. Suttle. 2001. *The Mineral Nutrition of Livestock*. CABI Publishing, New York. p. 624.

Undi, M., S. A. Moshtaghi-Nia, K. M. Wittenberg, and J. R. Ingalls. 1996. A comparative study on amino acid availability of moist heated canola meal for poultry versus ruminants. *Anim. Feed Sci. Technol.* 63:179–186.

USDA National Organic Program (NOP). 2008. Agricultural Marketing Service 7, Code of Federal Regulations (CFR), Part 205: the National Organic Program. Available at: http://ecfr.gpoaccess.gov/cgi/t/text/text-idx?c=ecfr&sid=65c6ded529144df1a6c1cba52ad42c61&tpl=/ecfrbrowse/Title07/7cfr205_main_02.tpl.

USDA. 2010. 7 CFR Part 205. National Organic Program, Amendment to the National List of Allowed and Prohibited Substances (Livestock). *Fed. Regist.* 75:51919–51924.

van Boekel, M. A. J. S. 1998. Effect of heating on Maillard reactions in milk. *Food Chem.* 62:403–414.

van Keulen, H. and W. Stol. 1991. Quantitative aspects of nitrogen nutrition in crops. *Fert. Res.* 27:151–160.

Vasanthakumar, T., R. A. Rajini, and C. Sarath. 2009. Yolk colour carotenoid and retinol level in the eggs of rural, urban range reared and cage reared chicken in two seasons. *Indian J. Poult. Sci.* 44:0974–8180.

Viveros, A., C. Centeno, A. Brenes, R. Canales, and A. Lozano. 2000. Phytase and acid phosphatase activities in plant feedstuffs. *J. Agric. Food Chem.* 48:4009–4013.

Viveros, A., C. Centeno, I. Arija, and A. Brenes. 2007. Cholesterol-lowering effects of dietary lupin (*Lupinus albus* var Multolupa) in chicken diets. *Poult. Sci.* 86:2631–2638.

Waguespack, A. M., S. Powell, T. D. Bidner, and L. L. Southern. 2009. The glycine plus serine requirement of broiler chicks fed low-crude protein, corn-soybean meal diets. *J. Appl. Poult. Res.* 18:761–765.

Wang, T. L., C. Domoney, C. L. Hedley, R. Casey, and M. A. Grusak. 2003. Can we improve the nutritional quality of legume seeds? *Plant Physiol.* 131:886–891.

Webb, K. E., Jr. 1990. Intestinal absorption of protein hydrolysis products: a review. *J. Anim. Sci.* 68:3011–3022.

Wilson, S. 2003. Feeding animals organically-the practicalities of supplying organic animal feed. In: P. C. Garnsworthy and J. Wiseman (eds) *Recent Advances in Animal Nutrition*. University of Nottingham Press, Nottingham. pp. 161–172.

Wu, Z., L. D. Satter, A. J. Blohowiak, R. H. Stauffacher, and J. H. Wilson. 2001. Milk production, estimated phosphorus excretion, and bone characteristics of dairy cows fed different amounts of phosphorus for two or three years. *J. Dairy Sci.* 84:1738–1748.

Yen, J. T., J. E. Wells, and D. N. Miller. 2004. Dried skim milk as a replacement for soybean meal in growing-finishing diets: effects on growth performance, apparent total-tract nitrogen digestibility, urinary and fecal nitrogen excretion, and carcass traits in pigs. *J. Anim. Sci.* 82:3338–3345.

Zijlstra, R. T., K. Lopetinsky, B. Dening, G. S. Bégin, and J. F. Patience. 2004. The nutritional value of zero-tannin faba bean for grower-finisher pigs. *Can. J. Anim. Sci.* 84:792–793.

11 Production of Forage Crops Suitable for Feeding Organically Raised Meat Animals

Ivan Manolov and Christina Yancheva

Abstract: This chapter briefly reviews how to produce sufficient amount of high-quality forage suitable for the feeding of organically raised meat animals. Possible solutions for securing of forage balance on the farm are well-designed crop rotations. These include rotations of cereals, legumes, and leys. Intercropping, undersowing, and green manuring are also recommended. Maintaining of soil fertility by measures for increasing fixed nitrogen in the soil, mobilization of soil phosphorus and potassium reserves, and manure application, is an important feature for sustainability of forage production in the farm. Pasture management is another important element that influences the sustainability of feeding the organic animals. The main leguminous crops (protein source) and grasses (energy source) are discussed as well. Whole crop forage (fresh or conserved) complements forage balance of an organic stock-breeding farm.

Keywords: organic forage crops; cereals; crop rotations; soil fertility; leys; pastures

11.1 INTRODUCTION

Organic meat can be produced successfully by farmers only when they are provided with sufficient amounts of high-quality feedstuffs for livestock. Supplying sufficient energy and the required protein level for a livestock farm is a key factor in the management of organic systems (Weller & Jones, 2002). Feedstuffs should ensure enough energy, mainly in the form of carbohydrates and cellulose, proteins, and minerals for high-quality meat production. Forage crops that are rich in carbohydrates are considered primary energy sources. They include cereal grains and their by-products, such as barley (*Hordeum vulgare* L.), triticale (X *Triticosecale* Wittmack), rye (*Secale cereale*), oats (*Avena sativa*), wheat (*Triticum aestivum* L.), spelt (*Triticum spelta*), grain, and fodder maize (*Zea mays* L.). The major protein sources are legumes, namely soybeans (*Glycine max*), vetch (*Vicia* sp.), field peas (*Pisum* sp.), alfalfa (*Medicago sativa*), red clover (*Trifolium pratense* L.), white clover (*Trifolium repens*), faba beans (*Vicia faba* L.), and lupine (*Lupinus albus*) (Blair, 2007).

However, pastures are the foundation of organic livestock production as they enable herbivorous animal nutrition in harmony with the environment and welfare (Haddad & Alves, 2002). Although, to a lesser extent, they are also considered necessary for the feeding of nonherbivore animals such as pigs and poultry when permanent access to open air areas is

Organic Meat Production and Processing, First Edition. Edited by Steven C. Ricke, Ellen J. Van Loo, Michael G. Johnson, and Corliss A. O'Bryan.

desired, preferably pasture (Council regulation (EC) No 834/2007). During critical periods of the year (winter in temperate regions and wet seasons in tropics), silage and hay are important forages. International and national organic farming regulations require all feedstuffs to be produced organically preferably on the same farm where food animals are reared or at least be grown or available locally (US National Organic Program Regulations 2000; Codex Alimentarius—Organically Produced Foods, 2001; Blair, 2007; Council regulation (EC) No 834/2007). Therefore, moving to 100% organic rations demands that farmers attempt to produce more feed from their own resources. When growing cereals and forages, consideration should be given to organic crop rotations, weed management, green manure (GM), cover crops (CCs), undersown crops, intercropping, and pasture management. The criteria for a successful cropping system includes sustaining soil fertility, providing a balance between nitrogen (N_2)-fixing and N-demanding crops, producing high-quality feed, and improving the efficiency of nutrient utilization to meet the annual requirements of the livestock farm.

11.2 CROP ROTATIONS

Organic farming systems are based on organically manured, legume-based multiannual crop rotations utilizing organic fertilizers from livestock raised on organic-based farms. Rotations should be capable of producing sufficient forage for livestock and preserving soil fertility without the use of synthetic fertilizers. Growing legumes is the main method for enriching soil with nutrients, especially for N in organic farming (Talgre *et al.*, 2009). Crop rotations ensuring forages for meat production usually include more than 30%–50% of the area as permanent grasses and arable land for cereals, seed legumes, and other forages (Lampkin, 1999; Watson *et al.*, 2008; Möller, 2009). For meat production farms, establishment of high-input rotations with a grass–clover crop and catch crops (fast-growing crop that is grown simultaneously with, or between, successive plantings of a main crop) versus low-input cereal rotations without catch crops is recommended (Vinther *et al.*, 2004). These same researchers reported high N inputs into the soil ranging from 113 to 168 kg N/ha/year at high-input rotations mainly due to N_2 fixation by clover, while the same parameter was considerably lower at low-input cereal rotations varying from 36 to 39 kg N/ha/year (Table 11.1).

A typical crop rotation on a mixed organic farm in Northwest Europe and Australia includes 3-year grass–clover ley (arable land put down to grass, clover, etc., for a single season or a limited number of years, in contrast to permanent pasture land) (Watson *et al.*, 2008; Thomas *et al.*, 2009). The N storage from the ley could support 2 or 3 years of arable cropping. In addition, soil N reserves could be extended by including N-fixing cash crop, such as beans, or by including a short period N_2-fixing GM, such as vetch, between cash crops. Nutrient and particularly N availability may also be enhanced by soil cultivation since it stimulates microbial activity and organic matter breakdown (Silgram & Shepherd, 1999). Examples of crop rotations suitable for mix crop–livestock farms in different countries with various conditions are presented in Table 11.1.

The producers of organic meat increase the amount and quality of produced feedstuff from a rotation by the inclusion of intercropping practices, GM, CCs, and undersowing (Table 11.1).

Table 11.1 Examples of high-input crop rotations for mixed crop–livestock farms.

Crop rotation	Country	Reference
Spring barley/grass–clover ley[a], winter wheat/grass, pea/barley–grass	Denmark	Vinther *et al.*, 2004
Winter wheat: grass, pea/barley: grass, spring barley: grass–clover ley		
Grass–clover, potatoes, field beans, winter wheat, spring wheat, winter rye (cereals were undersown with red clover–grass mixture)	Germany	Zaller & Köpke, 2004
Winter wheat, oat–clover intercrop, sunflower, spelt, grass–clover, silage maize	Switzerland	Berner *et al.*, 2008
Barley, grass–clover, grass–clover, barley/pea, winter wheat, fodder beets	Denmark	Askegaard & Eriksen, 2002
3-years grass–clover ley, winter wheat, winter triticale	UK beef farm	Berry *et al.*, 2003
2-years grass–clover ley, winter wheat, spring cereal, winter wheat	UK pigs/sheep farm	
2-years grass–clover ley, winter wheat, spring wheat, winter cereal	UK pigs farm	
2-years grass–clover ley, winter wheat, winter oats, winter beans, winter wheat, spring barley	UK chickens	
3-years grass–clover ley, winter wheat, winter oats, winter beans, winter wheat, spring oats	UK beef/sheep farm	
Pea, winter wheat, undersown with clover–grass, set-aside, winter rape, field bean, winter wheat, undersown with clover–grass, set-aside, winter rye	Sweden	Helander, 2004
Potato–spring barley intercrop, clover–grass, clover–grass, winter wheat/aftercrop	Poland	Stalenga, 2007
Oat + insown green manure (GM), GM, spring wheat, oat + insown GM, GM, potato 1 rye (cover crop (CC))	Sweden	Torstensson *et al.*, 2006
Clover–grass ley, clover–grass ley, winter wheat + CC, silage maize, winter rye + CC	Germany	Möller, 2009
Barley + insown red clover, red clover, winter wheat + insown red clover, beans + insown red clover, potato, peas + insown ryegrass	Sweden	Kirchmann *et al.*, 2007

[a]Arable land put down to grass, clover, etc., for a single season or a limited number of years, in contrast to permanent pasture.
CC, cover crop; GM, green manure.

11.3 INTERCROPPING

Intercropping of two or more crops represents an optimal strategy for stabilization of yields, better exploitation of natural resources such as light, water, and nutrients, and increase of agro-biodiversity. The optimum intercrop plant densities may be greater than the optimum density for sole crops because intercrop components might utilize the growth resources more efficiently than sole crops (Hauggaard-Nielsen *et al.*, 2006). A yield advantage of intercropping occurs as a result of complementary use of the resources of an intercropping system composed of two or more components. For example, combinations of cereals and legumes demonstrate increased use of symbiotically fixed N_2, increased field productivity, protein quality, and digestibility of the forages without compromising grain yield (Bulson *et al.*, 1997; Jensen, 1996, 2006; Möller *et al.*, 2008; Ksiezak & Staniak, 2009; Lauk & Lauk, 2009). Legume–cereal intercrops facilitate direct transfer of fixed N from legumes to cereals. In field conditions, by using an ^{15}N-labeled technique, Patra *et al.* (1986) determined that 28% of N uptake by maize (21.2 kg N/ha) was obtained by transfer of N fixed by cowpea.

Intercropping of grain legumes with cereals might be a method to improve the crops utilization of soil inorganic N. It can improve yield potentials, and at the same time minimize N leaching during autumn and winter as a result of the cereal root system and residual chemical composition. Intercropping is especially valuable for the production of protein-rich crops on land with weed problems. It enhances the grain protein concentration in cereals to levels typical for cereals heavily fertilized with animal manure (Jensen, 2006). Yield advantage of cereal–cereal and cereal–legume intercropping systems was confirmed by Song *et al.* (2007) in an experiment carried out in China with wheat, maize, and faba bean grown in the field solely and intercropped (wheat–faba bean, wheat–maize, and maize–faba bean). The intercrops generated up to 30% higher grain yields compared to sole crops. The intercropping of cereals with legumes, considerably increased protein yields (approximately 570 kg/ha) in comparison to sole cereal crops (approximately 400 kg/ha) (Möller *et al.*, 2008; Lauk & Lauk, 2009).

Cultivation of legume–cereal intercrops reduces competition among cereal plants for N, which leads to increases in N levels, and as a result, protein contents. Hauggaard-Nielsen *et al.* (2009a) found that pea–barley intercrops accumulate 25%–30% more N in their grains compared to respective sole crops in simultaneous experiments conducted in Denmark, United Kingdom, France, Germany, and Italy. When the amount of cereals is higher compared to that of legumes, increased competition for soil N was observed (Karpenstein-Machan & Stuelpnagel, 2000).

11.4 GREEN MANURE AND COVER CROPS

GM and CCs are important additional forage sources for meat-producing farms. The terms GM and CC are often used synonymously (Baggs *et al.*, 2000). The authors determined CCs as the crops grown during the winter in arable rotation aiming to reduce soil erosion, leaching of nutrients, and gaseous N emissions. The main purpose of GM crops is to improve soil fertility by increasing soil organic matter and other nutrients incorporated into the soil with plant biomass, particularly N fixed by leguminous crops. The most widely used GM legume crops are clovers (*Trifolium* sp.), alfalfa (*M. sativa*), trefoils (*Lotus* sp.) and other medics (*Medicago* sp.), vetches (*Vicia* sp.), and lupins (*Lupinus* sp.). Rye (*S. cereale*), oat (*A. sativa*), wheat (*Triticum* spp.), oil radish (*Raphanus sativus*), mustards (*Brassica* spp.), and buckwheat (*Fagopyrum esculentum Moench*) are the most often seeded nonlegume crops.

GM and CC reduce soil erosion and nutrients losses by leaching when they are used in place of fallow periods in crop rotations, when weather is unfavorable for optimal production of more economical crops (Dapaah & Vyn, 1998; Baggs *et al.*, 2000; Cherr *et al.*, 2006; Möller *et al.*, 2008; Hauggaard-Nielsen *et al.*, 2009b). Cultivation of leguminous or biculture legume and nonlegume CCs increases N inputs in the soil and decreases leaching of nitrates (Sainju *et al.*, 2007; Möller *et al.*, 2008; Sainju & Singh, 2008). The delay of the CC incorporation until winter diminishes the risk of nitrate leaching, even when legumes were used as CC, while a CC incorporation in autumn will increase the nitrate-leaching risk dramatically (Baggs *et al.*, 2000; Mitchell *et al.*, 2000; Möller *et al.*, 2008). Grazing or harvesting of CC prior to its incorporation will decrease nitrate-leaching risk. Bergström and Kirchmann (2004) found that interseeded ryegrass (*Lolium perenne* L.) and red clover (*T. pratense* L.) decrease leaching loads from spring barley (*H. vulgare* L.). Leguminous CCs are usually sown after the main grain crop decreases residual soil N levels after harvesting.

GM and CCs are good management tools for suppression of weed development (Bàrberi, 2002; Blaser *et al.*, 2006; Perry & Galatowitsch, 2006). Usually, the increased density of interseeded crops increases suppression of weeds. GM or CC is typically plowed into the soil. When the farm is specialized in meat production, above-ground biomass from CCs can be used for animal feeding. Möller *et al.* (2008) reported 18% higher N content in the dry matter from legumes than nonlegume crops. Therefore, a higher percent of leguminous crops in CC composition will increase overall N content in the above-ground biomass, and as a result, the protein content. The legume crop in the mixture contains more phosphorus (P) in biomass dry matter while nonlegume crops take up more potassium (K) from the soil. Therefore, leguminous crops can mobilize higher amounts of inorganic P from the soil, and a greater proportion will be available for the next crop after slow mineralization of ploughed biomass into the soil. Incorporation of fresh organic matter in the soil has a favorable effect on the soil biota and the biological activity, which improves mineral nutrition of the ensuing crop (Cherr *et al.*, 2006). The timing and amount of N mineralization from residues following GM generally have major effects on subsequent crop yields.

GM could be cultivated as pure crops, intercrops, or undersown crops or natural vegetation (weeds) can even be used for this purpose (Cherr *et al.*, 2006; Sainju & Singh, 2008). In this situation, the amount of N storage in vegetation, particularly incorporation into the soil is the greatest in legume and legume–cereal crops followed by cereals and the natural weeds, which have the least impact on N accumulation. In an experiment conducted in Georgia, USA, Sainju and Singh (2008) estimated that N input varied between 70 and 108 kg N/ha/year in vetch and vetch–rye GM compared to N contents in winter weeds. In an independent study performed in Scotland, no significant differences in N utilization and N recovered between naturally regenerated vegetation and sown CCs were established (Baggs *et al.*, 2000).

11.5 UNDERSOWING

Undersowing is practiced to ensure additional forage for livestock. After harvesting the main crop, it protects the soil from erosion and adds additional amounts of fixed N into the soil. Legumes such as clover, trefoil, or alfalfa are the most utilized crops for undersowing winter cereals (barley, wheat, oats, etc.) because they increase N availability in soil for the main cereal crop (Möller *et al.*, 2008). Undersowing of clover into cereals is a common practice for establishing leys (Taylor *et al.*, 2001). Ryegrasses are another option for undersowing crops. However, cultivation of ryegrasses can increase competition of the next crop for N because of the immobilization of the element during the decomposition of the ryegrass in the soil. Helander (2004) noted a 25% decrease of grain yield of oats when undersown with wheat/ryegrass. Significant positive effects were detected for oats grain yield from undersown clover (white and red) that reached an approximately 1500 kg/ha higher yield than the control.

Legumes and grasses are sown in the spring by drilling or simply by spreading seeds over growing cereal crop. The seeding rates of undersowings are reduced by half compared to the normal ones typically utilized for pure crops (Talgre *et al.*, 2009). The optimum seeded rates for triticale and wheat in the North Central regions of the United States, from 300 to 400 seeds m^{-2}, or seeds per m^2 while seeded densities of red clover must be above 120 plants per m^2. At that seeding rate, sufficient photosynthetically active radiation for the growth of red clover is transmitted throughout the grain canopy of cereals (Blaser *et al.*, 2006). Interseeded legume CCs—red clover and chickling vetch (*Lathyrus sativus* L.)—into corn fields in Michigan, USA, did not affect corn yield among any of the corn densities. In addition, increasing of corn plants per unit area did not influence germination and height of CCs but decreased their dry mass (Baributsa *et al.*, 2008).

After harvesting cereals, the undersown crop works as a CC during the autumn and winter prior to sowing of the following crop. Cereal harvest improves conditions for faster growth of undersowings since no competition for nutrients and water in soil and no overshadowing over the ground occurs, thus ensuring feed for grazing livestock. The success of undersown crop requires sufficient precipitation and soil moisture during seed germination. Consequently, this practice is more popular in temperate regions of the world.

Undersown crops (particularly ryegrass) are one of the most effective measures to control nitrate leaching, particularly on sandy soils. Eriksen *et al.* (2008) observed reduction of leached N from 60 to 9 kg/ha after plowing grazed unfertilized grass–clover and growing spring barley with undersown ryegrass. However, legume undersown crops are valuable N sources for grain production on a coarse sandy soil. White and red clover provided 120 and 103 kg N/ha, respectively, for the following unfertilized spring barley (Askegaard & Eriksen, 2007).

11.6 WEED MANAGEMENT

The prohibition of herbicide application generates serious problems with weed management in organic farming. Long-term effects of herbicides must be replaced by preventive approaches that will restrict dissemination and propagation of weeds (Christoffoleti *et al.*, 2007) and involve cultural weed management practices, most of which have a shorter term effect on weed control (Blackshaw *et al.*, 2007). The weed problem has to be examined from different angles that are focused on optimizing the whole cropping system. Emphasis has to be put on methods preventing weed infestation of fields. The most effective ones include but are not limited to a suitable crop sequence in more diverse crop rotation, soil tillage and smothering CCs, inclusion of grass–clover mixtures, cultivation of crops with increased competitiveness to weeds (quickly germinating and growing crops, and taller varieties that shade the weeds), balanced application of organic fertilizers, and harvesting of silage crops before weeds produce mature seeds (Bàrberi, 2002; Pekrun & Claupein, 2006; Blackshaw *et al*., 2007; Lundkvist *et al*., 2008; Østergård *et al*., 2008; Gruber & Claupein, 2009; Uchino *et al*., 2009). Cereals such as rye, barley, wheat, spelt, and triticale represent some of the best crops for suppressing weed growth because these cereal crops tend to produce a greater root biomass and have a higher root to shoot ratio at the beginning of the crop cycle (Jensen, 2006). Effective strategy against weeds includes maximum diversification of the cropping system.

11.7 SOIL FERTILITY

Long-term maintenance of soil fertility is a considerable challenge for organic producers. Breeding of livestock and the application of manure in the mixed farms help to close the circle of nutritional elements to a large extent compared to crop only-based farms. The most problematic nutritional element is N. Fixing N from legumes and manure are the main N inputs for organic system.

N, in the form of crude protein, is a critical component of the feed for livestock, as well as an important nutritional element for pasture and forages (Kleinman & Soder, 2008). Kirchmann *et al*. (2007a, 2007b) outlined N deficiency as a major yield-limiting factor in organic farming. In meat-producing farms, this problem is solvable because organic grassland–ruminant systems enable a high input of N to the soil that determines approximately equal yields from organic and conventional grass–clover forage crops (Eltun *et al*., 2002). Inclusion of more legume crops, grass–legumes leys, leguminous cover and GM crops, undersowing of some field crops with trefoil, and intercropping cereal–legume mixtures in the crop rotations increases the amount of N and its uniform distribution in the frame of rotation. In six out of seven livestock farms in United Kingdom studied by Berry *et al*. (2003), positive N balances due to large proportion of grass–clover leys and other leguminous crops in the rotation were found. The best N fixers were perennial legumes—red and white clover fixing up to 250 kg N/ha/year (Kristensen *et al*., 1995; Schmidt *et al*., 1999) and alfalfa, up to 500 kg N/ha/year (Spiertz & Sibma, 1986). In temperate environments, winter-hardy legumes such as vetch, clover, and medics are capable of accumulating large amounts of biomass (7–10 t/ha) and N (150–250 kg N/ha), and delivering substantial N benefit to subsequent spring-planted crops. Long-term management of soil under organic conditions increased soil total N content of up to 21% mainly due to simultaneous increase of soil organic matter varying from 6% to 34% (Erhart & Hartl, 2009).

The translocation of nutritional elements among the farmland subcompartments, arable land and permanent grassland, occurs through the manure (Haas *et al*., 2002). The nutrient composition of the manure depends mainly on the forages that are utilized. Incoming nutrients from fodder are mixed in the manure and afterwards reallocated to the fields. The concentrations of N and K in roughages harvested on grassland are very high, but is relatively low in P (8.69 kg N per kg P and 7.88 kg K per kg P), whereas concentrates derived from arable land are characterized by high P levels, intermediate or high N, and relatively low K contents (1.25–2.36 kg K per kg P) (Möller, 2009). The application of the greater part of manure produced on the farm on arable land leads to withdrawal and transfer of nutrients (mostly N and K) from grassland to areas with field crops. Manure is preferably applied to field crops before deep tillage, which helps the incorporation of manure into the soil and reduces losses of N as NH_3. Long-term application of manure has been shown to significantly increase water-stable macroaggregates, potentially mineralizable N and soil preseeding NO_3–N levels (Nyiraneza *et al.*, 2009). Surface application of manure to pastures temporarily decreases the palatability of forages and substantially increases environmental losses of nutrients (Kleinman & Soder, 2008). The amount of applied manure and slurry in Europe is limited to 170 kg N/ha/year by EU Regulation 2092/91.

The studies on soil phosphorus status of organic farms in conditions of short- and long-term experiments have revealed negative P budget and depletion of available phosphorus (Oehl *et al.*, 2002; Deria *et al.*, 2003; Gosling & Shepherd, 2005; Torstensson *et al.*, 2006; Goulding *et al.*, 2008; Miller *et al.*, 2008; Welsh *et al.*, 2009). The P budgets from "DOK" (bio-Dynamic, bio-Organic, and "Konventionell") system comparison trial, which is based on a ley rotation in Switzerland were negative after a 21-year period for both organic (ORG) and bio-dynamic (DYN) systems with decreases of 5.7 kg P/ha/year and 7.8 kg P/ha/year, respectively (Oehl *et al.*, 2002). By comparing organic rotations, Welsh *et al.* (2009) determined greater P export from forage–grain rotation because of the higher P requirements of leguminous crops. Deria *et al.* (2003) determined that 3–10 kg P/ha was removed from the wheat fields without being replaced by other sources. The low contents of available phosphorus in some environments restrict the development of N_2-fixing crops as they are sensitive to P deficiency and therefore, decrease the quality of forages for livestock. Higher levels of available P in organic soils could be maintained by increased or more frequent manure applications and/or fine-ground rock phosphates.

The effect of rock phosphates on crop P nutrition has a delayed action (Kitchen *et al.*, 2003; Ryan *et al.*, 2004). Low-soluble phosphates should preferably be allocated on grasslands fields of crop rotation where a long growing period (root exudates) and improved biological activity of the soil have a beneficial effect on P mobility (Möller, 2009). The combined application of ground rock phosphates and elemental sulfur increased P reserves, the amount of available P, soil productivity of permanent pastures, and cereal–pasture rotations (Evans *et al.*, 2006; Vidrih *et al.*, 2007). The application of large quantities of basic slag and apatite on soil with a very low P fertility status and pH 5–6 in organic system in Sweden increased the amount of soil-exchangeable P from 10 mg P/kg soil at the control level to 40 mg P/kg soil for a period of 5 years (Kirchmann *et al.*, 2007a). The farmers can apply acceptable P mineral sources only when they can convince the certifying body of the crucial need for P fertilization by submission of soil analysis or by presentation of a nutrient budget.

Nutrient acquisition of P and some micronutrients for plants are facilitated by arbuscular mycorrhizal fungi (AMF) (Smith & Read, 1997; Douds & Millner, 1999; Lampkin, 1999; Jansa *et al.*, 2002; Oehl *et al.*, 2004; Ryan & Tibbett, 2008). Comparatively low amounts of available P in organic managed soils are another stimulus for better development of AMF. Inclusion of cover/GM crops with deep rooting systems such as lupin, common vetch, and others in the rotation has been shown to additionally increase the quantity of available P for the crops that follow (Möller *et al.*, 2008).

Data generated from organic long-term experiments usually demonstrate negative potassium nutritional balances (Berry *et al.*, 2003; Kirchmann *et al.*, 2007a). Despite the large quantities of total K in soil, K reserves (exchangeable and fixed K) become depleted after long-term cropping (Øgaard *et al.*, 2001). Berry *et al.* (2003) found that only three of the nine organic farms examined in the United Kingdom exhibited a positive K budget from 9 to 28 kg K/ha/year for a period of 5 years. A positive K budget was achieved due to substantial inputs of manure to the fields obtained from livestock fed with imported concentrates. Negative balances for the other farms were within the range of 21–52 kg/ha/year. The study revealed that K inputs via manure are not sufficient to balance the removal in silage and other crops in livestock organic farms. Increasing farm stocking rate in a mixed system from 0.9 to 1.4 livestock units (LU) per hectare changed the K budget from negative to positive because of the K return derived from increased manure application (Askegaard & Eriksen, 2002). According to Kayser *et al.* (2009), application of 15–30 t/ha manure per rotation is considered to be the best approach for compensating the large K offtake caused by harvested crops and is a comparatively easily achieved goal for livestock farms.

11.8 CEREAL CROPS

Cereal crops are grown for grain and straw production. Availability of winter and spring varieties makes cereals appropriate for incorporation in various crop rotations. Cereal grain is a high-energy and low-protein feed suitable for inclusion in concentrated mixtures for different livestock groups. Grain can be stored for a relatively long period after drying to a moisture content less than 14%. Cereal grains have a reliable nutritive quality and a higher energy value than forage crops. The digestibility of organic cereals varies from 72% for oats to 92% for wheat and triticale. They contain 10%–11% crude protein and 60%–68% starch (Hancock *et al.*, 2003). Straw can be used for bedding and in small proportions be included in the ration of ruminants.

The problem in an organic system is the lower yields of cereals (Table 11.2) as compared to yields from conventional production. The low organic wheat grain yields can be explained by several factors, such as N deficiency, water deficit, weeds, pests and diseases, or compacted soil structure (Taylor *et al.*, 2001). Organic wheat production is also characterized by variable yields and grain protein contents ranging from 1 to 6.5 t/ha and from 7 to 13 g per 1000 g, respectively, in a single region (David *et al.*, 2004). Most of the studies conducted thus far have revealed an approximately 24% yield reduction in mixed crop–animal farms. The yield decrease is much higher (47%) when the farm is specialized in crop production only (Kirchmann *et al.*, 2007a). The productivity of crops under organic or conventional conditions depends on the type of the crop. Usually, the annual and perennial leguminous crops and grass–clover mixtures are comparatively more productive in comparison to cereals and other forage crops under organic cultivation (Kirchmann *et al.*, 2007a).

After the review of a few studies, Kirchmann *et al.* (2007a) identified N deficiency as the main reason for the diminished yield in organic systems. The N inputs in organic crop rotation are not equally distributed and not well adapted for the following crop. Years of high N inputs following N_2-fixing crops can be alternated with years of little or no N inputs after a cereal crop growing cycle (Kirchmann & Ryan, 2004). Only a system with a very large proportion of N_2-fixing forage crops such as an organic grassland–ruminant system is capable of maintaining sufficient N input to the soil and is able to sustain high yields from cultivated crops (Eltun *et al.*, 2002). The increased proportion of ley in the crop rotations combined with reduced cropping intensity lessens environmental damage on the organic system as it reduces soil erosion, N and P runoff losses. It is easier to maintain relatively high yield levels in mixed farming systems with livestock than in a stockless system.

Low productivity of organic managed crops could be due to the cultivation of varieties that were selected for conventional agriculture to be grown with high inputs of mineral fertilizers. Screening of varieties has been based on grain yield as the main criterion, which sometimes leads to greater differences in quality between varieties within a cereal species than among

Table 11.2 Mean yield (t/ha) and yield decrease (%) of cereals in organic and conventional farming systems.

Crop	Organic yield (t/ha)	Conventional yield (t/ha)	Yield decrease (%)	Remarks	Country	Reference
Wheat	5.6	9.3	40		France	Pelosil *et al.*, 2009
Wheat	4.1	4.6	11	DOK trial, 21 years	Switzerland	Mader *et al.*, 2002
Wheat	2.95	3.14	6	Mean for 7 years	Maryland, USA	Teasdale *et al.*, 2007
Maize	4.89	7.06	31			
Spring wheat	1.12	3.28	66	Grain crop rotation	Canada	Welsh *et al.*, 2009
Spring wheat	2.02	3.51	42	Forage–grain crop rotation		
Wheat	4.33	6.26	31	Variety Kobra, 3 years	Poland	Stalenga, 2007
Wheat	3.44	5.86	41	Variety Juma		
Wheat	4.20	6.08	31	18 years	Sweden	Kirchmann *et al.*, 2007b
Barley	2.10	3.75	44			
Barley	2.13	4.06	49	6 years' crop rotation	Sweden	Torstensson *et al.*, 2006
Oats	4.43	5.39	18			
Maize	9.94	11.79	16	Mean from 9 lines	Germany	Burger *et al.*, 2008

DOK, bio-Dynamic, bio-Organic, and "Konventionell."

species (Hancock *et al.*, 2003). Recently, breeding has been focused on developing varieties suitable for organic farming with efficient use of nutrients and water, weed stability, resistance to disease and pests, stable yields, and good quality of grain (Charmet *et al.*, 2005; Burger *et al.*, 2008; Hoad *et al.*, 2008; Wolfe *et al.*, 2008).

It is possible that there may be larger differences in quality between varieties within a cereal species than between species, as screening of varieties has been based on grain yield as the main criterion. Therefore, a further study on the forage potential of the currently available cereal species and varieties could lead to farmers being able to select and grow crops that can be classified as high-energy feeds rather than moderate energy sources.

11.9 FODDER CROPS

11.9.1 Legumes

Forage legumes play an important role in organic farming not only as forage but also as a valuable source of N and means of preventing weeds development and soil-borne diseases in the following arable crop (Younie & Hermansen, 2000). Forage legumes have been the foundation of meat production for centuries (Russelle, 2001). When properly managed, they are rich sources of protein, fiber, and energy. Meat production in developing countries is almost solely dependent on forage legumes and grasses. Inclusion of legumes is critical for sustainable meat and dairy production on the infertile savannah soils of tropics and subtropics (Graham & Vance, 2003).

White clover (*T. repens*) is the most appropriate and widely used forage legume for organic farming systems in regions with temperate maritime climates because of its adaptability to a range of management and soil fertility conditions. It is persistent as well as tolerant to a wide range of soil pH and drainage conditions. It can be used for different management regimes ranging from continuous sheep grazing where varieties with small leaves are most suitable, to lax defoliation including cutting where varieties with larger leaves would be more appropriate. Depending on the level of self-sufficiency, clover has the potential to provide 50%–80% of the total N input to the organic livestock system. Pure stands of white clover grown organically on lowland farms yielded from 5 to 8 t DM/ha (Halling *et al.*, 2001). However, growing white clover alone is not a practical option for organic farmers because of its relatively low yields compared to other forages (35%–37% lower than red clover and alfalfa), high risks of bloat to the grazing animal, and an increase in the weed dock (*Rumex crispus*) population. In addition, white clover is not a good competitor against weeds. The most popular practice is to grow white clover in mixed swards with grass species (Hancock *et al.*, 2003).

Red clover (*T. pratense*) is less persistent than white clover but does fix more N. Red clover tends to be used in short-term conservation leys (up to 3 years), primarily for cutting. Red clover is not a demanding legume in terms of soil conditions but it is also not persistent for grazing. Red clover exhibits high yields of digestible organic matter, metabolized energy, and N (Halling *et al.*, 2002; Abberton *et al.*, 2006). If compared to grasses and legumes, red clover when included in diets stimulated meat production and the content of beneficial fatty acids in meat. These diets improved the efficiency of feed N utilization as well (Lee *at al.*, 2006).

Birdsfoot trefoil (*Lotus corniculatus*) is a long-lived perennial legume suitable for different soils and climatic conditions. It does not cause bloat as many other commonly used legumes do. Using birdsfoot trefoil into grass pastures leads to 30% higher daily gains of grazing animals and a more uniform distribution of forage production. In a series of trials in four different countries, Halling *et al.* (2001) reported yields of 4.3–8.4 t DM/ha when the crop was grown on a range of soils, with the lowest and highest yields from crops grown on sandy and loam soils, respectively. The yield of the birdsfoot trefoil declined after three cuts, which demonstrated poor persistence. Higher intakes and digestibility values have been recorded with birdsfoot trefoil silage when compared to other legume silages (Fraser *et al.*, 2000). Birdsfoot was also found to improve protein utilization by the ruminant animals and to reduce methane emission and diffused N pollution when consumed as a major part of the ruminant diet (Abberton *et al.*, 2006).

Alfalfa (*M. sativa*) is the prevalent forage legume in temperate climates (Russelle, 2001) because of high nutritional quality, high yields, excellent summer regrowth, persistence, and drought tolerance. Alfalfa contains between 15% and 22% crude protein. It is also an excellent source of vitamins and minerals, and has considerable potential as a crop for conservation. The crop is most often harvested as hay, but can also be made into silage, or fed as greenchop. Alfalfa could be used for grazing as well. A good strategy for producers practicing continuous stocking, rotational stocking, and/or haymaking is to use grazing-tolerant cultivars rather than cultivars not selected for grazing tolerance (Bouton & Gates, 2003).

Sainfoin (*Onobrychis viciaefolia*) can be both grazed and conserved for winter forage. Typically, it provides a heavy cut for silage or hay in June and then regrows for grazing during summer. It is bloat-free and drought resistant. Stock performance when fed sainfoin is superior compared to other forages due to its very high palatability leading to high voluntary intake and higher protein assimilation by animals.

11.9.2 Grasses

Grass, as a major energy source, is vital to organic beef production. The main species of grass that can be used are perennial ryegrass, Italian ryegrass, hybrid ryegrass, timothy, and cocksfoot. Another group of grass species including meadow fescue, tall fescue, red fescue, and smooth-stalked meadow grass can also be used when specific cultivation circumstances are required, such as dry areas, wet and heavy soils, etc.

Perennial ryegrass (*L. perenne*) is very suitable for mild climatic conditions. However, it is too susceptible to the winter cold of continental climates where timothy and meadow fescue are more commonly used. Although perennial ryegrass has been criticized as being more appropriate for intensive high N systems than for organic systems, it is undoubtedly the most suitable species for ley farming in temperate maritime conditions because of its easy establishment, yield potential, persistence, and high-quality forage characteristics (Younie *et al.*, 2001). The continued reliance on the ryegrasses is because of the wide availability of varieties, which encompass early flowering erect varieties suitable for early grazing and cutting to late flowering prostrate and persistent varieties suitable for long-term intensive grazing management. Whilst there are very strong arguments in favor of the ryegrasses being the main components of most mixtures for grass leys, other grass species have significant attributes that should be exploited for specific situations.

Timothy (*Phleum pratense*) is suitable for heavy, wet soils and should always be considered where the sward is primarily intended for cutting management. Timothy has good palatability and very good winter hardiness.

Cocksfoot (*Dactylis glomerata*) is more suitable for dry sandy soils than perennial ryegrass and should always be included in the mixtures for grass leys. In addition, cocksfoot rapidly recovers after grazing.

Tall fescue (*Festuca pratensis*) is adapted to a wide range of soil and climatic conditions, but performs best on well-drained clay soils.

Red fescue (*Festuca rubra*) is suitable for very dry conditions on poor fertility but the fine rolled leaves reduce its palatability.

Italian ryegrass (*Lolium multiflorum*) is easy to establish, very productive in fertile soils, with early spring growth, high digestibility, palatable, and easy to ensile because of high carbohydrates content. It is not persistent (generally considered a 2-year crop) and susceptible to winter kill during continental climatic conditions.

11.9.3 Herbs

Herb species including chicory (*Cichorium intybus*) and ribgrass plantain (*Plantago lanceolata*) can significantly increase the availability of minerals in a sward (Weller & Bowling, 2001), improve animal performance (Fraser & Rowarth, 1996), and reduce parasitic problems (Deane *et al.*, 2002; Moorhead *et al.*, 2002; Marley *et al.*, 2003, 2006). Establishing and maintaining the herbs in the sward can be difficult due to overgrazing and winter hardiness.

Different climatic conditions will also have a marked influence on the type of herbs found in permanent pastures. When herbs are grown in a mixed sward management practices, both grazing intensity and cutting frequency influence their persistency and contribute to the total yield of the sward. Generally, herbal pastures are sown in a separate paddock reserved for this purpose or sown as herbal strips within a paddock. These pastures are not used as the primary grazing land. Instead, they are considered a health-based supplement to the diet for a few hours grazing or are reserved for haymaking.

11.9.4 Grass–clover leys

Grass–clover leys are important components of organic production systems. They provide forage for ruminants and aid with the retention of soil fertility as well as promote nutrient cycling through manure and slurry (Boller *et al.*, 2008). Doyle and Topp (2004) reported that the economic advantage of organic systems using forage legumes will depend on new advances and management techniques that decrease the risk associated with more difficult crop establishment and higher nitrate leaching of forage legumes compared to grasses. According to these authors, from an economic perspective, forage legumes are probably best grown in a mixture with grass. In organic farming, fodder production still relies considerably on newly established grass–white clover leys often combined with a second fodder crop to supply energy to animals (Van Eekeren, 2000).

Herbage production of grass–clover leys depends on ley age, soil properties, and geographical location (Riesinger & Herzon, 2008). Two-year-old leys produced the highest yield (6.5 t/ha DM), while 3-year-old leys (4.9 t/ha DM) resulted in the lowest yield. No significant differences between 1- and 2-year-old leys were observed (Nykänen *et al.*, 2000). Organically grown grass/white clover swards yielded between 5 and 12 t/ha DM (Weller & Jones, 2002) with the yield influenced not only by soil type and climatic conditions but also by management practices. The clover content (20%–80% of DM) benefited from using mixed swards due to their persistence, palatability, digestibility, and availability for extended grazing. However, the clover content of a mixed sward can vary between seasons. A limitation of white clover-based swards is the erratic supply of protein with the protein content of both grass and clover increasing during the growing season (Weller & Copper, 2001). The late spring growth of clover also shortens the grazing season and reduces the protein content of herbage cut for silage in the late spring period. By early summer, the protein content of the sward is higher than the level required by the grazing animals. During the summer period when the clover content of the sward is high, the fiber content of the grazed herbage is less than 40% (Chamberlain & Wilkinson, 1998).

The following type of grass–clover mixtures and seed rates can be used for different purposes (Philipps, 2001):

- Short term (15 kg/ha Italian ryegrass + 10 kg/ha red clover).
- Medium/long-term leys for cattle grazing (20 kg/ha perennial ryegrass + 8 kg/ha Italian ryegrass/timothy + 4 kg/ha white clover).
- Medium/long-term leys for sheep grazing (18 kg/ha perennial ryegrass + 4 kg/ha meadow fescue + 4 kg/ha Cocksfoot + 4 kg/ha white clover).

In many dryland areas, alfalfa–grass mixtures are grown to help fill voids in the stand and reduce weed pressure. Alfalfa–grass mixtures should also be considered for fields that will be left in hay for 5 years or more, fields with excessively wet or shallow soils, and fields that will be grazed. Grasses are often interseeded as the alfalfa stand thins out, but grass establishment will usually be improved by seeding at the same time as alfalfa. In irrigated mixtures, blends that include timothy, orchard grass, and improved (endophyte free) tall fescue are often planted. Under dryland conditions, the choice of grasses depends upon rainfall zone, competitiveness of the grass, and the types of animals that will be fed (Fuerst *et al.*, 2009).

11.9.5 Whole crop forage

While forage for grazing and conservation is primarily produced from grass–clover swards, other crops have the potential to increase the nutritional quality of the forage. In an organic livestock system, where the aim is to achieve a high level of feed self-sufficiency, growing of other crops within the rotation would be required if concentrated feed needs to be replaced. Introducing cereal crops into the rotation increases the flexibility of feed supplies. Cereals can be conserved as whole crop silage or stored as a high-energy grain feed. As energy is a limiting factor during the winter period, the inclusion of either forage maize or fodder beet in the rotation can increase the energy density of the diets. Whole crop cereal silage is a moderate energy feed (9.5–10.0 MJ of ME/kg [megajoules of metabolizable energy] DM) with a low protein content (<10%) (Hancock *et al.*, 2003).

Whole crop forage as an annual crop, which may be conserved, fits well into the mixed organic farm system. The whole crop forage as nurse crops for undersown grass reseeds provides a good opportunity for a break crop by adding diversity within the cropping season and arable rotation (Weller *et al.*, 2004). They can also provide high energy and less acid silage that complements grass silage and grazed grass, potentially improves intakes, reduces concentrate feed requirements, and enhances digestion, cow health, and performance.

Unlike a combined crop, harvesting cereals and/or legumes as a whole crop is neither a new practice nor is it confined to organic production. It allows greater flexibility of harvest dates within one harvest window (season) in contrast to multi cuts of grass silage. Choosing an earlier harvest date allows growing cereals in locations where climate may make combined cropping problematic. Recent research has focused on bi-cropping whole crop cereals with a variety of mixtures including legume species to increase the protein content to overcome this problem. Inclusion of a legume to create a mixture tends to increase the otherwise low protein of whole crop cereals. There is evidence that both undersowing legumes and including legumes in the forage mixture can also improve the efficiency of N uptake by the cereal component of the mixture. Whole crops can increase milk yield or live weight gain when fed as a supplement to grass silage (Chamberlain & Wilkinson, 1998; Dawson, 2006; Keady *et al.*, 2006). Whole crop cereals and cereal bi-crops can be successfully used to feed lambs and beef (Adesogan & Jones, 2000; Dawson, 2006; Keady *et al.*, 2006; Marley *et al.*, 2007).

11.10 PASTURES

On February 12, 2010 (7154 Federal Register), new organic regulations incorporating quantifiable measurements for tracking the pasturing of ruminant animals were introduced in the

United States. Livestock must actively graze on a daily basis during the grazing season, have access to outdoors, and not be confined during the nongrazing season. In order to provide at least 30% dry matter intake from pasture, animals must graze at least 120 days. Furthermore, grazing provides biodiversity, a clean environment, and premium quality meat (Wood *et al.*, 2007). Meat from pastured cows and lambs has less fat, more vitamin E (Yang *et al.*, 2002; Descalzo *et al.*, 2005), and higher levels of beneficial compounds, such as omega-3 and conjugated linoleic fatty acids, than grain-finished products (Scollan *et al.*, 2002; Fraser *et al.*, 2004).

The grazing management will ration out forage according to animal requirements, allowing full plant recovery while minimizing forage waste (Murphy, 1995). According to this author, the elements of a sustainable grazing management system includes a proper timing and intensity of grazing and recovery time after grazing. A rotational grazing system is preferred in organic production. The system optimizes pasture use, maintains pasture composition, maximizes grass production, reduces equipment requirements, and controls parasites. Rotational grazing can also improve soil structure and fertility recovery. Grazing for short periods followed by long recovery can help to prevent preferential grazing and allows the pasture composition to remain mixed, particularly for perennial species.

A system known as cell grazing (also known as holistic grazing, crash grazing, time control grazing, or high-density, short-duration grazing) has been designed to maximize time available for pasture plants to regenerate. The paddock number depends on the time for grass regrowth and number of days that animals have to graze in one paddock. Small paddocks are necessary for effective intensive management and this is often achieved using portable or fixed electric fencing, or by redesigning the farm with the use of additional fencing and shelterbelts. Success requires close monitoring of pasture conditions to determine optimum grazing pressure. It is essential for paddocks not to be grazed prematurely after rain so that the perennial pastures can be allowed to get a head start. In regions with wet and dry seasons, this will require disciplined management.

When the grazing pressure is excessive, the carrying capacity of the pasture (overgrazing) occurs and forage quality decreases. To avoid overgrazing, animals must not remain on a pasture for too long. The pastures that are continually overgrazed are weakened. Many productive species die, water runoff increases, soil temperature increases, and overall pasture quality and quantity decreases.

According to results of Defra project OF0328, undertaken by the Agricultural Development and Advisory Service (ADAS) (Anon, 2004), to improve forage utilization farmers should avoid depending on a single large cut of forage (which would compromise quality) and adopt one or more of the following procedures:

- Graze the sward before shutting up for silage.
- Make an earlier, smaller first cut of high-quality forage and a larger, higher protein second cut.
- Introduce a second high-protein forage for silage, such as whole crop cereals with vetches, peas, alfalfa, or lupins to reduce reliance on bought-in proteins.
- Put different forages into single silage clamp (or put different forages into different clamps).
- Use a second forage in the system that make it more agronomical robust and also result in an increase in forage dry matter intake.
- Always analyze conserved forages so that a balanced winter ration can be made up.

11.11 CONCLUSION

Production of organic meat cannot be a successful task without ensuring livestock farms with sufficient qualitative forage. The supply of organic feedstuffs on the market is limited, which leads to production of most of the necessary forages on the farm. A well-designed crop rotation includes a mixture of leguminous "fertility building" crops, grass–clover leys, cereals, and cash crops based on the production of forages. This achieves well-balanced feed rations to serve as sources of energy, proteins, and minerals for all types of animals. Inclusion of GM, CCs and undersowing of legumes, and grasses to cereals in a rotation manner increases land productivity and availability of roughages and pasture for livestock. The high-input crop rotation represents a means for overcoming decreased productivity of cereals in an organic system. Conserved cereals as whole crop silage improve feed rations of livestock during winter or dry seasons. Grass–clover leys and pastures are a mandatory part of rotations because of the requirement for outdoors access by all livestock groups. Pastures play an essential part of feeding of organic livestock. High-quality organic meat from ruminants can be achieved from well-tended, preferably cell-grazed pastures.

REFERENCES

Abberton, M. T., M. Forthergill, R. P. Collins, and A. H. Marshall. 2006. Breeding forage legumes for sustainable and profitable farming systems. *Aspects Appl. Biol.* 80:81–87.

Adesogan, A. T. and G. J. Jones. 2000. The effect of fermented whole crop cereal species on botanical composition, chemical composition, in vivo digestibility and live weight change in sheep. Available at: http://www.bsas.org.uk/downloads/annlproc/Pdf99/081.pdf (accessed February, 2010).

Anon. 2004. Optimising the production and utilisation of forage for organic livestock. *Final Report of Defra project OF0328* by ADAS (1/8/02 to 31/8/04). Available at: http://randd.defra.gov.uk/Document.aspx?Document=OF0328_2461_FRP.doc (accessed March, 2010).

Askegaard, M. and J. Eriksen. 2002. Exchangeable potassium in soil as indicator of potassium status in an organic crop rotation on loamy sand. *Soil Use Manag.* 18:84–90.

Askegaard, M. and J. Eriksen. 2007. Growth of legume and nonlegume catch crops and residual-N effects in spring barley on coarse sand. *J. Plant Nutr. Soil Sci.* 170:773–780.

Baggs, E., C. Watson, and R. Rees. 2000. The fate of nitrogen from incorporated cover crop and green manure residues. *Nutr. Cycl. Agroecosyst.* 56:153–163.

Bàrberi, P. 2002. Weed management in organic agriculture: are we addressing the right issues? *Weed Res.* 42:177–193.

Baributsa, D. N., E. F. Foster, K. D. Thelen, A. N. Kravchenko, D. R. Mutch, and M. Ngouajio. 2008. Corn and cover crop response to corn density in an interseeding system. *Agron. J.* 100:981–987.

Bergström, L. and H. Kirchmann. 2004. Leaching and crop uptake of nitrogen from nitrogen-15-labeled green manures and ammonium nitrate. *J. Environ. Qual.* 33:1786–1792.

Berner, A., I. Hildermann, A. Fließbach, L. Pfiffner, U. Niggli, and P. Mäder. 2008. Crop yield and soil fertility response to reduced tillage under organic management. *Soil Tillage Res.* 101:89–96.

Berry, P. M., E. A. Stockdale, R. Sylvester-Bradley, L. Philipps, K. A. Smith, E. I. Lord, C. A. Watson, and S. Fortune. 2003. N, P and K budgets for crop rotations on nine organic farms in the UK. *Soil Use Manag.* 19:112–118.

Blackshaw, R. E., R. L. Anderson, and D. Lemerle. 2007. Cultural Weed Management. In: M. K. Upadhyaya, and R. E. Blackshaw (eds). *Non-Chemical Weed Management: Principles, Concepts and Technology.* 1st edn. CABI Publishing, Cambridge, MA. pp. 36–47.

Blair, R. 2007. *Nutrition and Feeding of Organic Pigs.* 1st edn. CABI Publishing, Cambridge, MA. p. 322.

Blaser, B. C., L. R. Gibson, J. W. Singer, and J. L. Jannink. 2006. Optimizing seeding rates for winter cereal grains and frost-seeded red clover intercrops. *Agron. J.* 98:1041–1049.

Boller, B., P. Tanner, and F. X. Schubiger. 2008. Breeding forage grasses for organic conditions. *Euphytica.* 163:459–467.

Bouton, J. H. and R. N. Gates. 2003. Grazing-tolerant alfalfa cultivars perform well under rotational stocking and hay management. *Agron. J.* 95:1461–1464.

Bulson, H. A. J., R. W. Snaydonet, and C. E. Stopes. 1997. Effects of plant density on intercropped wheat and field beans in an organic farming system. *J. Agr. Sci.* 128:59–71.

Burger, H., M. Schloen, W. Schmidt, and H. H. Geiger. 2008. Quantitative genetic studies on breeding maize for adaptation to organic farming. *Euphytica* 163:501–510.

Chamberlain, A. T. and J. M. Wilkinson. 1998. *Feeding the Dairy Cow.* Chalcombe Publications, Lincoln. pp. 153–158.

Charmet, G., F. X. Oury, and C. Ravel. 2005. Use of molecular markers to improve wheat quality in organic farming systems. In: E.T. Lammerts van Bueren, I. Goldringen, and H Ostergard (eds). *Proc. COST SUSVAR/ECO-PB Work shop Org Plant Breeding Strateg Use Mol Markers.* January 17th–19th. Driebergen. pp. 36–41.

Cherr, C. M., J. M. S. Scholberg, and R. McSorley. 2006. Green manure approaches to crop production. *Agron. J.* 98:302–319.

Christoffoleti, P. J., S. J. P. Carvalho, M. Nicolai, D. Doohan, and M. Van Gessel. 2007. Prevention strategies in weed management. In: M. K. Upadhyaya, and R. E. Blackshaw (eds). *Non-Chemical Weed Management: Principles, Concepts and Technology.* 1st edn. CABI Publishing, Cambridge, MA. pp. 3–21.

Codex Alimentarius—Organically Produced Foods. Joint FAO/WHO Food Standards Programme, Food and Agriculture Organization of the United Nations. 2001. Available at: http://www.fao.org/DOCREP/005/Y2772E/Y2772E00.HTM (accessed February, 2010).

Council Regulation (EC) No 834/2007, Official Journal of the European Union, 20.7.2007, L 189/1. Available at: http://eurlex.europa.eu/LexUriServ/LexUriServ.do?uri=OJ:L:2007:189:0001:0023:EN:PDF (accessed February, 2011).

Dapaah, H. K. and T. J. Vyn. 1998. Nitrogen fertilization and cover crop effects on soil structure stability and corn performance. *Commun. Soil Sci. Plant Anal.* 129:2557–2569.

David, C., P. Viaux, and J. M. Meynard. 2004. Les enjeux de la production biologique en France. *Le Courrier de l'Environnement de l'INRA* 51:43–53.

Dawson, L. 2006. Options for feeding beef cattle this winter—key issue for Hillsborough beef open day Agri-Food and Biosciences Institute Hillsborough. NI. Available at: http://www.afbini.gov.uk/index/news/news-releases/news-releases-archive-2006.htm?newsid=6907 (accessed April, 2010).

Deane, J. C., J. Warren, L. Findlay, M. Dagleish, S. Cork, F. Jackson, and R. Keatinge. 2002. The effect of *Cichorium intybus* and *Lotus corniculatus* on nematode burdens and production in grazed lambs. In: Powell, Jane, *et al.* (eds). *Proc. UK Org. Res. 2002 Conf.* pp. 89–92.

Deria, A. M., R. W. Bell, and G. W. O'Hara. 2003. Organic wheat production and soil nutrient status in a Mediterranean climatic zone. *J. Sustain. Agric.* 21:21–47.

Descalzo, A. M., E. M. Insani, A. Biolatto, A. M. Sancho, P. T. Garcia, and N. A. Pensel. 2005. Influence of pasture or grain-based diets supplemented with vitamin E on antioxidant/oxidative balance of Argentine beef. *Meat Sci.* 70:35–44.

Douds, D. D. and P. Millner. 1999. Biodiversity of arbuscular mycorrhizal fungi in agroecosystems. *Agric. Ecosyst. Environ.* 74:77–93.

Doyle, C. J. and C. F. E. Topp. 2004. The economic opportunities for increasing the use of forage legumes in north European livestock systems under both conventional and organic management. *Renew. Agric. and Food Syst.* 19:15–22.

Eltun R., A. Korsaeth, and O. Nordheim. 2002. A comparison of environmental, soil fertility, yield, and economic effects in six cropping systems based on an 8-year experiment in Norway. *Agric. Ecosyst. Environ.* 90:155–168.

Erhart, E. and W. Hartl. 2009. Soil protection through organic farming: A review. In: E. Lichtfouse (eds). *Organic Farming, Pest Control and Remediation of Soil Pollutants.* Springer, Heidelberg. pp. 203–226.

Eriksen, J., M. Askegaard, and K. Søegaard 2008. Residual effect and nitrate leaching in grass-arable rotations: effect of grassland proportion, sward type and fertilizer history. *Soil Use Manag.* 24:373–382.

Evans, J., L. McDonald, and A. Price. 2006. Application of reactive phosphate rock and sulphur fertilizers to enhance the availability of soil phosphate in organic farming. *Nutr. Cycl. Agroecosyst.* 75:233–246.

Fraser, M. D., H. M. Speijers, V. J. Theobald, R. Fychan, and R. Jones. 2004. Production performance and meat quality of grazing lambs finished on red clover, lucerne or perennial ryegrass swards. *Grass Forage Sci.* 59:345–356.

Fraser, M. D., R. Fychan, and R. Jones. 2000. Voluntary intake, digestibility and nitrogen utilization by sheep fed ensiled forage legumes. *Grass Forage Sci.* 55:271–279.

Fraser, T. J. and J. S. Rowarth. 1996. Legumes, herbs or grass for lamb performance. *Proc. N. Z. Grassl. Assoc.* 58:49–52.

Fuerst, P. E., R. T. Koenig, J. Kugler, K. Painter, M. Stannard, and J. Goldberger. 2009. *EB2039E Organic Alfalfa Management Guide*. Washington State University Extension, Pullman, WA. pp. 5–6.

Gosling, P. and M. Shepherd. 2005. Long-term changes in soil fertility in organic arable farming systems in England, with particular reference to phosphorus and potassium. *Agric. Ecosyst. Environ.* 105:425–432.

Goulding, K., E. Stockdale, and C. Watson. 2008. Plant Nutrients in Organic Farming. In: *Organic Crop Production—Ambitions and Limitations*. Springer, Heidelberg. pp. 73–88.

Graham, P. and C. P. Vance. 2003. Legumes: importance and constraints to greater use. *Plant Physiol.* 131:872–877.

Gruber, S. and W. Claupein. 2009. Effect of tillage intensity on weed infestation in organic farming. *Soil Tillage Res.* 105:104–111.

Haas, G., B. Caspari, and U. Kopke. 2002. Nutrient cycling in organic farms: stall balance of a suckler cow herd and beef bulls. *Nutr. Cycl. Agroecosyst.* 64:225–230.

Haddad, C. M. and F. V. Alves. 2002. Organic feeds for beef cattle supplementation. *First Virtual Global Conference on Organic Beef Cattle Production*, September 2–October 15, 2002. Available at: http://www.cpap.embrapa.br/agencia/congressovirtual/pdf/ingles/03en05.pdf (accessed April 2010).

Halling, M., A. Hopkins, A. Nissinen, O. Paul, C. Tuori, and M. Soelter, 2001. Forage legumes—productivity and composition. *Landbauforschung Völkenrode* 234:5–15.

Halling, M. A., A. Hopkins, O. Nissinen, C. Paul, M. Tuori, and U. Soelter. 2002. Nutritive quality of forage legumes grown in northern Europe. Grassland sci. in Europe, Multi-Function Grasslands: Quality Forages, Animal Products and Landscapes. *Proc. 19th General Meeting EGF*, La Rochelle, France. 7:126–127.

Hancock, J., R. Weller, H. McCalman. 2003. 100% Organic livestock feeds—preparing for 2005. A report prepared for Organic Centre WalesBy Soil Association Producer Services and assisted by IGER. Available at: http://orgprints.org/10831/1/feeds2005.pdf (accessed February, 2011).

Hauggaard-Nielsen, H., M. Gooding, P. Ambus, G. Corre-Hellou, Y. Crozat, C. Dahlmann, A. Dibet, P von Fragstein, A. Pristeri, M. Monti, and E. S. Jensen. 2009a. Pea–barley intercropping and short-term subsequent crop effects across European organic cropping conditions. *Nutr. Cycl. Agroecosyst.* 85: 141–155.

Hauggaard-Nielsen, H., M. K. Andersen, B. Jornsgaard, and E. S. Jensen. 2006. Density and relative frequency effects on competitive interactions and resource use in pea-barley intercrops. *Field Crops Res.* 95:256–267.

Hauggaard-Nielsen, H., S. Mundus, and E. S. Jensen. 2009b. Nitrogen dynamics following grain legumes and subsequent catch crops and the effects on succeeding cereal crops. *Nutr. Cycl. Agroecosyst.* 84:281–291.

Helander, C. A. 2004. Residual nitrogen effects on a succeeding oat (*Avena sativa L.*) Crop of clover species and ryegrass (*Lolium perenne L.*) undersown in winter wheat (*Triticum aestivum L.*). *Acta. Agric. Sc. and B—Pl. Soil Sci.* 54:67–75.

Hoad, S., C. Topp, and K. Davies. 2008. Selection of cereals for weed suppression in organic agriculture: a method based on cultivar sensitivity to weed growth. *Euphytica* 163:355–366.

Jansa, J., A. Mozafar, T. Anken, R. Ruh, I. R. Sanders, and E. Frossard. 2002. Diversity and structure of AMF communities as affected by tillage in a temperate soil. *Mycorrhiza* 12:225–234.

Jensen, E. S. 1996. Grain yield, symbiotic N_2 fixation and interspecific competition for inorganic N in pea-barley inter-crops. *Plant Soil* 182:25–38.

Jensen, E. 2006. Intercropping of cereals and grain legumes for increased production, weed control, improved product quality and prevention of N-losses in European organic farming systems. Final report: Quality of Life and management of living resources (QLK5-CT-2002–02352). Available at: http://www.intercrop.dk/Publications.htm (accessed June, 2010).

Karpenstein-Machan, M. and R. Stuelpnagel. 2000. Biomass yield and nitrogen fixation of legumes monocropped and intercropped with rye and rotation effects on a subsequent maize crop. *Plant Soil* 218:215–232.

Kayser, M., J. Müller, and J. Isselstein. 2009. Nitrogen management in organic farming: comparison of crop rotation residual effects on yields, N leaching and soil conditions. *Nutr. Cycl. Agroecosyst.* 87:21–31.

Keady, T. W. J., F. L. Lively, D. J. Kilpatrick, and B. W. Mosse. 2006. Effects of replacing grass silage with either maize or whole-crop wheat silages on the performance and meat quality of beef cattle offered two levels of concentrates. *Int. J. of Anim. Biosci.* 1:613–623.

Kirchmann, H., L. Bergström, T. Kätterer, L. Mattsson, and S. Gesslein. 2007b. Comparison of long-term organic and conventional crop—Livestock systems on a previously nutrient-depleted soil in Sweden. *Agron. J.* 99:960–972.

Kirchmann, H. and M. H. Ryan. 2004. Nutrients in organic farming—Are there advantages from the exclusive use of organic manures and untreated minerals? *Proc. 4th Int. Crop Sci. Congr.* September 26th–October 1st, 2004. Brisbane, Australia. Available at: www.cropscience.org.au (accessed February, 2010).

Kirchmann, H., M. H. Ryan, and L. Bergström. 2007a. Plant nutrient use efficiency in organic farming—consequences of exclusive use of organic manures and untreated minerals. *CAB Reviews:* Perspectives in Agriculture, Veterinary Science, Nutrition and Natural Resources 2007 2, No. 076. pp. 12. Available at: http://www.cabi.org/cabreviews/ (accessed February, 2010).

Kitchen, J. L., G. K. McDonald, K. W. Shepherd, M. F. Lorimer, and R. D. Graham. 2003. Comparing wheat grown in South Australian organic and conventional farming systems. I. Growth and grain yield. *Aust. J. Agric. Res.* 54:889–901.

Kleinman, P. J. A. and K. Soder. 2008. The impact of hybrid dairy systems on air, soil and water quality: focus on nitrogen and phosphorus cycling. In: R. W. McDowell (ed.). *Environmental Impacts of Pasture-Based farming.* CABI Publishing. Cambridge, MA. pp. 249–277.

Kristensen, E. S., H. Hogh-Jensen, and I. S. Kristensen. 1995. Estimation of biological N_2 fixation in a clover grass system by the 15N dilution method and total N difference method. *Biol. Agric. Hortic.* 11: 203–219.

Ksiezak, J. and M. Staniak. 2009. Evaluation of legume-cereal mixtures in organic farming as raw material for silage production. *J. Res. Appl. Agr. Eng.* 54:157–163.

Lampkin, N. 1999. *Organic Farming.* Farming Press, Miller Freeman House, London. p. 715.

Lauk, R. and E. Lauk. 2009. Dual intercropping of common vetch and wheat or oats, effects on yields and interspecific competition. *Agron. Res.* 7:21–32.

Lee, R. F., Michael, P. L. Connelly, J. K. S. Tweed, R. J. Dewhurst, R. J. Merry, and N. D. Scollan. 2006. Effects of high-sugar ryegrass silage and mixtures with red clover silage on ruminant digestion. 2. Lipids. *J. Anim. Sci.* 84:3061–3070.

Lundkvist, A., L. Salomonsson, L. Karlsson, and A. D. Gustavsson. 2008. Effects of organic farming on weed flora composition in a long term perspective. *Eur. J. Agron.* 28:570–578.

Mader, P., A. Fliebbach, D. Dubois, L. Gunst, P. Fried, and U. Niggli. 2002. Soil fertility and biodiversity in organic farming. *Science* 296:1694–1697.

Marley, C. L., M. D. Fraser, R. Sanderson, and R. Jones. 2007. Effects of feeding different ensiled forages on the productivity and nutrient-use efficiency of finishing lambs. *Grass Forage Sci.* 62:1–12.

Marley, C. L., R. Cook , R. Keatinge, J. Barrett, and N. H. Lampkin. 2003. The effect of birdsfoot trefoil (*Lotus corniculatus*) and chicory (*Cichorium intybus*) on parasite intensities and performance of lambs naturally-infected with helminth parasites. *Vet. Parasitol.* 112:147–155.

Marley, C. L., R. Cook, J. Barrett, R. Keatinge, and N. H. Lampkin. 2006. The effects of birdsfoot trefoil (*Lotus corniculatus*) and chicory (*Cichorium intybus*) when compared with perennial ryegrass (*Lolium perenne*) on ovine gastrointestinal parasite development, survival and migration. *Vet. Parasitol.* 138:280–290.

Miller, P. R., D. E. Buschena, C. A. Jones, and J. A. Holmes. 2008. Transition from intensive tillage to no-tillage and organic diversified annual cropping systems. *Agron. J.* 100:591–599.

Mitchell, R. D. J., R. Harrison, K. J. Russell, and J. Webb. 2000. The effect of crop residue incorporation date on soil inorganic nitrogen, nitrate leaching and nitrogen mineralization. *Biol. Fertil. Soils* 32:294–301.

Möller, K. 2009. Inner farm nutrient flows between arable land and permanent grassland via the stable in organic cropping systems. *Eur. J. Agron.* 31:204–212.

Möller, K., W. Stinner, and G. Leithold. 2008. Growth, composition, biological N_2 fixation and nutrient uptake of a leguminous cover crop mixture and the effect of their removal on field nitrogen balances and nitrate leaching risk. *Nutr. Cycl. Agroecosyst.* 82:233–249.

Moorhead, A. J. E., H. G. Judson, and A. V. Stewart. 2002. Liveweight gain of lambs grazing 'Ceres Tonic' plantain (*Plantago lanceolata*) or perennial ryegrass (*Lolium perenne*). *Proc. N. Z. Soc. Anim. Prod.* 62:171–173.

Murphy, B. 1995. Pasture management to sustain agriculture. In: M. Altieri (ed.). *Agroecology: The Science of Sustainable Agriculture.* 2nd edn. Westview Press, Boulder, CO. pp. 321–347.

Nyiraneza, J., M. H. Chantigny, A. N'Dayegamiye, and M. R. Laverdière. 2009. Dairy cattle manure improves soil productivity in low residue rotation systems. *Agr. J.* 101:207–214.

Nykänen, A., A. Granstedt, A. Laine, and S. Kunttu. 2000. Yields and clover contents of leys of different age in organic farming in Finland. *Biol. Agric. Hortic.* 18:55–66.

Oehl F., A. Oberson, H. U. Tagmann, J. M. Besson, D. Dubois, P. Mäder, H. R. Roth, and E. Frossard. 2002. Phosphorus budget and phosphorus availability in soils under organic and conventional farming. *Nutr. Cycl. Agroecosyst.* 62:25–35.

Oehl, F., E. Sieverding, P. Mäder, D. Dubois, K. Ineichen, T. Boller, and A. Wiemken. 2004. Impact of long-term conventional and organic farming on the diversity of arbuscular mycorrhizal fungi. *Oecologia* 138:574–583.

Øgaard, A. F., T. Krogstad, and A.-K. Løes. 2001. Potassium uptake by grass from a clay and a silt soil in relation to soil tests. *Acta. Agric. Scand. B* 51:97–105.

Østergård, H., K. Kristensen, H. O. Pinnschmidt, P. K. Hansen, and M. S. Hovmøller. 2008. Predicting spring barley yield from variety-specific yield potential, disease resistance and straw length, and from environment-specific disease loads and weed pressure. *Euphytica* 163:391–408.

Patra, D., M. Sachdev, and B. Subbiah. 1986. ^{15}N studies on the transfer of legume-fixed nitrogen to associated cereals in intercropping systems. *Biol. Fertil. Soils* 2:165–171.

Pekrun, C. and W. Claupein. 2006. The implication of stubble tillage for weed population dynamics in organic farming. *Weed Res.* 46:414–423.

Pelosil, C., M. Bertrand, and J. Roger-Estrade. 2009. Earthworm community in conventional, organic and direct seeding with living mulch cropping systems. *Agron. Sustain. Dev.* 29:287–295.

Perry, L. G. and S. M. Galatowitsch. 2006. Light competition for invasive species control: A model of cover crop—weed competition and implications for *Phalaris arundinacea* control in sedge meadow wetlands. *Euphytica* 148:121–134.

Philipps, L. 2001. A guide to organic grassland. *Elm Farm Research Centre Bulletin*. Berkshire. No. 56: 9–10.

Riesinger, P. and I. Herzon. 2008. Variability of herbage production in mixed leys as related to ley age and environmental factors: a farm survey. *J. Sci. Food Agric.* 17:394–412.

Russelle, M. 2001. Alfalfa. *Am. Sci.* 89:252–259.

Ryan, M. H., J. W. Derrick, and P. R. Dann. 2004. Grain mineral concentrations and yield of wheat grown under organic and conventional management. *J. Sci. Food Agric.* 84:207–216.

Ryan, M. H. and M. Tibbett. 2008. The role of arbuscular mycorrhizas in organic farming. In: H. Kirchmann, and L. Bergstrom (eds). *Organic crop production—Ambitions and Limitations*. Springer Science+Business Media B.V. pp. 189–229.

Sainju, U. M. and B. P. Singh. 2008. Nitrogen storage with cover crops and nitrogen fertilization in tilled and nontilled soils. *Agron. J.* 100:619–627.

Sainju, U. M., B. P. Singh, W. F. Whitehead, and S. Wang. 2007. Accumulation and crop uptake of soil mineral nitrogen as influenced by tillage, cover crops, and nitrogen fertilization. *Agron. J.* 99: 682–691.

Schmidt, H., L. Philipps, J. P. Welsh, and P. von Fragstein. 1999. Legume breaks in stockless organic farming rotations: Nitrogen accumulation and influence on the following crops. *Biol. Agric. Hortic.* 17:159–170.

Scollan, N. D., M. Enser, R. I. Richardson, and J. D. Wood. 2002. Effect of forage legumes on the fatty acid composition of beef. *Proc. Nutr. Soc. 61 (3a): 99A*. Available at: http://journals.cambridge.org/action/displayIssue?jid=PNS&volumeId=61&seriesId=0&issueId=3a (accessed May, 2010).

Silgram M. and M. A. Shepherd. 1999. The effects of cultivation on soil nitrogen mineralization. *Adv. Agron.* 65:267–311.

Smith, E. and D. J. Read. 1997. *Mycorrhizal Symbiosis*. Academic Press Inc., San Diego, CA. p. 803.

Song, Y. N., F. S. Zhang, P. Marschner, F. L. Fan, H. M. Gao, X. G. Bao, J. H. Sun, and L. Li. 2007. Effect of intercropping on crop yield and chemical and microbiological properties in rhizosphere of wheat (*Triticum aestivum* L.), maize (*Zea mays* L.), and faba bean (*Vicia faba* L.). *Biol. Fertil. Soils* 43:565–574.

Spiertz J. H. J. and L. Sibma. 1986. Dry matter production and nitrogen utilization in cropping systems with grass, lucerne and maize. 2. Nitrogen yield and utilization with various cropping systems and their after-effects. *Neth. J. Agr. Sci.* 34:37–47.

Stalenga, J. 2007. Applicability of different indices to evaluate nutrient status of winter wheat in the organic system. *J. Plant Nutr.* 30:351–365.

Talgre, L., E. Lauringson, H. Roostalu, A. Astover, and A. Makke. 2009. Phytomass formation and carbon amount returned to soil depending on green manure crop. *Agron. Res.* 7:517–521.

Taylor, B. R., C. A. Watson, E. A. Stockdale, R. G. McKinlay, D. Younie, and S. A. Cranstoun. 2001. Current practices and future prospects for organic cereal production: survey and literature review. (HGCA). Available at: http://www.hgca.com/publications/documents/cropresearch/RR45_complete_final_report.pdf (accessed June, 2010).

Teasdale, J. R., C. B. Coffman, and R. W. Mangum. 2007. Potential long-term benefits of no-tillage and organic cropping systems for grain production and soil improvement. *Agron. J.* 99:1297–1305.

Thomas, A., C. Dalal, J. Weston, J. Lehane, J. King, N. Orange, J. Holes, and B. Wildermuth. 2009. Pasture-crop rotations for sustainable production in a wheat and sheep-based farming system on a vertosol in south-west Queensland, Australia. *Anim. Prod. Sci.* 49:682–695.

Torstensson, G., H. Aronsson, and L. Bergström. 2006. Nutrient use efficiencies and leaching of organic and conventional cropping systems in sweden. *Agron. J.* 98:603–615.

Uchino, H., K. Iwama, Y. Jitsuyama, T. Yudate, and S. Nakamura. 2009. Yield losses of soybean and maize by competition with interseeded cover crops and weeds in organic-based cropping systems. *Field Crops Res.* 113:342–351.

USA National Organic Program Regulations. 2000. Available at: http://www.nationalaglawcenter.org/readingrooms/organicprogram/ (accessed May, 2010).

Van Eekeren, N. 2000. Balancing summer rations of dairy cows by means of urea content in bulk tank milk. In: K. Soegaard, C. Ohlsson, J. Sehested, N. J. Hutchings, and T. Kristensen (eds). Grassland Farming: Balancing environmental and economic demands. *Proc. 18th Gen. Meet. Eur/ Grassland Fed.* pp. 555–557.

Vidrih, M., A. Vidrih, and S. Trdan. 2007. Direct drilling of winter cereals on pastures for increasing the early spring herbage production. *Zbor. Rad. Per. Sci. Res. Field & Veg. Crops.* 44:161–168.

Vinther, F., E. Hansen, and J. Olesen. 2004. Effects of plant residues on crop performance, N mineralisation and microbial activity including field CO_2 and N_2O fluxes in unfertilized crop rotations. *Nutr. Cycl. Agroecosyst.* 70:189–199.

Watson, C., E. A. Stockdale, and R. Rees. 2008. Assessment and maintenance of soil fertility in temperate organic agriculture. *CAB Rev.: Perspect. Agric., Vet. Sci., Nutr. Nat. Resour.* 3(219):1–11.

Weller, R. and A. Cooper. 2001. Seasonal changes in the crude protein concentration of mixed swards of white clover/perennial ryegrass grown without fertilizer N in an organic farming system in the UK. *Grass Forage Sci.* 56:92–95.

Weller, R. and E. Jones. 2002. An overview on the role and potential of forage production on lowland organic livestock farm. *Proc. Cor Conf.* Aberystwyth. pp. 81–84. Available at: http://orgprints.org/8297/1/weller_Forage_lowland_farms.pdf (accessed February, 2010).

Weller, R. F. and P. J. Bowling, 2001. The yield and quality of plant species grown in mixed organic swards. *Eur. Ass Anim. Prod.* Netherlands. 106:177–180.

Weller, R. F., P. J., Bowling, and J. Valentine. 2004. The potential value of different nurse crops for organic systems and their influence on the undersown swards. In: A. Hopkins (ed.). *BGS Occas. Symp., Org. Farming: Sci. Pract. Profitable Livestock Cropping.* 37:125–128.

Welsh, C., M. Tenuta, D. N. Flaten, J. R. Thiessen-martens and M. H. Entz. 2009. High yielding organic crop management decreases plant-available but not recalcitrant soil phosphorus. *Agron. J.* 101:1027–1035.

Wolfe, M. S., J. P. Baresel, D. Desclaux, I. Goldringer, S. Hoad, G. Kovacs, F. Löschenberger, T. Miedaner, H. Østergård, and E. T. Lammerts van Bueren. 2008. Developments in breeding cereals for organic agriculture. *Euphytica* 163:323–346.

Wood, J. D., R. I. Richardson, N. D. Scollon, A. Hopkin, R. Dunn, H. Butler, and F. M. Wittington. 2007. Quality of meat from biodiverse grassland. In: J. Hopkin, A. J. Duncan, D. I. McCracken, S. Peel, and S. J. R. B. Tallowin (eds). *High Value Grassland: Providing Biodiversity, a Clean Environment and Premium Products. Occasional symposium.* British Grassland Society 38:107–116.

Yang, A., M. C. Lanari, M. Brewster, and R. K. Tume. 2002. Lipid stability and meat colour of beef from pasture- and grain-fed cattle with or without vitamin E supplement. *Meat Sci.* 60:41–50.

Younie, D., A. P. Umrani, D. Gray, and M. Coutts. 2001. Effect of chicory or perennial ryegrass diets on mineral status of lambs. In: J. Isselstein, G. Spatz, and M. Hoffman (eds). *Organic Grassland Farming.* Grassland Science in Europe 6:278–280.

Younie, D. and J. Hermansen. 2000. The role of grassland in organic livestock farming. In: K. Soegaard, C. Ohlsson, J. Sehested, N.J. Hutchings, and T. Kristensen (eds). *Grassland Farming: Balancing Environmental and Economic Demands.* Grassland Science in Europe 5:493–513.

Zaller, J. and U. Köpke. 2004. Effects of traditional and biodynamic farmyard manure amendment on yields, soil chemical, biochemical and biological properties in a long-term field experiment. *Biol. Fertil. Soils* 40:222–229.

Section III

Processing, Sensory, and Human Health Aspects of Organic Meats

12 Slaughter Options for Organic Meat Producers in the United States

Corliss A. O'Bryan, Kristen E. Gibson, Philip G. Crandall, and Steven C. Ricke

Abstract: This chapter briefly discusses the standards for slaughter of organic meat animals in the United States. It touches upon the traditional fixed facility, and then explores the options of on-farm processing and the use of mobile processing units. The chapter briefly discusses the waste management for the traditional fixed facility and examines the alternatives for on-farm processing and mobile processing units.

Keywords: mobile processing units; waste management; slaughter

12.1 INTRODUCTION

As of this writing, there are no worldwide regulations governing the processing of organic meat. The global community often relies on the Codex Alimentarius (Codex) and the International Organization for Standardization (ISO) as the basis for production, processing, and labeling principles (Codex Alimentarius, 2010; International Organization for Standardization, 2010). The Codex Alimentarius Commission has developed a guidance document, *Guidelines for the Production, Processing, Labeling and Marketing of Organically Produced Foods* in an attempt to facilitate trade and prevent misleading claims (Codex Alimentarius, 2010). Many countries have promulgated their own organic food standards, such as the United States with the National Organic Program (NOP) or Australia with the National Standard for Organic and Bio-Dynamic Products (DAFF, 2009). The forthcoming sections provide a brief discussion of processing options for organic livestock as laid out in the United States Department of Agriculture (USDA) NOP (Agricultural Marketing Service, 2011). Environmental impact and waste management during organic processing are also outlined.

12.2 FIXED FACILITIES

In the United States, processing of meat is highly regulated. Individual slaughter plants may offer either state or federal inspection, but plants that process organic meat must also go through a certification process. Any producer that sells meat across state lines must have the animals slaughtered in a federally inspected plant, and certified organic meat must additionally be slaughtered in a certified organic slaughter facility. Processors of organic

Organic Meat Production and Processing, First Edition. Edited by Steven C. Ricke, Ellen J. Van Loo, Michael G. Johnson, and Corliss A. O'Bryan.

meats can still process nonorganic meats, but the organic product must be segregated from nonorganic, not only physically, but documented through record keeping. Contamination with prohibited substances, especially pesticides, must also be prevented.

12.2.1 Record keeping

Animals to be slaughtered must be certified organic, and be traceable by individual animal or flock. Processors must continue the record keeping paper trail that was begun by the producer to ensure the integrity of the final organic product.

For fixed facilities, a map showing the facility perimeter and buildings, all equipment, and areas used for receiving, raw material storage, processing, packaging, finished product storage, and shipping should be prepared and kept with the records. Flow charts should include equipment used in each step or stage of the process and show the flow of products through the facility—from receiving of raw ingredients to shipping of the final product.

Sources of ingredients and processing aids must be recorded. A copy of the organic certificate from the supplier of any organic ingredient or processing aid, showing that it is certified to NOP standards must be kept, along with the level of certification that supports the label claim made. For example, for a label claim of 100% organic, all ingredients and processing aids must be documented as certified 100% organic. Nonorganic ingredients and processing aids must be documented as: (1) the ingredient or aid is not commercially available as organic; (2) it does not contain prohibited ingredients and has not been produced using prohibited methods, such as genetic engineering; (3) it has not been treated with ionizing radiation; and (4) it is not produced from a crop grown using sewage sludge.

Pest management documentation should include procedures, maps, logs, service reports, and incident records. All materials used should be documented with product labels or Material Safety Data Sheet (MSDS) pages on file. If substances not on the National List of Allowed and Prohibited Substances (7 CFR 205.601–606) are used, documentation must be provided for how organic products and materials are protected from contamination during pest control applications.

For sanitation, documentation must be provided for standard operating procedures, equipment cleaning, equipment purge logs, and residue testing. Residue test procedures must be appropriate for the sanitation materials used. For instance, if chlorine is used as a sanitizer, a chlorine test strip with sensitivity in the low (0–10 ppm) range must be used to show that the level of chlorine remaining in the effluent is below 4 ppm, the level allowed in NOP Section 205.606. Materials that are not listed as allowed sanitizers may be used, but if they are used, they must be completely removed before running organic products. For example, if acid or alkaline sanitizers are used, a pH test with a neutral result (or one that matches the plain water used in the facility) indicates that the sanitizer material has been washed off. Quaternary ammonia must be completely removed, such that there are no detectable residues and residues do not contaminate organic products. Records must be maintained for each area or production line where organic processing occurs, demonstrating how organic products and packaging materials are protected from contamination by conventional product residues and/or sanitation chemicals on food contact surfaces.

12.2.2 Segregation of product

When slaughter takes place in a facility that also slaughters nonorganic animals, all incoming animals should be segregated into separate pens. Individual animals should be tagged, or in the case of small animals or poultry, identified by flock or herd. The slaughter facility can have a dedicated line for only organic, or they may use the same equipment for organic and conventional. If using the same equipment, slaughter of organically grown animals should be first in the day, or on a specific day of the week to make sure equipment has no residues from conventional animals. Organic meat must be separated from nonorganic, which can be accomplished by having a separate storage area or hanging rail in the cooler designated for organic use only. The same cleaning products may be used for organic processing lines as on nonorganic, but all equipment must be thoroughly washed with clean, 180°F (82.2°C) water before it comes into contact with organic product. Records should be maintained that document all cleaning processes and all organic processing.

12.2.3 Prevention of contamination with prohibited substances

Pesticides should be used only on the outside of a facility, but not in the vicinity of pens holding organic livestock. Rather than using pesticides within the plant, removing pest habitats, food sources, and breeding areas is advised. Preventing access of pests is critical, and manage other factors, such as temperature and lighting, to control pest reproduction should they gain access. Using mechanical or physical controls, such as lures, traps, and repellants before resorting to chemical pesticides is recommended. If chemical pesticides are used, producers must keep records of application and confirm that the organic product is not contaminated.

12.2.4 Locating a certified processing facility

Finding an organically certified USDA meat processor can be a difficult task. Meat production in the United States has been consolidating for many years leading to large farms supplying large slaughter operations and a simultaneous decline in small and medium-size processors. The poultry industry began consolidating in the mid-1950s, and in the 1980s, the cattle and swine industry followed suit, leading to fewer processing facilities available to independent farmers (Barkema et al., 2001). Once a facility is located, it may require a trip of several miles for animals that can lead to stress and loss of meat quality (Minka and Ayo, 2010). An option now available to some farmers is the mobile slaughter unit (MSU).

12.3 MOBILE SLAUGHTER UNITS

In May of 2010, the USDA announced compliance guidelines for MSUs (Food Safety and Inspection Service (FSIS), 2010). Among the advantages of an MSU versus a fixed structure are the lower costs and reduced stress on animals. As of this writing, there are approximately eight MSUs approved by USDA for red meats, and an equivalent number for poultry (eXtension, 2011). The first step in putting an MSU in operation is to contact the District Office of the USDA for a Grant of Inspection. Among the requirements for an MSU are letters indicating an approved water source as well as an approved sewage system. Units must also have a Sanitation Standard Operating Procedure (SSOP) as well as a Hazard Analysis Critical Control Point (HACCP) plan, just as fixed facilities are required to have. Most units are equipped to only slaughter, dress out, wash, and store the carcasses. The carcasses are then transported to a centralized cut and wrap stationary establishment or to a farmer designated meat locker or other such processing plant.

The live animal must be inspected prior to killing, and this usually takes place in outside pens supplied by the farm and adjacent to the MSU. There must be adequate restraint provided so that the inspector can examine the animal's teeth if necessary, and there must be a means of humane stunning. Depending on the number of animals slaughtered and how often the site is used, the pen flooring may be grass, gravel, or concrete. After stunning and killing, the animal is placed in the MSU on a rail, skinned, inspected, and washed. Inedible offal and heads are also inspected before being properly disposed of. Some states allow on-farm composting of the offal.

12.3.1 Grounds and facility requirements

Walls, floors, and ceilings must be durable, easily cleanable, and nonabsorbent. There should be a pest management program that includes the location where slaughter occurs, as well as where the MSU is stored when not in use. When in operation, the MSU must be placed in an area where pooling water and blood can be controlled so as to not allow adulteration of the product. The MSU should also be situated such that chemicals and odors do not become a problem for neighbors. Hand washing and toilet facilities must be located within a "reasonable distance" of the MSU as determined by the District Officer. The USDA inspector must have access to the equivalent of an office space as outlined in 9 CFR 307.1 that includes a cooler or space in a cooler for storage of any residue or microbiological samples taken.

12.3.2 Sanitation Standard Operating Procedures

Sanitation Standard Operating Procedures (SSOPs) must be developed and implemented by the operators of the MSU. These are written procedures that are meant to prevent direct contamination or adulteration of product. Daily records must also be maintained that document what has been done during that time frame. The establishment is required to maintain these written procedures on file, and they must be available to FSIS upon request. If product becomes adulterated or contaminated, there must be corrective actions that include how the contaminated product will be disposed of, how sanitary conditions will be restored, and any measures taken to ensure a repeat does not occur. An individual SSOP should include the equipment or area to be cleaned, identified by common name, the tools to be used, any disassembly instructions necessary, and the actual method of cleaning and sanitizing.

12.3.3 Approved sanitizers

As is the case during the growing stage of the livestock, all products used during processing must adhere to NOP regulations. Approved sanitizers appear on the National List of Allowed and Prohibited Substances. Materials that are not listed may be used, but must be completely removed as documented by testing before running organic products.

Chlorine is approved as an algicide, disinfectant, and sanitizer. Residual chlorine levels in the water at the discharge or effluent point is restricted to 4 mg/L (4 ppm). However, the levels of chlorine used at the beginning of disinfection of tools, equipment, product, or food contact surfaces may be higher than 4 mg/L, but care must be taken to ensure that the effluent water does not exceed the limit. Chlorine is quickly bound by soil, debris, and other organic matter and becomes ineffective for disinfection (Dychdala, 2001).

Ozone is an oxidizer with comparable disinfecting power to chlorine. Ozone treatments do not form by-products that chlorine treatments do, such as trihalomethane, chloroform, and other dangerous compounds. Ozone is faster acting than chlorine, allows for adequate disinfection with short-term contact, but it requires a greater capital investment and ongoing operating costs since it is unstable and must be generated on-site.

Peroxyacetic acid or peracetic acid (PAA), a combination of acetic acid and hydrogen peroxide, is also allowed in organic production. The disinfection capacity of PAA is comparable to chlorine and ozone, but produces a strong odor, so should be used in well-ventilated areas.

12.3.3.1 *HACCP*

In 1996, the Pathogen Reduction (PR) , HACCP Systems, Final Rule was published (Federal Register, 1996). This system is a science-based system designed to improve food safety by preventing the introduction of pathogens into food intended for human consumption, and consists of the following seven principles:

(i) Hazard analysis (to determine possible risks).
(ii) Identification of points where hazards can be controlled (critical control points).
(iii) Establishment of critical limits.
(iv) Monitoring of the critical control points (CCPs).
(v) Corrective actions to be taken in the event of a failure at a CCP.
(vi) Verification.
(vii) Documentation.

A written HACCP plan tailored to the MSU must be developed before the MSU is granted federal inspection. Some hazards typically found for slaughter include: (1) contamination by feces, stomach contents, or milk; (2) food-borne pathogens, such as *Escherichia coli* O157:H7 or *Salmonella*; and (3) residues of chemicals, pesticides, or drugs.

12.4 ON-FARM POULTRY PROCESSING

Some small poultry producers are opting to slaughter on farm and sell directly to the customer who comes to the farm or sell their product at local farmers markets. In the absence of a local, small slaughter operation or access to an MSU, slaughter and processing may take place on the farm. When poultry is killed and processed on the farm, it is a batch process, and farmers usually do one activity at a time, with each worker performing a variety of tasks. Little or no special equipment is required for on-farm processing. Regardless of site of slaughter, there must be arrangements in place for waste management.

During processing at any of the facilities or units described in the previous sections, waste products must be handled and disposed of according to the regulations outlined in the United States Environmental Protection Agency (USEPA) Effluent Limitations Guidelines and New Source Performance Standards for the Meat and Poultry Products Point Source Category (40 CFR Part 432) (USEPA 2004). Basic waste management practices are outlined in Section 12.5.

12.5 WASTE MANAGEMENT

12.5.1 Fixed facility

The management of liquid and solid waste products during meat and poultry processing in fixed facilities is similar for both conventional and organic processing systems, which is often a shared facility with segregated processing for conventional and organic. Meat and poultry processing wastes described here refer to slaughtering (i.e., first processing). The liquid and solid waste streams are separated throughout each processing step and are managed in different ways. Solid wastes (i.e., feathers, hair, hide, head, feet, offal, etc.) may be sold or disposed of (e.g., landfill or incineration) depending on the demand for such products. Liquid wastes (i.e., blood not collected, suspended solids, solubilized fat, proteins, urine, and other compounds), or wastewater, are discharged into either (1) the municipal waste stream for further treatment at a municipal wastewater treatment plant, or (2) directly into the environment after on-site treatment at the processing facility. In order to discharge wastewater from a fixed facility, operators are required to obtain wastewater permits from the appropriate regulatory agency. Permits required for discharge of wastewater from fixed facilities are covered in Sections 12.5.2 and 12.5.3.

12.5.2 Discharge permit and limits: municipal waste stream

Wastewater discharged into a publicly owned treatment works requires a permit from the city or municipality receiving the wastewater. In order to discharge into a municipal waste stream, processing facilities are required to install screen and oil-and-grease separation units to further remove suspended solids. After the processed wastewater enters the municipal waste stream, it receives the same treatment as domestic sewage, including primary (e.g., physical removal of floating and settleable solids); secondary (e.g., biological removal of organic matter); and tertiary (e.g., advanced physiochemical for removal of nutrients and suspended solids) treatment. Treatment technologies and requirements vary by state; therefore, operators of fixed facilities should refer to their respective state's and county's regulations.

12.5.3 Discharge permit and limits: surface water

Effluent wastewater receiving on-site treatment at the processing facility may be discharged directly into surface (navigable) waters under the authority of a National Pollutant Discharge Elimination System (NPDES) permit. Options for the on-site treatment of wastewater from first processing facilities with direct discharge into surface waters are outlined in the USEPA Final Rule on Effluent Limitations Guidelines for Meat and Poultry Products (40 CFR Part 432, Section 9) (USEPA 2004). Typically, facilities with an NPDES permit apply both primary and secondary treatments to generated wastewaters. Table 12.1 lists the distribution of the types of wastewater treatment found in first processing facilities.

Pollutants selected for regulation of meat and poultry processing facilities directly discharging into surface waters include Biochemical Oxygen Demand (BOD), total suspended solids (TSS), hexane extractable materials (oil and grease), fecal coliforms, nitrogen in the form of ammonia, and total nitrogen. These selected pollutants represent the constituents characteristic of meat and poultry processing wastewater, and are key indicators of the efficacy and performance of treatment processes in place at fixed facilities. Effluent limitations

Table 12.1 Distribution of wastewater treatment options in meat and poultry processing industry.

Treatment category	Treatment option	Percentage of direct discharging facilities having the treatment option in place
Primary treatment	Screen	98
	Oil and grease removal	83
	Dissolved air floatation	81
	Flow equalization	75
Secondary and tertiary treatment	Biological treatment[a]	100
	Filtration	23
	Disinfection	92

Source: Table adapted from EPA Effluent Limitations Guidelines and New Source Performance Standards for the Meat and Poultry Products Point Source Category, Final Rule (40 CFR Part 432).
[a]Biological treatment includes any combination of the following: aerobic lagoon, anaerobic lagoon, facultative lagoon, any activated sludge process, and/or other biological treatment processes (e.g., trickling filter).

for these pollutants differ between meat and poultry processing and depend on the size and type of facility (e.g., simple or complex slaughterhouses). The current regulation also requires facilities to maintain a pH between 6.0 and 9.0 in effluent wastewater at all times. More detailed information on pollutant effluent limitations can be found in the EPA Final Rule on Effluent Limitations Guidelines for Meat and Poultry Products (40 CFR Part 432, Appendix C) (USEPA, 2004).

12.5.4 On-farm processing and mobile slaughter units

On-farm processing and MSU for the slaughter of meat and poultry products generate both liquid and solid waste products similar to fixed facilities, though on a smaller scale. The options for management of on-farm processing and MSU waste products will depend primarily on individual state and county regulations regarding waste. The state agency charged with regulatory authority over meat and poultry processing wastes from on-farm processing and MSUs will vary by state, and may fall under either the state's department of environmental protection or department of agriculture resources or an equivalent agency. In general, there are three primary options for the management of liquid and solid wastes from on-farm processing and MSUs, and these treatment options are outlined in Table 12.2.

Table 12.2 Waste management options for on-farm or mobile meat and poultry slaughter units.

Type of waste	Treatment option
Solid waste	Incineration
	Sell to rendering company
	Farm-scale composting
Liquid waste	Discharge into municipal waste stream entering treatment plant[a]
	Hold in septic tank, pump out, and transport to treatment facility
	Apply to fields with nonfood growing crops

[a]Pretreatment with a screen and an oil separation unit is required.

12.6 CONCLUSIONS

In this chapter, the standards for slaughter of organic meat animals in the United States are discussed. It touches upon the traditional fixed facility, and then explores the options of on-farm processing and the use of mobile processing units. Waste management is discussed for the traditional fixed facility and the alternatives for on-farm processing and mobile processing units are examined.

ACKNOWLEDGMENT

Support in writing this chapter is gratefully acknowledged from National Integrated Food Safety Initiative (NIFSI) (2008–51110–04339) and USDA Food Safety Consortium grants to author Steven C. Ricke, and an NIFSI (2010–51110–21004) to author Philip G. Crandall.

REFERENCES

Agricultural Marketing Service. 2011. National Organic Program. Available at: http://www.ams.usda.gov/AMSv1.0/nop (accessed January 07, 2011).

Barkema, A., M. Drabenstott, and N. Novack. 2001. The new U.S. meat industry. *Econ. Rev.* Second Quarter. 33–56.

Codex Alimentarius. 2010. Available at: http://www.codexalimentarius.net/web/index_en.jsp (accessed January 11, 2011.

DAFF. 2009. National standard for organic and biodynamic produce. Available at: http://www.daff.gov.au/_data/assets/pdf_file/0018/126261/national-standard.pdf (accessed January 11, 2011).

Dychdala, G. R. 2001. Chlorine and chlorine compounds. In: S. S. Blocked (ed.) *Disinfection, Sterilization and Preservation.* Lippincott Williams & Wilkins, Philadelphia, PA. pp. 131–151.

eXtension. 2011. Mobile slaughter/processing units currently in operation. Available at: http://www.extension.org/pages/Mobile_slaughter%2Fprocessing_units_currently_in_operation (accessed January 11, 2011).

Federal Register. 1996. Pathogen reduction; hazard analysis and critical control point HACCP) systems; final rule. 61:38805–38989.

Food Safety and Inspection Service (FSIS). 2010. Mobile slaughter unit compliance guideline. Available at: http://www.fsis.usda.gov/PDF/Compliance_Guide_Mobile_Slaughter.pdf (accessed January 11, 2011).

International Organization for Standardization. 2010. Available at: http://www.iso.org/iso/home.html (accessed January 11, 2011).

Minka, N. S. and J. O. Ayo. 2010. Physiological responses of food animals to road transportation stress. *African J. Biotechnol.* 9:6601–6613.

USEPA. 2004. Technical development document for the final effluent limitations guidelines and standards for the meat and poultry products point source category (40 cfr 432) *in* effluent guidelines: meat and poultry products (mpp), final rule. Available at: http://water.epa.gov/scitech/wastetech/guide/mpp/index.cfm (accessed November 01, 2010).

LIST OF RESOURCES

Codex Alimentarius – Organically Produced Foods. Available at: http://www.fao.org/DOCREP/005/Y2772E/Y2772E00.HTM.

Guidelines for the Production, Processing, Labeling and Marketing of Organically Produced Foods. Available at: www.fao.org/organicag/doc/glorganicfinal.pdf.

The National Organic Program including National List of Allowed and Prohibited Substances. Available at: http://www.ams.usda.gov/AMSv1.0/nop.

FSIS Database on Federally Inspected Plants. Available at: http://www.fsis.usda.gov/PDF/MPI_Directory_Establishment_Name.pdf.
State Meat and Poultry Processor Listing. Available at: http://www.extension.org/pages/State_Listings.
Local Harvest Niche Meat Processor Directory. Available at: http://www.localharvest.org/.
HACCP and Pathogen Reduction. Available at: http://www.fsis.usda.gov/Science/hazard_analysis_&_pathogen_reduction/index.asp.
Small Scale Poultry Processing. Available at: http://attra.ncat.org/attra-pub/poultryprocess.html.
Niche Meat Processing Webinars by the Cooperative extension System. Available at: http://www.extension.org/pages/Archived_Niche_Meat_Processor_Webinars.
Guidelines for National Pollutant Discharge Elimination System (NPDES) permits. Available at: http://cfpub.epa.gov/npdes/.

13 Alternatives to Traditional Antimicrobials for Organically Processed Meat and Poultry

T. Matthew Taylor, Rolf Joerger, Enrique Palou, Aurelio López-Malo, Raúl Avila-Sosa, and Thelma Calix-Lara

Abstract: The selection of a food antimicrobial for use in organic meat and poultry processing can be a complex decision. Several factors influence the activity of a food antimicrobial on a processed meat or poultry carcass, or on a further processed product. Antimicrobials should be chosen so that their presence results in inhibition of relevant spoilage and/or pathogenic microorganisms likely present in the food, but should not compromise the palatability of the food. This chapter discusses antimicrobials approved for use in organic food manufacture, including weak organic acids, chlorine and oxidizing compounds, microbially produced antimicrobial substances, and biopreservation technologies. Antimicrobial mechanisms, microorganisms sensitive to their presence and the levels at which sensitivity are observed, factors influencing their efficacy in differing foods, and the levels required for microbial inhibition are reviewed. Food antimicrobials present processors with options for preservation of food quality and safety in addition to physical processing.

Keywords: organic acids; chlorine; peroxyacetic acid; peroxide; ozone; nisin; lysozyme; biopreservation; meat; poultry

13.1 INTRODUCTION

The microbiological quality and safety of meat and poultry may begin to deteriorate almost at the point of animal slaughter (i.e., harvest), and will continue through consumption. The rates of safety and quality loss are influenced by a host of processes, some inherent to the conversion of muscle into meat and others the result of processing. Some factors that speed up the loss of microbiological safety and quality include, but are not limited to: (1) breakdown of meat by naturally occurring enzymes (proteases, lipases) into compounds more easily utilized by microbes; (2) contamination of meat with microorganisms that metabolize peptides and lipids into compounds releasing off-flavors, odors, or colors in meat (e.g., microbial synthesis of biogenic amines and H_2S-induced greening of meat); (3) contamination of meat by microbial pathogens; and (4) postprocess storage under abusive conditions allowing accelerated growth of microbial contaminants. Conversely, the application of physical and chemical interventions, and combinations thereof, can extend meat and poultry quality and safety by the inhibition or inactivation of microbes. Examples in the

Organic Meat Production and Processing, First Edition. Edited by Steven C. Ricke, Ellen J. Van Loo, Michael G. Johnson, and Corliss A. O'Bryan.

manufacture of organic meat and poultry products include the use of hot water and/or organic acid rinses for pathogen inactivation on carcass surfaces, vacuum packaging, refrigerated storage of carcasses and fabricated cuts, use of approved substances in further processed and/or ready-to-eat (RTE) products, and low-temperature storage, distribution, and retail display.

Food antimicrobials have been successfully applied to carcasses for inhibition of spoilage and pathogenic microorganisms, incorporated into brine and marinating solutions and injected into nonintact products for deep-tissue decontamination, added into formulations of, or applied topically to surfaces of processed products to inhibit surface-contaminating microbes (Acuff *et al.*, 1987; Hardin *et al.*, 1995; Byelashov *et al.*, 2010a, 2010b; Schirmer & Langsrud, 2010). Food antimicrobials are varied with respect to their sources, mode of action, and spectrum of antimicrobial activity. Most food antimicrobials exert bacteriostatic or fungistatic activity; multiple log-cycle reductions in populations of contaminating microbes are not likely to be observed. Indeed, use of a food antimicrobial will not result in the indefinite lengthening of a food's shelf life but only delay its loss.

13.1.1 Food antimicrobial classification

Although food antimicrobials are defined as chemical preservatives, they are distinguished from other preservatives in the US Code of Federal Regulations (CFR) by their antibacterial or antifungal activity that results in extension of product shelf life by preventing microbial growth (21CFR101.22(a)(5); 21CFR170.3(o)(2)). This chapter discusses those antimicrobials approved for use in the manufacture of organic meat and poultry products as set forth by the US Department of Agriculture—Agriculture Marketing Service (USDA-AMS) National Organic Program for the production of products labeled "organic," "100% organic," or "made with organic (specified ingredients or food groups)" (7CFR205.102). Readers are cautioned that, while the antimicrobials reviewed herein are currently approved for use in organic food manufacture, antimicrobials may gain or lose approval as federal standards are revised. Furthermore, given the requirements for certification of organic food processors, often accomplished by third-party auditors, it is not impossible for different interpretations of allowable antimicrobial usage to exist amongst auditors. Processors must, therefore, utilize antimicrobials that are approved or are likely to be approved for use in the production of organic meat or poultry.

Food antimicrobials are identified as either natural or traditional; many fall into both groupings based on the antimicrobial's history of use, sources, and regulatory approvals (Davidson & Taylor, 2007). Examples of natural antimicrobials include hen egg white lysozyme (HEWL), nisin produced by *Lactococcus lactis*, microbially produced acetic or lactic acid from food substrates, and organic acids synthesized by various fruiting plants (e.g., citric acid). Examples of traditional antimicrobials include acetic and benzoic acid, nisin, lysozyme, and natamycin. Many traditional antimicrobials possess natural sources, but can also be produced in large quantities by industrial synthesis or fermentation. The reader is directed to recent reviews of interventions available for organic food production and processing for more information (O'Bryan *et al.*, 2008; Sirsat *et al.*, 2009).

13.1.2 Factors affecting food antimicrobial efficacy

The factors that affect the activity of an antimicrobial in a food are diverse, related to the food product, its processing, the antimicrobial, and the targeted microorganisms (Drosinos *et al.*, 2009; Naidu, 2000). Gould (1989) identified four types of factors that impact the utility of antimicrobials applied to foods: (1) implicit (hereafter referred to as microbial), (2) intrinsic, (3) extrinsic, and (4) process-related. Intrinsic factors are the characteristics of the food's inherent chemical and physical nature; extrinsic factors are those associated with the process environment and postprocess storage of the product. Intrinsic factors include the product's physical structure, pH and presence of acidulants, moisture content and water activity (a_w), oxidation/reduction (REDOX) potential (E_h), and the presence of other compounds or ingredients that may synergize with or antagonize the antimicrobial. Of primary interest is product pH; pH is likely have the greatest impact on microbial survival and growth. Food pH is also likely to significantly influence the activity of many of the antimicrobials approved for use in meat and poultry (Eklund, 1983; Abee *et al.*, 1995; Gill & Badoni, 2004). Drying or dehydration of a food lowers a_w, reducing the available water for use by microorganisms and inhibiting enzyme activity and microbial growth in and on meat/poultry (Koutsoumanis *et al.*, 2004; Kinsella *et al.*, 2006; 2009). Reduction of E_h may result in growth of microaerophilic or anaerobic microorganisms as dissolved oxygen is consumed, generally resulting in microbial spoilage by the lactic acid bacteria (LAB), the Clostridia, and *Brochothrix* spp. Food-borne pathogens that may compromise the safety of vacuum-packaged or low-E_H meat and poultry include *Clostridium perfringens*, *Clostridium botulinum*, and various facultative anaerobes (Nychas *et al.*, 2007). Finally, interaction of various types of antimicrobials (e.g., phenolic essential oil (EO) compounds, bacteriocins) with food constituents (e.g., protein, lipid, multivalent cations) has been shown to reduce the activity of several food antimicrobials (Aureli *et al.*, 1992; Jung *et al.*, 1992).

Significant extrinsic factors affecting the activity of food antimicrobials include the temperature of postprocessing storage, relative humidity and gaseous atmosphere of the storage environment, and characteristics of packaging material used. Lowering of storage temperature inhibits microbial growth and proliferation for many microorganisms, and in some cases will halt replication altogether. Lowering of environmental relative humidity will result in slowing of microbial growth by the reduction of moisture available for growth of surface-contaminating microbes. Likewise, a decline in environmental oxygen will select for anaerobic and facultative microbes at the expense of aerobes, changing the nature and evidence of spoilage. In many products, the package used and its characteristics exert great influence over these extrinsic factors. Finally, process factors include the application of heat to the product (intensity of thermal processing, method of heating) and product drying or dehydration. Numerous studies have been published detailing the combination of thermal processing, dehydration, or high-pressure processing with differing food antimicrobials for the inhibition of spoilage and pathogenic microbes on meat and poultry products.

Davidson and Branen (2005) divided the microbial grouping into those factors directly related to the microorganisms contaminating the food and those related to the physicochemistry of the antimicrobial. Microbial factors include the inherent resistance of the targeted microbe(s) to the antimicrobial, the number of cells present at the time of antimicrobial application, the rate of cellular growth/replication, the life cycle phase of cells at the point of antimicrobial application (i.e., lag, log, stationary, or death phase), the potential for other contaminating microbes to antagonize targeted microorganisms, presence of sublethally injured cells, and the ability of microbes to form capsules or biofilms that aid survival in the presence of the antimicrobial (Gould, 1989). Antimicrobial-related factors predict the functionality of the compound as a result of its physico-chemistry. Key factors include the antimicrobial's inherent polarity and hydrophilic/lipophilic balance (HLB). Polarity and HLB have been suggested to be the primary factors influencing antimicrobial efficacy due to the need for an antimicrobial to possess some degree of water solubility to allow interaction with contaminating microorganisms present in the aqueous phase of a food, while simultaneously having some lipophilicity to allow interaction with the microbial membrane (Branen *et al.*, 1980; Robach, 1980). Directly related to the antimicrobial's structure is its mode of action, or method by which it exerts its antimicrobial activity. Although the mechanisms of some food antimicrobials have been elucidated, the exact mode of action of many of the naturally occurring and/or traditional antimicrobials remains only partially understood (Breukink & de Kruijff, 1999).

13.1.3 Selection of a food antimicrobial

While multiple substances are approved for use in the processing of organic meat and poultry, not all are appropriate for all products. Selection and application of an antimicrobial must be preceded by careful consideration of the factors that impact antimicrobial efficacy in the food in question Section 13.1.2.

Of interest are concerns related to the incorporation of an antimicrobial into the manufacturing process, the antimicrobial's likely efficacy against targeted microorganisms, and its potential impacts on the sensory attributes of the product. Few food antimicrobials are broadly active, but are inhibitory to only a focused group of microorganisms. For example, the lytic enzyme lysozyme is inhibitory toward multiple Gram-positive genera (e.g., *Staphylococcus*, *Micrococcus*, *Clostridium*) but is generally ineffective against Gram-negative bacterial organisms except when combined with a membrane-destabilizing agent (Branen & Davidson, 2004; Johnson & Larson, 2005). The strong oxidizers are likely to display antimicrobial activity against both Gram-positive and Gram-negative bacteria, fungi, and some viruses. Their utility, however, is limited by rapid decomposition and difficulty in obtaining application conditions that can be adequately controlled (Pohlman *et al.*, 2002a, 2002b). Processors wishing to implement the use of a food antimicrobial should consider the factors that will affect the activity of the antimicrobial in the food product, identifying optimal antimicrobial(s) for control of food-borne pathogenic and spoilage microbes and those conditions that are optimal for antimicrobial application.

13.2 WEAK ORGANIC ACIDS AND ASSOCIATED SALTS

The pH of a food is one of the most critical factors affecting the metabolic activity, growth, and proliferation of microorganisms in foods, including meat and poultry (Davidson & Taylor, 2007). Most microbial contaminants in foods grow optimally at neutral or near-neutral pH, though a growing body of work has reported many organisms capable of survival in foods with pH ranging from 4–9, and in some instances outside these limits (Doores, 2005). Though multiple organic acids have been successfully applied for the preservation of quality and safety of foods, the forthcoming discussion centers on those organic acids that are commonly applied to the processing of meat and poultry products approved for use in organic foods manufacture.

13.2.1 Organic acid mechanisms of action

Organic acids and their salts are considered weak acids, meaning they do not fully dissociate in water but do so in a pH-dependent manner. Their protonation state is predicted by the

Henderson–Hasselbalch equation, which relates the dissociation of an organic acid in solution as a function of system pH and concentrations of dissociated and protonated acid as:

$$\mathrm{pH} = \mathrm{p}K_\mathrm{a} + \log[\mathrm{A}^-]/[\mathrm{HA}],$$

where $\mathrm{p}K_\mathrm{a}$ is the acid dissociation constant and $[\mathrm{A}^-]$ and [HA] are the concentrations of dissociated and protonated acids, respectively. Consequently, the antimicrobial activity of organic acids is enhanced as the pH of the food is lowered to that of, or below, the $\mathrm{p}K_\mathrm{a}$ of the acid. Reduction in pH results in a greater concentration of protonated acid, decreasing the polarity of the molecule and increasing opportunity for diffusion of acid across the membrane and into the cytoplasm. However, the substitution of the donatable proton with a monovalent (Na^+, K^+) or multivalent (Ca^{+2}) cation significantly increases the solubility of organic acid in aqueous systems. Thus, a balance should be struck between the need to maintain acid solubility with the need to achieve maximal activity via pH reduction. Subsequent discussion of organic acids focuses on the antimicrobial efficacy of the undissociated acid, as the mode of action is thought to arise primarily from the activity of the protonated acid on the cell.

Organic acids are thought to effect microbial inhibition by two primary mechanisms: (1) cytoplasmic acidification with subsequent uncoupling of energy production and regulation, and (2) cytoplasmic accumulation of dissociated acid anion to toxic levels. Diffusion of an undissociated acid through a microbial membrane in a food where the pH of the cellular cytoplasm is higher than that of the surrounding environment favors the establishment of a transmembrane gradient (Gould, 1989). As protonated acid diffuses across the membrane, an environment is encountered favoring the dissociation of the acid into the acid anion and free proton (Eklund, 1983). The cell reacts by working to efflux the proton, in some cases exchanging the proton for some other cation (e.g., Na^+, K^+); this is the basis of the chemiosmotic theory (Mitchell, 1961; Mitchell & Moyle, 1969). It is proposed that the microbial membrane is impermeable to protons, requiring active transport to efflux protons and maintain pH homeostasis in the cellular interior (Brul & Coote, 1999; Hirshfield *et al.*, 2003). Freese *et al.* (1973, 1978) revised this theory, suggesting an uncoupling of electron transport from oxidative respiration. Neither theory completely explains the mode of action of the organic acids, as in any system where organic acids are used there is always some concentration of dissociated anion, and the presence of an uncoupling effect suggests the potential for other inhibitory effects to occur following the diffusion of the undissociated acid across the cell membrane.

Cherrington *et al.* (1991) identified possible modes of inhibition by organic acids resulting from accumulation of acid anion in the microbial cytoplasm. In addition to shifting the internal pH out of range for optimal enzymatic activity, protein and DNA/RNA syntheses are negatively affected by the presence of organic acids at elevated levels (Cherrington *et al.*, 1990). Shelef (1994) reported that sodium lactate exerted antimicrobial activity not only through the lowering of a_w. Russell (1992) reported accumulation of acid anion to be the driving force for inhibition of cells, as the accumulation of anion eventually negates the proton motive force (PMF) and inhibits the microbe's ability to realkalinize its cytoplasm. Others have reported that lactic acid permeabilizes the Gram-negative membrane and releases lipopolysaccharide (LPS) from the outer membrane (Alakomi *et al.*, 1999). Most recently, research has demonstrated that the interplay of all these mechanisms likely drives inhibition of microbes by organic acids (Figure 13.1) (Koczoń, 2009).

13.2.2 Citric acid and the citrates

Citric acid (CAS No. 77–92-9) is a hydroxy tricarboxylic acid produced naturally by various plants. It is water soluble, approved for direct addition to multiple foods, is affirmed as

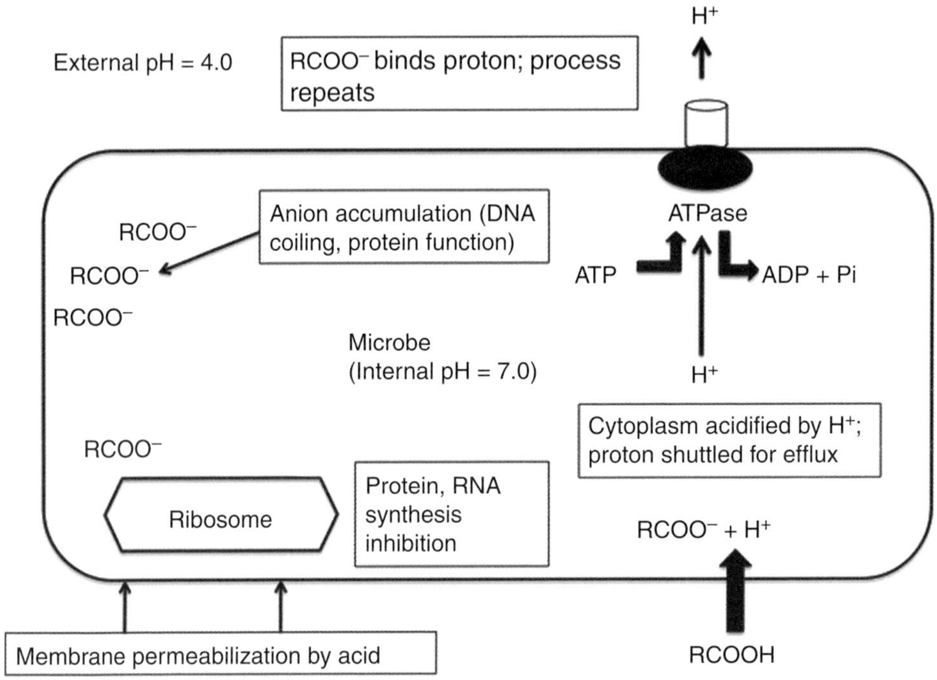

Figure 13.1 Theorized interrelationship between antimicrobial mechanisms of weak organic acids for inhibition of microbial pathogens. (Adapted from Davidson & Taylor, 2007.)

generally recognized as safe (GRAS) (21CFR184.1033), and is approved for use in the manufacture of fresh and processed meats and poultry at concentrations specific to its purpose (USDA-FSIS, 2010). In addition, its calcium, potassium, and sodium salts are also approved for use in organic food processing. In addition to the antimicrobial mechanisms discussed previously, citric acid is known to inhibit cells through metal chelation, in some instances resulting in enhanced pathogen inhibition versus the monocarboxylic acids (e.g., lactic acid) (Miller *et al.*, 1993).

The antimicrobial efficacy of citric acid against food-borne microorganisms in fluid media has been repeatedly documented. Oh and Marshall (1994) reported that the minimum inhibitory concentration (MIC) of citric acid against *Listeria monocytogenes* Scott A in tryptic soy broth supplemented with 0.6% yeast extract (TSB-YE) was 26 mM. Citric acid at 0.27% lengthened the lag phase of food-borne pathogenic and spoilage microbes, decreased the maximum growth rate, and increased the time required for bacterial populations to attain stationary phase (del Río *et al.*, 2007, 2008). Similar effects (i.e., lag phase lengthening, decreased total population) were observed against species of *Lactobacillus*, *Pseudomonas*, *Brochothrix*, and *Serratia* in brain heart infusion (BHI) and deMan, Rogosa, and Sharpe (MRS) broth supplemented with citric acid up to 1% (Ouattara *et al.*, 1997). *Salmonella enterica* serovar Typhimurium generation times in BHI containing citric acid (pH 6.4, 5.4, 4.5) increased from 0.36 hours to 0.54 hours, and finally to 0.75 hours, respectively (Álvarez-Ordóñez *et al.*, 2009). Citric acid, however, has also been shown to be less inhibitory than other organic acids and to contribute some level of protection to microbes during processing. Blackburn *et al.* (1997) reported the $D_{62.5^\circ C}$ for *Salmonella* Enteriditis and *Escherichia coli* O157:H7 to be 43.8 and 61.5 minutes, respectively, in TSB containing 1% w/w citric acid (pH 4.8) and 3.5% w/w NaCl. By comparison, $D_{62.5°C}$ for these pathogens in TSB containing 1% w/w acetic acid (pH 4.8) and 3.5% NaCl were 13.3 and 48.8 minutes, respectively, for *Salmonella* and *E. coli* O157:H7. This phenomenon is thought to result from the capacity of the organic acid to act as a buffering agent given its triprotic nature (Abdul-Raouf *et al.*, 1993).

Multiple recent studies have examined the antimicrobial efficacy of citric acid against food-borne microbes on fresh meat, poultry, and processing surfaces. Tamblyn and Conner (1997) compared antimicrobial effects of multiple organic acids applied at points along the poultry slaughter continuum for the reduction of *S.* Typhimurium on chicken skin. Citric acid (4%) in a simulated chiller tank resulted in a 1.9 $\log_{10}$ reduction compared to controls for attached *S.* Typhimurium cells. By comparison, 4% lactic acid application resulted in 0.7 $\log_{10}$ reduction under similar conditions (Tamblyn & Conner, 1997). Conversely, reduction of *E. coli* O157:H7 on lean and adipose tissues of beef carcasses with intact fascia treated with 1%, 3%, or 5% lactic, acetic, or citric acid were similar; acid type did not significantly influence observed pathogen inhibition (Cutter & Siragusa, 1994). Spray application of the commercial antimicrobial intervention Chicxide (Birko Corp., Denver, CO, USA), containing blended lactic and citric acids, to surfaces of broiler carcasses resulted in a 1.3 $\log_{10}$ CFU/mL reduction in inoculated salmonellae, whereas immersion of broiler carcasses in the antimicrobial intervention for up to 20 seconds resulted in 2.3 $\log_{10}$ CFU/mL reduction in inoculated *Salmonella* (Laury *et al.*, 2009). A similar product, Beefxide (Birko Corp., Denver, CO, USA), produced *E. coli* O157:H7 and *Salmonella* reductions of 1.4 and 1.1 $\log_{10}$ CFU/100 cm^2 on inoculated fresh beef (Laury *et al.*, 2009). González-Fandos *et al.* (2009) evaluated citric acid inhibition of *L. monocytogenes* on refrigerated (4°C) poultry skin; significant reductions in *L. monocytogenes* ($p < 0.05$) were realized on citric acid-treated samples as compared to controls. However, pathogen recovery was also observed during product storage, with no differences in surviving populations on treated and control samples remaining after 8 days of storage (González-Fandos *et al.*, 2009). Citric acid at 3% did not produce unacceptable odors and color acceptability was retained for longer periods versus untreated controls (González-Fandos *et al.*, 2009).

In addition to fresh meat and poultry applications, citric acid has been studied for the decontamination of processed meat and poultry products. Palumbo and Williams (1994) inoculated frankfurters with *L. monocytogenes* and then dipped franks in solutions of 1% organic acid (lactic, acetic, citric, citric + acetic) for 2 minutes. Following 80 days of 5°C vacuum storage, *L. monocytogenes* counts on citric acid-treated surfaces were ~2.0 $\log_{10}$ less than inoculated nontreated controls; maximal reductions were observed on franks treated with a mixture of citric + acetic acid (~4.0 $\log_{10}$ less than controls) (Palumbo & Williams, 1994). Miller *et al.* (1993) applied organic acid salts to *C. botulinum*-inoculated uncured turkey roll; use of 6% citrate resulted in delay of *C. botulinum* toxin detection to 18 days postinoculation (control samples had detectable toxin at 1 day postinoculation). Citric acid and its salts have demonstrated efficacy for pathogen control in both fresh and processed meat and poultry, but their usage is potentially limited by possible negative sensorial impacts and the need for low-pH maintenance for optimum antimicrobial activity.

13.2.3 Lactic acid and the lactates

Lactic acid (2-hydroxypropanoic acid) is a monocarboxylic acid (pK_a 3.79) produced during anoxic respiration or by fermentation by several bacterial microorganisms, including the LAB (Axelsson, 1998). It occurs in two isomeric forms (D-, L-); it has been reported that the

L-isomer is more effective for pathogen inhibition (McWilliam & Stewart, 2002a, 2002b). Lactate is used as a flavoring agent, color stabilizer, and can be useful for its effects on retardation of lipid oxidation and subsequent off-odor development (Papadopoulos *et al.*, 1991a, 1991b). It is allowed for direct addition to various foods and is affirmed as GRAS (21CFR184.1061).

Lactic acid is approved as an antimicrobial for application to animal carcasses (pre- and postchilling; $\leq 5\%$ acid solution), subprimal cuts and trimmings (2.0%–3.0% acid, $\leq 55°C$), and to beef heads and tongues (2.0%–2.8% in washing systems) (USDA-FSIS, 2010). Its use in the meat industry is widespread and many researchers have documented its efficacy for the reduction of enteric pathogens on the surfaces of carcasses and derived cuts (Dixon *et al.*, 1987; Castillo *et al.*, 1998, 1999, Ellebracht *et al.*, 1999; Delmore *et al.*, 2000, 2001; Baird *et al.*, 2006; Bosilevac *et al.*, 2006). It is used in manufacturing further processed and RTE meats, added on its own or in combination with other antimicrobials, or through incorporation into antimicrobial interventions such as acidic calcium sulfate (Nuñez de Gonzalez *et al.*, 2004). Nevertheless, while lactic acid currently bears USDA approval for use in organic foods manufacture, the lactates (e.g., sodium lactate, potassium lactate) are not explicitly approved, despite wide acceptance and use in the industry for both preservation of product safety and quality.

Oh and Marshall (1994) reported lactic acid in TSB-YE inhibited *L. monocytogenes* at 56 mM, while others have reported that 75 mM acid was required for *Salmonella*, *L. monocytogenes*, and *E. coli* O157:H7 inhibition (Over *et al.*, 2009). After 30 minutes of incubation at 21°C, *S.* Enteriditis populations in water were reduced by 1.5 $\log_{10}$ following 0.5% lactic acid (pH 2.6) application; combination of 0.5% acid with 0.05% sodium dodecyl sulfate reduced the pathogen to below detectable limits (Zhao *et al.*, 2009). Likewise, 0.4% lactic acid in Mueller–Hinton broth (MHB) effected a 3.0 $\log_{10}$ decrease in *S.* Typhimurium populations over 24 hours of incubation at 37°C (Zhou *et al.*, 2007). A 1.0–1.5 $\log_{10}$ reduction was observed against *Salmonella* serovars (Enteriditis, Typhimurium, Indiana) following 0.75% lactic acid application after only 5 minutes. (Kanellos & Burriel, 2005). Sodium lactate (2%, pH 5.5) reduced *L. monocytogenes* populations by ~3.0 $\log_{10}$ after 72 hours of incubation at 37°C (Apostolidis *et al.*, 2008).

The efficacy of lactic acid and lactates has been studied intensely with regard to their antimicrobial utility on animal carcasses and derived cuts. Castillo *et al.* (1998, 1999) reported that carcass rinses consisting of hot water spray followed by lactic acid were more effective at inhibiting enteric pathogens on beef carcasses than either water or acid alone. Carlson *et al.* (2008a), in studies testing the efficacy of lactic acid for beef carcass hide decontamination, determined that application of 10% lactic acid at 23°C for 7 seconds was more effective at reducing *E. coli* O157:H7 than 5% sodium metasilicate, 4% NaOH, or 10% acetic acid applied similarly. Subsequent work by this group demonstrated that *E. coli* O157:H7 and *Salmonella* were reduced on hides by 3.4 and 2.8 $\log_{10}$, respectively (10% lactic acid, 55°C) (Carlson *et al.*, 2008b). Heller *et al.* (2007) observed 1.0–1.1 $\log_{10}$ reduction of *E. coli* O157:H7 inoculated on beef outside rounds treated with 55°C-tempered lactic acid at 2.5% and 5.0%, respectively. Spraying of 2.0% and 4.0% lactic acid on beef trim surfaces resulted in reductions of *E. coli* O157:H7 and *Salmonella* of ~2.0 and 1.5 $\log_{10}$, respectively (Harris *et al.*, 2006). These reductions were subsequently observed when trimmings were ground and stored under refrigeration for up to 5 days post-lactic acid application, with only a ~0.5 $\log_{10}$ increase in *E. coli* O157:H7 and *Salmonella* populations on acid-treated beef (Harris *et al.*, 2006). Lactic acid has been successfully applied for decontamination of poultry carcasses and pieces. Hwang and Beuchat (1995) demonstrated that a 0.5% lactic acid/0.05% sodium benzoate solution was able to reduce *Salmonella* on chicken wings to nondetectable levels after 4–6 days at 4°C. *Campylobacter jejuni* and *E. coli* O157:H7 populations were reduced by 1.5 and 1.8 $\log_{10}$, respectively, following exposure to the acid wash (Hwang & Beuchat, 1995). In similar trials, disinfection of turkey carcasses using a 4.25% lactic acid rinse produced reductions of 3.0 and 5.5 $\log_{10}$ in aerobic and coliform plate counts, respectively (Bautista *et al.*, 1997). These authors also reported that *Salmonella* was significantly reduced on carcass surfaces; muscle tissue, however, was discolored at elevated acid concentrations (1.24% lactic acid).

Lactic acid and the lactates are also effective for the disinfection of surfaces of processed meat and poultry products. While the lactates are not explicitly approved for use in organic food processing, their discussion is nonetheless beneficial. Nuñez de Gonzalez *et al.* (2004) determined the efficacy of antimicrobial dips on frankfurters formulated with or without lactates, reporting that 30 seconds' dips of franks in 3.4% lactic acid solution followed by refrigerated storage controlled *L. monocytogenes* growth on surfaces and that pathogen numbers were ~2.0 $\log_{10}$ less than controls over 12 weeks. Incorporation of 1.8% sodium lactate and 0.125%–0.25% sodium diacetate into frankfurters, followed by dipping in 2.5% lactic acid solution, resulted in *L. monocytogenes* levels being maintained at or below 1.0 $\log_{10}$ CFU/cm^2 over 40 days of vacuum storage at 10°C (Barmpalia *et al.*, 2004). As with citric acid, processors wishing to use lactic acid must confirm not only the antimicrobial efficacy for the product(s) of interest, but also account for potential undesirable effects on product quality and sensorial attributes.

13.2.4 Tartaric acid, malic acid, and their salts

Tartaric and L-malic acids are dicarboxylic organic acids found naturally in various fruiting plants and berries (Table 13.1). In addition to being approved for use in the manufacture of organically processed foods, both are approved for direct addition to foods and affirmed as GRAS (21CFR184). Doores (2005) reported that both function as antimicrobials by acidification, though their inhibitory potential is not as great as that of other organic acids. In addition to reported antibacterial activity, both have been shown to possess antifungal activity.

Álvarez-Ordóñez *et al.* (2009) determined the ability of *S.* Typhimurium to activate its acid tolerance response (ATR) phenotype following growth in BHI acidified to pH 4.5 with various acids, including malic acid. Compared to controls, malic acid exposure resulted in longer generation times and lag phase ($p < 0.05$), but not in time to achieve stationary phase, indicating the pathogen was able to adapt to the acid. Interestingly, the pathogen's heat resistance also increased following incubation in the acid-containing medium, increasing the time to inactivate 1.0 $\log_{10}$ from 9.6 minutes to >42 minutes (Álvarez-Ordóñez *et al.*, 2009). Incubation of *C. jejuni* in BHI containing 0.5% v/v malic or tartaric acid resulted in a 6.0 $\log_{10}$ reduction following 24 hours of incubation at 4°C (Birk *et al.*, 2010). Likewise, in broth supplemented with 75-mM tartaric or malic acid, populations of *L. monocytogenes*, *E. coli* O157:H7, or *S.* Typhimurium were reduced by 4.0 $\log_{10}$ following 12 hours of storage at 4°C (Over *et al.*, 2009).

Following 3 days of refrigeration, *C. jejuni* populations on chicken medallions exposed to 4% or 6% w/v tartaric acid (pH 2.4) were reduced by 1.0 or 1.5 $\log_{10}$, respectively (Birk *et al.*, 2010). *L. monocytogenes* on frankfurter surfaces were 2.0 $\log_{10}$ less than controls following dipping in a 1% tartaric acid solution and storing for 80 days under refrigerated

Table 13.1 Physicochemical properties of weak organic acids approved for use in organic meat and poultry processing.

Name	Molecular weight (g/mol)	Synonym	Chemical formula	Melting point (°C)	Density (g/mL)	pK_a(s) at 25°C
L-Ascorbic acid	176.12	Vitamin C	$C_6H_8O_6$	191	1.65 at 25°C	4.04, 11.70
Citric acid	192.12	2-Hydroxy-1,2,3-propanetricarboxylyic acid	$C_6H_8O_7$	153	1.67 at 20°C[a]	3.13, 4.76, 6.40
D-, L-Lactic acid[a]	90.08	D-, L-2-Hydroxypropanoic acid	$C_3H_6O_3$	53	1.21 at 21°C	3.86[b]
Malic acid	134.09	Hydroxybutanedioic acid	$C_4H_6O_5$	132	1.60 at 20°C	3.40, 5.11
D-, L-Tartaric acid[c]	150.09	2,3-Dihydroxybutanedioic acid	$C_4H_6O_6$	206	1.76 at 20°C[d]	2.98, 4.34[e]

Source: Lide (2010).
[a]Properties given are identical for isomers of lactic acid except where indicated.
[b]Value listed is specific for D-,L-lactic acid.
[c]Properties given are identical for isomers of tartaric acid except where indicated.
[d]Density given is specific to D-tartaric acid.
[e]Listed pK_as are specific to L-tartaric acid.

vacuum (Palumbo & Williams, 1994). Riedel *et al.* (2009) demonstrated that *C. jejuni* in chicken skin rinsate was reduced by 0.9 $\log_{10}$ following a 1 minute dip in 2% w/v tartaric acid (pH 2.6). Following a 1 minute dip and 24 hours of incubation under refrigeration, levels of the pathogen were reduced by 3.7 $\log_{10}$ compared to the water-treated control samples, where only ~1.0 $\log_{10}$ reduction was observed (Riedel *et al.*, 2009). Nonetheless, tartaric and malic acid utility may be limited by factors similar to those discussed previously for citric and lactic acid.

13.3 CHLORINE AND THE OXIDIZING ANTIMICROBIALS

According to 7CFR205.605, the oxidizing agents, chlorine materials, and hydrogen peroxide, ozone, and peroxyacetic acid are permitted, with certain restrictions, for disinfection and surface sanitation in organic food production. With respect to the "chlorine materials," three compounds are listed: the sodium and calcium salts of hypochlorous acid, and chlorine dioxide. Inclusion of chlorous compounds in organic agriculture and food processing has not been without reservations, as seen in a statement by the National Organic Standards Board (NOSB) (2003), which reads

> It was the NOSB's opinion that while chlorine needs to be allowed in the handling of organic food out of concern for public health and safety, its use needs to be minimized and operators need incentives and clear guidance to develop viable alternatives that protect the public as effectively as chlorine, but are less harmful to food handlers and the environment.

13.3.1 Chlorine disinfection and intermediate formation

The use of chlorine in food processing is common, but not universal. For example, while its use in poultry chillers is ubiquitous in the United States, it is not allowed in the European Union (EC Regulation No. 853/2004 Article 3). The efficacy of antimicrobial chlorine treatment is influenced, to a large extent, by the inevitable presence of organic matter alongside microorganisms. Adequate chlorination can be achieved by dissolving chlorine gas (Cl_2) in water where it forms hypochlorous acid (HOCl) according to the reaction:

$$Cl_2 + H_2O \leftrightarrow HOCl + Cl^- + H^+$$

Alternatively, dissolving sodium or calcium hypochlorite in water can generate HOCl via:

$$NaOCl + H_2O \leftrightarrow HOCl + Na^+ + OH^-$$

$$Ca(OCl)_2 + 2H_2O \leftrightarrow 2\,HOCl + Ca^{+2} + 2\,OH^-$$

Hypochlorous acid can dissociate into OCl^- and H^+, but between pH 2 and 7, HOCl is the predominant species (Deborde & von Gunten, 2008). This acid is capable of reacting with numerous inorganic and organic compounds by oxidation, addition to unsaturated bonds, or electrophilic substitution at nucleophilic sites (Deborde & von Gunten, 2008). Such reactions diminish or abolish the function of vital molecules on or in microorganisms, leading to cell injury or death. Presumably, chlorination affects multiple cellular targets. Changes in membrane permeability have been observed, but their contributions to cell death are not fully understood (Arana *et al.*, 1999). Unsaturated fatty acids are potential targets for chlorination, but the reaction can be slow or negligible (Gibson *et al.*, 1986). However, chlorination of sulfur- or amine-containing compounds is much more efficient, and in its nonionized state, HOCl is expected to pass easily through the membrane of cells and react with internal proteins and nucleic acids (Deborde & von Gunten, 2008). Hypochlorous acid not only inactivates fungal and bacterial agents, but also viruses (Page *et al.*, 2010; Sobsey *et al.*, 1998). Damage to proteins involved in the binding of the virus to host cells and of proteins involved in processes following binding of virus to hosts appears sufficient for inactivation of viruses (Nuanualsuwan & Cliver, 2003).

In virtually all cases, the reaction of HOCl with organic and some inorganic molecules leads to formation of chlorinated intermediate products. The formation of such compounds (e.g., trihalomethanes) is at the heart of concerns surrounding the use of chlorine for the purpose of water disinfection, and perhaps even more so for use on foods. Some of these compounds are known carcinogens, possess genotoxicity, and may affect reproduction (Richardson *et al.*, 2007; Legay *et al.*, 2010). Ever since such compounds were detected in drinking water in the 1970s, concerns over chlorination and its ecological and health consequences have been raised (Rook, 1974). The extent of danger posed by chlorinated drinking water is still not known, and even fewer data are available relating potential hazards associated with consuming foods exposed to chlorine (Nieuwenhuijsen *et al.*, 2009). Vizzier-Thaxton *et al.* (2010) tested chicken leg quarters for the presence of trihalomethanes but were unable to detect such compounds in samples exposed to 50 ppm of chlorine. Only exposure to 70 and 100 ppm resulted in the detection of minute amounts of trihalomethanes at levels below allowable limits in drinking water.

13.3.2 Chlorine disinfection of meat and poultry

Strict maintenance of permitted chlorine levels in processing environments is necessary to achieve satisfactory levels of microbial inactivation. Loretz *et al.* (2010) recently reported that spray or immersion application of chlorinated water to poultry carcasses resulted in reductions of naturally occurring aerobic bacteria, coliforms, and *E. coli* of 0.2–1.2 $\log_{10}$ CFU. Higher reductions were seen on poultry carcasses artificially contaminated with *C. jejuni* (0.5–3.0 $\log_{10}$) (Kim *et al.*, 2005; Li *et al.*, 2002; Park *et al.*, 2002) and *E. coli* (1.9–2.1 $\log_{10}$) (Northcutt *et al.*, 2005). When immersed in chlorinated hot water, lamb carcasses experienced reductions of 1.6 $\log_{10}$ total aerobes (James *et al.*, 2000). Calcium hypochlorite (pH 5.0) at 60, 70, and 100 mg/L reduced the viability of *L. monocytogenes* on artificially contaminated chicken breast meat by 2.8–4.4 $\log_{10}$ (Gonçalves *et al.*, 2005). The need for relatively high ($>$ 500 ppm) levels of chlorine (supplied as NaOCl) to achieve about 1.0 $\log_{10}$ reduction of *E. coli* on beef carcass surfaces was also demonstrated by Cutter and Siragusa (1995).

One antimicrobial treatment that should, at least theoretically, produce few chlorinated compounds, is chlorine dioxide (ClO_2). It is viewed as being at least as effective as HOCl, though its cost of use is higher than the hypochlorides, and it has the disadvantage that it must be generated on-site. Chlorine dioxide is an explosive gas that cannot be compressed and is thus not easily transported. It has to be generated from sodium chlorite ($NaClO_2$) by reaction with gaseous Cl_2, hypochlorite, or hydrochloric acid (HCl) (EPA, 1999). Its decomposition does not lead to chlorinated by-products because its reaction mechanism differs from that of other chlorous compounds, specifically HOCl. The compound is a radical capable of easily accepting electrons:

$$ClO_2\,(aq) + e^- \leftrightarrow ClO_2^-,$$

thereby oxidizing other compounds. In practice, certain pH conditions or the presence of contaminating HOCl can lead to the formation of unwanted chloro-organic compounds or of chlorate (ClO_3^-) (EPA, 1999). However, in a mutagenicity assay based on the Ames test, only "negligible mutagenicity" of chlorine dioxide-treated poultry chiller water was detected when even ClO_2 was applied at four times the level required for disinfection (Tsai *et al.*, 1997).

Though ClO_2 has been extensively studied for its antimicrobial utility on fresh produce, studies on meat and poultry are rare. Pohlman *et al.* (2002a) studied a combined ClO_2/trisodium phosphate (TSP) treatment of beef and realized a reduction in $\log_{10}$ CFU of only 0.6, 0.4, and 0.3 for *E. coli*, coliforms, and total aerobes, respectively. The authors mentioned possible pH effects of TSP for the mild levels of bacterial inactivation, but it has been demonstrated that ClO_2 efficacy is compromised by macromolecular constituents of meat (protein, lipid) (Vandekinderen *et al.*, 2009). Valderrama *et al.* (2009) observed divalent cation-induced reductions in antilisterial activity of ClO_2 in chilled brines used to rapidly cool RTE meats prior to packaging, determining that ClO_2 concentrations of 3–30 ppm were ineffective in spent brines. Finally, as with HOCl, chlorine dioxide reacts with multiple targets in microorganisms and viruses. The amino acids cysteine, tryptophan, and tyrosine in proteins are oxidized, as are nucleic acids and lipids (Gordon & Rosenblatt, 2005). Most likely, cumulative impairment of enzymatic, genetic, and membrane functions results in microbial death (Bernarde *et al.*, 1967; Jeng & Woodworth, 1990; Huang *et al.*, 1997; Young & Setlow, 2003).

13.3.3 Antimicrobial utility of ozone

Like chlorine dioxide, ozone is a gas that must be generated on-site for application as a disinfectant or sanitizer, but generation of ozone does not involve use of toxic chemicals. When oxygen in air or in its molecular form is subjected to ultraviolet irradiation or bombarded with electrons by corona discharge, some oxygen (O_2) molecules are split into free radicals that are able to react with intact O_2 to form ozone (O_3). With air as a feed gas, ~1%–3% ozone may be generated; with pure oxygen as starting material, up to 6% O_3 may be produced. Ozone is unstable and cannot be stored; it can be applied as a gas for disinfection or sanitation purposes, but it may also be applied aqueously. Ozone is more water soluble than molecular oxygen, but its solubility decreases with increasing temperature; virtually no solubility is observed at 60°C. Maintaining an effective concentration of O_3 in open aqueous systems, such as chillers, is challenging. In water, as in air, ozone is reactive and decays rapidly. The formation of oxy-radicals confers antimicrobial properties to ozone; these radicals oxidize cell components on or within the cell surface. Sulfhydryl (–SH) groups are easily oxidized, and amino acid oxidation results in breakdown of protein structure and function. Unsaturated fatty acids are also targets for oxidation, and physical disruption of membrane integrity can lead to leakage and death. Nucleic acids are attacked, and viruses may be inactivated by destruction of nucleic acid or coat proteins. The oxidation products of O_3 are not considered a health concern, and any ozone in a processing environment ultimately decays, leaving no residue. Wastewater treated with O_3 exhibited no toxicity in a bioluminescence assay 2 hours posttreatment, whereas the same water treated with chlorine still exhibited toxicity (Arana *et al.*, 1999). Though ozone has received GRAS status (21CFR184.1563) for treatment of water, this compound causes lung damage and precautions must be taken to minimize inhalation (Lippmann, 1989).

Few studies have utilized ozone for disinfection of meat and poultry products. Moore *et al.* (2000) reported that meat-derived broth reduced the antimicrobial efficacy of ozone substantially. When beef carcasses were sprayed with ozonated water (95 mg/L), the antimicrobial did not reduce *E. coli* O157:H7 and *S.* Typhimurium populations beyond what water alone could achieve (Castillo *et al.*, 2003). Reductions in aerobic bacteria and coliforms on pork loins of 0.5–1.0 $\log_{10}$ CFU were achieved by exposure to ozone gas, and similar reductions of inoculated *Salmonella* Infantis and *Pseudomonas aeruginosa* were observed on chicken skin exposed to >2000-ppm gaseous O_3 (Al-Haddad *et al.*, 2005; Jeong *et al.*, 2007). Sheldon and Brown (1986) previously observed a reduction of <1.0 $\log_{10}$ CFU on poultry carcasses treated with gaseous ozone. Somewhat higher reductions were observed on focally contaminated beef brisket samples sprayed with water followed by spraying with 0.5% O_3 in spray cabinet (Gorman *et al.*, 1995). Novak and Yuan (2004a) reported that O_3 sensitized *C. perfringens* on beef surfaces to subsequent heating. Ozone in conjunction with heat (60°C) was shown to decrease vegetative cells of *C. perfringens* by 2.1 $\log_{10}$ CFU/g and spores by 1.2 $\log_{10}$ CFU/g (Novak & Yuan, 2004a, 2004b).

13.3.4 Hydrogen peroxide and peroxyacetic acid

Hydrogen peroxide is another agent approved for use in organic food processing. It has found application for disinfecting hatching eggs in the poultry industry (Cox *et al.*, 2007). Unlike ozone and chlorine dioxide, it does not have to be generated on-site; it is industrially produced and can be safely shipped in aqueous solution. Hydrogen peroxide is a weak acid and a strong oxidizer. Its antimicrobial activity is more pronounced at acidic pH since the fully protonated form is more reactive. When dissolved in pure water, it slowly decomposes into water and molecular oxygen:

$$2\,H_2O_2 \rightarrow 2\,H_2O + O_2$$

In the presence of susceptible molecules, H_2O_2 rapidly reacts to form oxidized products and water. Of interest is the generation of highly reactive hydroxyl radicals in the presence of ions such as ferrous iron (Fe^{+2}) (Goldstein *et al.*, 1993).

When applied as a 3% solution for spray washing *Salmonella*-inoculated lean and adipose beef tissue, H_2O_2 reduced the pathogen by less than 1.0 $\log_{10}$ compared to water spray (Bell *et al.*, 1997). Combining H_2O_2 with 1% acetic acid resulted in increased pathogen reduction, producing decreases of just over 1.0 $\log_{10}$ (Bell *et al.*, 1997). Only very low levels of reductions in aerobic bacteria were observed when a mixture of H_2O_2 and sodium bicarbonate were sprayed on poultry carcasses (Fletcher *et al.*, 1993). When applied immediately after inoculation, ~3.0 $\log_{10}$ reduction of *E. coli* on beef-fat fascia was achieved by spraying with 5% H_2O_2; pathogen reductions were decreased considerably when the antimicrobial containing spray was applied 24 hours postinoculation (Cabedo *et al.*, 1996).

The controlled reaction of hydrogen peroxide with acetic acid in the presence of sulfuric acid as a catalyst produces peroxyacetic acid (i.e., peracetic acid (PAA)). PAA bears approval for use in organic food environments for decontamination/disinfection of wash/rinse waters and food process surfaces. Unlike hydrogen peroxide that is formed naturally during aerobic metabolism and which can be enzymatically destroyed, PAA cannot be inactivated enzymatically. It decomposes to acetic acid and H_2O_2 in water as equilibrium products, and the resulting peroxide can lead to oxidation of cellular structures/components. PAA can also break down to form acetic acid and oxygen; acetic acid likely contributes to the antimicrobial mechanism of PAA via acidification. The maximum allowable concentrations for PAA and H_2O_2 in antimicrobial applications in poultry processing are 220 and 120 ppm, respectively (21CFR173.370). At 85 ppm, a 91.8% reduction in *Salmonella*-positive carcasses was observed in a chiller application; this reduction was higher than that achieved with 30-ppm chlorine (Bauermeister *et al.*, 2008). Nonetheless, 200 ppm of PAA sprayed onto distal surfaces of pieces of brisket from chilled beef carcass quarters produced only a small reduction of aerobes, coliforms, or *E. coli* (Gill & Badoni, 2004). Limited effectiveness of spray chilling with 200 ppm PAA was also observed in beef carcass tissue inoculated with *E. coli* O157:H7 (Stopforth *et al.*, 2004). Relatively low levels of reductions in pathogenic bacteria ($<1 \log_{10}$) were observed in artificially contaminated chicken legs dipped in solution of 220 ppm peroxyacids (del Río *et al.*, 2007).

Ozone and peroxides have higher oxidation potentials than chlorine compounds. However, this reactivity does not necessarily result in more efficient killing of microbes and viruses. When oxidizers are used up rapidly in the wash water or on the surfaces of microbes, penetration to targets that are important for cell survival might not be achieved; organisms may thus be injured but not inactivated. For all oxidizing antimicrobials, implementing measures to reduce the organic load during processing will increase effectiveness and potentially allow the use of lower amounts of the antimicrobial compounds. Such reductions can lead to cost savings and, in the case of chlorination, to reduced levels of by-products entering the wastewater stream. The higher costs of chlorine dioxide-, ozone-, and peroxide-based compounds make it difficult to replace chlorination in large-volume operations. For applications at smaller scales and for products for which consumers are willing to pay the extra cost, oxidizers that leave minimal residues are viable alternatives to chlorine.

13.4 ANTIMICROBIAL POLYPEPTIDES AND BIOPRESERVATION

Traditional methods of food preservation will not preclude microbial hazards in the manufacture of organic meat and poultry products. Moreover, consumers are demanding more natural or mildly processed foods with long shelf life but without chemical incorporation (Aymerich *et al.*, 2006). Thus, a wide range of natural preservatives derived from animals, plants, and nonpathogenic microorganisms have been identified and are being studied. Naturally occurring antimicrobial compounds are abundant in the environment; while some of these antimicrobials are currently employed for food preservation, many are just being studied for use in foods (López-Malo *et al.*, 2000). Some antimicrobials utilized in the food industry as biopreservatives are antimicrobial polypeptides that are widely distributed in nature and are used by nearly all life forms as essential components of the host's nonspecific immune response. More than 800 polypeptides and enzymes have been identified from microorganisms, insects, amphibians, plants, and mammals. In addition to their ability to kill microbes directly, these compounds seem to be able to recruit and promote various elements of host immunity and are produced both constitutively and by induction, predominantly in exposed tissues (e.g., lungs, nasal tracts, skin). They are rapidly synthesized at low metabolic cost, easily stored in large amounts, and have a wide antimicrobial spectrum as well as promote synergistic activity, providing for broad-spectrum coverage against a variety of microbial pathogens (Rydlo *et al.*, 2006). Though the USDA does not explicitly allow the use of such antimicrobials for organic meat and poultry processing, their presence is indicated on products treated with fermentative cultures (e.g., LAB), which are allowed in organic foods manufacture.

13.4.1 Nisin and the bacteriocins

Some bacteria synthesize polypeptides with bactericidal activity (i.e., bacteriocins). Biochemically, two types of bacteriocins have been identified in the LAB: (1) those characterized by the presence of posttranslationally modified amino acids (e.g., dehydroalanine) and/or thioether bridges between residues (i.e., lanthionine rings), usually identified as the lantibiotics (Class I), and (2) those containing unmodified amino acids (Klaenhammer, 1988). Nonlantibiotics are divided into Classes II–IV depending on size and presence of nonprotein moieties. Both lantibiotics and nonlantibiotics are ribosomally synthesized. Intensive research into the bacteriocins produced by the LAB has been undertaken with the aim of improving the microbial quality and safety of fermented products (de Vuyst & Leroy, 2007). In this section, the Class IA lantibiotic nisin as a model for the LAB-derived bacteriocins is discussed.

Nisin is a heat-stable polycationic peptide produced by *L. lactis* subspecies *lactis* fermentation. Multiple forms of nisin are known to exist (e.g., nisin A and nisin Z; nisin A differs from nisin Z by the substitution of His at position 27 [nisin A] for Asn at position 27 [nisin Z]) (Abee *et al.*, 1994). Nisin is recognized as safe for application in foods by the World Health Organization (WHO); nisin bears GRAS approval by the US Food and Drug Administration (FDA) (FDA, 1988). Nisin is active against Gram-positive bacteria including *Listeria*, *Clostridium*, *Bacillus*, and *Staphylococcus* spp. (Sobrino-López & Martín-Belloso, 2006). Nisin electrostatically binds its cationic *C*-terminus to anionic lipids on the exterior of a bacterium, subsequently intimately interacting with the peptidoglycan (PG) precursor lipid II and ultimately establishing a pore that dissipates the transmembrane electrochemical gradient (Breukink *et al.*, 1997, 2003; Van Kraau *et al.*, 1997; van Heusden *et al.*, 2002). However, it exhibits little or no activity against Gram-negatives, yeasts, or molds unless the outer membrane is destabilized or removed (Branen & Davidson, 2004; Al-Nabulsi *et al.*, 2009). Nisin is commercially available and used in a variety of foods including processed cheeses, eggs, vegetables, meat, fish, beverages, and cereals (Delves-Broughton, 1990; Delves-Broughton *et al.*, 1996).

Recently, nisin has been studied widely, both alone and in combination with other antimicrobials, for its efficacy in the inhibition of bacterial spoilage and pathogenic microbes on fresh meat and poultry (Table 13.2) (Ye *et al.*, 2008; Kouakou *et al.*, 2009; Ercolini *et al.*, 2010). Long and Phillips (2003) reported that the combination of 500 IU/mL nisin with 2% sodium lactate resulted in 2.7 $\log_{10}$ reduction of *Arcobacter butzleri* inoculated on chicken skin incubated at 5°C. Govaris *et al.* (2010) evaluated the antimicrobial effect of oregano EO (0.6%, 0.9%) in combination with nisin (500, 1000 IU/g) against *S.* Enteriditis in minced sheep meat during storage at 4°C or 10°C for 12 days. While treatment of sheep meat with nisin at either level proved insufficient to act against *S.* Enteriditis, the combination of 0.6% oregano EO and 500 IU/g nisin exhibited stronger antimicrobial activity against the pathogen than the EO or bacteriocin alone. Solomakos *et al.* (2008) evaluated antimicrobial effects of thyme EO at 0.6% and nisin at 500 or 1000 IU/g against *L. monocytogenes*. Combined addition of thyme EO and nisin synergistically inhibited the pathogen, decreasing *L. monocytogenes* populations below the official acceptable limit imposed by the European Union (2.0 $\log_{10}$ CFU/g) during refrigerated storage. Recently, others have evaluated the antimicrobial effectiveness of lysozyme, nisin, and ethylenediaminetetraacetic acid (EDTA) combinations on bacterial growth on ostrich patties packaged in air, vacuum, and two differing modified atmospheres (Mastromatteo *et al.*, 2010).

Table 13.2 Selected previous applications of antimicrobial polypeptides for pathogen control in meat or poultry.

Polypeptide	Type of study	Microorganism inhibited	Food matrix	Reference
Unidentified bacteriocin	In combined effects of fat content and sodium nitrite	*Listeria monocytogenes*	Lean pork meat	Kouakou *et al.*, 2009
Nisin	Incorporated into chitosan-coated plastic films	*L. monocytogenes*	Ham steaks	Ye *et al.*, 2008
Nisin	In combination with thyme essential oil	*L. monocytogenes*	Minced beef	Solomakos *et al.*, 2008
Nisin	Incorporated into cellulose films	*L. monocytogenes*	Processed meats	Nguyen *et al.*, 2008
Nisin	Incorporated into alginate coating	*L. monocytogenes*	Deli turkey products	Juck *et al.*, 2010
Nisin	In combination with oregano essential oil	*Salmonella* Enteriditis	Minced sheep meat	Govaris *et al.*, 2010
Nisin	Incorporated in antimicrobial packaging	*Carnobacterium*, lactic acid bacteria, and *Brochothrix thermosphacta*	Beef	Ercolini *et al.*, 2010
Nisin/lysozyme	In combination with EDTA	*Pseudomonas* spp.	Packed ostrich patties	Mastromatteo *et al.*, 2010
Nisin/lysozyme	In package pasteurization combined with prior-surface application	*L. monocytogenes*	Ready-to-eat turkey bologna	Mangalassary *et al.*, 2008

Nisin has also been studied for its ability to protect the quality and safety of various processed meat and poultry products (Gomes *et al.*, 2009). Application of 5000 IU/g nisin plus 2.5% acetic acid reduced *L. monocytogenes* counts on ham and bologna held at 10°C to nondetectable levels by the end of a 48-day storage period (Geornaras *et al.*, 2005). Mangalassary *et al.* (2008) investigated the efficacy of surface application of nisin and/or lysozyme combined with in-package pasteurization for the reduction and prevention of subsequent recovery and growth of *L. monocytogenes* during refrigerated storage on the surfaces of low-fat turkey bologna. In-package pasteurization combined with nisin or nisin–lysozyme treatments was effective in reducing the population below detectable levels for 2–3 weeks of refrigerated storage (Mangalassary *et al.*, 2008). Likewise, incorporation of nisin or pediocin into beef brining solutions resulted in ~0.5–0.6 $\log_{10}$ reductions in *E. coli* O157:H7 populations on cooked beef as compared to controls, indicating reduced thermal resistance of the pathogen in the presence of bacteriocin (Byelashov *et al.*, 2010a). More recently, application of a 0.5% nisin solution to surfaces of RTE turkey ham achieved a nearly 3.0 $\log_{10}$ decrease in *L. monocytogenes* following 60 days of refrigeration (Ruiz *et al.*, 2010).

13.4.2 Egg white lysozyme

Lysozyme (1,4-β-*N*-acetylmuramidase) (CAS No. 9001–63-2) belongs to the muramidases; lysozyme cleaves the glycosidic bond between C-1 of *N*-acetylmuramic acid (NAM) and C-4 of *N*-acetylglucosamine (NAG) in the bacterial peptidoglycan (PG). The lysozymes are distributed throughout nature, where they constitute a natural defense mechanism against bacterial pathogens. Many bacteriophages also produce lysozymes that locally hydrolyze the PG to facilitate cell lysis at the end of the phage replication cycle. Lysozymes are divided into families and types on the basis of similarities in amino acid sequences and three-dimensional structure. Extensive hydrolysis of PG by lysozyme results in cell lysis and death in hypo-osmotic environments. Some lysozymes can kill bacteria by stimulating autolysin activity upon interaction with the cell surface. In addition, a nonlytic bactericidal mechanism involving membrane damage without PG hydrolysis has been reported for *C*-type lysozymes, including human and HEWL. The most studied lysozyme and the only one bearing approval for direct addition to foods is HEWL, also considered GRAS (FDA, 1998). Although Gram-positive bacteria are generally sensitive to HEWL because their PG is directly exposed, some are innately immune due to modified PG structure (Conte *et al.*, 2006). Gram-negative bacteria are generally resistant to lysozymes due to the outer membrane that shields the PG from lysozyme attack (Song *et al.*, 2002). In general, little is known about the antimicrobial activity and specificity of different lysozymes, though some have been reported active against Gram-negative bacteria, even in the absence of a membrane-destabilizing agent (Branen & Davidson, 2004).

Nakimbugwe *et al.* (2006b) evaluated six lysozymes under ambient and high pressure on a panel of five Gram-positive ((1) *Enterococcus faecalis*, (2) *Bacillus subtilis*, (3) *Listeria innocua*, (4) *Staphylococcus aureus*, and (5) *Micrococcus lysodeikticus*) and Gram-negative ((1) *Yersinia enterocolitica*, (2) *Shigella flexneri*, (3) *E. coli* O157:H7, (4) *P. aeruginosa*, and (5) *S.* Typhimurium) bacteria. At ambient pressure, each Gram-positive organism displayed a specific pattern of sensitivity to the six lytic enzymes, but no Gram-negative microbe was sensitive to any of the studied lysozymes. High-pressure application (130–300 MPa, 15 minutes, 25°C) sensitized several Gram-positive and Gram-negative genera for one or more lysozymes (Nakimbugwe *et al.*, 2006a). *Micrococcus* and *Pseudomonas* cells became sensitive to every lysozyme tested under high pressure. In general, these modifications may also explain the poor activity of the other lysozymes against these bacteria at ambient pressure (Nakimbugwe *et al.*, 2006a).

13.4.3 Biopreservation of meat and poultry

In addition to the antimicrobials discussed already, the use of protective cultures for the inhibition or antagonism of microbial spoilers and pathogens has been heavily studied and multiple applications of differing competitive microbes for meat and poultry safety have been published (Sirsat *et al.*, 2009). The use of such cultures to inhibit food-borne microbial pathogens, either by competition for nutrients or by synthesis of metabolites that possess antimicrobial activity is often referred to as "biopreservation" (Lücke, 2000). Biopreservation systems such as various bactericidal LAB and/or their bacteriocins have been developed and commercialized (USDA-FSIS, 2010). Multiple LAB genera and species commonly associated with meats demonstrate antagonism toward pathogenic and spoilage organisms (Katikou *et al.*, 2005; Kostrzynska & Bachand, 2006). The inhibition of pathogens results from acid fermentation and synthesis of antimicrobial compounds (bacteriocins, peroxides), and from competition for nutrients with other microbes (Amézquita & Brashears, 2002; Katikou *et al.*, 2005).

Numerous studies have detailed the efficacy of LAB-derived biopreservation technologies for the inhibition of Gram-negative and Gram-positive microbes on fresh and processed meat and poultry (Stiles, 1996; De Martinis *et al.*, 2002). Bredholt *et al.* (1999) reported that *E. coli* O157:H7 on cooked hams stored at 10°C for 28 days under vacuum were reduced by 2.0–3.0 $\log_{10}$ compared to controls. Fresh ground beef inoculated with *Lactobacillus reuteri* at 3.0 or 6.0 $\log_{10}$ CFU/g suppressed *E. coli* O157:H7 populations to nondetectable levels after 20 days of refrigerated vacuum storage (Muthukumarasamy *et al.*, 2003). Maragkoudakis *et al.* (2009) described a *Lactobacillus fermentum* isolate that exerted antimicrobial activity against *S.* Enteriditis on fresh chicken meat. After 7 days of incubation at 8°C–10°C, *Salmonella* on chicken inoculated with *L. fermentum* were ~1.5 $\log_{10}$ lower than controls. Ruby and Ingham (2009) reported that *E. coli* O157:H7 and *Salmonella* in beef broth stored at 10°C for 9 days were reduced by ~2.5 $\log_{10}$ per mL in the presence of a *Lactobacillus sakei* isolate inoculated at 8.0 $\log_{10}$ CFU/mL. These authors also reported that no significant increases in populations of *E. coli* O157:H7 were observed in inoculated beef stored for 6 days at 5°C. Key factors to consider in the application of a protective culture or blend of cultures are: (1) targeted pathogen(s) and likely inhibition provided by the protective cultures; (2) ability of cultures to suppress pathogens under storage conditions and loss of antimicrobial effects; (3) potential for spoilage to occur by proliferation of protective cultures during storage; and (4) levels of cultures required for effective pathogen inhibition and resulting changes in plate counts of indicator organisms following application of competitive/protective cultures (Gomes *et al.*, 2009).

13.5 CONCLUDING REMARKS

Despite the variety of food antimicrobials available for the preservation of food safety and quality, only a limited number of antimicrobials are currently approved for use in the manufacture of organic meat and poultry. The application of one or more antimicrobials in a process must be preceded by thoughtful consideration of the factors affecting the activity of chosen antimicrobials. The organic acids and their salts, the oxidizers, and the antimicrobial polypeptides inhibit microorganisms by differing mechanisms and will yield different effects on the ultimate product safety and quality based on the microorganisms targeted, intrinsic and extrinsic factors specific to the product of interest, and the conditions of antimicrobial application. Nonetheless, antimicrobial utilization can provide significant enhancement of microbiological safety and quality of fresh and processed organic meat and poultry.

REFERENCES

Abdul-Raouf, U. M., L. R. Beuchat, and M. S. Ammar. 1993. Survival and growth of *Escherichia coli* O157:H7 in ground, roasted beef as affected by pH, acidulants, and temperature. *Appl. Environ. Microbiol.* 59:2364–2368.

Abee, T., L. Krockel, and C. Hill. 1995. Bacteriocins: modes of action and potentials in food preservation and control of food poisoning. *Int. J. Food Microbiol.* 28:169–185.

Abee, T., F. M. Rombouts, J. Hugenholtz, G. Guihard, and L. Letellier. 1994. Mode of action of nisin Z against *Listeria monocytogenes* Scott A grown at high and low temperatures. *Appl. Environ. Microbiol.* 60:1962–1968.

Acuff, G. R., C. Vanderzant, J. W. Savell, D. K. Jones, D. B. Griffin, and J. G. Ehlers. 1987. Effect of acid decontamination of beef subprimal cuts on the microbiological and sensory characteristics of steaks. *Meat Sci.* 19:217–226.

Al-Haddad, K. S. H., R. A. S. Al-Qassemi, and R. K. Robinson. 2005. The use of gaseous ozone and gas packaging to control populations of *Salmonella infantis* and *Pseudomonas aeruginosa* on the skin of chicken portions. *Food Cont.* 16:405–410.

Al-Nabulsi, A. A., T. M. Osaili, M. A. Al-Holy, R. R. Shaker, M. M. Ayyash, A. N. Olaimat, and R. A. Holley. 2009. Influence of desiccation on the sensitivity of *Cronobacter* spp. to lactoferrin or nisin in broth and powdered infant formula. *Int. J. Food Microbiol.* 136:221–226.

Alakomi, H.-L., E. Skyttä, M. Saarela, T. Mattila-Sandholm, K. Latva-Kala, and I. M. Helander. 1999. Lactic acid permeabilizes gram-negative bacteria by disrupting the outer membrane. *Appl. Environ. Microbiol.* 66:2001–2005.

Álvarez-Ordóñez, A., A. Fernández, A. Bernardo, and M. López. 2009. Comparison of acids on the induction of an acid tolerance response in *Salmonella typhimurium*, consequences for food safety. *Meat Sci.* 81:65–70.

Amézquita, A. and M. M. Brashears. 2002. Competitive inhibition of *Listeria monocytogenes* in ready-to-eat meat products by lactic acid bacteria. *J. Food Prot.* 65:316–325.

Apostolidis, E., Y.-I. Kwon, and K. Shetty. 2008. Inhibition of *Listeria monocytogenes* by oregano, cranberry and sodium lactate combination in broth and cooked ground beef systems and likely mode of action through proline metabolism. *Int. J. Food Microbiol.* 128:317–324.

Arana, I., A. Santorum, A. Muela, and I. Barcina. 1999. Chlorination and ozonation of waste-water: comparative analysis of efficacy through the effect on *Escherichia coli* membranes. *J. Appl. Microbiol.* 86:883–888.

Aureli, P., A. Costantini, and S. Zolea. 1992. Antimicrobial activity of some plant essential oils against *Listeria monocytogenes*. *J. Food Prot.* 55:344–348.

Axelsson, L. T. 1998. Lactic acid bacteria. In: S. Salminen and A. Wright (eds). *Lactic Acid Bacteria*. Marcel Dekker, Inc., New York. pp. 1–63.

Aymerich, T., M. Garriga, A. Jofré, B. Martín, and J. M. Monfort. 2006. The use of bacteriocins against meat-borne pathogens. In: L. M. L. Nollet and F. Toldrá (eds). *Advanced Technologies for Meat Processing*. Taylor & Francis Group, LLC, Boca Raton, FL. pp. 371–399.

Baird, B. E., L. M. Lucia, G. R. Acuff, K. B. Harris, and J. W. Savell. 2006. Beef hide antimicrobial interventions as a means of reducing bacterial contamination. *Meat Sci.* 73:245–248.

Barmpalia, I. M., I. Geornaras, K. E. Belk, J. A. Scanga, P. A. Kendall, G. C. Smith, and J. N. Sofos. 2004. Control of *Listeria monocytogenes* on frankfurters with antimicrobials in the formulation and by dipping in organic acid solutions. *J. Food Prot.* 67:2456–2464.

Bauermeister, L. J., J. W. J. Bowers, J. C. Townsend, and S. R. McKee. 2008. Validating the efficacy of peracetic acid mixtures as an antimicrobial in poultry chillers. *J. Food Prot.* 71:1119–1122.

Bautista, D. A., N. Sylvester, S. Barbut, and M. W. Griffiths. 1997. The determination of efficacy of antimicrobial rinses on turkey carcasses using response surface designs. *Int. J. Food Microbiol.* 34:279–292.

Bell, K. Y., C. N. Cutter, and S. S. Sumner. 1997. Reduction of foodborne micro-organisms on beef carcass tissue using acetic acid, sodium bicarbonate, and hydrogen peroxide. *Food Microbiol.* 14:439–448.

Bernarde, M. A., W. B. Snow, V. P. Olivieri, and B. Davidson. 1967. Kinetics and mechanism of bacterial disinfection by chlorine dioxide. *Appl. Microbiol.* 15:257–265.

Birk, T., A. C. Grønlund, B. B. Christensen, S. Knøchel, K. Lohse, and H. Rosenquist. 2010. Effect of organic acids and marination ingredients on the survival of *Campylobacter jejuni* on meat. *J. Food Prot.* 73:258–265.

Blackburn, C. d. W., L. M. Curtis, L. Humpheson, C. Billon, and P. J. McClure. 1997. Development of thermal inactivation models for *Salmonella enteriditis* and *Escherichia coli* O157:H7 with temperature, pH, and NaCl as controlling factors. *Int. J. Food Microbiol.* 38:31–44.

Bosilevac, J. M., X. Nou, G. A. Barkocy-Gallagher, T. M. Arthur, and M. Koohmarie. 2006. Treatments using hot water instead of lactic acid reduce levels of aerobic bacteria and Enterobacteriaceae and reduce the prevalence of *Escherichia coli* O157:H7 on preevisceration beef carcasses. *J. Food Prot.* 69:1808–1813.

Branen, A. L., P. M. Davidson, and B. Katz. 1980. Antimicrobial properties of phenolic antioxidants and lipids. *Food Technol.* 34:42–53, 63.

Branen, J. K. and P. M. Davidson. 2004. Enhancement of nisin, lysozyme, and monolaurin antimicrobial activities by ethylenediaminetetraacetic acid and lactoferrin. *Int. J. Food Microbiol.* 90:63–74.

Bredholt, S., T. Nesbakken, and A. Holck. 1999. Protective cultures inhibit growth of *Listeria monocytogenes* and *Escherichia coli* O157:H7 in cooked, sliced, vacuum- and gas-packaged meat. *Int. J. Food Microbiol.* 53:43–52.

Breukink, E. and B. de Kruijff. 1999. The lantibiotic nisin, a special case or not? *Biochim. Biophys. Acta* 1462:223–234.

Breukink, E., H. E. van Heusden, P. J. Vollmerhaus, E. Swiezewska, L. Brunner, S. Walker, A. J. Heck, and B. de Kruijff. 2003. Lipid II is an intrinsic component of the pore induced by nisin in bacterial membranes. *J. Biol. Chem.* 278:19898–19903.

Breukink, E., C. van Kraaij, R. A. Demel, R. J. Siezen, O. P. Kuipers, and B. de Kruijff. 1997. The C-terminal region of nisin is responsible for the initial interaction of nisin with the target membrane. *Biochemistry* 36:6968–6976.

Brul, S. and P. Coote. 1999. Preservative agents in foods. Mode of action and microbial resistance mechanisms. *Int. J. Food Microbiol.* 50:1–17.

Byelashov, O. A., J. M. Adler, I. Geornaras, K. Y. Ko, K. E. Belk, G. C. Smith, and J. N. Sofos. 2010a. Evaluation of brining ingredients and antimicrobials for effects on thermal destruction of *Escherichia coli* O157:H7 in a meat model system. *J. Food Sci.* 75:M209–M217.

Byelashov, O. A., H. Daskalov, I. Geornaras, P. A. Kendall, K. E. Belk, J. A. Scanga, G. C. Smith, and J. N. Sofos. 2010b. Reduction of *Listeria monocytogenes* on frankfurters treated with lactic acid solutions of various temperatures. *Food Microbiol.* 27:783–790.

Cabedo, L., J. N. Sofos, and G. C. Smith. 1996. Removal of bacteria from beef tissue by spray washing after different times of exposure to fecal material. *J. Food Prot.* 59:1284–1287.

Carlson, B. A., I. Geornaras, Y. Yoon, J. A. Scanga, J. N. Sofos, G. C. Smith, and K. E. Belk. 2008a. Studies to evaluate chemicals and conditions with low-pressure applications for reducing microbial counts on cattle hides. *J. Food Prot.* 71:1343–1348.

Carlson, B. A., J. Ruby, G. C. Smith, J. N. Sofos, G. R. Bellinger, W. Warren-Serna, B. Centrella, R. A. Rowling, and K. E. Belk. 2008b. Comparison of antimicrobial efficacy of multiple beef hide decontamination strategies to reduce levels of *Escherichia coli* O157:H7 and *Salmonella*. *J. Food Prot.* 71:2223–2227.

Castillo, A., L. M. Lucia, K. J. Goodson, J. W. Savell, and G. R. Acuff. 1998. Comparison of water wash, trimming, and combined hot water and lactic acid treatments for reducing bacteria of fecal origin on beef carcasses. *J. Food Prot.* 61:823–828.

Castillo, A., L. M. Lucia, K. J. Goodson, J. W. Savell, and G. R. Acuff. 1999. Decontamination of beef carcass surface tissue by steam vacuuming alone and combined with hot water and lactic acid sprays. *J. Food Prot.* 62:146–151.

Castillo, A., L. M. Lucia, I. Mercado, and G. R. Acuff. 2001. In-plant evaluation of a lactic acid treatment for reduction of bacteria on chilled beef carcasses. *J. Food Prot.* 64:738–740.

Castillo, A., K. S. McKenzie, L. M. Lucia, and G. R. Acuff. 2003. Ozone treatment for reduction of *Escherichia coli* O157:H7 and *Salmonella* serotype Typhimurium on beef carcass surfaces. *J. Food Prot.* 66:775–779.

Cherrington, C. A., M. Hinton, and I. Chopra. 1990. Effect of short-chain organic acids on macromolecular synthesis in *Escherichia coli*. *J. Appl. Bacteriol.* 68:69–74.

Cherrington, C. A., M. Hinton, G. S. Mead, and I. Chopra. 1991. Organic acid: chemistry, antibacterial activity and practical applications. *Adv. Microb. Physiol.* 32:87–108.

Conte, A., M. Sinigaglia, and M. A. Del Nobile. 2006. Antimicrobial effectiveness of lysozyme immobilized on polyvinylalcohol-based film against *Alicyclobacillus acidoterrestris*. *J. Food Prot.* 69:861–865.

Cox, N. A., L. J. Richardson, R. J. Buhr, M. T. Musgrove, M. E. Berrang, and W. Bright. 2007. Bactericidal effects of several chemicals and hatching eggs inoculated with *Salmonella* serovar Typhimurium. *J. Appl. Poult. Res.* 16:623–627.

Cutter, C. N. and G. R. Siragusa. 1994. Efficacy of organic acids against *Escherichia coli* O157:H7 attached to beef carcass tissue using a pilot scale model carcass washer. *J. Food Prot.* 57:97–103.

Cutter, C. N. and G. R. Siragusa. 1995. Application of chlorine to reduce populations of *Escherichia coli* on beef. *J. Food Saf.* 15:67–75.

Davidson, P. M. and A. L. Branen. 2005. Food antimicrobials – an introduction. In: P. M. Davidson, J. N. Sofos, and A. L. Branen (eds) *Antimicrobials in Food*. 3rd edn. CRC Press, New York. pp. 1–10.

Davidson, P. M. and T. M. Taylor. 2007. Chemical preservatives and natural antimicrobial compounds. In: M. P. Doyle and L. R. Beuchat (eds) *Food microbiology: Fundamentals and Frontiers*. 3rd edn. ASM Press, Inc., Washington, DC. pp. 713–745.

De Martinis, E. C. P., V. F. Alves, and B. D. G. M. Franco. 2002. Fundamentals and perspectives for the use of bacteriocins produced by lactic acid bacteria in meat products. *Food Rev. Int.* 18:191–208.

de Vuyst, L. and F. Leroy. 2007. Bacteriocins from lactic acid bacteria: production, purification, and food applications. *J. Mol. Microbiol. Biotechnol.* 13:194–199.

Deborde, M. and U. von Gunten. 2008. Reactions of chlorine with inorganic and organic compounds during water treatment-kinetics and mechanisms: a critical review. *Water Res.* 42:13–51.

Delmore, R. J., Jr., J. N. Sofos, G. R. Schmidt, K. E. Belk, W. R. Lloyd, and G. C. Smith. 2000. Interventions to reduce microbiological contamination of beef variety meats. *J. Food Prot.* 63:44–50.

del Río, E., B. González de Caso, M. Prieto, C. Alonso-Calleja, and R. Capita. 2008. Effect of poultry decontaminants concentration on growth kinetics for pathogenic and spoilage bacteria. *Food Microbiol.* 25:888–894.

del Río, E., R. Muriente, M. Prieto, C. Alonso-Calleja, and R. Capita. 2007. Effectiveness of trisodium phosphate, acidified sodium chlorite, citric acid, and peroxyacids against pathogenic bacteria on poultry during refrigerated storage. *J. Food Prot.* 70:2063–2071.

Delves-Broughton, J. 1990. Nisin and its uses as a food preservative. *Food Technol.* 44:110–117.

Delves-Broughton, J., P. Blackburn, R. J. Evans, and J. Hugenholtz. 1996. Applications of the bacteriocin, nisin. *Antonie van Leeuwenhoek* 69:193–202.

Dixon, Z. R., C. Vanderzant, G. R. Acuff, J. W. Savell, and D. K. Jones. 1987. Effect of acid decontamination of beef loin steaks on microbiological and sensory characteristics. *Int. J. Food Microbiol.* 5:181–186.

Doores, S. 2005. Organic acids. In: P. M. Davidson and A. L. Branen (ed) *Antimicrobials in Foods*. 3rd edn. Marcel Dekker, Inc., New York. pp. 95–136.

Drosinos, E. H., P. N. Skandamis, and M. Mataragas. 2009. Antimicrobials treatment. In: F. Toldrá (ed) *Safety of Meat and Processed Meat*. Springer, New York. pp. 255–298.

EC. 2004. Regulation (EC) No. 853/2004 of the European Parliament and of the Council of 29 April 2004 Laying Down Specific Hygiene Rules for on the Hygiene of Foodstuffs. Available at: www.fsai.ie/uploadedFiles/Reg853_2004(1).pdf (accessed February 9, 2011).

Eklund, T. 1983. The antimicrobial effect of dissociated and undissociated sorbic acid at different pH levels. *J. Appl. Bacteriol.* 54:383–389.

Ellebracht, E. A., A. Castillo, L. M. Lucia, R. K. Miller, and G. R. Acuff. 1999. Reduction of pathogens using hot water and lactic acid on beef trimmings. *J. Food Sci.* 64:1094–1099.

EPA. 1999. Alternative disinfectants and oxidants guidance manual. Ch. 4: Chlorine dioxide (EPA 815-R-99–014). Available at: http://www.epa.gov/ogwdw000/mdbp/alternative_disinfectants_guidance.pdf (accessed September 3, 2010).

Ercolini, D., I. Ferrocino, A. La Storia, G. Mauriello, S. Gigli, P. Masi, and F. Villani. 2010. Development of spoilage microbiota in beef stored in nisin activated packaging. *Food Microbiol.* 27:137–143.

FDA. 1988. Direct food substances affirmed as generally recognized as safe: nisin preparation. *Fed. Reg.* 53:11247–11251.

FDA. 1998. Direct food substances affirmed as generally recognized as safe: egg white lysozyme. *Fed. Reg.* 63:12421–12426.

Fletcher, D. L., S. M. Russell, J. M. Walker, and J. S. Bailey. 1993. An evaluation of a rinse procedure using sodium bicarbonate and hydrogen peroxide on the recovery of bacteria from broiler carcasses. *Poult. Sci.* 72:2152–2156.

Freese, E. 1978. Mechanism of growth inhibition by lipophilic acids. In: J. J. Kabara (ed) *The Pharmacological Effect of Lipids*. American Oil Chemists Society, Champaign, IL. pp. 123–131.

Freese, E., C. W. Sheu, and E. Galliers. 1973. Function of lipophilic acids as antimicrobial food additives. *Nature* 241:321–327.

Geornaras, I., K. E. Belk, J. A. Scanga, P. A. Kendall, G. C. Smith, and J. N. Sofos. 2005. Postprocessing antimicrobial treatments to control *Listeria monocytogenes* in commercial vacuum-packaged bologna and ham stored at 10°C. *J. Food Prot.* 68:991–998.

Gibson, T. M., J. Haley, M. Righton, and C. D. Watts. 1986. Chlorination of fatty acids during water treatment disinfection: reactivity and product identification. *Environ. Technol. Lett.* 7:365–372.

Gill, C. O. and M. Badoni. 2004. Effects of peroxyacetic acid, acidified sodium chlorite or lactic acid solutions on the microflora of chilled beef carcasses. *Int. J. Food Microbiol.* 91:43–50.

Goldstein, S., D. Meyerstein, and G. Czapski. 1993. The Fenton reagents. *Free Radical Biol. Med.* 15:435–445.

Gomes, B. C., L. K. Winkelströter, F. B. dos Reis, and E. C. P. De Martinis. 2009. Biopreservation. In: F. Toldrá (ed). *Safety of Meat and Processed Meat*. Springer, New York. pp. 297–312.

Gonçalves, A. C., R. C. C. Almeida, M. A. O. Alves, and P. F. Almeida. 2005. Quantitative investigation on the effects of chemical treatments in reducing *Listeria monocytogenes* populations on chicken breast meat. *Food Cont.* 16:617–622.

González-Fandos, E., B. Herrera, and N. Maya. 2009. Efficacy of citric acid against *Listeria monocytogenes* attached to poultry skin during refrigerated storage. *Int. J. Food Sci. Technol.* 44:262–268.

Gordon, G. and A. A. Rosenblatt. 2005. Chlorine dioxide: the current state of the art. *Ozone Sci. Eng.* 27:203–207.

Gorman, B. M., J. N. Sofos, J. B. Morgan, G. R. Schmidt, and G. C. Smith. 1995. Evaluation of hand-trimming, various sanitizing agents, and hot water spray-washing as decontamination interventions for beef brisket adipose tissue. *J. Food Prot.* 58:899–907.

Gould, G. W. 1989. *Mechanisms of Action of Food Preservation Procedures*. Elsevier Applied Science, London. p. 448.

Govaris, A., N. Solomakos, A. Pexara, and P. S. Chatzopoulou. 2010. The antimicrobial effect of oregano essential oil, nisin and their combination against *Salmonella* Enteriditis in minced sheep meat during refrigerated storage. *Int. J. Food Microbiol.* 137:175–180.

Hardin, M. D., G. R. Acuff, L. M. Lucia, J. S. Oman, and J. W. Savell. 1995. Comparison of methods for decontamination from beef carcass surfaces. *J. Food Prot.* 58:368–374.

Harris, K., M. F. Miller, G. H. Loneragan, and M. M. Brashears. 2006. Validation of the use of organic acids and acidified sodium chlorite to reduce *Escherichia coli* O157 and *Salmonella* Typhimurium in beef trim and ground beef in a simulated processing environment. *J. Food Prot.* 69:1802–1807.

Heller, C. E., J. A. Scanga, J. N. Sofos, K. E. Belk, W. Warren-Serna, G. R. Bellinger, R. T. Bacon, M. L. Rossman, and G. C. Smith. 2007. Decontamination of beef subprimal cuts intended for blade tenderization or moisture enhancement. *J. Food Prot.* 70:1174–1180.

Hirshfield, I. N., S. Terzulli, and C. O'Byrne. 2003. Weak organic acids: a panoply of effects on bacteria. *Sci Prog.* 86:245–269.

Huang, J., L. Wang, N. Ren, F. Ma, and J. Ma. 1997. Disinfection effect of chlorine dioxide on bacteria in water. *Water Res.* 31:607–613.

Hwang, C.-A. and L. R. Beuchat. 1995. Efficacy of a lactic acid/sodium benzoate wash solution for reducing bacterial contamination of raw chicken. *Int. J. Food Microbiol.* 27:91–98.

James, C., J. A. Thornton, L. Ketteringham, and S. J. James. 2000. Effect of steam condensation, hot water or chlorinated hot water immersion on bacterial numbers and quality of lamb carcasses. *J. Food Eng.* 43:219–225.

Jeng, D. K. and A. G. Woodworth. 1990. Chlorine dioxide gas sterilization under square-wave conditions. *Appl. Environ. Microbiol.* 56:514–519.

Jeong, J.-Y., C.-R. Kim, K.-H. Kim, S.-J. Moon, K. Kook, and S.-N. Kang. 2007. Microbial and physico-chemical characteristics of pork loin cuts treated with ozone gas during storage. *Kor. J. Food Sci. Animal Res.* 27:80–86.

Johnson, E. A. and A. E. Larson. 2005. Lysozyme. In: P. M. Davidson, J. N. Sofos, and A. L. Branen (eds) *Antimicrobials in Foods*. 3rd edn. CRC Press, New York. pp. 361–388.

Juck, G., H. Neetoo, and H. Chen. 2010. Application of an active alginate coating to control the growth of *Listeria monocytogenes* on poached and deli turkey products. *Int. J. Food Microbiol.* 142:302–308.

Jung, D.-S., F. W. Bodyfelt, and M. A. Daeschel. 1992. Influence of fat and emulsifiers on the efficacy of nisin in inhibiting *Listeria monocytogenes* in fluid milk. *J. Dairy Sci.* 75:387–393.

Kanellos, T. S. and A. R. Burriel. 2005. The *in vitro* bactericidal effects of the food decontaminants lactic acid and trisodium phosphate. *Food Microbiol.* 22:591–594.

Katikou, P., I. Ambrosiadis, D. Georgantelis, P. Koidis, and S. A. Georgakis. 2005. Effect of *Lactobacillus*-protective cultures with bacteriocin-like inhibitory substances' producing ability on microbiological, chemical and sensory changes during storage of refrigerated vacuum-packed sliced beef. *J. Appl. Microbiol.* 99:1303–1313.

Kim, C., Y.-C. Hung, and S. M. Russell. 2005. Efficacy of electrolyzed water in the prevention and removal of fecal material attachment and its microbiocidal effectiveness during simulated industrial poultry processing. *Poult. Sci.* 84:1778–1784.

Kinsella, K. J., D. M. Prendergast, M. S. McCann, I. S. Blair, D.A. McDowell, and J. J. Sheridan. 2009. The survival of *Salmonella enterica* serovar Typhimurium DT104 and total viable counts on beef surfaces at different relative humidities and temperatures. *J. Appl. Microbiol.* 106:171–180.

Kinsella, K. J., T. A. Rowe, I. S. Blair, D. A. McDowell, and J. J. Sheridan. 2006. Survival and recovery of *Salmonella enterica* serovar Typhimurium DT104 at low temperature and water activity in a broth system. *Foodborne Pathog. Dis.* 3:375–383.

Klaenhammer, T. R. 1988. Bacteriocins of lactic-acid bacteria. *Biochimie* 70:337–349.

Koczoń, P. 2009. Growth inhibition mode of action of selected benzoic acid derivatives against the yeast *Pichia anomala*. *J. Food Prot.* 72:791–800.

Kostrzynska, M. and A. Bachand. 2006. Use of microbial antagonism to reduce pathogen levels on produce and meat products: a review. *Can. J. Microbiol.* 52:1017–1026.

Kouakou, P., H. Ghalfi, J. Destain, R. Dobois-Dauphin, P. Evrard, and P. Thonart. 2009. Effect of curing sodium nitrite additive and natural meat fat on growth control of *Listeria monocytogenes* by the bacteriocin-producing *Lactobacillus curvatus* strain CWBI-B28. *Food Microbiol.* 26:623–628.

Koutsoumanis, K. P., P. A. Kendall, and J. N. Sofos. 2004. Modeling the boundaries of growth of *Salmonella* Typhimurium in broth as a function of temperature, water activity, and pH. *J. Food Prot.* 67:53–59.

Laury, A. M., M. V. Alvarado, G. Nace, C. Z. Alvarado, J. C. Brooks, A. Echeverry, and M. M. Brashears. 2009. Validation of a lactic acid- and citric acid-based antimicrobial product for the reduction of *Escherichia coli* O157:H7 and *Salmonella* on beef tips and whole chicken carcasses. *J. Food Prot.* 72: 2208–2211.

Legay, C., M. J. Rodriguez, J. B. Sérodes, and P. Levallois. 2010. Estimation of chlorination by-products presence in drinking water in epidemiological studies on adverse reproductive outcomes: a review. *Sci. Total Environ.* 408:456–472.

Li, Y., H. Yang, and B. L. Swem. 2002. Effect of high-temperature inside-outside spray on survival of *Campylobacter jejuni* attached to prechill chicken carcasses. *Poult. Sci.* 81:1371–1377.

Lide, D. R. 2010. *CRC Handbook of Chemistry and Physics: Internet Version*. 90th edn. CRC Press, Boca Raton, FL. p. 2804.

Lippmann, M. 1989. Health effects of ozone, a critical review. *J. Am. Air Pollut. Cont. Assoc.* 39:672–695.

Long, C. and C. A. Phillips. 2003. The effect of sodium citrate, sodium lactate and nisin on the survival of *Arcobacter butzleri* NCTC 12481 on chicken. *Food Microbiol.* 20:495–502.

López-Malo, A., S. M. Alzamora, and S. Guerrero. 2000. Natural antimicrobials from plants. In: S. M. Alzamora, M. S. Tapia, and A. López-Malo (eds) *Minimally Processed Fruits and Vegetables: Fundamental Aspects and Applications*. Aspen Publishers, Inc., Gaithersburg, MD. pp. 237–264.

Loretz, M., R. Stephan, and C. Zweifel. 2010. Antimicrobial activity of decontamination treatments for poultry carcasses: a literature survey. *Food Cont.* 21:791–804.

Lücke, F. K. 2000. Utilization of microbes to process and preserve meat. *Meat Sci.* 56:105–115.

Mangalassary, S., I. Han, J. Rieck, J. Acton, and P. Dawson. 2008. Effect of combining nisin and/or lysozyme with in-package pasteurization for control of *Listeria monocytogenes* in ready-to-eat turkey bologna during refrigerated storage. *Food Microbiol.* 25:866–870.

Maragkoudakis, P. A., K. C. Mountzouris, D. Psyrras, S. Cremonese, J. Fischer, M. D. Cantor, and E. Tsakalidou. 2009. Functional properties of novel protective lactic acid bacteria and application in raw chicken meat against *Listeria monocytogenes* and *Salmonella enteriditis*. *Int. J. Food Microbiol.* 130:219–226.

Mastromatteo, M., A. Lucera, M. Sinigaglia, and M. R. Corbo. 2010. Use of lysozyme, nisin, and EDTA combined treatments for maintaining quality of packed ostrich patties. *J. Food Sci.* 75:M178–M186.

McWilliam Leitch, E. C. and C. S. Stewart. 2002a. *Escherichia coli* O157 and non-O157 isolates are more susceptible to L-lactate then to D-lactate. *Appl. Environ. Microbiol.* 68:4676–4678.

McWilliam Leitch, E. C. and C. S. Stewart. 2002b. Susceptibility of *Escherichia coli* O157 and non-O157 isolates to lactate. *Lett. Appl. Microbiol.* 35:176–180.

Miller, A. J., J. E. Call, and R. C. Whiting. 1993. Comparison of organic acid salts for *Clostridium botulinum* control in an uncured turkey product. *J. Food Prot.* 56:958–962.

Mitchell, P. 1961. Coupling of phosphorylation to electron and hydrogen transfer by a chemiosmotic type of mechanism. *Nature* 191:144–148.

Mitchell, P. and J. Moyle. 1969. Estimation of membrane potential and pH difference across the cristae membrane of rat liver mitochondria. *Eur. J. Biochem.* 7:471–484.

Moore, G., C. Griffith, and A. Peters. 2000. Bactericidal properties of ozone and its potential application as a terminal disinfectant. *J. Food Prot.* 63:1100–1106.

Muthukumarasamy, P., J. H. Han, and R. A. Holley. 2003. Bactericidal effects of *Lactobacillus reuteri* and allyl isothiocyanate on *Escherichia coli* O157:H7 in refrigerated ground beef. *J. Food Prot.* 66:2038–2044.

Naidu, A. S. 2000. Overview. In: A. S. Naidu (ed.) *Natural Food Antimicrobial Systems*. CRC Press, Boca Raton, FL. pp. 1–16.

Nakimbugwe, D., B. Masschalck, G. Anim, and C. W. Michiels. 2006a. Inactivation of gram-negative bacteria in milk and banana juice by hen egg white and lambda lysozyme under high hydrostatic pressure. *Int. J. Food Microbiol.* 112:19–25.

Nakimbugwe, D., B. Masschalck, M. Atanassova, A. Zewdie-Bosüner, and C. W. Michiels. 2006b. Comparison of bactericidal activity of six lysozymes at atmospheric pressure and under high hydrostatic pressure. *Int. J. Food Microbiol.* 108:355–363.

Nguyen, V. T., M. J. Gidleyb, and G. A. Dykes. 2008. Potential of a nisin-containing bacterial cellulose film to inhibit *Listeria monocytogenes* on processed meats. *Food Microbiol.* 25:471–478.

Nieuwenhuijsen, M. J., R. Smith, S. Golfinopoulos, N. Best, J. Bennett, G. Aggazzotti, E. Righi, G. Fantuzzi, L. Bucchini, S. Cordier, C. M. Villanueva, V. Moreno, C. La Vecchia, C. Bosetti, T. Vartiainen, R. Rautiu, M. Toledano, N. Iszatt, R. Grazuleviciene, and M. Kogevinas. 2009. Health impacts of long-term exposure to disinfection by-products in drinking water in Europe: HIWATE. *J. Water Health* 7:185–207.

Northcutt, J. K., D. P. Smith, M. T. Musgrove, K. D. Ingram, and A. Hinton Jr. 2005. Microbiological impact of spray washing broiler carcasses using different chlorine concentrations and water temperatures. *Poult. Sci.* 84:1648–1652.

NOSB. 2003. Measuring effluent: clarification of chlorine contact with organic food. Available at: http://www.ams.usda.gov/AMSv1.0/getfile?dDocName = STELDEV3104548 (accessed September 10, 2010).

Novak, J. S. and J. T. C. Yuan. 2004a. Increased inactivation of ozone-treated *Clostridium perfringens* vegetative cells and spores on fabricated beef surfaces using mild heat. *J. Food Prot.* 67:342–346.

Novak, J. S. and J. T. C. Yuan. 2004b. The fate of *Clostridium perfringens* spores exposed to ozone and/or mild heat treatment on beef surfaces followed by modified atmosphere packaging. *Food Microbiol.* 21:667–673.

Nuanualsuwan, S. and D. O. Cliver. 2003. Capsid functions of inactivated human picornaviruses and feline calicivirus. *Appl. Environ. Microbiol.* 69:350–357.

Nuñez de Gonzalez, M. T., J. T. Keeton, G. R. Acuff, L. J. Ringer, and L. M. Lucia. 2004. Effectiveness of acidic calcium sulfate with propionic and lactic acid and lactates as postprocessing dipping solutions to control *Listeria monocytogenes* on frankfurters with or without potassium lactate and stored vacuum packaged at 4.5°C. *J. Food Prot.* 67:915–921.

Nychas, G.-J.E., D. L. Marshall, and J. N. Sofos. 2007. Meat, poultry, and seafood. In: M. P. Doyle and L. R. Beuchat (eds) *Food Microbiology: Fundamentals and Frontiers*. 3rd edn. ASM Press, Washington, DC. pp. 105–140.

O'Bryan, C. A., P. G. Crandall, and S. C. Ricke. 2008. Organic poultry pathogen control from farm to fork. *Foodborne Path. Dis.* 5:709–720.

Oh, D. H. and D. L. Marshall. 1994. Enhanced inhibition of *Listeria monocytogenes* by glycerol monolaurate with organic acids. *J. Food Sci.* 59:1258–1261.

Ouattara, B., R. E. Simard, R. A. Holley, G.J.-P. Piette, and A. Bégin. 1997. Inhibitory effect of organic acids upon meat spoilage bacteria. *J. Food Prot.* 60:246–253.

Over, K. F., N. Hettiarachchy, M. G. Johnson, and B. Davis. 2009. Effect of organic acids and plant extracts on *Escherichia coli* O157:H7, *Listeria monocytogenes*, and *Salmonella* Typhimurium in broth culture model and chicken meat systems. *J. Food Sci.* 74:M515–M521.

Page, M. A., J. L. Shisler, and B. J. Mariñas. 2010. Mechanistic aspects of adenovirus serotype 2 inactivation with free chlorine. *Appl. Environ. Microbiol.* 76:2946–2954.
Palumbo, S. A. and A. C. Williams. 1994. Control of *Listeria monocytogenes* on the surface of frankfurters by acid treatment. *Food Microbiol.* 11:293–300.
Papadopoulos, L. S., R. K. Miller, G. R. Acuff, L. M. Lucia, C. Vanderzant, and H.R. Cross. 1991a. Consumer and trained sensory comparisons of cooked beef top rounds treated with sodium lactate. *J. Food Sci.* 56:1141–1146.
Papadopoulos, L. S., R. K. Miller, G. R. Acuff, C. Vanderzant, and H. R. Cross. 1991b. Effect of sodium lactate on microbial and chemical composition of cooked beef during storage. *J. Food Sci.* 56:341–347.
Park, H., Y.-C. Hung, and R. E. Brackett. 2002. Antimicrobial effect of electrolyzed water for inactivating *Campylobacter jejuni* during poultry washing. *Int. J. Food Microbiol.* 72:77–83.
Pohlman, F. W., M. R. Stivarius, K. S. McElyea, Z.B. Johnson, and M.G. Johnson. 2002a. The effects of ozone, chlorine dioxide, cetylpyridinium chloride and trisodium phosphate as multiple antimicrobial interventions on microbiological, instrumental color, and sensory color and odor characteristics of ground beef. *Meat Sci.* 61:307–313.
Pohlman, F. W., M. R. Stivarius, K. S. McElyea, Z. B. Johnson, and M.G. Johnson. 2002b. Reduction of microorganisms in ground beef using multiple intervention technology. *Meat Sci.* 61:315–322.
Richardson, S. D., M. J. Plewa, E. D. Wagner, R. Schoeny, and D. M. Demarini. 2007. Occurrence, genotoxicity, and carcinogenicity of regulated and emerging disinfection by-products in drinking water: a review and roadmap for research. *Mutat. Res.* 636:178–242.
Riedel, C. T., L. Brøndsted, H. Rosenquist, S. N. Haxgart, and B.B. Christensen. 2009. Chemical decontamination of *Campylobacter jejuni* on chicken skin and meat. *J. Food Prot.* 72:1173–1180.
Robach, M. C. 1980. Use of preservatives to control microorganisms in food. *Food Technol.* 34: 81–84.
Rook, J. J. 1974. Formation of haloforms during chlorination of natural waters. *Water Treat. Exam.* 23:234–243.
Ruby, J. R. and S. C. Ingham. 2009. Evaluation of potential for inhibition of growth of *Escherichia coli* O157:H7 and multidrug-resistant *Salmonella* serovars in raw beef by addition of a presumptive *Lactobacillus sakei* ground beef isolate. *J. Food Prot.* 72:251–259.
Ruiz, A., S. K. Williams, N. Djeri, A. Hinton Jr., and G. E. Rodrick. 2010. Nisin affects the growth of *Listeria monocytogenes* on ready-to-eat turkey ham stored at four degrees Celsius for sixty-three days. *Poult. Sci.* 89:353–358.
Russell, J. B. 1992. Another explanation for the toxicity of fermentation acids at low pH: anion accumulation versus uncoupling. *J. Appl. Bacteriol.* 73:363–370.
Rydlo, T., J. Miltz, and A. Mor. 2006. Eukaryotic antimicrobial peptides: promises and premises in food safety. *J. Food Sci.* 71:R125–R135.
Schirmer, B. C. and S. Langsrud. 2010. Evaluation of natural antimicrobials on typical meat spoilage bacteria *in vitro* and it vacuum-packed pork meat. *J. Food Sci.* 75:M98–M102.
Sheldon, B. W. and A. L. Brown. 1986. Efficacy of ozone as a disinfectant for poultry carcasses and chill water. *J. Food Sci.* 51:305–309.
Shelef, L. A. 1994. Antimicrobial effects of lactates: a review. *J. Food Prot.* 57:445–450.
Sirsat, S. A., A. Muthaiyan, and S. C. Ricke. 2009. Antimicrobials for foodborne pathogen reduction in organic and natural poultry production. *J. Appl. Poult. Res.* 18:379–388.
Sobrino-López, A. and O. Martín-Belloso. 2006. Enhancing inactivation of *Staphylococcus aureus* in skim milk by combining high-intensity pulsed electric fields and nisin. *J. Food Prot.* 69:345–353.
Sobsey, M. D., D. A. Battigelli, G.-A. Shin, and S. Newland. 1998. RT-PCR amplification detects inactivated viruses in water and wastewater. *Water Sci. Technol.* 38:91–94.
Solomakos, N., A. Govaris, P. Koidis, and N. Botsoglou. 2008. The antimicrobial effect of thyme essential oil, nisin, and their combination against *Listeria monocytogenes* in minced beef during refrigerated storage. *Food Microbiol.* 25:120–127.
Song, Y., E. E. Babiker, M. Usui, A. Saito, and A. Kato. 2002. Emulsifying properties and bactericidal action of chitosan-lysozyme conjugates. *Food Res. Int.* 35:459–466.
Stiles, M. E. 1996. Biopreservation by lactic acid bacteria. *Anton. Leeuw. J. Microbiol.* 70:331–345.
Stopforth, J. D., Y. Yoon, K. E. Belk, J. A. Scanga, P. A. Kendall, G. C. Smith, and J. N. Sofos. 2004. Effect of simulated spray chilling with chemical solutions on acid-habituated and non-acid-habituated *Escherichia coli* O157:H7 cells attached to beef carcass tissues. *J. Food Prot.* 67:2099–2106.
Tamblyn, K. C. and D. E. Conner. 1997. Bactericidal activity of organic acids against *Salmonella typhimurium* attached to broiler chicken skin. *J. Food Prot.* 60:623–633.

Tsai, L.-S., R. Wilson, and V. Randall. 1997. Mutagenicity of poultry chiller water treated with either chlorine dioxide or chlorine. *J. Agric. Food Chem.* 45:2267–2272.

USDA-FSIS. 2010. Safe and suitable ingredients used in the production of meat and poultry products. Directive 7120.1, Rev. 2. Available at: www.fsis.usda.gov/OPPDE/rdad/FSISDirectives/7120.1Rev2.pdf (accessed June 6, 2010).

Valderrama, W. B., E. W. Mills, and C. N. Cutter. 2009. Efficacy of chlorine dioxide against *Listeria monocytogenes* in brine chilling solutions. *J. Food Prot.* 72:2272–2277.

van Heusden, H. E., B. de Kruijff, and E. Breukink. 2002. Lipid II induces a transmembrane orientation of the pore-forming peptide lantibiotic nisin. *Biochemistry* 41:12171–12178.

Van Kraau, C., E. Breukink, H. S. Rollema, R. Siezen, R. A. Demel, B. De Kruijkk, and O. P. Kuipers. 1997. Influence of charge differences in the C-terminal part of nisin on antimicrobial activity and signaling capacity. *Eur. J. Biochem.* 247:114.

Vandekinderen, I., F. Devlieghere, J. Van Camp, B. Kerkaert, T. Cucu, P. Ragaert, J. De Bruyne, and B. De Meulenaer. 2009. Effects of food composition on the inactivation of foodborne microorganisms by chlorine dioxide. *Int. J. Food Microbiol.* 131:138–144.

Vizzier-Thaxton, Y., M. L. Ewing, and C. M. Bonner. 2010. Generation and detection of trihalomethanes in chicken tissue from chlorinated chill water. *J. Appl. Poult. Res.* 19:169–173.

Ye, M., H. Neetoo, and H. Chen. 2008. Control of *Listeria monocytogenes* on ham steaks by antimicrobials incorporated into chitosan-coated plastic films. *Food Microbiol.* 25:260–268.

Young, S. B. and P. Setlow. 2003. Mechanisms of killing of *Bacillus subtilis* spores by hypochlorite and chlorine dioxide. *J. Appl. Microbiol.* 95:54–67.

Zhao, T., P. Zhao, and M. P. Doyle. 2009. Inactivation of *Salmonella* and *Escherichia coli* O157:H7 on lettuce and poultry skin by combinations of levulinic acid and sodium dodecyl sulfate. *J. Food Prot.* 72:928–936.

Zhou, F., B. Ji, H. Zhang, H. Jiang, Z. Yang, J. Li, J. Li, Y. Ren, and W. Yan. 2007. Synergistic effect of thymol and carvacrol combined with chelators and organic acids against *Salmonella* Typhimurium. *J. Food Prot.* 70:1704–1709.

14 Nutritional Value of Organic Meat and Potential Human Health Response

Ewa Rembiałkowska and Maciej Badowski

Abstract: The nutritional quality of organically produced meat is affected by organic diet and breeding conditions. A number of studies confirm the lower total fat content of organic farm meat and this means a decreased calorific value. Carcasses from organically reared animals have usually higher content of intramuscular fat, positively affecting sensory properties. Moreover, the fatty acids' profile in organic meat is considered to be more beneficial to human health. Organic meat contains more favourable proportions of unsaturated fatty acids, including *n*-3 acids, and decreased level of saturated ones. The reason for this is a longer period of pasturage applied to organically reared animals. Therefore, consumption of organic meat can impose anti-cancer effects, stimulate the immune system and prevent coronary heart diseases.

Keywords: meat quality; organic meat; organic farming; organic breeding; fatty acids; animal nutrition; nutritional value

14.1 INTRODUCTION

The global demand for organic food is systematically growing. More and more people are becoming convinced that such food is healthier than the mass-produced food, which is commonly available on the conventional food market (Anderson, 2000; Harper & Makatouni, 2002; Magnusson *et al.*, 2003; Yiridoe *et al.*, 2005). Different kinds of disease epidemics, such as Bovine spongiform encephalopathy (BSE) or foot-and-mouth disease (FMD), have illustrated how changes in demand for a specific raw material or even a meat product are influenced by the sense of danger, which may be foreseen and estimated.

For a modern consumer, who has a developed organic awareness, a basic discriminant of valuable food products is quality. This term has a complex nature and is a sum of many factors. The first factor is public health and safety specified by hygienic and toxicological indicators (level of environmental impurities, mycotoxins, bacteria and parasites). There are also other factors such as sensory discriminates (taste, smell, colour and tenderness), physiological and nutritional parameters and technological features of meat (Andersen, 2000; Cooper *et al.*, 2007). The meat quality is conditioned by factors such as: type of fodder for animals, breeding system (including a physical activity), breed, age and sex of animals as well as the course of post-slaughter activities (bleeding out, steaming and cooling of carcasses) (Troy, 1995; Kerry *et al.*, 2000).

Organic Meat Production and Processing, First Edition. Edited by Steven C. Ricke, Ellen J. Van Loo, Michael G. Johnson, and Corliss A. O'Bryan.

The focus of the present chapter is the nutritional value, an important meat quality attribute. It is a combination of separate mineral components and organic compounds, which have a direct or indirect impact on consumer's health – usually on a long-term basis. The nutritional value is defined as the usefulness of products and food groups to cover the consumer's needs related to metabolism. The nutritional value is, therefore, a function of content, balance and bioavailability of nutritional components. Organic raw materials include higher amounts of bioactive substances, which are considered very desirable from the health point of view. It also relates to meat, which is confirmed by the studies presented later in this chapter.

Meat and meat products are one of the most important protein sources of high nutritional value, i.e. protein of favourable amino acid composition, including all essential amino acids in appropriate proportions. With respect to different species, the protein content in muscular tissue ranges from 15% to 20%, and in the meat exposed to heat treatment – due to a lower content of water – it is higher per product mass. Offal has much less protein (11%–17%); however, it is richer in vitamins and mineral components, but high content of cholesterol reduces its use in human diet (Bartnikowska, 2005).

Amino acid composition of meat proteins is well balanced, i.e. they include all amino acids necessary for synthesis of systemic proteins, which cannot be produced in a human organism (essential amino acids). Only proteins of connective tissue are of a low biological value, since they contain little tryptophan and cysteine (essential amino acids). Technological processes, particularly thermal ones and those connected with reduction of water content in a product (e.g. drying), may significantly reduce amino acid bioavailability, which consequently causes a decrease in nutritional value of meat and meat products (Bartnikowska, 2005).

Protein is needed not only to restore spent cells, but to generate new cells as well. Therefore, fast-growing organisms such as children and youth need a much higher supply of protein of high nutritional value per body mass than adults. A diet rich in proteins is often nutritionally needed for people affected with a variety of illnesses, for example people suffering from maldigestion and malabsorption syndromes, inflammatory bowel disease (large intestine), hyperthyroidism, weakened patients, mostly with cancer or those chronically undernourished due to other reasons (Bartnikowska, 2005).

The energy value of meat and meat products depends on the content of fat and water and the content of fat varies significantly, depending on species, section of the carcass and type of product. Due to the high-energy value of fat, its high content in meat is considered unfavourable from a health point of view. Therefore, most consumers – following the nutritionists' instructions – shop for the meat and the meat products of low-fat content. However, fat is a carrier of taste and olfactory substances and its presence is recommended up to specific limits (Pospiech *et al.*, 2006). However, fat is also associated with the cholesterol content. With regard to atherosclerosis prophylaxis, its level in blood serum may be taken into consideration as a risk indicator of blood vessel and heart diseases. It is believed that low-fat food products contain lower amounts of cholesterol.

In terms of fat content, meat from different species is characterised by high variability. This fraction consists of approximately 95% triacylglycerols, which in turn are composed of fatty acids with chain length and saturation levels providing a definitive nutritional value for the animal fat. Saturated acids are regarded as a factor with a negative impact on human health, since they contribute to the development of atherosclerosis (Pfeuffer & Schrezenmeir, 2000) and an increase of the cholesterol level in blood, which fosters circulatory system diseases (Haug *et al.*, 2007). Among polyunsaturated fatty acids (PUFA), *n*-3 PUFA, having the first double bond at C-3 from the methyl end (e.g. linolenic acid 18:3), have a beneficial influence on humans. They positively affect the nervous system function and decrease the risk of diabetes and heart and circulatory diseases (Horrobin, 1993; Hu *et al.*, 1999). The ratio of *n*-3 acids to *n*-6 acids is crucial as well. If the content of the second group of PUFA *n*-6, having first double bond at C-6 from methyl end (e.g. linoleic acid (LA) 18:2), is too high, the risk of inflammatory conditions, thrombus and autoimmune symptoms increases. The composition of fatty acids is a very important factor in contributing to nutritional value. Originally, meat consumed by humans is naturally rich in *n*-3 acids, which are beneficial in terms of anti-inflammatory properties as well as reducing the risk of heart attack (Bucher *et al.*, 2002), breast, prostate and colorectal cancers (Deckere, 1999; Augustsson *et al.*, 2003). The content of *n*-6 acids that increases the risk of atherosclerosis lesions and development of cancerous lesions was lower in meat eaten originally than in currently available meat of commercial animal breeds. In human diet, the ratio of *n*-6:*n*-3 acids amounted to approximately 1:1. Currently, due to industrial production of animal fodders (based on numerous grains including *n*-6 acids), this ratio comes to 30:1 (Berrisch-Hempen, 1995). *N*-6 fatty acids appear in quantities significantly higher than *n*-3 acids (Enser *et al.*, 1998), which have a negative impact on consumers' health status.

Among *n*-3 acids, alpha-linolenic acid (LNA) is of the greatest significance; however, among *n*-6 acids, the highest content belongs to LA. With respect to monounsaturated acids, the oleic acid content should also be considered since it represents one-fourth of the total fatty acids. Oleic acid plays a protective role with respect to *n*-3 and *n*-6 acids, preventing their oxidation as well as reducing cholesterol level and exhibits an anti-cancer effect (Ip, 1997; Kris-Etherton *et al.*, 1999; Mensink *et al.*, 2003).

Docosahexaenoic acid (DHA) belongs to the *n*-3 acid group and is important from dietary viewpoint. It influences reduction of death rate of brain cells in Alzheimer's disease (Uauy & Dangour, 2006), supports treatment of Parkinson disease, circulatory system diseases, chronic arthritis and attention deficit hyperactivity disorder (ADHD) (Haug *et al.*, 2007). Eicosapentaenoic acid (EPA) favourably influences the circulatory system, fighting symptoms of ADHD and has anti-inflammatory and anti-atherogenic effects (Calder, 2006). Both acids improve the skin condition in atopic dermatitis (Boelsma *et al.*, 2004).

A component of the fat fraction that is of particular interest is conjugated linoleic acid (CLA). Beside milk, beef is a fundamental source of the compound isomers in the human diet (Haug *et al.*, 2004). The most important CLA isomer, representing approximately 90% CLA, is *cis*-9 *trans*-11 isomer since it prevents tumour development, heart disease and has a stimulating effect on the immune system (Whigham *et al.*, 2000). It is called rumenic acid since the rumen is a primary origin for its synthesis from LA. Other CLA isomers (*trans*-7 *cis*-9, *trans*-10 *cis*-12 and *trans*-9 *cis*-11) counteract obesity (reducing adipose tissue and increasing muscle mass) and support diabetes treatment (Taylor & Zahradka, 2004). The content of CLA in animal fat is influenced by many factors. The most important factors are the type of fodder fed to the animals (Parodi, 1999), followed by seasonal fluctuations (Parodi, 1977), endogenous synthesis from *trans*-vaccenic acid (TVA) (Griinari *et al.*, 2000) and free radical oxidation of LA during processing (Ha *et al.*, 1989).

Meat and offal are good sources of many mineral components, such as iron (so-called haeme iron, well-absorbable) and zinc, copper, phosphorus and sulphur. Because of the high content of phosphorus and sulphur compounds, meat and meat products are classified as highly acid-forming compounds, which means that in order to maintain the acid–alkaline balance, it is necessary to consume alkaline-forming products (primarily vegetables) together with meat products (Bartnikowska, 2005).

Meat is a very good source of B-group vitamins, but the content of the separate vitamins significantly differs with regard to animal species, for example the highest content of thiamine is in pork, and niacin (vitamin PP) in veal. Only animal products are the source of vitamin B_{12}. In the average Western diet, meat and meat products meet about 70% of the demand for vitamin B_{12} and thus vegetarians who have consciously refrained from consuming animal products are recommended to supplement their diets with vitamin B_{12} supplements. During food processing, in particular thermal processes, a significant loss of B-group vitamins can take place (Bartnikowska, 2005).

Fat-soluble vitamins (mainly A and D) are stored in internal organs of animals; therefore, offal such as liver or kidneys are a primary source. As mentioned previously, these animal products include a very high level of cholesterol, which reduces their use in human nutrition. Vitamin E can also be found in meat and meat products, but its concentration is not sufficient to protect unsaturated fatty acids against oxidation processes (Bartnikowska, 2005).

14.2 BEEF

Ruminant meat is particularly valuable, since numerous studies have shown that the ratio of PUFA *n*-6:*n*-3 is much lower compared to other kinds of meat. In beef from organic farms, the proportions are even more favourable. It results from a high concentration of LNA (18:3, *n*-3), the high content of which can be found in grass (Wood *et al.*, 1999, 2003). This hypothesis has been confirmed by the results of the studies by Marmer *et al.* (1984) and Matthes and Pastushenko (1999), when diets of animals were switched from a pasture system to a grain mix. These dietary shifts caused a decrease in the content of PUFA (in particular LNA) and consequently an increase in the ratio of fatty acids *n*-6:*n*-3. The results of the studies by Pastushenko *et al.* (2000) were similar where a lower concentration of saturated fatty acids (22.4%) had been ascertained in organic beef compared to conventional beef sources (40.0%). In the meat of animals fed mother's milk and subsequently pastured and hay-fed during the winter, there was a lower content of saturated fatty acids compared to the meat of heifers fed milk-substitute mix, fodder concentrates and small amounts of hay. Differences in the content of *n*-6 and *n*-3 acids in both kinds of beef, reported by Pastushenko *et al.* (2000) and Enser *et al.* (1998), are presented in Figure 14.1.

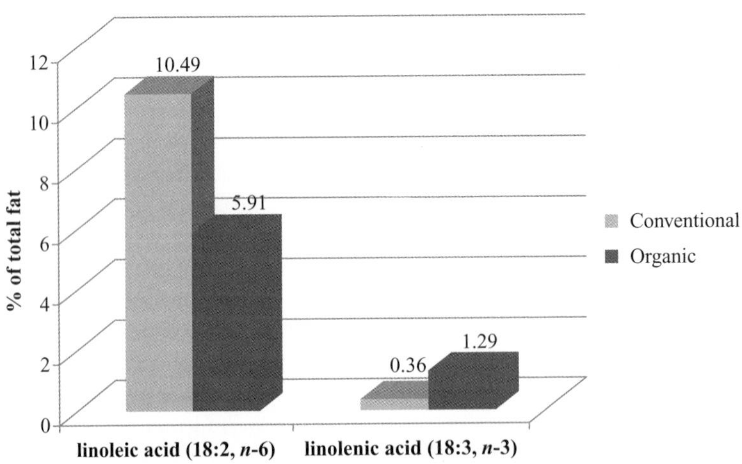

Figure 14.1 The content of unsaturated fatty acids in organic and conventional beef (% of total fat) (Enser *et al.*, 1998; Pastushenko *et al.*, 2000).

Results of the studies by Nuernberg *et al.* (2002) further confirmed that animal diet composition has the greatest impact on the proportion of fatty acids in meat. Cows grazing in fresh grass on pasture land exhibited four times higher content of LNA (18:3), an *n*-3 fatty acid that is considered very beneficial to human health, while simultaneously yielding a reduced content of oleic (18:1, *n*-9) and linoleic (18:2, *n*-6) acids when compared to the meat of grain concentrated-fed cows.

Similar findings were demonstrated in a veal study. Miotello *et al.* (2009) compared characteristics of veal from organic and conventional farms. The organic meat included significantly lower content of ether extract ($P < 0.001$) and cholesterol ($P < 0.05$). From nutritional viewpoint, this meat proved to be more valuable due to a higher content of *n*-3 acids, reduced *n*-6:*n*-3 ratio and a higher CLA level compared to conventional veal.

Walshe *et al.* (2006) compared bull meat from organic and conventional farming. According to their studies, organic meat samples have a considerably higher content of fat and lower moisture content than conventional ones. However, there were no significant differences observed between both kinds of meat in terms of the content of proteins, ash, beta-carotene, alpha-tocopherol, retinol or the content of fatty acids. However, the conventional meat exhibited a better storage quality since its samples had better colour and lipid stability during storage compared to the organic meat samples. It is probably the result of the higher fat content in organic bull meat, which caused more intensive oxidation of the unsaturated fatty acids in the fat of these samples (Walshe *et al.*, 2006). This was confirmed by a high content of the substances reacting with thiobarbituric acid reactive substances (TBARS) found in organic meat.

A lower lipid stability of such meat may also be characterised by a higher content of metal ions (in particular total and haeme iron), which are catalysts of peroxidation processes (Fukozawa & Fuji, 1992). The physical activity of an organism increases the level of haeme iron (Hoffmann, 1995), particularly in muscles, which undergo the greatest oxidation (Petersen *et al.*, 1997). Due to an obligatory access to outdoor runs, animals from organic farming are involved in more physical activities than intensive farming stock; therefore, this conclusion may be warranted (Braghieri & Napolitano, 2009).

14.3 MUTTON AND LAMB

Mutton yields similar proportions of fatty acids as beef. Fisher *et al.* (2000) stated that both in organic and conventional farming, slaughter yields can be similar but organic mutton yields more favourable quality properties. Three sheep breeds were compared: (1) Welsh Mountain, (2) Soay and (3) Suffolk. It was demonstrated that appropriate breed selection was also responsible for gaining desirable yield.

Angood *et al.* (2008) compared the composition of fatty acids' pool and the nutritional quality of organic and conventional lamb products, offered on the British market. The study presented significant differences, i.e. organic meat had a higher content of *n*-3 PUFA and displayed a better nutritional quality in terms of juiciness, tastiness and general acceptability compared to the conventional lamb available on the market. The greater juiciness resulted from a higher content of intramuscular fat in the organic lamb chops. The greater tastiness, preferred by the British consumers, was attributed to the difference in the fatty acids composition, in particular a higher content of linolenic acid (18:3) and total content of *n*-3 acids in organic meat chops (as a result of pasture feeding). Conventional meat contained more LA (18:2, *n*-6), which probably resulted from predominant dietary concentrate in the animal diet.

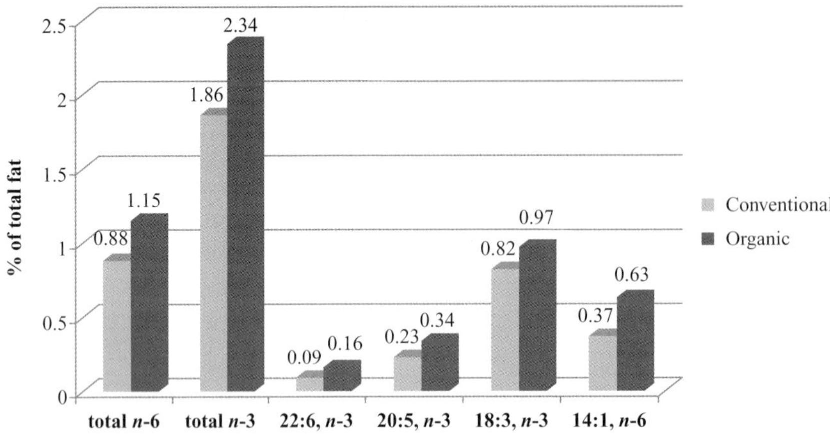

Figure 14.2 Some fatty acids in intramuscular fat in longissimus dorsi muscle (% of total fat). (After Morbidini *et al.*, 1999.)

Both kinds of meat, however, exhibited a favourable proportion of *n*-6:*n*-3 acids' quantity (Angood *et al.*, 2008).

Feeding lambs with fresh green fodder has a considerable impact on the quality of their meat, which was confirmed by the studies by Enser *et al.* (1998). The large amount of dietary fibre in the fodder and the considerable consumption of fresh grass led to a beneficial composition of fatty acids as well as the content of intramuscular and abdominal fat. According to Oksbjerg *et al.* (2005), the influence on the proportions of fatty acids may be dependent on dietary fibre supply as well as restrictively defined animal feeding, which consequently resulted in a lower total mass. In fact, lean meat possesses greater proportions of phospholipids and is richer in PUFA, in particular C20 and C22 acids (Elmore *et al.*, 1999). Due to pasture feeding, organic lamb meat showed a higher content of CLA, an indirect product of biohydrogenation in rumen, which is considered very beneficial to human health (Pariza *et al.*, 2001).

Comparative studies of mutton have also been conducted in Italy. Morbidini *et al.* (1999) analysed the quality of carcasses and intramuscular fat of sheep meat from a certified organic farm, comparing the results with those obtained in a control group. Conventional sheep meat was characteristic of a greater energetic value due to a higher fat content. However, a composition of intramuscular fat turned out to be more favourable in organic sheep meat (see Figure 14.2). Saturated fatty acids predominated in meat samples of both groups, but the content of *n*-3 and *n*-6 acids – the most valuable from a health viewpoint – was higher in organic meat.

14.4 PORK

Recently, the interest in pork meat from organic farms has been growing (Hamm & Gronefeld, 2004). However, there has been considerable discrepancy regarding the quality of the meat obtained due to the differentiation of fodders used within organic system, species changeability and slaughter methods (Rembiałkowska & Wiśniewska, 2010). In conjunction with these variables, some efforts to define optimal breeding conditions have been undertaken (under the guidelines by the International Federation of Organic Agriculture Movements) in an effort to allow an increase in yield as well as maintaining a high meat quality (Guy & Edwards, 2002).

An organoleptic assessment of pork – similarly to beef and mutton – is largely influenced by the content of intramuscular fat (Fernandez *et al.*, 1999). However, the results of studies comparing swine carcasses from organic and conventional farms are not unambiguous. Hansen *et al.* (2006) showed that animals from both systems reached a similar mass, yielded a similar fat-free body mass and content of intramuscular fat. Yet, Sundrum *et al.* (2000, 2003) and Millet *et al.* (2004) obtained different results, with marbling level of organic meat being higher. However, drawing definitive conclusions is hindered by animals being fed differently in these studies. The experiment conducted by Hansen *et al.* (2006) was based on animals receiving organic fodder, including 70% cereal grain concentrates and 30% silage. In the experiments by Sundrum *et al.* (2000, 2003) and Millet *et al.* (2004), diet of animals from organic farms was more diverse, since they contained grains of wheat and barley and seeds of broad bean, pea and lupine while the proportion of concentrates in fodder was less. Consequently, it was hypothesised that dietary composition was responsible for the differing contents of intramuscular fat of the pork meat samples. However, the studies by Olsson *et al.* (2003) revealed a lower content of intramuscular fat and lower fat-free body mass in the organically raised animals.

According to the results of the previous studies, organic pork is less tender than pork meat from conventional sources (Danielsen *et al.*, 1999). This may be the result of lower daily mass gains of pigs from organic farming. Therefore, while slaughtering, their meat presents a decreased proteolytic potential – its low value is considered to be the cause of decreased tenderness (Sather *et al.*, 1997).

Nevertheless, these studies unequivocally indicate an advantage for organic pork in terms of a high level of PUFA and a lower content of saturated fatty acids (Hansen *et al.*, 2000; Nilzen *et al.*, 2001). These differences have impact on reduced processing quality of organic meat, owing to increased lipid oxidation and content of soft fat. The results from analysis of the content of TBARS reveal a poorer storage quality of organic meat (Lopez-Bote *et al.*, 1998; Warnants *et al.*, 1999; Nilzen *et al.*, 2001).

A concept of higher fat quality in organic pork is also confirmed by the studies carried out by Kim *et al.* (2009) on Korean native black pigs. A level of saturated and monounsaturated fatty acids, less beneficial to health, turned out to be higher in conventional meat, whereas the content of valuable *n*-6 and *n*-3 polyunsaturated acids was considerably higher in organic meat (see Figure 14.3). The proportion *n*-6:*n*-3 was lowered in organic pork by more than half (see Figure 14.4).

In addition, the chemical composition of the pork of outdoor run animals (with pasture access) was compared with the meat of pigs raised only in farm buildings (Hansen *et al.*, 2000; Nilzen *et al.*, 2001). In the first group animals' meat, a higher concentration of vitamin E and alpha-tocopherol – antioxidant compounds – was detected. However, their level was not high enough to balance a great susceptibility of organic meat to lipid oxidation, resulting from a high content of PUFA.

Pork carcasses from organic farms also have a higher fat-free body mass (Sundrum & Acosta, 2003; Bee *et al.*, 2004), which is why they have a higher estimated wholesale pricing – compared to conventional carcasses; they exhibit a higher mass of loin and gammon (Sundrum & Acosta, 2003). It is believed that appropriate selection of animal breeds and adaptation to local conditions should sort out the problem of lower daily mass gains, such as is the case with organic swine.

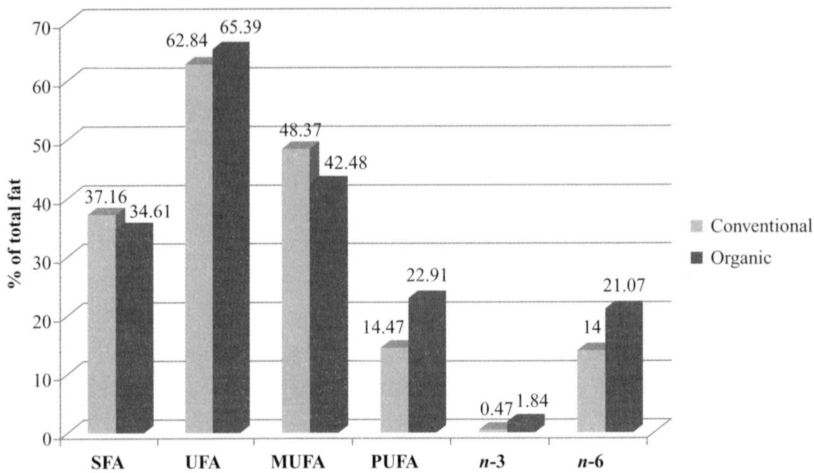

Figure 14.3 Comparison of fatty acid composition (%) of longissimus muscle between conventional ($n =$ 30) and organic ($n = 30$) pork from Korean black pigs. (After Kim *et al.*, 2009.) SFA, saturated fatty acids; UFA, unsaturated fatty acids; MUFA, monounsaturated fatty acids; PUFA, polyunsaturated fatty acids.

Similar results were obtained by Hogberg *et al.* (2002), who detected a higher content of vitamin E in organic pork, resulting from the higher vitamin content in organic fodder. Significant differences in content of other vitamins between organic and conventional pork were not reported.

An organic breeding system also promotes greater quantity of protein in meat and reduced water/protein index (Dworschak *et al.*, 1995; Entfält *et al.*, 1997; Olsson *et al.*, 2003). Such animal farming also results in higher content of ash and mineral components, such as zinc, copper and iron (Dworschak *et al.*, 1995; Olsson *et al.*, 2003). The ability of binding metal ions by protein particles is most likely higher in muscle cells of more physically active

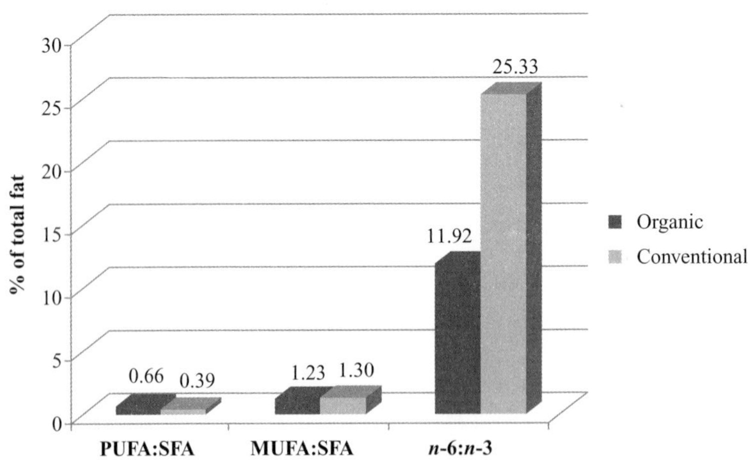

Figure 14.4 Comparison of fatty acid profile of M. longissimus muscle between conventional ($n = 30$) and organic ($n = 30$) pork from Korean black pigs. (After Kim *et al.*, 2009.) PUFA, polyunsaturated fatty acids; MUFA, monounsaturated fatty acids; SFA, saturated fatty acids.

swine, raised with access to an outdoor run. Moreover, higher content of iron may be correlated with greater myoglobin level, which is observed in animals raised in conditions providing opportunities for the animals to participate in recreational activities (Shorthose & Harris, 1991).

14.5 POULTRY

Castellini *et al.* (2002) and Combes *et al.* (2003) analysed quality parameters of poultry from organic and conventional farms. Broilers raised under these different conditions were the focus of their studies. In organic farming, broilers had access to a specific run area, whereas in conventional system they were raised in cages where space per one broiler was very limited. Fodder for organic chickens consisted of organic raw materials, while traditional farm fodder was used in conventional farming. The breast and thigh meat were taken as the experimental samples. Smaller mass gains and lower post-slaughter mass were observed with respect to organic chickens. However, they were approximately two to three times less fatty in abdominal and breast parts; the percentage fat content of meat in breast and thigh parts was higher as well. The fat content in the thigh part was 1.8 times lower in the organic birds. The meat included a higher level of iron compared to conventional ones. In the researchers' opinion, a lower total body mass and smaller mass gains of organic chickens were caused by their greater physical activity (confirmed by the behavioural observations) and larger energy expenditures related to thermoregulation. Greater physical activity of the organic chickens was also responsible for a lower content of abdominal adipose tissue and better developed breast and thigh muscles, which would increase the commercial value of the meat (Lei & Van Beek, 1997; Lewis *et al.*, 1997). A higher level of iron was most likely conditioned by greater physical activity, since oxidation of muscle tissue raises the level of haeme iron in the organism (Hoffmann, 1995).

Analyses of chemical composition of poultry from both production systems illustrated a greater content of saturated and PUFA as well as a lower content of monounsaturated fatty acids in organic meat. Significant differences were found especially in the level of *n*-3 acids, including DHA, which was twofold greater than in conventional meat (Figure 14.5). Most likely, it was caused by the grass present in the animals' diet.

Among unfavourable properties of organic poultry, it should also be mentioned that there was an increased TBARS concentration, which indicates intensive fat oxidation processes in organic meat. Lower pH values and poorer water-binding capability in organic poultry meat were linked to greater losses while cooking. The above-mentioned dependence between a farming method and meat pH was not confirmed in the studies conducted by Combes *et al.* (2003). However, in both experiments, a lower content of total fat in organic meat was noted, which attests to its higher nutritional value (Castellini *et al.*, 2002; Combes *et al.*, 2003).

14.6 RABBIT MEAT

The quality of rabbit meat is quite rarely brought up in analytical studies. It is difficult to quantify, since – according to many authors (Combes, 2004; Dalle Zotte, 2004; Combes & Dalle Zotte, 2005; Hernández & Gondret, 2006) – rabbit meat yields great nutritional value when compared against other kinds of meat. Since there are many rabbit breeds as well as breeding methods, a reliable comparison of the meat quality of rabbits from organic and

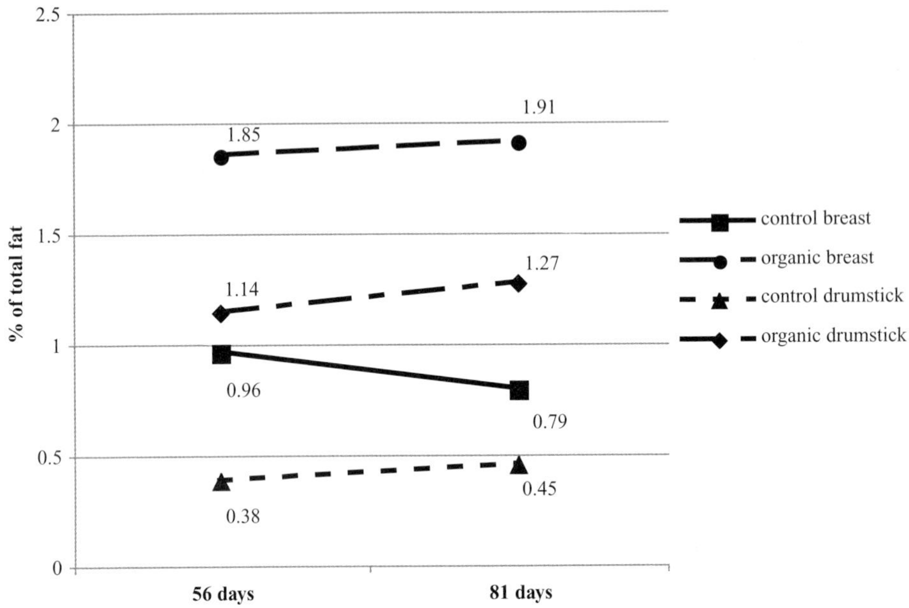

Figure 14.5 Content of docosahexaenoic acid (22:6, *n*-3) in organic and conventional chicken breast and drumstick (% of total fat.) (After Castellini *et al.*, 2002.)

conventional farms is quite complicated. Consequently, it comes as no surprise that there are very few publications on this topic.

In the studies by Lebas *et al.* (2002), same age rabbits were slaughtered from both systems. The analysis revealed a slightly higher ratio of the cooled carcass weight to the pre-slaughter weight and more alkaline reaction of muscles and higher fat content in organic rabbit meat. The results seem to be surprising because – like other animal species from organic farms – organic rabbits also have more opportunities for physical activities due to run access during their production cycle.

Combes *et al.* (2003) compared organic and conventional meat of the rabbit of similar weight but different ages. Organic animals had a lower content of meat in the back part of the carcass and a lower fat content. They were also characterised by smaller amounts of intramuscular fat.

Pla *et al.* (2007) analysed the composition of fatty acids in rabbit meat from both farming systems. Back leg meat of organic rabbits contained a lower level of monounsaturated fatty acids and a higher content of PUFA compared to the same parts of conventional animals. The concentration of saturated fatty acids turned out to be similar for both types of meat. From a nutritional viewpoint, organic rabbit meat has the advantage of having a higher ratio of polyunsaturated to saturated fatty acids. These differences in chemical composition were explained by the dissimilar production systems where a crucial role is most likely played by the diet of the hindgut fermentation of the mammal as well as the age at slaughter (Pla *et al.*, 2007), given that the animal breeds were the same in both groups.

In follow-up research, Pla (2008) compared organic and conventional rabbits of different slaughter age and weight, raised under the national organic standard in Spain. According to the authors, the age differentiation did not have any impact on differences found during the studies. The differences were primarily associated with the organic carcasses being leaner and their indicator of meat to bone weight being lower. Moreover, organic meat included less protein and a lower content of fatty acids. Conventional meat contained more saturated and monounsaturated fatty acids but less polyunsaturated and *n*-6 and *n*-3 fatty acids. In organic rabbit meat, the ratio of polyunsaturated to saturated fatty acids was higher, so theoretically it was better; but the proportion of *n*-6:*n*-3 fatty acids was higher as well, and therefore it was less favourable from a nutritional point of view. Organic meat protein was richer in methionine and cysteine. These particular amino acids represent sulphur exogenous amino acids, not synthesised in humans, where the available amount depends entirely on their availability in diets. These amino acids are necessary for synthesis of keratin – a basic structural protein of hair and nails as well as other functions.

14.7 SUMMARY

Analysing study results discussed in this chapter, some conclusions may be drawn. Factors such as organic diet and breeding conditions have a significant influence on the nutritional quality of organic farm meat. A frequently – although not always – stated effect of the use of this system in combination with animal breeding is the lower total fat content in meat, regardless of the source of meat. The legal framework of organic farming imposed on the farmer's management obligation is to provide animals with free motion opportunities by the means of run access. It supports not only animal's well-being, i.e. physical and mental comfort, but is also positively reflected in meat quality. Lower carcass fat, and consequently lower calorific value, is a factor considered to be very beneficial to consumers. Such characteristics prevent obesity, development of atherosclerosis or type 2 diabetes, i.e. civilisation diseases of the 21st century, where diet is most often considered to be the cause. Besides fat content in the carcass, there are also observed differences in fat distribution in body – organic meat has a higher content of intramuscular fat, which is defined as 'marbling' due to its appearance. It is a favourable property and is considered a sensory assessment, since it relates to higher meat juiciness and tastiness.

Diets fed to organic animals also have a fundamental impact on the composition of meat fatty acids. A significantly longer period of pasturage in vegetation season leading to a greater share of fresh grass in the daily animal diet followed by a winter season where there is more hay and hay silage, in conjunction with lower levels of other cereal silages and concentrates compared to conventional system, appears to have a major impact on meat quality and composition. The reasons described include considerable differences in meat composition – organic raw materials possess much more favourable proportions of unsaturated fatty acids, including *n*-3 acids, all considered important for optimal human health. Among the human health attributes are their anti-cancer effect, as well as their ability to act preventively against coronary heart diseases via reduction in cholesterol levels in blood, stimulating the operation of immune system and favourably influencing the operation of nervous system (De Deckere, 1999; Bucher *et al.*, 2002; Augustsson *et al.*, 2003).

Several studies – though much less numerous – allow for the assumption that organic animal meat includes greater mineral content, in particular haeme iron, which ensures proper operation of circulatory system and improves resistance to infections and stimulates brain development.

It is difficult to conclude whether organic meat has a higher protein content; however, there are assumptions that in such raw materials this fraction is considered more valuable from human nutritional viewpoint. It may include greater amounts of essential amino acids.

Table 14.1 The comparison of selected quality indicators of organic and conventional meat.

The kind of meat	The examined indicators	The obtained results	The author
Beef, mutton	Composition of fatty acids	Higher level of *n*-3 PUFA and lower level of *n*-6 PUFA in ORG meat; higher DHA and EPA content in ORG meat	Enser *et al.*, 1998
Beef	Carcass mass	Lower in ORG breeding	Woodward & Fernandez, 1999
	Fat content	Higher in ORG meat	
	Content of intramuscular fat	Higher in ORG meat	
Beef	Carcass classification under EUROP system	Higher marks for ORG breeding carcasses	Hansson *et al.*, 2000
	Content of total fat	Lower in ORG meat	
Beef	Content of total fat	Higher in ORG meat	Walshe *et al.*, 2006
	Content of protein	Comparable in both groups	
	Content of ash	Comparable in both groups	
	Content of beta-carotene	Comparable in both groups	
	Content of alpha-tocopherol	Comparable in both groups	
	Content of retinol	Comparable in both groups	
	Composition of fatty acids	Comparable in both groups	
	Storage quality	Worse in case of ORG meat due to colour and lipid stability	
Beef	Composition of fatty acids	Lower SFA level in ORG meat; lower MUFA level in ORG meat higher *n*-3 PUFA level and lower *n*-6 MUFA level in ORG meat	Pastushenko *et al.*, 2000
Pork	Fat-free body mass	Higher in ORG meat	Sundrum and Acosta, 2003
	Content of intramuscular fat	Higher in ORG meat	
Pork	Content of intramuscular fat	Higher in ORG meat	Millet *et al.*, 2004
Pork	Composition of fatty acids	Lower SFA level and more PUFA in ORG meat	Hansen *et al.*, 2006
	Content of TBARS	Higher in ORG meat	
Pork	Composition of fatty acids	Higher PUFA level and lower SFA level in ORG meat	Hansen *et al.*, 2000
Pork	Composition of fatty acids	Higher PUFA level and lower SFA level in ORG meat	Nilzen *et al.*, 2001
	Content of TBARS	Higher in ORG meat	
Pork	Content of TBARS	Higher in ORG meat	Lopez-Bote *et al.*, 1998; Warnants *et al.*, 1999
Pork	Fat-free body mass	Higher in ORG meat	Bee *et al.*, 2004
Pork	Fat-free body mass	Lower in ORG meat	Olsson *et al.*, 2003
	Content of total fat	Higher in ORG meat	
	Content of intramuscular fat	Lower in ORG meat	

Table 14.1 (*Continued*)

The kind of meat	The examined indicators	The obtained results	The author
Pork	Composition of fatty acids	Higher PUFA level (including *n*-3 acids) in ORG meat	Kim *et al.*, 2009
Mutton	Composition of fatty acids	Higher *n*-3 PUFA level and lower *n*-6 PUFA level in ORG meat	Fisher *et al.*, 2000
	Content of intramuscular fat	Higher in ORG meat	
	Content of total fat	Lower in ORG meat	
	Content of TBARS	Higher in ORG meat	
Lamb	Composition of fatty acids	Higher *n*-3 PUFA level (in particular linolenic acid) in ORG meat; comparable *n*-6:*n*-3 ratio	Angood *et al.*, 2008
Poultry	Content of abdominal fat	Lower in ORG meat	Castellini *et al.*, 2002; Combes *et al.*, 2003
	Content of iron	Higher in ORG meat	
	Content of breast and thigh meat	Higher in ORG breeding	
	Composition of fatty acids	Higher PUFA and SFA levels and lower MUFA level in ORG meat; higher *n*-3 acids level (in particular DHA) in ORG meat	
	Content of TBARS	Higher in ORG meat	
	Meat pH	Lower in ORG meat	
	Content of total fat	Lower in ORG meat	
Rabbit meat	Content of total fat	Higher for ORG rabbits	Lebas *et al.*, 2002
	Meat pH	Higher for ORG rabbits	
Rabbit meat	Content of total fat	Lower for ORG rabbits	Combes *et al.*, 2003
Rabbit meat	Composition of fatty acids	Lower MUFA level and higher PUFA level in ORG rabbit meat; comparable SFA concentration; PUFA:SFA ratio higher in ORG rabbit meat	Pla *et al.*, 2007
Rabbit meat	Content of total fat	Lower for ORG rabbits	Pla, 2008
	Content of protein	Lower for ORG rabbits	
	Composition of fatty acids	More *n*-6 and *n*-3 PUFA and lower SFA and MUFA levels in ORG rabbit meat; PUFA:SFA ratio higher in ORG rabbit meat; *n*-6:*n*-3 acids ratio higher in ORG meat	
	Composition of protein	More Met and Cys in ORG rabbit meat	

ORG, organic meat; MUFA, monounsaturated fatty acids; PUFA, polyunsaturated fatty acids; SFA, saturated fatty acids; TBARS, thiobarbituric acid reactive substances: Met, methionine; Cys, cysteine.

Because humans are unable to synthesise them, it is recommended to supply them in diet as exogenous amino acid supplements.

The list of the most essential nutritional properties of meat, representing a quality comparison of organic and conventional raw materials on the basis of the previous studies, is presented in Table 14.1.

REFERENCES

Andersen, H. J. 2000. What is pork quality? In: H. J. Andersen (ed) *Quality of Meat and Fat Affected by Genetics and Nutrition*. Proceedings of the Joint Session of the EAAP Commissions on Pig Production, Animal Genetics and Animal Nutrition. EAAP Publication No. 100, Zurich, Switzerland. pp. 15–26.

Anderson, W. A. 2000. The future relationship between the media, the food industry and the consumer. *Br. Med. Bull.* 56:254–268.

Angood, K. M., J. D. Wood, G. R. Nute, F. M. Whittington, S. I. Hughes, and P. R. Sheard. 2008. A comparison of organic and conventionally-produced lamb purchased from three major UK supermarkets: price, eating quality and fatty acid composition. *Meat Sci.* 78:176–184.

Augustsson, K., D. S. Michaud, E. B. Rimm, M. F. Leitzmann, M. J. Stampfer, W. C. Willett, and E. Giovannucci. 2003. A prospective study of intake of fish and marine fatty acids and prostate cancer. *Cancer Epidemiol. Biomarkers Prev.* 12:64–67.

Bartnikowska, E. 2005. Aspekty zdrowotne związane ze spożywaniem mięsa. Available at: http://docs7.chomikuj.pl/623867855,PL,0,0,Aspekty-zdrowotne-zwi%C4%85zane-ze-spo%C5%BCywaniem-mi%C4%99sa.doc.

Bee, G., G. Guex, and W. Herzog. 2004. Free-range rearing of pigs during the winter: adaptations in muscle fiber characteristics and effects on adipose tissue composition and meat quality traits. *J. Anim. Sci.* 82:1206–1218.

Berrisch-Hempen, D. 1995. Fettsäurezusammensetzung von Wildfleisch – Vergleich zum Fleisch schlachtbarer Haustiere. *Fleischwirtschaft* 75:809–813.

Boelsma, E., H. F. J. Hendriks, and L. Roza. 2004. Nutritional skin care: health effects of micronutrients and fatty acids. *Am. J. Clin. Nutr.* 73:853–864.

Braghieri, A. and F. Napolitano. 2009. Organic meat quality. In: J. P. Kerry and D. Ledward (eds) *Improving the Sensory and Nutritional Quality of Fresh Meat*. Woodhead Publishing, Cambridge. pp. 387–417.

Bucher, H. C., P. Hengstler, Ch. Schindler, and G. Meier. 2002. n-3 polyunsaturated fatty acids in coronary heart disease: a meta-analysis of randomized controlled trials. *Am. J. Med.* 112:298–304.

Calder, P. C. 2006. n-3 Polyunsaturated fatty acids, inflammation, and inflammatory diseases. *Am. J. Clin. Nutr.* 83:1505S–119S.

Castellini, C., C. Mugnai, and A. Dal Bosco. 2002. Effect of organic production system on broiler carcass and meat quality. *Meat Sci.* 60:219–225.

Combes, S. 2004. Valeur nutritionnelle de la viande de lapin. *INRA Prod. Anim.* 17:373–383.

Combes, S. and A. Dalle Zotte. 2005. La viande de lapin: valeur nutritionnelle et particularités technologiques. In: *Proc. 11èmes Journées Recherche Cunicole*, World Rabbit Science Association, November, Paris, France, pp. 167–180.

Combes, S., F. Lebas, L. Lebreton, T. Martin, N. Jehl, L. Cauquil, B. Darche, and M. A. Corboeuf. 2003. Comparaison lapin ≪ Bio ≫/ lapin standard: Caractéristique des carcasses et composition chimique de 6 muscles de la cuisse. In: *Proc. 10èmes Journées de la Recherche Cunicole*, World Rabbit Science Association, pp. 133–136.

Cooper, J., U. Niggli, and C. Leifert. 2007. *Handbook of Organic Food Safety and Quality*. Woodhead Publishing Limited, Cambridge. p. 521.

Dalle Zotte, A. 2004. Dietary advantages: rabbit must tame consumers. *Viandes et Produits Carnés* 23:161–167.

Danielsen, V., L. L. Hansen, F. Möller, C. Bejerholm, and S. Nielsen. 1999. Production results and sensory meat quality of pigs fed different amounts of concentrate *ad lib.* Clover grass or clover grass silage. In: J. E. Hermansen, V. Lund, and R. Thuen (eds) *Proc. Semin. No. 303: Ecol. Anim. Husbandry Nordic Countries*. Horsens. pp. 79–86.

De Deckere, E. A. 1999. Possible beneficial effect of fish and fish n-3 polyunsaturated fatty acids in breast and colorectal cancer. *Eur. J. Cancer Prev.* 8:213–221.

Dworschak E., E. Barna, A. Gergely, P. Czuczy, J. Hovari, M. Kontraszti, O. Gaal, L. Radnoti, G. Biro, and J. Kaltenecker. 1995. Comparison of some components of pigs kept in natural (free-range) and large-scale condition. *Meat Sci.* 39:79–86.

Elmore, J. S., D. S. Mottram, M. Enser, and J. D. Wood. 1999. Effect of the polyunsaturated fatty acid composition of beef muscle on the profile of aroma volatiles. *J. Agric. Food Chem.* 47:1619–1625.

Enfält, A. C., K. Lundstrom, I. Hansson, N. Lundeheim, and P. E. Nystrom. 1997. Effects of outdoor rearing and sire breed (Duroc or Yorkshire) on carcass composition and sensory and technological meat quality. *Meat Sci.* 45:1–15.

Enser, M., K. G. Hallett, B. Hewett, G. A. J. Fursey, J. D. Wood, and G. Harrington. 1998. Fatty acid content and composition of UK beef and lamb muscle in relation to production system and implications for human nutrition. *Meat Sci.* 49:329–341.

Fernandez, X., G. Monin, A. Talmant, J. Mourot, and B. Lebret. 1999. Influence of intramuscular fat content on the quality of pig meat. *Meat Sci.* 53:67–72.

Fisher, A. V., M. Enser, R. I. Richardson, J. D. Wood, G. R. Nute, E. Kurt, L. A. Sinclair, and R. G. Wilkinson. 2000. Fatty acid composition and eating quality of lamb types derived from four diverse breed × production system. *Meat Sci.* 55:141–147.

Fukozawa, K. and T. Fuji. 1992. Peroxide dependent and independent lipid peroxidation sitespecific mechanism of initiation by chelated iron inhibition by α-tocopherol. *Lipids* 27:227–233.

Griinari, J. M., B. A. Corl, S. H. Lacy, P. Y. Chouinard, K. V. V. Nurmela, and D. E. Bauman. 2000. Conjugated linoleic acid is synthesized endogenously in lactating dairy cows by Δ9-desaturase. *J. Nutr.* 130:2285–2291.

Guy, J. H. and S. A. Edwards. 2002. Consequences for meat quality of producing pork under organic standards. *Pigs News Inf.* 23:75–80.

Ha, Y. L., N. K. Grimm, and M. W. Pariza. 1989. Newly recognized anticarcinogenic fatty acids: identification and quantification in natural and processed cheese. *J. Agric. Food Chem.* 37:75–81.

Hamm, U. and F. Gronefeld. 2004. *The European Market for Organic Food: Revised and Updated Analysis, Organic Marketing Initiatives and Rural Development Series*, vol. 5. School of Business and Management. University of Wales, Aberystwyth. p. 165.

Hansen, L. L., C. Bejerholm, M. C. Claudi, and H. J. Andersen. 2000. Effects of organic feeding including roughage on pig performance, technological meat quality and the eating of the pork. *Proc. 13th IFOAM Int. Sci. Conf.*, Basel, Switzerland. p. 288.

Hansen, L. L., C. Cludi-Magnussen, S. K. Jensen, and H. J. Andersen. 2006. Effect of organic production system on performance and meat quality. *Meat Sci.* 74:605–615.

Hansson, J., C. Hamilton, T. Ekman, and K. Forslund. 2000. Carcass quality in certified organic production compared with conventional livestock production. *J. Vet. Med.* 47:111–120.

Harper, G. C. and A. Makatouni. 2002. Consumer perception of organic food production and farm animal welfare. *Br. Food J.* 104:287–99.

Haug, A., A. T. Hostmark, and O. M. Harstad. 2007. Bovine milk in human nutrition-a review. *Lipids Health Dis.* 6:25. Available at: www.lipidworld.com/content/6/I/25.

Haug, A., O. Taugbol, E. S. Olsen, A. S. Biong, and O. M. Harstad. 2004. Milk fat in human nutrition. Studies in dairy cows with special reference to CLA. *Anim. Sci. Pap. Rep.* 3:381–390.

Hernández, P. and F. Gondret. 2006. Rabbit meat quality. In: L. Maertens and P. Coudert (eds) *Recent Advances in Rabbit Sciences*. ILVO, Merelbeke. pp. 269–290.

Hoffmann, G. 1995. Sport medical aspects of iron metabolism. *J. Inorg. Biochem.* 59:237.

Hogberg, A., J. Pickova, J. Babol, K. Andersson, and P. C. Dutta. 2002. Muscle lipids, vitamins E and A and lipid oxidation as affected by diet and RN genotype in female and castrated male Hampshire crossbreed pigs. *Meat Sci.* 60:411–420.

Horrobin, D. F. 1993. Fatty acid metabolism in health and disease: the role of delta-6-desaturase. *Am. J. Clin. Nutr.* 57:732S–736S.

Hu, F. B, M. J. Stampfer, J. E. Manson, E. B. Rimm, A. Wolk, G. A. Colditz, C. H. Hennekens, and W. C. Willett. 1999. Dietary intake of {alpha}-linolenic acid and risk of fatal ischemic heart disease among women. *Am. J. Clin. Nutr.* 69:890–897.

Ip, C. 1997. Review of the effects of trans fatty acids, oleic acid, n-3 polyunsaturated fatty acids, and conjugated linoleic acid on mammary carcinogenesis in animals. *Am. J. Clin. Nutr.* 66:1523–1529.

Kerry, J. P., D. J. Buckley, and P. A. Morrissey. 2000. Improvement of oxidative stability of beef and lamb with vitamin E. In: E. Decker, C. Faustman, C. J. Lopez-Bote (eds) *Antioxidants in Muscle Foods, Nutritional Strategies to Improve Quality*. John Wiley & Sons, New York. pp. 229–261.

Kim, D. H., P. N. Seong, S. H. Cho, J. H. Kim, J. M. Lee, C. Jo, and D. G. Lim. 2009. Fatty acid composition and meat quality traits of organically reared Korean native black pigs. *Livest. Sci.* 2009: 96–102.

Kris-Etherton, P. M., T. A. Pearson, Y. Wan, R. L. Hargrove, K. Moriarty, V. Fishell, and T. D. Etherton. 1999. High-monounsaturated fatty acid diets lower both plasma cholesterol and triacylglycerol concentrations. *Am. J. Clin. Nutr.* 70:1009–1015.

Lebas, F., L. Lebreton, and T. Martin. 2002. Statistics on organic production of rabbits on grassland. *Cuniculture*. 164:74–80.

Lei, S. and G. Van Beek. 1997. Influence of activity and dietary energy on broiler performance. Carcass yield and sensory quality. *Br. Poult. Sci.* 38:183–189.

Lewis, P. D., G. C. Perry, L. J. Farmer, and R. L. S. Patterson. 1997. Responses of two genotypes of chicken to the diets and stocking densities typical of UK and label rouge production system. I. Performance, behaviour and carcass composition. *Meat Sci.* 4:501–516.

Lopez-Bote, C. J., A. Diestre, and J. M. Monfort, 1998. Sustained utilization of the Iberian pig breed. *Meat Sci.* 49:17–27.

Magnusson, M. K., A. Arvola, U. K. Hursti, L. Aberg, and P. O. Sjoden. 2003. Choice of organic foods is related to perceived consequences for human health and to environmentally friendly behaviour. *Appetite* 40:109–117.

Marmer, W. N., R. J. Maxwell, and J. E. Williams. 1984. Effects of dietary regimen and tissue site on bovine fatty acid profiles. *J. Animal Sci.* 59:109–121.

Matthes, H. D. and V. Pastushenko. 1999. Einflußder landwirtschaftlichen Produktionsweise auf den Fettsäurengehalt des Fleisches. *Ernährungs-Umschau* 46:335–338.

Mensink, R. P., P. L. Zock, A. D. Kester, and M. B. Katan. 2003. Effects of dietary fatty acids and carbohydrates on the ratio of serum total to HDL cholesterol and on serum lipids and apolipoproteins: a meta-analysis of 60 controlled trials. *Am. J. Clin. Nutr.* 77:1146–1155.

Millet, S., M. Hesta, M. Seynaeve, E. Ongenae, S. De Smet, J. Debraekeleer, and G. P. J. Janssens. 2004. Performance, meat and carcass traits of fattening pigs with organic versus conventional hausing and nutrition. *Livest. Prod. Sci.* 87:109–119.

Miotello, S., V. Bondesan, F. Tagliapietra, S. Schiavon, and L. Bailon. 2009. Meat quality of calves obtained from organic and conventional farming. *Ital. J. Anim. Sci.* 8:231–215.

Morbidini, L., D. M. Sarti, P. Pollidori, and A. Valigi. 1999. Carcass, meat and fat quality in Italian Merino derived lambs obtained with "organic" farming system. *FAO-CIHEAM Network on sheep and goat*, Molina de Segura, Murcia, Spain, September 23rd–25th.

Nilzen, V., J. Babol, P. C. Dutta, N. Lundeheim, A. C. Enfält, and K. Lundstrom. 2001. Free range rearing of pigs with access to pasture grazing - effect on fatty acid composition and lipid oxidation products. *Meat Sci.* 58:267–275.

Nuernberg, K., G. Nuernberg, K. Ender, S. Lorenz, K. Winkler, R. Rickert, and H. Steinhart. 2002. N-3 fatty acid and conjugated linoleic acids of longissimus muscle in beef cattle. *Eur. J. Lip. Sci. Technol.* 104:463–471.

Oksbjerg, N., K. Strudsholm, L. Gunilla, J. Gunilla, and E. J. Hermansen. 2005. Meat quality of fully or partly outdoor reared pigs in organic production. *Agric. Scand.* 55:106–112.

Olsson, V., K. Andersson, I. Hansson, and K. Lundström. 2003. Differences in meat quality between organically and conventionally produced pigs. *Meat Sci.* 3:287–297.

Pariza, M. W., Y. Park, and M. E. Cook. 2001. The biological active isomers of conjugated linoleic acid. *Prog. Lipid Res.* 40:283–298.

Parodi, P.W. 1977. Conjugated octadecadienoic acids of milk fat. *J. Dairy Sci.* 60:1550–1553.

Parodi, P. W. 1999. Conjugated linoleic acid and other anticarcinogenic agents of bovine milk fat. *J. Dairy Sci.* 82:1339–1349.

Pastushenko, V., H. D. Matthes, T. Hein, and Z. Holzer. 2000. Impact of cattle grazing on meat fatty acid composition in relation to human health. *Proc. 13th IFOAM Int. Sci. Conf.*, Basel, Switzerland, pp. 293–296.

Petersen, J. S., P. Berge, P. Henckel, and M. T. Sorensen. 1997. Collagen characteristics and meat texture of pigs exposed to different levels of physical activity. *J. Muscle Foods.* 8:47–61.

Pfeuffer, M. and J. Schrezenmeir. 2000. Bioactive substances in milk with properties decreasing risk of cardiovascular disease. *Br. J. Nutr.* 84:155–159.

Pla, M. 2008. A comparison of the carcass traits and meat quality of conventionally and organically produced rabbits. *Livest. Sci.* 115:1–12.

Pla, M., P. Hernandez, B. Arino, J. A. Ramirez, and I. Diaz. 2007. Prediction of fatty acid kontent in Rabbit meat and discrimination between conventional and organic production systems by NIRS methodology. *Food Chem.* 100:165–170.

Pospiech, E., A. Łyczyński, and K. Borzuta. 2006. Problemy jakości mięsa wieprzowego. *Referat. Materiały Konferencji Surowcowej: "Problemy gospodarki surowcowej w przemyśle mięsnym" Skorzęcin, październik*. p. 24.

Rembiałkowska E. and K. Wiśniewska. 2010. Jakość mięsa z produkcji ekologicznej. *Medycyna Wet.* 66:188–191.

Sather, A. P., S. D. M. Jones, A. L. Schaefer, J. Colyn, and W. M. Robertson. 1997. Feedlot performance, carcass composition and meat quality of free-range reared pigs. *Can. J. Anim. Sci.* 77:225–232.

Shorthose, R. W. and P. V. Harris. 1991. Effects of growth and composition on meat quality. In: A. M. Pearson and T. R. Dutson (eds) *Advances in Meat Research.* Elsevier, London. p. 515.

Sundrum, A. and A. Y. Acosta. 2003. Nutritional strategies to improve the sensory quality and food safety of pork while improving production efficiency within organic framework conditions. Report of EU-project, Improving Quality and Safety and Reduction of Costs in the European Organic and "Low Input" Supply Chain, CT-2003-506358.

Sundrum, A., L. Butfering, M. Henning, and K. H., Hoppenbrock. 2000. Effects of on-farm diets for organic pig production on performance and carcass quality. *J. Anim. Sci.* 78:1199–1205.

Taylor, G. C. and P. Zahradka. 2004. Dietary conjugated linoleic acid and insulin sensitivity and resistance in rodent models. *Am. J. Clin. Nutr.* 79:1164–1168.

Troy, D. J. 1995. Modern methods to improve and control meat quality. In: D. J. Troy (ed.) *International Developments in Process Efficiency and Quality in the Meat Industry.* Teagasc. National Food Centre, Dublin Castle. pp. 57–72.

Uauy, R. and A. D. Dangour. 2006. Nutrition in brain development and aging: role of essential fatty acids. *Nutr. Rev.* 64(5 Pt 2):S24–S33.

Walshe, B. E., E. M. Sheehan, C. M. Delahunty, P. A. Morrissey, and J. P. Kerry. 2006. Composition, sensory and shelf life stability analyses of *Longissimus dorsi* muscle from steers reared under organic and conventional production systems. *Meat. Sci.* 73:319–325.

Warnants, N., M. J. Van Oeckel, and C. V. Boucque. 1999. Incorporation of dietary polyunsaturated fatty acids into pork fatty tissues. *J. Anim. Sci.* 77:2478–2490.

Whigham, L. D., M. E. Cook, and R. L. Atkinson. 2000. Conjugated linoleic acid: implications for human health. *Pharmacol. Res.* 42:503–510.

Wood J. D., M. Enser, A. V. Fisher, G. R. Nute, R. I. Richardson, and P. R. Shed. 1999. Manipulating meat quality and composition. *Proc. Nutr. Soc.* 58:363–370.

Wood, J. D., R. I. Richardson, G. R. Nute, A. V. Fisher, M. M. Campo, E. Kasapidou, P. R. Sheard, and M. Enser. 2003. Effects of fatty acids on meat quality: a review. *Meat Sci.* 66:21–32.

Woodward, B. W. and M. I. Fernandez. 1999. Comparison of conventional and organic beef production systems II. Carcass characteristics. *Livest. Prod. Sci.* 61:225–231.

Yiridoe, E. K., S. Bonti-Ankomah, and R. C. Martin. 2005. Comparison of consumer perceptions and preference toward organic versus conventionally produced foods: a review and update of the literature. *Renew. Agric. and Food Syst.* 20:193–205.

15 Sensory Assessment of Organic Meats

Lydia J. Rice and Jean-François Meullenet

Abstract: Discriminative, descriptive, and affective testing methods are the three categories of sensory methods used to evaluate meat products that are discussed. Discriminative analysis can be used to determine if there are perceivable differences between products. Descriptive analysis techniques such as the Sensory Spectrum™ method and quantitative descriptive analysis (QDA®) are used to determine if significant differences in product attributes exist. Descriptive analysis generally involves a group of 10–12 trained panelists who evaluate a series of products according to a lexicon of sensory attributes. To measure the acceptability of a product, affective testing measures with large samples of consumers are conducted to determine how much a product is liked by consumers. The use of affective testing can also address the appropriateness of attribute levels or purchase intent by consumers. Instrumental measures of sensory quality and the limited sensory evaluation that has been performed on organic pork, beef, pork, poultry, and lamb are discussed.

Keywords: descriptive; discriminative; affective; sensory; statistics; sensory evaluation; consumer; panelist; panel; triangle test

15.1 INTRODUCTION

The role of the sensory scientist in the assessment of organic meats is to quantify the entire human sensorial response in terms of all five senses: (1) sight, (2) touch, (3) hearing, (4) smell, and (5) taste. This can be accomplished through discriminative testing, descriptive analysis, affective testing, or any combination thereof. In terms of organic meat, discriminative testing can be used to establish sensorial differences either between organic and conventional meat or between different treatments of organic products. After differences among products are established, experimenters can use descriptive analysis to describe the sensorial attributes of each product in objective terms. If one wants to know how product attributes affect consumer liking, one can employ affective testing. The differing foci of discriminative, descriptive, affective testing dictate unique experimental approaches.

Organic Meat Production and Processing, First Edition. Edited by Steven C. Ricke, Ellen J. Van Loo, Michael G. Johnson, and Corliss A. O'Bryan.

15.2 TYPES OF SENSORY TESTING

15.2.1 Discriminative testing

Discriminative testing addresses the fundamental question: "Are two products different?" Once this question is answered, one can profile the differences in descriptive testing or use affective testing to determine how the differences affect consumer liking.

There are two basic categories of discriminative testing: (1) similarity testing, where one wants to determine if two products are similar enough to be used interchangeably, and (2) difference testing, where one wants to determine if differences can be detected. Important distinctions between these types of tests must be understood for experimenters to correctly design a discriminative test. These distinctions are largely based on underlying statistical principles. In difference testing, one should minimize the chances for a Type I error, in which one rejects the null hypothesis when in fact it is true. For any discriminative test, the null hypothesis, or H_o, is always A = B. Therefore, a Type I error is committed if one states that a difference exists between A and B when in fact they are the same. The chances for a Type I error are referred to as the α-risk; so for a difference test, one strives to minimize α. In similarity testing, one should minimize the chances for a Type II error, in which one fails to reject H_o when it is false. If one commits a Type II error in a discriminative test, one concludes that two products are the same when in fact they are different. The chances for a Type II error are referred to as β-risk; so, for a similarity test, one strives to minimize β. When designing a discriminative test, the sample size needed depends on desired sensitivity levels in terms of α and β and the proportion of the population that can detect a difference (p_d). Individuals can have varying degrees of sensory sensitivity; plausibly, more sensitive individuals are better at discrimination. Charts, such as the one in Appendix A made for triangle tests, can be used to design sensory tests with appropriate parameters.

Suppose, for example, a company wants to determine the shelf life of raw rib eye steak. The company will test the fresh product (A) and the stored product (B) with a triangle test at specified time points (e.g., 1,3,5,7 days). Sensory scientists want to determine when consumers can identify a difference. For this test, α is minimized most commonly at 0.05, the p_d is 20%, and β should be controlled but not necessarily as small as α. Examining the chart in Appendix A, the researcher sees that 117 subjects are sufficient to set α at 0.05, β at 0.10, and p_d at 20%. Based on these minimum requirements and budgetary constraints, the researcher decides to include 120 panelists at each time point. If there are 120 subjects and α is 0.05, Appendix B shows that 50 correct answers are needed to declare that a significant difference between the two products exists.

If, for example, the research and development department of a bread company wants to determine if slightly less expensive flour from a new supplier is similar enough to flour from the current supplier for the two to be used interchangeably, sensory scientists would then design a study slightly different from the previous example. In this example, when one looks at the chart from Appendix A, if p_d is still 20%, β should be minimized at 0.05, while α may increase. β is minimized in this example because it is a similarity test.

Among discriminative methods for both similarity and difference testing, the triangle test is the most common. In this method, three samples are identified with random 3-digit codes. Two of the products are the same, and one is different. The panelist is instructed to taste the samples from left to right and determine the odd sample. If a sufficient number of correct answers are elicited from enough panelists, then a significant difference is said to exist between the two samples. The number of correct answers needed to determine if there is a difference depends on the overall sample size. Charts such as the one in Appendix B can be used to determine if a significant difference exists on the basis of the number of correct answers.

During a triangle test, presentation order is randomized for each panelist. The possible presentation orders for the triangle test are: AAB, ABB, ABA, BAB, BAA, BBA, where A is product x and B is product y. Randomization is essential to prevent the error of central tendency, where subjects tend to choose the middle sample more often than the ones on the ends. Randomization for all sensory tests is important to prevent order of presentation effects from influencing the results.

Additional discriminative methods include two out of five, duo trio, and simple difference tests. The two out of five test most closely resembles the triangle test; a subject must select the two odd samples out of five total. In the duo trio test, subjects are given a reference and two additional samples, and they must select which of the two additional samples matches the reference. Subjects used in the simple difference test are asked if two products are the same or different. Bias for stating that products are different is created by merely asking the difference question. Thus, the analysis of this test is designed to correct this bias. Meilgaard *et al.* (2007) provide good explanations for the design and analyses of these tests; thus, they will not be discussed further here.

15.2.2 Descriptive testing

Descriptive analysis allows experimenters to describe the characteristics of a food product in qualitative and quantitative terms. The method has come a long way from the development of the Flavor Profile Method in the late 1940s (Caul, 1957). This early method utilized a 7-point scale and panel consensus to measure the intensity of aroma, flavor, and aftertaste attributes. Unfortunately, the 7-point scale was too limited to readily allow panelists to discriminate among samples. Additionally, problems surfaced with panel consensus. When panel leaders develop a consensus, panelists with stronger personalities tend to dominate the final result. The International Organization for Standardization (ISO) details a method similar to the Flavor Profile Method in their *Sensory Analysis—Methodology* materials. Besides the previously discussed issues that may arise with limited scale ranges and consensus (though consensus is optional in their guidelines), ISO also suggests that overall impression be measured with other descriptive attribute intensities (ISO 6564, 1985). Although descriptive panelists sometimes assess "total intensity of aroma or flavor" or attribute "balance" (Meilgaard *et al.*, 2007), the hedonic score for overall impression should be measured with large consumer panels, which is discussed further in Section "Affective Testing."

More recently developed methods such as Quantitative Descriptive Analysis (QDA®) (Stone *et al.*, 1980), Sensory Spectrum® (SS) (Muñoz & Civille, 1992, 1998), and free-choice profiling (Williams & Arnold, 1984) were made to correct deficiencies of earlier methods. QDA® and SS are both well structured, while free-choice profiling offers more flexibility. The decision to follow one methodology over another should begin with careful consideration of each technique's advantages and disadvantages. QDA® utilizes moderately trained panelists to rate the intensities of attributes on a 15-cm line scale. Due to the lack of training with a universal scale, the results are dependent on an individual's use of the scale and are, therefore, relative. The SS method may utilize a 15-cm line scale or a nonvisual 0–15 interval. This scale is anchored at certain points to specific references. For example, the universal scale, used to identify absolute intensities of aromatics and feeling factors, is referenced through the intensity of particular attributes of multiple food products (Meilgaard *et al.*, 2007). The universal scale as it would appear with a line scale is illustrated in Figure 15.1.

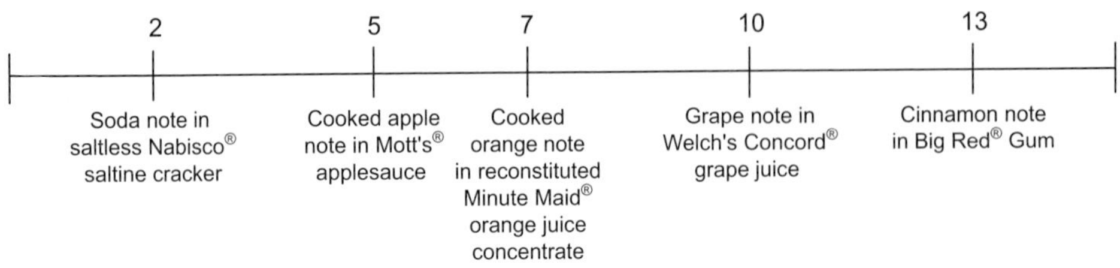

Figure 15.1 Spectrum universal scale.

The SS method has also defined the intensities of all predominant attributes of a number of food products such as Pepperidge Farm® Cookies, Bordeaux® Cookies, Lay's® Classic Potato Chips, and Heinz® Tomato Ketchup. The extensive use of references in the SS method necessitates extensive initial and continual training of panelists. While the reference training provides absolute intensity measures, it also makes SS panels expensive to maintain. QDA®, on the other hand, requires less panelist training; therefore, panel maintenance is less expensive. One must be aware, however, that moderately trained or untrained panelists may use attribute terms incorrectly. For example, artificial vanilla, vanillin, may be erroneously reported as vanilla. If panel training and calibration are sufficient, only 10–12 panelists are necessary (Lawless & Heymann, 1998). Careful consideration of experimental objectives and budget constraints will aid the experimenter in choosing the method that can best meet his or her needs.

These descriptive methods (QDA® and SS) can be adapted to suit particular products. For example, organic beef may need specific references for umami or warmed-over flavor. Umami is debatably the fifth basic taste and means "deliciousness" in Japanese. Warmed-over flavor is the resulting off-taste that occurs from lipid oxidation when cooked meat is stored in the refrigerator prior to reheating. A customized ballot can be generated from the descriptive panel prior to data collection. If the need for additional references arises at this time, the panel leader can collect them before testing proceeds.

Subjects used in free-choice profiling generate their own attribute terms to create individualized ballots. They are free to use as many descriptors as needed to describe a category of products. The experimenter utilizes General Procrustes Analysis (Gower, 1975), a multivariate statistical tool, to combine descriptors that consumers use similarly. The experimenter then uses his or her judgment to determine the meaning of the descriptors. Free-choice profiling has two main advantages over QDA® and SS: (1) minimal panel training is required and (2) the lack of training means subjects can still be considered naive consumers. However, the experimenter's extensive role in determining the meaning of the results may allow the experimenter's subjectivity to too heavily influence final interpretations (Meilgaard *et al.*, 2007).

15.2.3 Affective testing

Affective testing is perhaps the approach that remains closest to the consumer. Here, the objective is to identify consumer inclinations in terms of like and dislike. For organically produced meat, affective testing can be employed to answer:

- Is organically produced meat liked significantly more than conventionally produced meat?
- Can different strategies of organic farming change the end product enough to significantly affect consumer liking?

What is your overall impression of this product?

Dislike extremely	Dislike very much	Dislike moderately	Dislike slightly	Neither like nor dislike	Like slightly	Like moderately	Like very much	Like extremely
1	2	3	4	5	6	7	8	9

Figure 15.2 Nine-point verbal hedonic scale.

Most commonly used to measure overall consumer liking is the 9-point verbal hedonic scale, which is presented in Figure 15.2.

For identifying significant differences in liking among products, the principal measurement is the mean response to the overall impression question across all consumers. The overall impression must be presented in a straightforward, nonbiased way. "What is your overall impression of this product?" is more experimentally correct than "How much do you like this product?" because the latter suggests that the consumer likes the product even before the question is answered. Consumers can also be questioned about their liking for broad category attributes (e.g., texture, flavor).

Statistical mean separation is performed to determine if a significant difference in liking among two or more products exists. Often, if two or more products are compared directly, consumers may be asked to rank the samples in order of their most preferred to least preferred. Friedman analysis can determine if a significant difference occurs between ranked products. The decision to utilize ranking data, liking data, or both rests on the experimenter's objectives.

Besides hedonic scales, affective testing also employs diagnostic scales, which can be used to determine if attributes are at appropriate levels. For example, in the case of organic chicken breast, experimenters may want to know if moistness, chicken flavor, and tenderness are sufficient. A balanced just-about-right (JAR) scale is useful for this task. Figure 15.3 demonstrates the use of this scale with the previously mentioned attributes.

Consumers have heterogeneous needs and preferences. Presumably, not all consumers will agree that every attribute is JAR. Generally, if 70% of consumers fall within the JAR category, then reformulation is not needed (Chambers & Wolf, 1996).

What is your opinion of the **moistness** of this product?

Much too dry	Too dry	Just about right	Too moist	Much too moist
1	2	3	4	5

What is your opinion of the **chicken flavor** of this product?

Much too weak	Too weak	Just about right	Too strong	Much too strong
1	2	3	4	5

What is your opinion of the **tenderness** of this product?

Much too tough	Too tough	Just about right	Too tender	Much too tender
1	2	3	4	5

Figure 15.3 Just-about-right scales.

The heterogeneity of consumers necessitates the use of large panels in affective tasting ($n = 75–600$). In most situations, consumers recruited to participate in affective tests should represent the target demographic of the product. Occasionally, consumers are recruited randomly from the population.

Though the 9-point hedonic scale and the 5-point JAR scales are most commonly used, at times, liking and appropriateness of attribute levels are tested through anchored line scales. For example, a hedonic line scale may have "dislike extremely" and "like extremely" as anchors, while a diagnostic line scale may be anchored by "less moist" on the left and "moist" on the right.

Regardless of the scale used, affective testing is best at gauging utility consumers derived from evident sensory characteristics: appearance, texture, and flavor. However, consumers derive utility from characteristics outside of this narrow scope, especially with value-added products such as organic meats, which consumers may believe are healthier or better for the environment (AMS-USDA, 1995; Magnusson *et al.*, 2003). Plausibly, the perception that an organic product has additional nutritional benefits or was produced through more sustainable agriculture affects consumer liking. In order to quantify the utility derived from characteristics outside of sensory attributes, experimental auctions and choice-based experiments can be employed. Experimental auctions can elicit willingness to pay through a nonhypothetical exchange of cash for a value-added product. In choice-based experiments, the consumer sees a series of different attribute combinations and chooses their most preferred combination. The utility for a combination of attributes is subsequently calculated. These methods blend sensory science methodology, economics, and marketing. Further discussion of these methods appears in Chapter 6. The future of sensory assessment of organic products should utilize conventional affective methodology as well as methods that consider nonsensorial attributes that affect consumer liking.

15.2.4 Instrumental analyses

When determining quality of meat products, instrumental analyses measure physical characteristics, which contribute to the sensorial experience. Among the most commonly measured are shear force, water-holding capacity, ultimate pH, colorimeter values (L*, a*, b*, hue, chroma), and fatty acid content.

To establish instrumental tenderness, shear force is often measured. The Warner–Bratzler Shear Blade and the Allo–Kramer Compression-Shear Device are two well-established machines for measuring shear force, which is associated with tenderness. The general principle of this procedure is to measure the force required to cut through meat. As the measured shear force values increase, tenderness decreases. However, since texture is a multidimensional characteristic, fully quantifying it with either of these devices is difficult; thus, eventually the Texture Profile Analysis method was created. Additional discussion of this method is provided by Owens and Meullenet (2010).

Water-holding capacity can also be an important instrumental measure because juiciness, a sensory attribute, is an indicator of it. Juiciness is important to the sensorial experience because it provides lubrication for chewing (Owens & Meullenet, 2010). Aaslyng (2002) explains:

> The water holding capacity can be determined by three fundamentally different principles: (i) using external forces to drive out the water like the filter press method and by centrifugation, (ii) by letting the water drip out of the raw meat in a standardized way over a certain time period

like the 'Honikel bag method' or the EZ-drip loss, or (iii) by heating the meat and measuring the cooking loss.

Honiker, 1989; Honiker & Hamm, 1994; Rasmussen & Andersson, 1996; Christensen, 2003

Colorimeter measures (L*, a*, b*, hue, chroma) quantify a product's appearance, which can be important in setting consumers' sensorial expectations. These expectations may differ depending on a consumer's context. For example, the yellow color of corn-fed chicken was found to imply freshness and superior quality in North America, while North Ireland consumers considered it "unnatural" and "unfamiliar" (Sunde, 1992; Kennedy *et al.*, 2005). Recording colorimeter values also allows data to be compared among different studies.

Ultimate pH is measured by inserting an electrode into the meat or into a homogenized sample and can affect both water-holding capacity and reflectance (thus, colorimeter measurements) of meat products. Lower ultimate pH causes contractile fibers of the muscle to shrink, and this decreases water-binding ability, which thus increases light scattering (Warris, 2000; Castellini *et al.*, 2002).

Meat fatty acid profiles may have nutrition and sensory ramifications. Higher percentages of unsaturated fatty acids may promote cardiovascular health (Ascherio, 2002; Wahrburg, 2004), though this can also lead to faster development of rancid flavor and thus shorter shelf life. Gas chromatography can aid in the identification and quantification of fatty acids.

15.3 SENSORY RESEARCH ON ORGANIC MEAT

Research comparing the sensory quality of organically and conventionally produced food presents special considerations. In the case of certain types of produce, research suggests that cultivar, harvest date, irrigation levels, and chemical treatment can affect sensory attributes (Berenguer *et al.*, 2006; Clayton *et al.*, 2006; Kihlberg *et al.*, 2006; Sinesio *et al.*, 2007; Obenland *et al.*, 2009; Ulrich *et al.*, 2009; Perez-Sanchez *et al.*, 2010). Similarly, inputs such as type of animal, animal diet, and postslaughter processes (e.g., aging and storage) can affect final sensory profiles of meat products (for a review on diet effects, see Melton (1990)) (de Huidobro *et al.*, 2003; Rodriguez-Perez *et al.*, 2003; Hopkins *et al.*, 2005; Brewer & Novakofski, 2008).

Assessing the changes that occur in the farming system itself, inputs within the farming system, and postslaughter processes remains a challenge for studies that attempt to compare the attributes of conventionally produced and organically produced meat. Some studies attempt to account for farming conditions, while others use products readily available to consumers without accounting for production practices. In the latter case, researchers are testing the differences in products readily available to consumers, not necessarily the differences resulting from inputs of the farming system. Arguably, this question is an equally valid point of research.

15.3.1 Sensory research on organic pork

When one considers the sensory quality of either conventionally or organically produced pork, one has to be aware of all the factors that may affect the sensory attributes of pork. For example, some pigs carry the RN^- allele, which leads to a high glycogen content and potential affectation of sensory attributes. Sensorially, the loin muscle of carriers may have differences in flavor (greater acidulous and meat taste) and texture (more tender and juicy)

(Lundstrom *et al.*, 1996, 1998; Johansson *et al.*, 1999; Jonsall *et al.*, 2000; 2002). Gender also sometimes affects the final sensorial attributes; meat from castrates has been found to be more tender and juicy than meat from gilts (Enfalt *et al.*, 1997; Jonsall *et al.*, 2001). Outdoor rearing, more commonly found in organic farming systems, may produce meat lower in juiciness and acidulous taste than meat from indoor-reared animals (Enfalt *et al.*, 1997; Jonsall *et al.*, 2001), though some studies have shown that exercise does not affect the sensory quality of meat (Vanderwal *et al.*, 1993; Petersen *et al.*, 1997).

Considering these factors, Jonsall *et al.* (2002) compared conventionally and organically raised meat with consumer preference testing and descriptive analysis following ISO standards. Through descriptive analysis, these researchers found that conventionally produced loins were juicier and less crumbly than their organic counterparts. Animals positive for the RN^- allele produced loins that were more tender, juicy, acidulous, and crumbly. Gender also produced significant differences in descriptive analysis; loins from castrated males had more off-odor and were less juicy than loins from females. Consumer preference tests showed consumers preferred RN^- carriers but no significant preference was found for either organic or conventional loins (Jonsall *et al.*, 2002).

Additional differences between organically and conventionally produced pork have been addressed through instrumental analyses. Olsson *et al.* (2003) found that some of the ways quality between organic and conventional pork differed were increased shear force and drip loss of organically raised noncarriers (of the RN^- allele) compared to conventionally raised noncarriers. Reduced water-holding capacity (measured by increased drip loss) may be indicative of higher glycogen content, which could potentially bind water (Fernandez, 1991). Increased rates of exercise in organically raised animals may contribute to different postmortem rates of glycolysis (Olsson *et al.*, 2003). Higher shear forces may be indicative of lower levels of intramuscular fat, slower protein turnover, or differences in intramuscular collagen (Wood *et al.*, 1992; Petersen *et al.*, 1997; Olsson *et al.*, 2003). Although reduced water-holding capacity and higher shear rates are considered properties of lower quality meat, one cannot conclude that humans would necessarily perceive sensorial differences without sensory analysis, which was not performed in this study (Olsson *et al.*, 2003).

Millet *et al.* (2004) compared organic housing and nutrition with conventional housing and nutrition. Pigs were subdivided according to a 2×2 factorial design including housing and nutrition factors. Meat produced from animals given organic nutrition exhibited redder meat, higher intramuscular fat, and lower ultimate pH of the ham and loin. Organic housing contributed to thicker back fat, more muscle, and redder meat. Millet *et al.* (2004) concluded that both organic housing and nutrition have implications for final quality.

Hansen *et al.* (2006) further explored the implications for differences in nutrition in the quality of organically and conventionally raised pigs. Organic pigs were subdivided into three groups and fed 100% concentrate, 70% concentrate with 30% organic barley/pea silage, or 70% concentrate with 30% organic clover grass silage. Conventional pigs were also given 100% concentrate. Organic pigs fed 70% concentrate generated leaner carcasses; however, descriptive analysis with eight trained assessors showed this meat was also harder, less tender, and had less meat flavor (measured on an unstructured scale transformed to a numeric scale with values 0–15). Organic pigs fed 70% concentrate and 30% clover grass silage also had less meat flavor than conventional pigs (Hansen *et al.*, 2006). This study suggests that not only must one compare "organic" and "conventional" products, one must also question how the interaction effects of housing and feeding within the organic system promote optimum quality. Previous authors have discussed that final eating quality is highly variable, and depends on feeding management and housing conditions (Lebret, 2008; Bonneau & Lebret, 2010).

Much of the discussion thus far has centered on laboratory measures of pork quality (e.g., shear force, water-holding capacity, ultimate pH). Not yet discussed and perhaps more important are quality characteristics that serve as cues for consumer purchase decisions. European consumers from four different countries (France, Denmark, Sweden, and United Kingdom) were shown computer-modified photos of pork chops and asked to select their preferred levels (of two options) of fat cover, color, marbling, and drip. When evaluating their predilections, consumers put more emphasis on color and fatness. Selections for color were heterogeneous, but consumers from the same country tended to agree with higher frequency. French preferences leaned toward the darker pork, while the British and Danes favored the paler pork. Leaner pork was more popular across all countries. To study the effect of additional label information, images for marbling and no drip were stabilized while two levels of information for place of production (imported and home country) and production system (indoor and outdoor) were added to the choices of the computer-modified images. When considering only those consumers who consistently chose their responses, over 90% preferred pork labeled as being from their home country. A majority also preferred the outdoor "system of production" label (Dransfield *et al.*, 2005).

Tasting trials in Britain and France did not find significant differences in taste for the indoor- and outdoor-raised pork, although when these products were labeled as outdoor or home country-produced, overall appreciation (equivalent to overall impression) scores (on a 0–10 scale) increased for both groups. These labels also increased willingness to pay. One can conclude that though taste may not be significantly different between outdoor- and indoor-raised pork, other factors drive consumer liking (Dransfield *et al.*, 2005).

Studies comparing sensory properties of conventional and organic pork have returned somewhat inconsistent results, most likely due to confounding of variables that contribute to final sensory quality. Future research should attempt to test specific aspects of the organic system (e.g., outdoor rearing, organic concentrate) to identify the contribution of each. Affective tests that offer complete profiles of diagnostic and hedonic scores would also be a worthwhile contribution to the literature.

15.3.2 Sensory research on organic beef

While domestic studies of organic beef are limited, several have been performed in Europe. A study performed by the National University of Ireland obtained six organic and six conventional carcasses from a slaughterhouse. Unfortunately, the diets of these animals were confidential and could not be discussed in the study. Researchers harvested the longissimus dorsi, more commonly called the rib eye or loin eye (Calkins & Sullivan, 2007), from the steers and compared them with instrumental quality and descriptive sensory analyses. Instrumental analyses showed that organic meat had higher fat and lower moisture content. The authors also studied the effectiveness of different packaging types. Conventional samples had the best color stability when packaged with modified atmosphere packaging (MAP) and the best lipid stability when overwrapped with cling film; therefore, conventional samples had longer shelf life than organic samples.

The descriptive panel, composed of nine women and one man, generated a ballot that included the relevant and important terms for odor (sulfur cabbage, strength of beef, cold cooked beef fat), texture (moistness, chewiness, juice development in mouth), and flavor (strength of beef, roasted beef, sweet, cold beef fat). No significant differences were found between steaks from organically and conventionally reared animals in terms of these descriptors. This study demonstrates that though instrumentally measured differences (e.g., fat content, moisture content) may be found, these differences do not necessarily affect the human sensorial experience in a significant way (Walshe *et al.*, 2006).

The previously discussed article focused on comparing conventional and organic beef. In other reports, the focus has been on differences *within* organic rearing practices that may affect final meat quality. Here, we discuss an Italian study that explored forage-to-concentrate ratio and a Danish study that reviewed many potential factors affecting organic beef quality. Forage-to-concentrate ratio refers to the relative amounts of forage, generally fibrous plant material, and concentrate, a developed blend of food materials much more calorically dense than forage. Twenty Podolian bulls were divided into two groups: one given high concentrate (60/40 forage-to-concentrate ratio) and the other given low concentrate (70/30 ratio). The carcasses were then aged either 15 or 21 days. Eight descriptive panelists evaluated the resulting steaks for flavor and tenderness using a 10-cm unstructured line scale with 0 corresponding to absent and 10 corresponding to very strong. The higher forage diet (70/30 ratio) generated more tender steaks after 15 days of aging. However, after 21 days, the differences in meat tenderness between high concentrate-fed and low concentrate-fed animals disappeared. In both forage-to-concentrate ratio groups, aging improved the flavor of meat significantly. Interestingly, this study also found a significant correlation between sensorial tenderness and Warner–Bratzler Shear force (WBSF); thus, the authors suggested that WBSF may be utilized on uncooked meat to predict meat tenderness (Marino *et al.*, 2006), though not all studies have shown that shear force is a good predictor of sensory tenderness (Shackelford *et al.*, 1995). Danish researchers discussed many issues related to raising organic steers. Grazing and exercise, in particular, can affect the sensorial characteristics of beef. The meat can be darker or have less tenderness and flavor (Bowling *et al.*, 1978; Knight & Death, 1997; Vestergaard *et al.*, 2000a, 2000b). Though these potentially negative characteristics are present, grazing also increases unsaturated fatty acid content, which positively affects the healthfulness of the steak (Knight & Death, 1997). The extent to which the culminations of these characteristics affect consumer liking has not been tested; thus, their relative importance to the consumer cannot be determined at the present time.

For future improvement of organic beef, the authors suggested adjusting feeding regimens to increase conjugated linoleic acid (CLA) content in the final product, for example (Nielsen & Thamsborg, 2005). Additionally, one must be aware that the differences in fermentation activity in the rumen of beef animals can potentially impact the assessment of diet effects on meat quality.

Although to our knowledge, no studies have profiled the affective status of organic beef with domestic consumers, the University of Nebraska compared conventionally produced corn-fed domestic beef with Australian grass-fed beef and Canadian grain-fed beef. (Commonly, barley is the grain base in Canadian cattle diets.) Although these comparisons do not involve organic beef, they may provide insight into its potential consumer acceptance in the United States. As previously discussed, grazing and exercise, common in organic farming systems, can affect final beef quality. The majority of American consumers gave higher affective scores to domestic beef than to Australian grass-fed beef. Research has shown that off-notes and grassy flavor is more pronounced in grass-fed steers, and this may drive consumer responses (Xiong *et al.*, 1996). The Australian beef was also aged longer than domestic beef, and this could also have contributed to the development of off-flavors and the decrease of consumer acceptance (Campo *et al.*, 1999). When comparing domestic beef with Canadian grain-fed beef, the majority of American consumers again preferred domestic beef perhaps due to its higher fat content. These results indicate that the majority of Americans prefer domestic beef, though 19% preferred Australian and 26.7% preferred Canadian. This indicates that a niche market potentially exists for grass-fed or grain-fed beef. One must note that in this study, the sample with the significantly higher overall acceptability score based on an 8-point hedonic scale is defined as the preferred sample, rather than having the preferred sample being determined directly through responses to a preference question (Sitz *et al.*, 2005). One must also note that consumers in this study were potentially biased before rating overall appreciation; they were given a warm-up sample before testing, so that they could be familiarized with judging tenderness, juiciness, and flavor. While asking consumers how much they like these attributes or the appropriateness of the attribute intensities is suitable, asking consumers to judge their absolute intensity (without free-choice profiling) is not. The tasting training session, short though it was, could have made consumers more aware of certain attributes of the product, which potentially could have shifted their preferences. Also, a short training session is not sufficient to develop the palette of a descriptive panelist. Thus, the panelists in this study may no longer have been naive consumers, but they were also not trained panelists.

In general, the literature on sensory research of organic beef needs further effort from the scientific community. Expanding the literature on affective testing, especially domestically, could direct future research in meeting the needs, wants, and expectations of the consumer.

15.3.3 Sensory research on organic poultry

Among the organic meat products discussed in this chapter, organic poultry has the most published literature. Similar to other meats, organic poultry quality is dependent on a number of factors within the organic farming system. Genotype, rearing duration, access to outdoors, and gender have been reported as factors in final quality (Ristic, 2004; Horsted *et al.*, 2005; Fanatico *et al.*, 2007). Some genotypes or "strains" are slower growing, and their sensory properties (e.g., tenderness) can improve with longer rearing durations (Horsted *et al.*, 2005). Differences in positive sensory attributes (e.g., salt flavor, sweet corn flavor) (Horsted *et al.*, 2005) have been reported for different sexes. Fanatico *et al.* (2007) examined the effect of both genotype (fast-growing and slow-growing) and presence of outdoor access on descriptive sensory characteristics and consumer liking. Descriptive panelists (SS-trained) found that meat from the fast-growing genotype was saltier, while the slow-growing birds had greater dark meat fat flavor. Hedonic (i.e., liking) data from consumers ($n = 81$) showed no significant difference between the slow-growing genotype with outdoor access and the fast-growing genotype raised indoors, although diagnostic JAR data showed that many consumers considered breast meat from the outdoor access-raised slow-growing genotype too dry (Fanatico *et al.*, 2007). Nevertheless, some authors have suggested that slow-growing genotypes are more appropriate for organic farming (Castellini *et al.*, 2008).

Technological quality differences have been found between organic and conventional chicken. Castellini *et al.* (2002) found that organic broilers had increased cooking losses due to lower ultimate pH and water-holding capacity. Limited descriptive analysis found that organic breasts were juicier and higher in overall acceptability (on a 5-point scale), though only nine trained assessors were used (Castellini *et al.*, 2002). Overall acceptability should be measured by much larger consumer panels, rather than through biased descriptive panelists, because consumer preferences are heterogeneous and cannot be readily captured when using small sample sizes.

Jahan *et al.* (2005) employed two methods of descriptive analysis—(1) free-choice profiling and (2) conventional descriptive—to describe the differences among corn-fed, free-range, organic, and conventional chicken breasts. In both methods, products were scored on a 100-mm unstructured line scale anchored with *weak* and *strong* at 10 and 90 mm, respectively. Appearance and texture embodied the primary differences among the samples, though some assessors also discriminated on flavor and aroma. Corn-fed chicken was differentiated on the appearance attributes yellow and brown, which is expected due to the presence of maize xanthophylls (Farmer *et al.*, 1997). Conventional descriptive analysis showed that organic breasts were less tender (soft) than corn-fed or free-range breasts. Organic breasts scored higher for moistness than free-range breasts.

Brown *et al.* (2008) also compared conventional, organic, corn-fed, and free-range breasts. The samples in this study were collected from processors, rather than grocery stores as in the previously discussed study; thus, processing conditions that may contribute to final sensory quality were more consistent across farming systems. Instrumental analyses tested common quality parameters (weight, color, fat content, ultimate pH, water-holding capacity), and the sensory panel evaluated descriptive characteristics (texture, juiciness, chicken flavor, abnormal flavor), and hedonic scores (overall liking, flavor liking) on an 8-point category scale. The conventional breasts had the highest intramuscular fat content, while the organic breasts had the lowest. Ultimate pH was significantly higher in the conventional system, which Castellini *et al.* (2002) also found, though no differences in water-holding capacity were reported from the different production systems, unlike results reported by Castellini *et al.* (2002).

Sensory analysis found that meat from the organic system was the least tender, followed by free-range and corn-fed with conventional breasts being the most tender. The meat from the organic system was also the driest, while the meat from the conventional system was the juiciest. Overall liking (pooled across all production site visits) was significantly higher for conventional breasts than for organic or free-range (Brown *et al.*, 2008), though this finding was taken from the same panel performing descriptive analysis. Although the level of training for this panel is unclear, presumably, they do not represent naive consumers, who should be used to measure hedonic attributes.

A recent domestic study also compared organic, free-range, and conventional broilers with descriptive analysis among other parameters. Although descriptive differences in chicken breasts were not found in this study, organic chicken thighs were less tender and chewier than conventional thighs (Husak *et al.*, 2008).

Although the differences between organic and conventional poultry are not fully clear, organic chicken tends to be tougher. Part of the inconsistency most likely stems from the differing inputs of both the conventional and organic systems. Since these are not consistent across studies, we should not expect the results to be. Future research should focus on separating the contribution of each input into the system in terms of both descriptive analysis and affective testing.

15.3.4 Sensory research on organic lamb

British researchers compared conventionally and organically produced lamb purchased from three supermarkets and profiled them in terms of price, descriptive sensory characteristics, and fatty acid composition. Nine descriptive panelists created and agreed on terms to be included in the analysis. The measuring tool was a 100-mm line scale with anchors on either end. The anchors attached to zero were "very tender" and "very dry;" the opposite anchors were named "extremely tough" and "extremely juicy," respectively, and given a value of 100.

Between conventional and organic lamb chops, organic lamb scored significantly higher for lamb flavor and juiciness. The difference in lamb flavor was attributed to differences in fatty acid composition, and increased juiciness in the organic product was attributed to higher lipid content. Organic lamb also scored better in terms of overall impression; however, one must note that measuring overall impression is usually reserved for affective tests and should involve large consumer panels. As the authors acknowledge, one cannot readily assess consumer acceptance based on the overall impression scores from a limited number of biased assessors. Here, one should withhold conclusions about consumer preferences until they can be addressed in a well-structured affective test.

When comparing the lamb samples, more differences were found among grocery stores than between farming systems. The differences in quality attributes among the grocery stores was possibly due to differing "display until" dates, suggesting that freshness is an important factor to consider when measuring the quality of meat products (Angood *et al.*, 2008).

Revilla *et al.* (2009) subjected conventionally and organically produced suckling lamb meat to descriptive and affective testing. The dams of the suckling lambs had standardized diets, which were dependent upon their conventional or organic designation. Sixteen QDA®-trained panelists generated relevant descriptors to judge both raw and grilled meat on a 9-point scale. Raw organic suckling lamb meat appeared to possess more pink color, intramuscular fat, and be more fibrous ($\alpha = 0.10$). As for the odor profile, raw conventional lamb had greater intensity and suckling lamb aroma. In terms of texture, grilled conventional meat was more fibrous, had more fat sensation, and displayed more juiciness, while the flavor profile showed that grilled organic suckling lamb was sourer.

The affective part of this study required 35 volunteer families to assess the meat in their homes. The reported affective data is limited, but overall appreciation scores were higher for organic samples.

Organic lamb research thus far is extremely limited. Additional descriptive and affective data is necessary to draw meaningful conclusions about the characteristics of organic lamb and how these characteristics affect consumer preferences.

15.4 CONCLUSIONS

Sensory research on organic meat has many areas of deficiency. More collection of affective and descriptive data, both in the United States and abroad, is needed for all types of organic meat. Research on organic meats should adhere to scientifically acceptable sensory methodology. The ultimate objective of this research should be to accurately describe how sensory properties and the value-added attributes of organic products contribute to consumer preferences. Too much of the present discussion concerning sensory properties of organic products involves anecdotal evidence and potentially confounding experimental test conditions. Future discussions should revolve around unbiased empirical data.

REFERENCES

Aaslyng, M. D. 2002. Quality indicators for raw meat. In: J. Kerry and D. Ledward (eds) *Meat Processing Improving Quality*. CRC Press, Boca Raton, FL. pp. 157–174.

AMS-USDA. 1995. How to buy poultry. Available at: http://www.ams.usda.gov/AMSv1.0/getfile?dDocName=STELDEV3002966 (accessed September, 2010).

Angood, K. M., J. D. Wood, G. R. Nute, F. M. Whittington, S. I. Hughes, and P. R. Sheard. 2008. A comparison of organic and conventionally-produced lamb purchased from three major UK supermarkets: price, eating quality and fatty acid composition. *Meat Sci.* 78:176–184.

Ascherio, A. 2002. Epidemiologic studies on dietary fats and coronary heart disease. *Am. J. Med.* 113:9–12.

Berenguer, M. J., P. M. Vossen, S. R. Grattan, J. H. Connell, and V. S. Polito. 2006. Tree irrigation levels for optimum chemical and sensory properties of olive oil. *Hort. Sci.* 41:427–432.

Bonneau, M. and B. Lebret. 2010. Production systems and influence on eating quality of pork. *Meat Sci.* 84:293–300.

Bowling, R. A., J. K. Riggs, G. C. Smith, Z. L. Carpenter, R. L. Reddish, and O. D. Butler. 1978. Production, carcass and palatability characteristics of steers produced by different, management-systems. *J. Anim. Sci.* 46:333–340.

Brewer, S. and J. Novakofski. 2008. Consumer sensory evaluations of aging effects on beef quality. *J. Food Sci.* 73:S78-S82.

Brown, S. N., G. R. Nute, A. Baker, S. I. Hughes, and P. D. Warriss. 2008. Aspects of meat and eating quality of broiler chickens reared under standard, maize-fed, free-range or organic systems. *Br. Poult. Sci.* 49:118–124.

Calkins, C. R. and G. Sullivan. 2007. Ranking of beef muscles for tenderness. Available at: http://www.beefresearch.org/CMDocs/BeefResearch/Ranking%20of%20Beef%20Muscles%20for%20Tenderness.pdf (accessed September, 2010).

Campo, M. M., C. Sanudo, B. Panea, and P. Alberti, P. Santolaria. 1999. Breed type and ageing time effects on sensory characteristics of beef strip loin steaks. *Meat Sci.* 51:383–390.

Castellini, C., C. Berri, E. Le Bihan-Duval, and G. Martino. 2008. Qualitative attributes and consumer perception of organic and free-range poultry meat. *Worlds Poult. Sci. J.* 64:500–512.

Castellini, C., C. Mugnai, and A. Dal Bosco. 2002. Effect of organic production system on broiler carcass and meat quality. *Meat Sci.* 60:219–225.

Caul, J. F. 1957. The profile method of flavor analysis. *Adv. Food Res.* 7:1–40.

Chambers, E., IV, M.B. Wolf. 1996. *Sensory Testing Methods*. American Society for Testing and Materials, West Conshohocken, PA. p. 146.

Christensen, L. B. 2003. Drip loss sampling in porcine *m. longissimus dorsi*. *Meat Sci.* 63:469–477.

Clayton, M., W. V. Biasi, I. T. Agar, S. M. Southwick, and E. J. Mitcham. 2006. Sensory quality of 'Bing' sweet cherries following preharvest treatment with hydrogen cyanamide, calcium ammonium nitrate, or gibberellic acid. *Hort. Sci.* 41:745–748.

de Huidobro, F. R., E. Miguel, E. Onega, and B. Blazquez. 2003. Changes in meat quality characteristics of bovine meat during the first 6 days post mortem. *Meat Sci.* 65:1439–1446.

Dransfield, E., T. M. Ngapo, N. A. Nielsen, L. Bredahl, P. O. Sjoden, M. Magnusson, M. M. Campo, and G. R. Nute. 2005. Consumer choice and suggested price for pork as influenced by its appearance, taste and information concerning country of origin and organic pig production. *Meat Sci.* 69:61–70.

Enfalt, A. C., K. Lundstrom, I. Hansson, N. Lundeheim, and P. E. Nystrom. 1997. Effects of outdoor rearing and sire breed (Duroc or Yorkshire) on carcass composition and sensory and technological meat quality. *Meat Sci.* 45:1–15.

Fanatico, A. C., P. B. Pillai, J. L. Emmert, E. E. Gbur, J.- F. Meullenet, and C. M. Owens. 2007. Sensory attributes of slow- and fast-growing chicken genotypes raised indoors or with outdoor access. *Poultry Sci.* 86:2441–2449.

Farmer, L. J., G. C. Perry, P. D. Lewis, G. R. Nute, J. R. Piggott, and R. L. S. Patterson. 1997. Responses of two genotypes of chicken to the diets and stocking densities of conventional UK and Label Rouge production systems.2. Sensory attributes. *Meat Sci.* 47:77–93.

Fernandez, X. 1991. A review of the causes of variation in muscle glycogen content and ultimate pH in pigs. *J. Muscle Foods* 2:209–235.

Gower, J. C. 1975. Generalized procrustes analysis. *Psychometrika*. 40:33–51.

Hansen, L. L., C. Claudi-Magnussen, S. K. Jensen, and H. J. Andersen. 2006. Effect of organic pig production systems on performance and meat quality. *Meat Sci.* 74:605–615.

Honiker, K. O. and R. Hamm. 1994. Measurement of water-holding capacity and juiciness. In: A. M. Pearson and T. R. Dutson (ed) *Quality Attributes and Their Measurement in Meat, Poultry and Fish Products.* Blackie Academic & Professional; Chapman & Hall, Glasgow. pp. 125–161.

Honiker, K. O. 1989. The meat aspects of water and food quality. In: T. M. Hardman (ed.) *Water and Food Quality.* Elsevier Applied Science, London. pp. 277–303.

Hopkins, D. L., P. J. Walker, J. M. Thompson, and D. W. Pethick. 2005. Effect of sheep type on meat and eating quality of sheep meat. *Aust. J. Exp. Agric.* 45:499–507.

Horsted, K., J. Henning, and J. E. Hermansen. 2005. Growth and sensory characteristics of organically reared broilers differing in strain, sex and age at slaughter. *Acta Agric. Scand. A Anim. Sci.* 55:149–157.

Husak, R. L., J. G. Sebranek, and K. Bregendahl. 2008. A survey of commercially available broilers marketed as organic, free-range, and conventional broilers for cooked meat yields, meat composition, and relative value. *Poult. Sci.* 87:2367–2376.

ISO 6564. 1985. Sensory analysis – methodology – flavour profile methods.

Jahan, K., A. Paterson, and J. R. Piggott. 2005. Sensory quality in retailed organic, free range and corn-fed chicken breast. *Food Res. Int.* 38:495–503.

Johansson, L., K. Lundstrom, A. Jonsall, and T. Lundh. 1999. Effects of red clover silage and ageing time on sensory characteristics and cooking losses of loin (M-*longissimus dorsi*) from Hampshire crosses with and without the RN- allele. *Food Qual. Prefer.* 10:299–303.

Jonsall, A., L. Johansson, and K. Lundstrom. 2000. Effects of red clover silage and RN genotype on sensory quality of prolonged frozen stored pork (M-*Longissimus dorsi*). *Food Qual. Prefer.* 11:371–376.

Jonsall, A., L. Johansson, and K. Lundstrom. 2001. Sensory quality and cooking loss of ham muscle (M-biceps femoris) from pigs reared indoors and outdoors. *Meat Sci.* 57:245–250.

Jonsall, A., L. Johansson, K. Lundstrom, K. H. Andersson, A. N. Nilsen, and E. Risvik. 2002. Effects of genotype and rearing system on sensory characteristics and preference for pork (M-*Longissimus dorsi*). *Food Qual. Prefer.* 13:73–80.

Kennedy, O. B., B. J. Stewart-Knox, P. C. Mitchell, and D. I. Thurnham. 2005. Flesh colour dominates consumer preference for chicken. *Appetite* 44:181–186.

Kihlberg, I., A. Ostrom, L. Johansson, and E. Risvik. 2006. Sensory qualities of plain white pan bread: influence of farming system, year of harvest and baking technique. *J. Cereal Sci.* 43:15–30.

Knight, T. W. and A. F. Death. 1997. Is beef with yellow fat potentially healthier for you than beef with white fat? *Proc. N. Z. Soc. Anim. Prod.* 57:134–136.

Lawless, H. T. and H. Heymann. 1998. *Sensory Evaluation of Food: Principles and Practices.* Chapman & Hall, New York. p. 827.

Lebret, B. 2008. Effects of feeding and rearing systems on growth, carcass composition and meat quality in pigs. *Animal* 2:1548–1558.

Lundstrom, K., A. Andersson, and I. Hansson. 1996. Effect of the RN gene on technological and sensory meat quality in crossbred pigs with Hampshire as terminal sire. *Meat Sci.* 42:145–153.

Lundstrom, K., A. C. Enfalt, E. Tornberg, and H. Agerhem. 1998. Sensory and technological meat quality in carriers and non-carriers of the RN^- allele in Hampshire crosses and in purebred Yorkshire pigs. *Meat Sci.* 48:115–124.

Magnusson, M. K., A. Arvola, U. K. K. Hursti, L. Aberg, and P. O. Sjoden. 2003. Choice of organic foods is related to perceived consequences for human health and to environmentally friendly behaviour. *Appetite* 40:109–117.

Marino, R., M. Albenzio, A. Braghieri, A. Muscio, and A. Sevi. 2006. Organic farming: effects of forage to concentrate ratio and ageing time on meat quality of Podolian young bulls. *Livest. Sci.* 102:42–50.

Meilgaard, M., G. V. Civille, and B. T. Carr. 2007. *Sensory Evaluation Techniques.* CRC Press, New Boca Raton, FL. p. 416.

Melton, S. L. 1990. Effects of feeds on flavor of red meat – a review. *J. Anim. Sci.* 68:4421–4435.

Millet, S., M. Hesta, M. Seynaeve, E. Ongenae, S. De Smet, J. Debraekeleer, and G. P. J. Janssens. 2004. Performance, meat and carcass traits of fattening pigs with organic versus conventional housing and nutrition. *Livest. Prod. Sci.* 87:109–119.

Muñoz, A. M. and G. V. Civille. 1992. The Spectrum descriptive analysis method. In: R. C. Hootman (ed). *ASTM Manual Series MNL 13, Manual on Descriptive Analysis Testing.* ASTM International, West Conchohocken, PA. pp. 22–34.

Muñoz, A. M. and G. V. Civille. 1998. Universal, product and attribute scaling and the development of common lexicons in descriptive analysis. *J. Sens. Stud.* 13:57–75.

Nielsen, B. K. and S. M. Thamsborg. 2005. Welfare, health and product quality in organic beef production: a Danish perspective. *Livest. Prod. Sci.* 94:1–50.

Obenland, D., S. Collin, B. Mackey, J. Sievert, K. Fjeld, and M. L. Arpaia. 2009. Determinants of flavor acceptability during the maturation of navel oranges. *Postharvest Biol. Technol.* 52:156–163.

Olsson, V., K. Andersson, I. Hansson, and K. Lundstrom. 2003. Differences in meat quality between organically and conventionally produced pigs. *Meat Sci.* 64:287–297.

Owens, C. M. and J.-F. Meullenet. 2010. Poultry meat tenderness. In: I. Guerrero-Legarreta and Y. H. Hui (ed) *Handbook of Poultry Science and Technology*. John Wiley & Sons, Hoboken, NJ. pp. 491–514.

Perez-Sanchez, R., M. Angeles Gomez-Sanchez, and M. Remedios Morales-Corts. 2010. Description and quality evaluation of sweet cherries cultured in Spain. *J. Food Qual.* 33:490–506.

Petersen, J. S., P. Berge, P. Henckel, and M. T. Sorensen. 1997. Collagen characteristics and meat texture of pigs exposed to different levels of physical activity. *J. Muscle Foods* 8:47–61.

Rasmussen, A. J. and M. Andersson. 1996. New methods for determination of drip loss in pork muscles. In: *Meat for the Consumer, 42nd International Congress of Meat Science and Technology*, Lillehammer. pp. 286–287.

Revilla, I., M. A. Luruena-Martinez, M. A. Blanco-Lopez, A. M. Vivar-Quintana, C. Palacios, and P. Severiano-Perez. 2009. Comparison of the sensory characteristics of suckling lamb meat: organic vs conventional production. *Czech J. Food Sci.* 27:S267–S270.

Ristic, M. 2004. Meat quality of organically produced broilers. *World Poult.* 20:30–31.

Rodriguez-Perez, M. R., G. Zurera-Cosano, R. M. Garcia-Gimeno, E. Barco-Alcala, and A. M. Castillejo-Rodriguez. 2003. Sensory and microbiological quality evaluation of vacuum-packed sliced cooked chicken breast. Shelf-life estimation. *J. Food Qual.* 26:105–122.

Shackelford, S. D., T. L. Wheeler, and M. Koohmaraie. 1995. Relationship between shear force and trained sensory panel tenderness ratings of 10 major muscles from Bos Indicus and Bos Taurus Cattle. *J. Anim. Sci.* 73:3333–3340.

Sinesio, F., E. Moneta, and M. Peparaio. 2007. Sensory characteristics of traditional field grown tomato genotypes in Southern Italy. *J. Food Qual.* 30:878–895.

Sitz, B. M., C. R. Calkins, D. M. Feuz, W. J. Umberger, and K. M. Eskridge. 2005. Consumer sensory acceptance and value of domestic, Canadian, and Australian grass-fed beef steaks. *J. Anim. Sci.* 83:2863–2868.

Stone, H., J. L. Sidel, and J. Bloomquist. 1980. Quantitative descriptive analysis. *Cereal Foods World* 25:642–644.

Sunde, M. L. 1992. Symposium – the scientific way to pigment poultry products – introduction to the symposium. *Poult. Sci.* 71:709–710.

Ulrich, D., E. Hoberg, and C. Fischer. 2009. Diversity and dynamic of sensory related traits in different apple cultivars. *J. Appl. Botany and Food Quality – Angew Bot.* 83:70–75.

Vanderwal, P. G., G. Mateman, A. W. Devries, G. M. A. Vonder, F. J. M. Smulders, G. H. Geesink, and B. Engel. 1993. Scharrel (free range) pigs – carcass composition, meat quality and taste-panel studies. *Meat Sci.* 34:27–37.

Vestergaard, M., N. Oksbjerg, and P. Henckel. 2000a. Influence of feeding intensity, grazing and finishing feeding on muscle fibre characteristics and meat colour of semitendinosus, longissimus dorsi and supraspinatus muscles of young bulls. *Meat Sci.* 54:177–185.

Vestergaard, M., M. Therkildsen, P. Henckel, L. R. Jensen, H. R. Andersen, and K. Sejrsen. 2000b. Influence of feeding intensity, grazing and finishing feeding on meat and eating quality of young bulls and the relationship between muscle fibre characteristics, fibre fragmentation and meat tenderness. *Meat Sci.* 54:187–195.

Wahrburg, U. 2004. What are the health effects of fat? *Eur. J. Nutr.* 43:6–11.

Walshe, B. E., E. M. Sheehan, C. M. Delahunty, P. A. Morrissey, and J. P. Kerry. 2006. Composition, sensory and shelf life stability analyses of *Longissimus dorsi* muscle from steers reared under organic and conventional production systems. *Meat Sci.* 73:319–325.

Warris, P. D. 2000. *Meat Science. An Introductory Text.* CABI, New York. p. 310.

Williams, A. A. and G. M. Arnold. 1984. A new approach to sensory analysis of food and beverages. *4th Weurman Flavour Research Symposium*, Dourdan, May 9th–11th.

Wood, J. D., P. D. Warriss, and M. B. Enser. 1992. Effects of production factors on meat quality in pigs. In: D. E. Johnston and M. K. Knight (eds) *The Chemistry of Muscle-based Foods*. Royal Society of Chemistry, London. pp. 3–14.

Xiong, Y. L., W. G. Moody, S. P. Blanchard, G. Liu, and W. R. Burris. 1996. Postmortem proteolytic and organoleptic changes in hot-boned muscle from grass- and grain-fed and zeranol-implanted cattle. *Food Res. Int.* 29:27–34.

APPENDIX A MINIMUM NUMBER OF ASSESSMENTS IN A TRIANGLE TEST

Parameters (α, β, p_d) indicate sensitivity of test for sample size (n), which are table entries.

					β				
α		**0.50**	**0.40**	**0.30**	**0.20**	**0.10**	**0.05**	**0.01**	**0.001**
	$p_d = 50\%$								
0.40		3	3	3	6	8	9	15	26
0.30		3	3	3	7	8	11	19	30
0.20		4	6	7	7	12	16	25	36
0.10		7	8	8	12	15	20	30	43
0.05		7	9	11	16	20	23	35	48
0.01		13	15	19	25	30	35	47	62
0.001		22	26	30	36	43	48	62	81
	$p_d = 40\%$								
0.40		3	3	6	6	9	15	26	41
0.30		3	3	7	8	11	19	30	47
0.20		6	7	7	12	17	25	36	55
0.10		8	10	15	17	25	30	46	67
0.05		11	15	16	23	30	40	57	79
0.01		21	26	30	35	47	56	76	102
0.001		36	39	48	55	68	76	102	130
	$p_d = 30\%$								
0.40		3	6	6	9	15	26	44	73
0.30		3	8	8	16	22	30	53	84
0.20		7	12	17	20	28	39	64	97
0.10		15	15	20	30	43	54	81	119
0.05		16	23	30	40	53	66	98	136
0.01		33	40	52	62	82	97	131	181
0.001		61	69	81	93	120	138	181	233
	$p_d = 20\%$								
0.40		6	9	12	18	35	50	94	153
0.30		8	11	19	30	47	67	116	183
0.20		12	20	28	39	64	86	140	212
0.10		25	33	46	62	89	119	178	260
0.05		40	48	66	87	117	147	213	305
0.01		72	92	110	136	176	211	292	397
0.001		130	148	176	207	257	302	396	513
	$p_d = 10\%$								
0.40		9	18	38	70	132	197	360	598
0.30		19	36	64	102	180	256	430	690
0.20		39	64	103	149	238	325	439	819
0.10		89	125	175	240	348	457	683	1011
0.05		144	191	249	325	447	572	828	1178
0.01		284	350	425	525	680	824	1132	1539
0.001		494	579	681	803	996	1165	1530	1992

Source: Reprinted from Meilgaard *et al.* (2007), with permission.

APPENDIX B CRITICAL NUMBER OF CORRECT RESPONSE IN A TRIANGLE TEST (ENTRIES ARE $x_{\alpha,n}$)

Reject the null hypothesis (H_0: A = B) if correct responses are equal or greater to table entries, which are number of correct answers needed to declare that a significant difference exists at tabled α-level.

	α								α						
n	**.40**	**.30**	**.20**	**.10**	**.05**	**.01**	**.001**	***n***	**.40**	**.30**	**.20**	**.10**	**.05**	**.01**	**.001**
								31	12	13	14	15	16	18	20
								32	12	13	14	15	16	18	20
3	2	2	3	3	3	—	—	**33**	13	13	14	15	17	18	21
4	3	3	3	4	4	—	—	**34**	13	14	15	16	17	19	21
5	3	3	4	4	4	5	—	**35**	13	14	15	16	17	19	22
6	3	4	4	5	5	6	—	**36**	14	14	15	17	18	20	22
7	4	4	4	5	5	6	7	**42**	16	17	18	19	20	22	25
8	4	4	5	5	6	7	8	**48**	18	19	20	21	22	25	27
9	4	5	5	6	6	7	8	**54**	20	21	22	23	25	27	30
10	5	5	6	6	7	8	9	**60**	22	23	24	26	27	30	33
11	5	5	6	7	7	8	10	**66**	24	25	26	28	29	32	35
12	5	6	6	7	8	9	10	**72**	26	27	28	30	32	34	38
13	6	6	7	8	8	9	11	**78**	28	29	30	32	34	37	40
14	6	7	7	8	9	10	11	**84**	30	31	33	35	36	39	43
15	6	7	8	8	9	10	12	**90**	32	33	35	37	38	42	45
16	7	7	8	9	9	11	12	**96**	34	35	37	39	41	44	48
17	7	8	8	9	10	11	13	**102**	36	37	39	41	43	46	50
18	7	8	9	10	10	12	13	**108**	38	40	41	43	45	49	53
19	8	8	9	10	11	12	14	**114**	40	42	43	45	47	51	55
20	8	9	9	10	11	13	14	**120**	42	44	45	48	50	53	57
21	8	9	10	11	12	13	15	**126**	44	46	47	50	52	56	60
22	9	9	10	11	12	14	15	**132**	46	48	50	52	54	58	62
23	9	10	11	12	12	14	16	**138**	48	50	52	54	56	60	64
24	10	10	11	12	13	15	16	**144**	50	52	54	56	58	62	67
25	10	11	11	12	13	15	17	**150**	52	54	56	58	61	65	69
26	10	11	12	13	14	15	17	**156**	54	56	58	61	63	67	72
27	11	11	12	13	14	16	18	**162**	56	58	60	63	65	69	74
28	11	12	12	14	15	16	18	**168**	58	60	62	65	67	71	76
29	11	12	13	14	15	17	19	**174**	61	62	64	67	69	74	79
30	12	12	13	14	15	17	19	**180**	63	64	66	69	71	76	81

Source: Reprinted from Meilgaard *et al.* (2007), with permission.

16 Chemical Residues in Organic Meats Compared to Conventional Meats

Sergio Ghidini, Emanuela Zanardi, Mauro Conter, and Adriana Ianieri

Abstract: Organic agriculture is important for the lower environmental impact, although there is no clear scientific evidence that organic meat can better protect consumers from chemical contamination than can conventional meat, even if this is one of the major forces driving consumers to buy organic products. In particular, organic farming cannot protect consumers from environmental contaminants that are ubiquitous or whose sources cannot be controlled by organic farming. Outdoor grazing, frequently used in organic breeding, could expose the animals to contamination levels even higher than conventionally farmed animals, thus generating more contaminated meat.

Keywords: meat; organic; heavy metals; dioxin; contaminants; veterinary drugs; pesticides; polycyclic aromatic hydrocarbons

16.1 INTRODUCTION

The rapid pace of change in science and technology with the continuous introduction of new technologies in agriculture, the changes in legislation, and the current socioeconomic and sociodemographic realities have all had a marked impact on the food purchased currently. The intensification of farming, such as the use of fertilizers and pesticides, and the industrialization of food production, using additives and preservatives to improve taste, appearance, and shelf-life, for example, can be the causes for concern among many consumers. Furthermore, new technologies, used or under consideration, such as genetically modified organisms (GMOs), nanotechnology, and animal cloning, are modifying agriculture and food production in some fashion today, and are expected to produce great changes in the future. However, consumers tend to be more conservative in the field of food than in any other field. Even if they are increasingly more prepared to use technologies in other aspects of their lives, generally they do not want high levels of technology associated with their food (Grunert, 2002; Deliza *et al.*, 2003). In this context, the chemical safety of food products is probably the major concern for consumers all around the world.

The European Food Safety Authority (EFSA, 2010) recently published the results of a survey conducted among the consumers of the European Union (EU) regarding consumers' perceptions of food-related risks. The majority of Europeans associate food and eating with enjoyment. According to this survey, those who are concerned about possible food-related

Organic Meat Production and Processing, First Edition. Edited by Steven C. Ricke, Ellen J. Van Loo, Michael G. Johnson, and Corliss A. O'Bryan.

risks tend to worry more about chemical contamination of food rather than bacterial contamination or health and nutrition issues. No single widespread concern about food-related risks was mentioned spontaneously by a majority of respondents—19% cited chemicals, pesticides, and other substances as the major concerns. When subsequently prompted by a list of possible issues associated with food, respondents mentioned as risks to be "very worried" about: (1) chemical residues from pesticides in fruit, vegetables, and cereals (31%, up by 3 percentage points as compared to 2005); (2) antibiotics or hormones in meat (30%, up by 3 percentage points on 2005); (3) cloning animals for food products (30%); and (4) pollutants, such as mercury in fish and dioxins in pork (29%, up by 3 percentage points on 2005). Fewer people were "very worried" about bacterial contamination of foods (23%), and even fewer about possible nutritional risks, such as putting on weight (15%) or not having a healthy/balanced diet (15%). As a logical consequence, clearly summarized by Yiridoe *et al.* (2005) and Magkos *et al.* (2006), consumer concern over the quality and safety of conventional food has intensified, becoming the primary force for increasing demand for organically produced food, which is perceived as healthier and safer.

16.2 INORGANIC RESIDUES AND CONTAMINANTS

16.2.1 Lead and cadmium

Lead is a metal that has been largely used in industry, mainly in the manufacturing of pigments, coatings, containers, ointments, and electric batteries. Because of its toxicity, over the past decades, the lead level in food has decreased significantly, thanks to efforts to reduce the emission of lead and improvements in quality assurance of chemical analysis. The ban of leaded fuel (tetraethyl lead was used as an octane booster and as an antiknocking agent) in most industrialized countries has been the most effective action to reduce lead contamination (Thomas *et al.*, 1999). Lead is present in low concentrations in most foods. Offal and mollusks may contain higher levels. Contaminations of food during processing or food production in contaminated areas are the main reasons for enhanced lead intake via foodstuffs. Absorption of ingested lead may constitute a serious risk to public health. Some chronic effects of lead poisoning are colic, constipation, and anemia. It may also induce increased blood pressure and cardiovascular disease in adults. Lead affects the neurological development of fetus and reduces the learning capability of children. The Codex Alimentarius system and the EC regulations (EC, 2008) set the same maximum residue levels (MRLs) for lead in the meat of bovine animals, sheep, pig, and poultry (0.1 mg/kg) and for edible offals of these animals (0.5 mg/kg). In addition, the International Agency for Research on Cancer (IARC) has classified lead as category 2A carcinogen (IARC, 2006).

Cadmium is a heavy metal found as an environmental contaminant, both through natural occurrence, and from industrial and agricultural sources. Foodstuffs are the main source of cadmium exposure for the nonsmoking general population (cigarette smoke is rich in cadmium). Although cadmium absorption after dietary exposure in humans ranged between 3% and 5%, the exposure should not be a concern; nevertheless, cadmium is efficiently retained in the kidney and liver in the human body, with a very long biological half-life ranging from 10 to 30 years. The kidney is the primary target organ for cadmium since it accumulates in the proximal tubular cells and thus, can cause renal dysfunction. Cadmium can also cause bone demineralization, either through direct bone damage or indirectly as a result of electrolytic imbalance caused by renal dysfunction. After high exposure, the tubular damage may progress to decreased glomerular filtration rate, and eventually to renal failure. The IARC has classified cadmium as a group 1 human carcinogen (IARC, 1993). More recently, epidemiological data associated with high cadmium exposure in the population increased the risk of cancer in tissues such as the lung, endometrium, bladder, and breast (EFSA, 2009). The EU has set maximum levels (MLs) for cadmium in meat of bovine animals, sheep, pig, and poultry at 0.05 mg/kg wet weight and for edible offal of these animals at 0.5 mg/kg for liver and 1.0 mg/ kg for kidney (EC, 2008; EC, 2010). The Scientific Panel on Contaminants in the Food Chain (CONTAM) assessed the risks to human health related to the presence of cadmium in foodstuffs (EFSA, 2009). Approximately 140,000 data covering the time period from 2003 to 2007 on cadmium levels in various food commodities were considered. High cadmium concentrations were detected in seaweed, fish and seafood, chocolate, and foods that have been used for special dietary applications. In the food category "meat, meat products, and offal," the fractions of samples exceeding the MLs were: (1) bovine, sheep, and goat meat, 3.6%; (2) poultry and rabbit meat, none; (3) pork, 1.6%; (4) liver (bovine, sheep, pig, poultry, and horse), 3.7%; and (5) kidney (bovine, sheep, pig, poultry, and horse), 1.0%. The relative median values were 0.0050, 0.0030, 0.0050, 0.0430, and 0.1520 mg/kg respectively.

With a different approach (exploratory assessment) but with the same purpose, in 2003–2004, the US Department of Agriculture Food Safety and Inspection Service (FSIS) determined the levels of cadmium and lead in randomly collected samples of kidney, liver, and muscle tissues of mature chickens, boars/stags, dairy cows, and heifers (Pagan-Rodriguez *et al.*, 2007). The study found that none of the muscle samples contained cadmium or lead levels exceeding the MLs established by other countries or international organizations and that kidney concentrations were always higher than meat concentrations. There are sporadic cases in which liver samples from mature chickens and boars/stags contained elevated cadmium or lead levels; however, the 95th percentile and the mean concentrations for liver samples were below the tolerances established by other countries or international organizations. In boars/stags, mature chickens, and dairy cows, the cadmium levels for the 95th percentile (not for the mean) of kidney samples were above internationally accepted levels, exceeding the tolerance (1 mg/kg) established by the EU. The results of FSIS studies demonstrated that the incidence (percentage of positive samples) and levels of cadmium in kidney, liver, and muscle did not increase between 1985 and 2004. Since these organs are basically involved in accumulating contaminants, the age of the animals could be an important factor in determining consumer exposure via meat consumption. For this reason, Waegeneers *et al.* (2009) investigated the effect of animal age on concentrations of cadmium, and in bovine tissues (meat, kidney, and liver) sampled from animals reared in contaminated areas of Belgium. Cadmium concentrations in meat samples had an increasing trend with age. In addition, a significant positive linear relation was found between animal age and renal or hepatic cadmium levels. Also lead concentrations in kidneys and liver increased with age. They estimated that for 2-year-old animals from contaminated areas, the European ML of 1 mg/kg of cadmium in kidneys would be exceeded in zero to 5% of cases.

Andersson (1998) showed that emission of cadmium from fertilizers is an important issue in food production system. The use of phosphorous fertilizer in the conventional dairy life cycle is instead found on the arable farms that grow the crops for concentrate feed production. The import of feed, which is often substantial on a conventional milk farm, leads to a phosphorus inflow, most of which will stay on the farm and be spread on the arable land with the farmyard manure. The input of cadmium to the dairy farm can be traced to the feed import and the atmospheric deposition. Phosphorous fertilizers are very often contaminated by cadmium; therefore, the metal will remain on the farm in the manure since the output products (milk and meat) remove minute amounts of cadmium.

It seems evident that conventional farming systems with a large input of feed, and which only have animal output products, run a potential risk of accumulating heavy metals in the soil and thus of producing more contaminated products (Cederberg and Mattsson, 2000). The comparisons between heavy metals in organic and conventional milk are scarce, and not always consistent with this. Linden *et al.* (2001) found higher cadmium levels in kidneys from organically raised pigs than in kidneys from conventionally raised pigs, despite the lower level of cadmium in organic feed and similar daily feed allowances in both production systems. They explained the higher levels of cadmium in manure and kidneys from organic pigs compared to conventional pigs by the exposure of the organic pigs to cadmium from other sources than the feed, for example, soil ingestion through grubbing. Ghidini *et al.* (2005) undertook a comparison between the chemical safety of organic and conventional products of animal origin. Samples from organic farms were coupled to samples from conventional farms within maximum 3-km range in order to minimize environmental effects. Regarding lead and cadmium in meat, the levels detected were very low (all within the EU ML) and did not differ between organic and conventional products. Malmauret *et al.* (2002) conducted an extensive comparison between organic and conventional foodstuffs in France. Within this survey, six conventional and six organic samples of beef, pork, and poultry were analyzed. In all samples, cadmium and lead concentrations were below the limit of quantification (LOQ) of the adopted method (0.005 and 0.002 mg/kg, respectively).

Olsson *et al.* (2001) studied the impact of conventional and organic farming on cadmium levels in tissues from dairy cows. Kidney, liver, muscle, and mammary tissue samples were collected at slaughter from 67 cows, aged 30–95 months, in a project with conventional and organic production at the same farm. Significantly lower levels of cadmium in kidney and mammary tissue were found in cows from the organic system than from the conventional cows, while there were no differences in muscle tissue.

Blanco-Penedo *et al.* (2010) evaluated if differences in nonessential and essential trace element accumulation in the meat of beef cattle reared under different systems (including organic, conventional, and intensive management). Diaphragm muscles from 166 calves from nine farms were analyzed. Muscle cadmium concentrations were low (less than 10 μg/kg wet weight) and muscle lead levels were below the limits of detection (0.003 mg/kg) in most (97%) samples; there were no significant differences between farms. Toxic element concentrations detected in the muscle of the animals did not generally reflect differences in exposure. This is particularly relevant for animals reared in organic farms where cattle could be exposed to higher levels of toxic elements, probably due to soil ingestion when grazing, and does not confirm the tendency shown by Linden *et al.* (2001) in their study.

16.2.2 Arsenic and mercury

The maximum contribution to consumer exposure to arsenic and mercury is basically given by seafood and drinking water (for arsenic). Even if these elements can be detected by sensitive techniques (mainly ICP-MS) in meat and offals, the contribution of meat to the total intake of arsenic and mercury is negligible, and therefore, they will not be considered here.

16.3 ORGANIC RESIDUES AND CONTAMINANTS

16.3.1 PCDD/Fs and PCBs

Polychlorinated dibenzo-*p*-dioxins (PCDDs) and dibenzofurans (PCDFs) belong to the group of lipophilic and persistent organic contaminants. PCDDs and PCDFs are often referred to simply as "dioxins." Depending on the degree of chlorination (1–8 chlorine atoms) and the substitution pattern, one can distinguish between 75 PCDDs and 135 PCDFs, called "congeners." Although dioxins do not have any benefits and therefore, have not been produced specifically, these contaminants have meanwhile found a ubiquitous distribution due to their formation as unwanted and often unavoidable by-products in a number of industrial and thermal processes. The toxicity of dioxins differs considerably. In particular, those congeners, which are substituted in the 2,3,7,8-position, are of special importance. Thus, of the 210 theoretically possible congeners, only 17 are of toxicological concern. These compounds show a similar toxicity to that of the most toxic congener 2,3,7,8-tetrachlorodibenzo-*p*-dioxin (2,3,7,8-TCDD). In order to facilitate the comparison of analytical and exposure data, it has proven useful in the past to convert the analytical results of the determination of all 17 congeners of toxicological concern into one summarizing result, which is expressed as toxic equivalents (TEQ). This conversion is based on the assumption that all dioxin congeners show similar qualitative effects (binding to the same dioxin-receptor) but with different intensities. The different binding activity is expressed by toxic equivalency factors (TEF), estimated from the weaker toxicity of the respective congener in relation to the most toxic compound 2,3,7,8-TCDD, which is assigned the arbitrary TEF of 1. Moreover, it is assumed that the effects are neither synergistic nor antagonistic, but additive. By multiplying the amount of each congener present with the corresponding TEF, the TEQ value of a sample can be generated.

Polychlorinated biphenyls (PCBs) were largely produced in the past for their technological properties even if their production is currently discontinued. Over the 209 possible congeners, the planar ones show a dioxin-like mechanism of toxicity, thus have TEFs and are counted together with dioxins and furans to generate the TEQ value.

Recently, the EFSA (2010) published results of one of the largest monitoring efforts in the literature of dioxin levels in food and feed. A total of 7270 samples collected in the period 1999–2008 from 19 member states, Norway, and Iceland were analyzed. The percentage of samples below the LOQ varied considerably at the congener level. Overall, the percentages of censoring, defined as the proportion of nonquantified (less then LOQ) observations, varied sizably depending on how results were expressed: (1) on a fat basis (approximately 40%); (2) on a whole weight basis (approximately 30%), or (3) for feed on "12% moisture" basis (approximately 60%). The percentage of results exceeding different MLs for dioxins and dioxin-like PCBs set by legislation was on average 8% with a further 4% exceeding some action levels, but there were large variations among groups. It is important to bear in mind that a varying proportion of product testing reflects targeted and not random monitoring. This has the potential of introducing a degree of uncertainty and bias in the evaluation of background levels of dioxins and dioxin-like PCBs in food and feed, as higher total values are expected in targeted compared to random samples.

An extensive monitoring on meat and meat products was also conducted in Germany (Bundesministerium für Ernährung, Landwirtschaft und Verbraucherschutz, 2009) as reported by Sabine *et al.* (2010). The median dioxin content in beef samples was 0.9 ng/kg TEQ fat, which was in the range of the action level of 1.0 ng/kg fat. A clear correlation between dioxin contamination and age was also shown. Overall, it was concluded that an adult consumer with 70 kg bodyweight—solely with the consumption of meat and meat products—is approximately 3% of the acceptable weekly intake (TWI), which was set to 14 pg TEQ/kg bodyweight/week by the Scientific Committee on Food (SCF, 2001).

Between 2007 and 2008 a similar survey was conducted in the United States (FSIS, 2009). The results yielded mean total TEQ levels for beef, turkey, chicken, and pork at 0.66, 0.61, 0.17, and 0.16 ng/kg fat, respectively. A comparison with previous surveys indicates declining dioxin contamination in all slaughter classes over a 10-year period. The results are consistent with those found in the EU (except for samples from the Baltic Sea area showing higher values).

At present, no comparisons between dioxin contamination in conventional and organic meat are present in literature. Only a study on dioxin on organic eggs can be cited here. Kijlstra (2004) showed higher dioxin contamination in organic eggs than in conventional ones. These results were explained by the different breeding techniques adopted in organic farming where chickens have more outdoor space and are, as a result, exposed to dioxin contamination via air, soil, and worms and insect ingestion.

16.4 PESTICIDES

Among the pesticides, only organochlorinated pesticides can be regularly detected in meat since they are accumulating in fat tissue. Other pesticides do not persist in meat as residues, and meat is not considered as a possible means for exposure to organophosphates or other less lipophilic molecules (Lu *et al.*, 2008; Rawn *et al.*, 2004).

Organochlorine pesticides are currently banned but traces of these molecules (or their metabolites) in human serum may be explained by environmental exposure of the population. Foods are considered to represent a constant source of exposure, despite compliance with the maximum permitted residue levels (Rivas *et al.*, 2007). Only 60% of all the fruit or vegetables consumed by European citizens are free from pesticide residues. According to a recent study in Barcelona (Spain), organochlorine pesticides appeared to have completely disappeared from some foodstuffs, such as fruits; although, they continued to be detected in other types of food, such as vegetables, milk, and meat (Vicente *et al.*, 2004). In a recent analysis of organochlorine pesticides residues in US foods, it was found that even chemicals that had been banned for decades were consistently detected in food samples tested by the US Food and Drug Administration (FDA, 2008). This can be explained in part by the persistence of many organic contaminants in the environment (dieldrin and DDT breakdown products, for example, can remain in soil for decades) and in part by the importation or wind and water transportation of pesticides still used in other countries.

These findings show that organic meat also cannot be absolutely free from organochlorine pesticides residues. As a matter of fact, the only study that compared organochlorine residues in organic and conventional meat found no differences (Ghidini *et al.*, 2005).

16.5 POLYCYCLIC AROMATIC HYDROCARBONS (PAHs)

Polycyclic aromatic hydrocarbons (PAHs) constitute a large class of organic compounds that are composed of two or more fused aromatic rings. They are primarily formed by incomplete combustion or pyrolysis of organic matter and during various industrial processes. PAHs generally occur in complex mixtures that may consist of hundreds of compounds. Humans are exposed to PAHs by various pathways. While for nonsmokers, the major route of exposure is consumption of food; for smokers, the contribution from smoking may be significant. Food can be contaminated from environmental sources, industrial food processing, and certain home cooking practices. As shown by EFSA (2008), unprocessed meat is unlikely to be contaminated by PAHs while processed meat can be; in the EFSA opinion, barbecued meat had more than double the contamination level of grilled meat with smoked meat products in third place. Nevertheless, smoking is considered the critical point in meat contamination by PAHs. Fortunately, modern smoking methods are designed to avoid PAH contamination. This is one of the examples showing that "traditional" is not synonym of safe.

16.6 VETERINARY DRUGS

Within this category, many subcategories should be considered: antibiotics, hormones, antiparasitic agents, anti-inflammatory drugs, and so on. Within each category, many different molecules are comprised. The forthcoming paragraphs give a brief introduction and summary about the complexity of this topic.

In conventional animal breeding, veterinary drugs can be used both for therapy and as growth promoters. However, the use of veterinary drugs as growth promoters is becoming less extensive. For instance, antibiotics historically were used as feed additives in poultry and pigs but the concerns associated with the spread of antibiotic resistance led first to a limitation of their use and eventually to a ban. As a matter of fact, the EU in 2003 adopted Regulation 1831 (EC, 2003) stating that antibiotics, other than coccidiostats and histomonostats, might be marketed and used as feed additives only until December 31, 2005. Anticoccidial substances, such as antibiotics ionophores, will also be prohibited as feed additives in 2013. After this date, medical substances in animal feeds will be limited to therapeutic use by veterinary prescription. The FDA adopted a similar measure for the United States in 2005, essentially banning the use of antibiotics as feed additives. However, in the United States and Canada, three natural steroid hormones, (1) estradiol, (2) testosterone, and (3) progesterone, and three synthetic surrogates, (1) zeranol, (2) melengestrol, and (3) trenbolone, remain in widespread use by the US and Canadian beef cattle producers to promote growth and production (Stephany, 2001). The contemporary use of more than one steroid is allowed and practiced. It is widely acknowledged that the use of these hormone growth promoters results in residues in meat and the environment (Lange *et al.*, 2002); both the FDA and the Food and Agricultural Organization/World Health Organization's Joint Expert Committee on Food Additives (JECFA) have published acceptable daily intakes (ADIs) for all hormones in current use (Henricks *et al.*, 2001). Residues of these hormone growth promoters also persist for weeks to months in manure and in feedlot runoff, raising concerns about the added exogenous hormone load to the environment (Schiffer *et al.*, 2001).

Numerous countries have established legal limits for residues of veterinary drugs in food. According to the Agreement on the Application of Sanitary and Phytosanitary Measures (SPM) and Agreement on Technical Barriers to Trade, food safety standards for veterinary drug residues established by the Codex Alimentarius Commission are the reference points in international trade. Thus far, Codex has established MRLs for approximately 50 veterinary drugs. Once accepted, member states of Codex are expected to implement these MRLs in national (or community) law. Deviations from Codex MRLs are possible but must be substantiated with scientific proof of risk. MRLs for residues of veterinary drugs are set at the EU level. The European Medicines Agency (EMA) evaluates all applications for MRLs. The EU MRLs are listed in Table 1 of the Annex of Regulation (EU) 37/2010. Only veterinary drugs containing pharmacologically active substances for which MRLs have been set (or for which the conclusion was that no MRL is required) are allowed to be authorized by EU member states for use in food producing animals, and then only for use in those species for which MRLs were set.

Residues of veterinary drugs are generally monitored by national plans on residues worldwide. These data are usually public, but they are presented only as number of positive samples per therapeutic category; therefore, it is impossible to use these data for risk evaluation purposes. Scientific literature does not help in this case because, while it is very easy to find analytical methods, it is almost impossible to find surveys on veterinary drugs residues in meat.

Since it is difficult to assess the level of risk of veterinary drug residues in conventional meat, a complete comparison between conventional and organic meat cannot be performed. More importantly, organic meat should be virtually free of these residues since legislation on organic meat production generally prohibits the use of synthetic chemotherapy drugs while only phytotheraphy, vaccinations, and homeopathy are allowed. In this case, legislation in the United States is more conservative than the European legislation because in the United States, if an organic animal is treated only once with synthetic drugs it cannot be labeled as organic, while in the EU single treatment exceptions are allowed.

16.7 CONCLUSIONS

Organic agriculture is important for many reasons (less environmental impact, less use of energy, and other factors); however, even if the scientific literature in this field is scarce, there is no clear scientific evidence that organic meat can better protect consumers from chemical contamination than conventional meat, even if this is one of the major forces driving consumers to buy organic products. In particular, organic farming cannot protect consumers from environmental contaminants that are substantially ubiquitous or whose sources cannot be controlled by organic farming. The fact that organic farms buy less feed from outside sources rather than conventional ones is not protective against chemical contamination.

For certain sharable animal welfare reasons, organic livestock farming may adopt animal husbandry techniques (e.g., outdoor grazing) that could expose the animals to contamination levels even higher than conventionally farmed animals, thus generating more contaminated meat. Considering that organic animals are exposed to this route of contamination from the environment, it is crucial to carefully plan out the placement of organic farms. Since many chemicals accumulate in meat and offals as a direct function of time, the animal age factor should also be considered. Data on the age of the animals used to obtain organic meat in comparison to conventional ones is not generally possible. However, it is reasonable to suppose that: (1) animals bred organically only for meat production achieve lower production performances than convention ones and therefore, must live longer to reach market weight for harvesting; and (2) animals organically bred for purposes other than meat (i.e., lactating cows), whose meat is still used at the end of their production cycle, have productive careers longer than conventional ones.

Certainly, more research is needed in this field, but, at present, the advertising claims that consumption of organic meat rather than conventional meat can reduce the exposure to chemicals should be avoided to preclude any unwanted negative impacts on such an important sector as organic agriculture.

REFERENCES

Andersson, K. 1998. Life cycle assessment (LCA) of food products and production systems. PhD thesis. Chalmers University of Technology, Gothenburg.

Blanco-Penedo, I., M. López-Alonso, M. Miranda, J. Hernández, F. Prieto, and R. F. Shore. 2010. Non-essential and essential trace element concentrations in meat from cattle reared under organic, intensive or conventional production systems. *Food Addit. Contam. A.* 27:36–42.

Bundesministerium für Ernährung, Landwirtschaft und Verbraucherschutz. 2009. Statuserhebung zu Dioxinen und PCB in Futter- und vom Tier stammenden Lebensmitteln. Angewandte Wissenschaft Heft 522, Verlagsgesellschaft W.E. Weinmann e.K. or pdf-file. Available at: http://www.bmelv.de/cae/servlet/contentblob/793196/publicationFile/44180/Heft522.pdf.

Cederberg, C. and B. Mattsson. 2000. Life cycle assessment of milk production – a comparison of conventional and organic farming. *J. Clean. Prod.* 8:49–60.

Deliza, R., A. Rosenthal, and A. L. S. Silva. 2003. Consumer attitude towards information on non conventional technology. *Trends Food Sci. Tech.* 14:43–49.

EC. 2003. Commission Regulation 1831/2003/EC of 22 September 2003 on additives for use in animal nutrition. *Off. J. Eur. Union* 268:29–15.

EC. 2008. Commission Regulation 629/2008/EC of 2 July 2008 setting maximum levels for certain contaminants in foodstuffs. *Off. J. Eur. Union* 173:6–15.

EC. 2010. Commission Regulation 37/2010/EC of 22 December 2010 on pharmacologically active substances and their classification regarding maximum residue limits in foodstuffs of animal origin. *Off. J. Eur. Union* 173:6–15.

European Food Safety Authority (EFSA). 2008. Polycyclic aromatic hydrocarbons in food scientific opinion of the panel on contaminants in the food chain. *EFSA J.* 724:1–114.

European Food Safety Authority (EFSA). 2009. Scientific opinion of the panel on contaminants in the food chain on a request from the European Commission on cadmium in food. *EFSA J.* 980:1–139.

European Food Safety Authority (EFSA). 2010. New research results on EU consumers' perceptions of food-related risks. Available at: http://www.efsa.europa.eu/en/press/news/corporate101117.htm?WT.mc_id=EFSAHL01&emt=1 (accessed November, 2010).

European Food Safety Authority (EFSA). 2010. Results of the monitoring of dioxin levels in food and feed. *EFSA J.* 8:3–1385.

Food and Drug Administration (FDA). 2008. Pesticide Residue Monitoring Program FY 2004–2006. Available at: http://www.fda.gov/Food/FoodSafety/FoodContaminantsAdulteration/Pesticides/ResidueMonitoringReports/ucm125183.htm (accessed September, 2009).

Food Safety and Inspection Service (FSIS). 2009. Dioxins and dioxin-like compounds in the U.S. domestic meat and poultry supply. Available at: Fsis.usda.gov.http://www.fsis.usda.gov/PDF/Dioxin_Report_1009.pdf (accessed September, 2010).

Ghidini, S., E. Zanardi, A. Battaglia, G. Varisco, E. Ferretti, G. Campanini, and R. Chizzolini. 2005. Comparison of contaminant and residue levels in organic and conventional milk and meat products from Northern Italy. *Food Addit. Contam.* 22:9–14.

Grunert, K. 2002. Current issues in the understanding of consumer food choice. *Trends Food Sci. Tech.* 13:275–285.

Henricks D. M., S. L. Gray, and J. J. Owenby. 2001. Residues from anabolic preparations after good veterinary practice. *Acta Patholog. Microb.* 109:273–283.

IARC. 1993. *Beryllium, Cadmium, Mercury, and Exposures in the Glass Manufacturing Industry.* IARC monographs on the evaluation of carcinogenic risks to humans. Vol. 58. Lyon. p. 444.

IARC. 2006. *Inorganic and Organic Lead Compounds.* IARC monographs on the evaluation of carcinogenic risks to humans. Vol. 87. Lyon. p. 506.

Kijlstra, A. 2004. The role of organic and free range poultry production systems on the dioxin levels in eggs. *Proc. 3rd SAFO Workshop*, Falenty. pp. 83–89.

Lange, I. G., A. Daxenberger, B. Schiffer, H. Witters, D. Ibarreta, and H. H. D. Meyer. 2002. Sex hormones originating from different livestock production systems: fate and potential disrupting activity in the environment. *Anal. Chim. Acta.* 473:27–37.

Linden, A., K. Andersson, and A. Oskarsson. 2001. Cadmium in organic and conventional pig production. *Arch. Environ. Contam. Toxicol.* 40:425–431.

Lu, C., D. B. Barr, M. A. Pearson, and L. A. Waller. 2008. Dietary intake and its contribution to longitudinal organophosphorus pesticide exposure in urban/suburban children. *Environ. Health Persp.* 116:537–542.

Magkos, F., F. Arvaniti, and A. Zampelas. 2006. Organic food: buying more safety or just peace of mind? A critical review of the literature. *Crit. Rev. Food Sci.* 46:23–56.

Malmauret, L., D. Parent-Massin, J. L. Hardy, and P. Verger. 2002. Contaminants in organic and conventional foodstuffs in France. *Food Addit. Contam.* 19:524–532.

Olsson, I. M., S. Jonsson, and A. Oskarsson. 2001. Cadmium and zinc in kidney, liver, muscle and mammary tissue from dairy cows in conventional and organic farming. *J. Environ. Monit.* 3:531–538.

Pagan-Rodriguez, D., M. O'Keefe, C. Deyrup, P. Zervos, H. Walker, and A. Thaler. 2007. Cadmium and lead residue control in a hazard analysis and critical control point (HACCP) environment. *J. Agric. Food Chem.* 55:1638–1642.

Rawn, D. F., X. L. Cao, J. Doucet, D. J. Davies, W. F. Sun, R. W. Dabeka, and W. H. Newsome. 2004. Canadian Total Diet Study in 1998: pesticide levels in foods from Whitehorse, Yukon, Canada, and corresponding dietary intake estimates. *Food Addit. Contam.* 21:232–250.

Rivas, A., I. Cerrillo, A. Granada, M. Mariscal-Arcas, and F. Olea-Serrano. 2007. Pesticide exposure of two age groups of women and its relationship with their diet. *Sci. Total. Env.* 382:14–21.

Sabine, A., W. Jira, K. H. Schwind, H. Wagner, and F. Schwägele. 2010. Chemical safety of meat and meat products. *Meat Sci.* 86:38–48.

SCF. 2001. Scientific Committee on Food: opinion of the scientific committee on food on the risk assessment of dioxins and dioxin-like PCBs in food, 30 May 2001. Update based on new scientific information available since the adoption of the SCF opinion of 22nd November 2000, European Commission, CS/CNTM/DIOXIN/20 final.

Schiffer, B., A. Daxenberger, K. Meyer, and H. H. D. Meyer. 2001. The fate of trenbolone acetate and melengestrol acetate after application as growth promoters in cattle: environmental studies. *Environ. Health Persp.* 109:1145–1151.

Stephany, R. W. 2001. Hormones in meat: different approaches in the EU and in the USA. *Acta Patholog. Microb.* 109:S357–S363.

Thomas, V., R. Sokolow, J. Fanelli, and T. Spiro. 1999. Effects of reducing lead in gasoline: an analysis of the international experience. *Environ. Sci. Technol.* 33:3942–3948.

Vicente, A., J. F. Arques, J. R. Villalbi, F. Centrich, E. Serrahima, and X. Llebaria X. 2004. Pesticides in the diet: adding pieces to the puzzle. *Gac. Sanit.* 18:425–430.

Waegeneers, N., J. C. Pizzolon, M. Hoenig, and L. de Temmerman. 2009. The European maximum level for cadmium in bovine kidneys is in Belgium only realistic for cattle up to 2 years of age. *Food Addit. Contam. A.* 26:326–332.

Yiridoe, E. K., S. B. Ankomah, and R. C. Martin. 2005. Comparison of consumer perceptions and preference toward organic versus conventionally produced foods: a review and update of the literature. *Renew. Agr. Food Sys.* 20:193–205.

Section IV

The Current Food Safety Status of Organic Meats

17 Prevalence of Food-Borne Pathogens in Organic Beef

Megan E. Jacob, J. Trent Fox, and T. G. Nagaraja

Abstract: Consumers have cited safer food as one of the reasons for selecting organic or naturally labeled products; however, existing research does not support differences in either the prevalence or antimicrobial susceptibility of food-borne pathogens from organic or conventional cattle production systems. Still, the impact of these systems on food-borne pathogen epidemiology within the cattle populations has only been evaluated for the past 10 years, and there are relatively few studies available. In addition, little is known of the influence of these production systems on the contamination of feed or other products, which may be a source of the organisms to cattle. Food-borne pathogen prevalence and antimicrobial resistance are important to human and animal health, and risk factors for increased prevalence and resistance should be identified. The impact of production practices throughout the entire system needs further evaluation before large, far-reaching conclusions can be made.

Keywords: food-borne pathogens; organic; cattle; food safety; antimicrobial resistance

17.1 INTRODUCTION

The beef industry is an important component of the US economy, which, according to the United States Department of Agriculture Economic Research Service (USDA-ERS), has a retail value of $73 billion (USDA-ERS, 2010a). Further, the United States produced approximately 26 billion pounds of beef in 2009, of which 7.2% was exported (USDA-ERS, 2010a). Culled beef and dairy cattle, as well as commercial steers and heifers, all contribute to the total pool of beef products. Recently, niche markets, which are described as markets providing unique commodities to a subsector of the general market, have begun to emerge as important players in US beef production. There are two major niches in today's cattle production systems, natural and organic, although other smaller systems also exist (Fox *et al.*, 2008a; Rawls *et al.*, 2002). The impact of these niche marketing practices on cattle health and performance are important questions to researchers (Wileman *et al.*, 2009). In addition, there has been increasing interest in the safety of food produced in unconventional, niche marketing systems. Consumers perceive organic food as healthier, better tasting, more environmentally friendly, and safer than food produced conventionally (Brennan *et al.*, 2003;

Organic Meat Production and Processing, First Edition. Edited by Steven C. Ricke, Ellen J. Van Loo, Michael G. Johnson, and Corliss A. O'Bryan.

Table 17.1 Selected diet and management practice specifications for conventional and organic livestock production systems in the United States.

Production practice	Conventional production	USDA-certified organic programs
Vaccines	Allowed	Allowed
Prebiotic/probiotic use (generally recognized as safe by the FDA)	Allowed	Allowed
Vitamin/trace minerals	Allowed	Allowed
Antibiotic use	Allowed	Not allowed; sick animals must be treated and removed from the program
Ionophore use	Allowed	Not allowed
Hormone implant use	Allowed	Not allowed
Oxytocin, post parturition therapy	Allowed	Allowed
Lidocaine, local anesthetic	Allowed	Allowed
Procaine, local anesthetic	Allowed	Allowed
Synthetic parasiticides	Allowed	Allowed w/o routine use
Mammalian byproduct feedstuffs	Allowed, many restrictions	Not Allowed
100% certified organic feedstuffs, including continuous access to organic pasture	Not required	Required
Maintained under continuous organic management from the last third of gestation	Not required	Required
Certification	–	USDA certification
Regulation	–	USDA auditing

Source: USDA-AMS, 2008; Fox *et al.*, 2008a.
FDA , US Food and Drug Administration; USDA, United States Department of Agriculture Economic Research Service.

Hughner *et al.*, 2007; Magnusson *et al.*, 2003). Still, to date, there are few available scientific data to support these conclusions (Brennan *et al.*, 2003; Sofos, 2008).

Livestock management practice differences between organic and conventional systems (Table 17.1), particularly in the use of growth hormones and antimicrobials, have largely fueled these perceptions and the debate of food safety. There have been several studies evaluating the effect of management system on the prevalence of food-borne pathogens (Table 17.2) and antimicrobial susceptibility of fecal organisms in cattle, which will be the focus of this chapter.

Cattle are reservoirs for many bacterial food-borne pathogens including *Escherichia coli* O157 and non-O157 Shiga toxin-producing *E. coli* (STEC), *Salmonella enterica*, *Campylobacter jejuni*, and *Listeria monocytogenes* (Callaway *et al.*, 2006; Hannon *et al.*, 2009; Ho *et al.*, 2007). Often, these organisms are commensals of the gastrointestinal tract and are shed in the feces that serve as a source of contamination for food and water. Direct contamination of beef carcasses often occurs through feces-contaminated cattle hides that are removed at

Table 17.2 Summary of food-borne pathogen prevalence in organic or conventional cattle or beef products.

Food-borne pathogen	Study population	Results	Reference
Escherichia coli O157	Beef cattle	No difference in prevalence between organic and conventional cattle at harvest	Reinstein *et al.*, 2009
E. coli O157	Dairy cattle	No difference in percentage of positive samples in organic and conventional dairy farms	Cho *et al.*, 2006a
E. coli O157 and Shiga toxin-producing *E. coli* (STEC)	Dairy cattle[a]	No difference in prevalence or risk of carrying *E. coli* O157 or STEC	Kuhnert *et al.*, 2005
Salmonella	Dairy cattle	No difference in prevalence between organic and conventional dairy farms	Fossler *et al.*, 2005a, 2005b
Salmonella	Beefsteak	No positive samples detected	Miranda *et al.*, 2009
Campylobacter	Dairy cattle	No difference in prevalence between organic and conventional dairy farms	Sato *et al.*, 2004
Listeria monocytogenes	Beefsteak	No difference in percentage of positive samples between organic and conventional production	Miranda *et al.*, 2009

[a]Study conducted in Europe.

harvest, although carcass contamination can also occur during evisceration. Hides, which can be contaminated at the feedlot/dairy, during transportation, or during lairage, may come into contact with the beef carcass during hide removal, shortly after stunning. Hide and carcass prevalence with food-borne pathogens like *E. coli* O157 can be highly correlated (Bosilevac *et al.*, 2009; Jacob *et al.*, 2010). Indirect contamination of milk, watersheds, or other vehicles can occur by shedding these organisms into the surrounding environment. The ability of these organisms to persist within and between animals throughout the production process ultimately facilitates their spread to the consumer.

The Centers for Disease Control and Prevention (CDC) attributes 3,504,693 human cases of food-borne illness to these bacterial pathogens every year (USDA-ERS, 2010b), although not all cases result from exposure to cattle or beef products. Still, direct contamination of beef products or indirect contamination of other food products and water, by cattle shedding these organisms, is a significant concern to beef producers and consumers alike. Countless epidemiological investigations and experimental trials have been conducted to better understand management, environmental, and animal risk factors that are associated with the presence of these food-borne pathogens within cattle. Several studies have evaluated the presence of these organisms in organically raised cattle, often comparing fecal prevalence estimates to those from conventional systems. The presence of food-borne pathogens and antimicrobial resistant bacteria in food animals are two parameters used to ascertain the safety of food, particularly beef products. Several review articles have highlighted food-borne pathogen prevalence and antimicrobial susceptibility when only a single management factor (e.g., hormone implants) differs between cattle populations (Fox *et al.*, 2008a; Jacob *et al.*, 2008). Such studies are often experimental and will have treatment groups with and without a given management factor; these studies do not compare all the aspects of organic or conventional systems, and as such, studies will not be discussed in detail here.

17.2 *E. COLI* O157 AND NON-O157 STEC

E. coli O157 and other STEC are a significant public health concern. These organisms, which are characterized by their ability to produce at least one *Shigella*-like toxin, are commonly shed from the feces of cattle and other ruminants, which serve as asymptomatic reservoirs for the organisms. Contamination of cattle carcasses with STEC including *E. coli* O157 has been associated with the presence of the organism on cattle hides and in cattle feces (Elder *et al.*, 2000; Fox *et al.*, 2008b; Jacob *et al.*, 2010). *E. coli* O157 is commonly associated with beef products, having been the cause of several high-profile ground beef recalls (USDA-FSIS, 2010c) and is considered an adulterant by the United States Department of Agriculture Food Safety and Inspection Service under the Federal Meat Inspection Act (USDA-FSIS, 2010d), requiring a regulatory program to test for the presence of the pathogen after processing. Recent legislation to amend the Federal Meat Inspection Act (2010) has proposed adding several non-O157 STECs, common to cattle, to the list of adulterants as well. Although often associated with beef, other less conventional food products such as produce, also have emerged as important transmission vehicles of STEC (Berger *et al.*, 2010).

The fecal prevalence of *E. coli* O157:H7 in conventionally raised cattle has been estimated between 10%–40%, yet can vary dramatically by cattle operation, pen, diet, season, age of the animal, and other factors (Dewell *et al.*, 2008; Renter *et al.*, 2008; Sargeant *et al.*, 2003). Human infections with *E. coli* O157:H7 are not frequent, occurring at an estimated 0.99 cases per 100,000 people in the population (CDC, 2010), which highlights the ability of pre- and postharvest interventions to reduce the pathogen load and ultimately pathogen risk in beef products. Still, human infections can lead to serious illness, particularly in children and the elderly, which can develop the potentially fatal hemolytic uremic syndrome (HUS; Armstrong *et al.*, 1996). Antimicrobial therapy is not indicated for treating human *E. coli* O157 infections, so antimicrobial susceptibility patterns in these isolates have little clinical significance. However, even without the threat of therapy failure in treating STEC infections, these organisms still have the potential to be a source of resistant genetic elements for horizontal gene transfer to other environments, and reporting antimicrobial susceptibility patterns is useful.

The epidemiology and ecology of non-O157 STECs in cattle is not as well understood; multiple serotypes previously associated with illness in humans have been isolated from healthy cattle or beef carcasses. Previous serotypes isolated include, but are not limited to, O103, O113, and O121 (Arthur *et al.*, 2002; Bettelheim, 2007). Still, poor diagnostic capabilities and a lack of understanding of the public health consequence of specific serotypes have hampered progress. For example, although the prevalence of non-O157 STECs on beef carcasses has been reported as greater than 50% (prior to in-plant interventions), the proportion of isolates that could cause human infection is not known (Arthur *et al.*, 2002). Often, prevalence estimates for specific non-O157 serotypes in cattle feces are not available. In humans, surveillance results reported by the CDC (2010) estimate 0.57 cases of non-O157 STEC infection per 100,000 people in the US population. Again, severe complications including HUS can occur with non-O157 STEC serotypes.

There are few reports of *E. coli* O157 in beef cattle produced organically. In a 2009 study, Reinstein *et al.* evaluated the prevalence and antimicrobial susceptibilities of *E. coli* O157:H7 in commercial beef cattle produced in one of the three systems, (1) organic, (2) natural, or (3) conventional. In this study, cattle feces and swabs from the rectoanal mucosal junction were sampled for *E. coli* O157:H7 at harvest, and the total prevalence reported. The rectoanal mucosa has previously been identified as the preferential location of colonization within the cattle gastrointestinal tract (Naylor *et al.*, 2003). Prevalence of the organism in 553 certified organic cattle was 14.8%, while it was 14.2% in 506 naturally raised cattle (Reinstein *et al.*, 2009). These prevalence estimates are well within the typical range reported for conventionally raised animals (Dewell *et al.*, 2008; Renter *et al.*, 2008; Sargeant *et al.*, 2003). Prevalence estimates in organic cattle from this study could not be meaningfully compared to those from cattle produced conventionally because samples were taken from different processing plants, and the pen and feedlot management effects could not be accounted for. However, when the antimicrobial susceptibility of *E. coli* O157:H7 isolates from cattle produced organically or naturally were compared to isolates obtained from cattle produced conventionally, no major differences were observed (Reinstein *et al.*, 2009).

In a series of papers published by Cho *et al.* (2006a, 2006b, 2007), a study of STEC prevalence and antimicrobial susceptibility in organic and conventional Minnesota dairy farms was described. Their data indicated that herd size was different between organic (smaller) and conventional dairy herds, which along with other important management differences, may impact the fecal shedding of STEC. In a 2-year study with a relatively small number of organic farms (8), *E. coli* O157 was detected in 7.4% (year 1) and 13.3% (year 2) of dairy fecal samples (Cho *et al.*, 2006a). Both estimates were numerically higher than those obtained from conventional farms (3.6% and 2.0%, respectively); however, total prevalence was not assessed statistically. In addition, *E. coli* O157 was found on 50% of organic dairy farms (4 out of 8), indicating its widespread presence within Minnesota herds. The antimicrobial resistance and virulence gene profile of *E. coli* O157 isolates were not different between organic and conventional farms in this study (Cho *et al.*, 2006a). The occurrence of Shiga toxin-producing bacteria, as determined by the presence of at least one Shiga toxin-encoding gene in a colony sweep of a bacterial culture, also was numerically higher (6.6%) in fecal samples from organic dairy herds compared to conventional dairy herds (5.1%; Cho *et al.*, 2006b). Again, there were no differences in the virulence gene profiles of STEC isolated from organic or conventional herds. Eighty-three STEC isolates (43 *E. coli* O157; 40 non-O157 STEC) were evaluated for antimicrobial susceptibility patterns against a panel of 17 antimicrobial agents (Cho *et al.*, 2007). Results indicated that more STEC isolates from conventional herds (62%) were resistant to at least one antimicrobial agent than isolates from organic herds (48%); however, there was no significant difference in the recovery of multidrug resistant isolates and no differences between the isolates overall.

A European study evaluated the prevalence of STEC in fecal samples from organic and conventionally farmed dairy cattle and found no differences between the two management styles, for prevalence or risk for carrying *E. coli* O157 or STEC (Kuhnert *et al.*, 2005). In this study, farms were matched for size, zone, and other traits; still, European cattle production, whether organic or conventional, will have differences to the US systems and must be interpreted cautiously. Furthermore, many studies have evaluated specific risk factors (i.e., antimicrobial use, hormone implant use, etc.) associated with *E. coli* O157 and other food-borne pathogen prevalence and antimicrobial susceptibility in cattle (Edrington *et al.*, 2006; Looper *et al.*, 2003; McAllister *et al.*, 2006). Although these studies do not specifically address differences between organic and more conventional feeding systems, inferences between different management styles can be made. Rarely have consistent findings been found with regard to antimicrobial or growth promoting hormone implants and *E. coli* O157 or STEC prevalence. These, along with the data mentioned previously, would lead us to believe that we would expect a similar prevalence distribution of *E. coli* O157 and STEC in cattle raised organically to what we find in conventional production systems.

Fecal generic *E. coli* isolates, not specifically STEC, have been the focus of several studies evaluating antimicrobial susceptibility differences in organisms from organic and conventional cattle production systems (Berge *et al.*, 2010; Miranda *et al.*, 2009; Sato *et al.*, 2005; Walk *et al.*, 2007). Because it is not known how representative these isolates would be of STEC, they are not covered in detail here, but may be worth considering. Findings were generally a mixed bag that ranged from no differences in the susceptibility of *E. coli* isolates between cattle produced organically versus conventionally, to more resistance in the conventionally produced isolates (Berge *et al.*, 2010; Miranda *et al.*, 2009; Sato *et al.*, 2005; Walk *et al.*, 2007).

17.3 *SALMONELLA*

S. enterica is one of the most common, yet serious food-borne pathogens in the United States. The CDC estimated the incidence of *Salmonella* cases in 2009 as 15.2 for every 100,000 people (CDC, 2010). *S. enterica*, which comprise over 2500 serotypes, can be associated with clinical disease in livestock, but often the organisms are shed asymptomatically. Not all serotypes result in severe human disease, and different serotypes are often associated with different food products (Callaway *et al.*, 2008). Within cattle, the reported prevalence of *Salmonella* has been highly variable; in feedlot cattle, prevalence has ranged from quite low (3.8%; Callaway *et al.*, 2006) to quite high in cattle treated for respiratory disease (74%; Alam *et al.*, 2009). In the Dairy 2007 study reported by the National Animal Health Monitoring System (NAHMS), which represented almost 80% of US dairy operations, 13.7% of healthy cows and 39.7% of operations tested were positive for *Salmonella* (USDA-APHIS, 2009). Unlike STEC infections, human and animal salmonellosis infections can be treated with antimicrobials, although not all infections require treatment. Food-borne illnesses due to multidrug-resistant *Salmonella* are a growing concern (Alcaine *et al.*, 2007); therefore, studies evaluating the risk factors associated with multidrug resistance are important and warranted. In the 2007 dairy study, 92.8% of isolates were susceptible to all antimicrobials evaluated, whereas 5.5% of isolates were considered multidrug resistant (USDA-APHIS, 2009).

Within organic dairy farms, the prevalence of *Salmonella* has been evaluated; unfortunately, no data exists for *Salmonella* prevalence within organic feedlot cattle. In addition, there is no information on the serotype distribution of *Salmonella* in organic cattle. In a large, multistate study with 129 organic (32) and conventional (97) dairy farms, *Salmonella* species were isolated from approximately 5% of fecal samples from cows and 4% of calves; 88% of farms had at least one cow with a positive *Salmonella* fecal sample (Fossler *et al.*, 2005a, 2005b). The prevalence of *Salmonella* in cows or calves was not different between organic and conventional farms (Fossler *et al.*, 2005a, 2005b). The same group published a report on antimicrobial susceptibility differences in *Salmonella* isolates from the group of samples (Ray *et al.*, 2006). For most of the 14 antimicrobial agents evaluated in this study, there were no significant differences in the percentage of isolates showing resistance; however, for at least streptomycin, there were more resistant isolates from conventional compared to organic farms. In addition, the percentage of isolates resistant to five or more antimicrobial agents was higher in conventional isolates (24.6%) compared to organic (11.5%), but statistically, this was only a trend ($P = 0.12$; Ray *et al.*, 2006). As a whole, 21.1% of *Salmonella* isolates in this study were resistant to five or more antimicrobials, which is considerably higher than that reported by the NAHMS Dairy study.

Salmonella prevalence was also examined in 150 raw beefsteak samples from cattle produced conventionally (75) and organically (75) (Miranda *et al.*, 2009). There were no positive samples detected in this study, so differences could not be determined. In general, as with STEC, there does not appear to be a difference in the prevalence of *Salmonella,* and there is very little data to support major differences in antimicrobial susceptibility of isolates between conventionally and organically raised beef.

17.4 *CAMPYLOBACTER*

Along with *Salmonella*, *Campylobacter* species are frequent bacterial causes of human food-borne illness in the United States. Most often, this infection results in diarrhea, sometimes with blood; occasionally, severe complications including Guillain-Barré syndrome have been associated with *Campylobacter* infection (Rees *et al.*, 1995). The CDC estimates that *Campylobacter* cases occur in 13.0 of every 100,000 people in the United States each year (CDC, 2010). Typically, persons with *Campylobacter* recover without treatment, but specific antimicrobial therapy may be given at the discretion of the provider.

Often, poultry products are implicated as the major route of human infection, but *Campylobacter* species have been reported throughout the beef production system. The prevalence of *Campylobacter* in the feces of cattle can vary widely by sampling time (Besser *et al.*, 2005); however, they appear to be present on most dairies and feedlots (Englen *et al.*, 2007; Hannon *et al.*, 2009; USDA-APHIS, 2009). In a longitudinal study of 20 feedlot cattle pens at a large commercial feedyard, total fecal-pen prevalence of *C. jejuni* ranged from 1.6% to 62.2% across a 5-month period (Besser *et al.*, 2005). The pen-floor fecal prevalence of *Campylobacter* species was 87% over seven Canadian feedlots (Hannon *et al.*, 2009). In the 2007 NAHMS dairy study, 33.7% of cows and 92.6% of operations were positive for *Campylobacter* by fecal culture (USDA-APHIS, 2009). In addition, among *C. jejuni* isolates, 36.6% were susceptible to all antimicrobials evaluated, and 2.2% were resistant to two or more antimicrobials (USDA-APHIS, 2009).

Although more extensively evaluated in the swine and poultry literature, there have been several reports of the prevalence and/or antimicrobial susceptibility of *Campylobacter* isolates from dairy cattle raised organically. Sato *et al.* (2004) conducted a study of 30 organic and 30 conventional dairy farms and feces were collected from cows and calves. Fecal prevalence, which was not different between management systems, was 26.7% and 29.1% for organic and conventional farms, respectively (Sato *et al.*, 2004). These prevalence estimates fit within the broad range previously reported for conventionally raised cattle. In addition, this study evaluated the antimicrobial susceptibility of isolates to four antimicrobials; there was no widespread resistance to ciprofloxacin, gentamicin, or erythromycin. Although tetracycline resistance in this study was 45%, there were no differences in resistance between management systems (Sato *et al.*, 2004). An evaluation of the antimicrobial susceptibility of *Campylobacter* isolates obtained from dairy herds throughout the United States also showed little difference in the proportion of isolates resistant to any given antimicrobial (Halbert *et al.*, 2006). Only one of the antimicrobials evaluated, tetracycline, resulted in significant differences in the proportion of isolates resistant; again, widespread resistance was reported but isolates from organic cattle had proportionally less resistance (49.3%) than those from conventional cattle (58.3%). These data, like that for other food-borne pathogens, suggest that cattle production system is not associated with the fecal prevalence of *Campylobacter*, and only rarely with the antimicrobial resistance of isolates. There is no information on the prevalence of *Campylobacter* in retail organic beef products.

17.5 *LISTERIA MONOCYTOGENES*

L. monocytogenes is the causative agent of listeriosis in people, which is often characterized by fever, muscle aches, and diarrhea. Unfortunately, serious clinical disease and complications can occur in the elderly, immunocomprosmised, and pregnant individuals (Freitag *et al.*, 2009). Listeriosis has several unique properties including a high case-fatality rate (20%–30%), a long incubation period (can be greater than 30 days), and the ability to cause fetal distress or death in human neonates (Swaminathan & Gerner-Smidt, 2007). Many different food products have been associated with human outbreaks in the United States; ready to eat foods, especially cooked and refrigerated protein foods, are frequently associated with *L. monocytogenes* cases (Lianou & Sofos, 2007). Still, human infections with *Listeria* are not frequent, occurring at an estimated 0.34 cases per 100,000 people in the US population in 2009 (CDC, 2010).

Listeria species are ubiquitous in the environment (Freitag *et al.*, 2009) and have been previously reported in the feces of both feedlot and dairy cattle (Callaway *et al.*, 2006; Ho *et al.*, 2007; Nightingale *et al.*, 2004). Previous work by Ho *et al.* (2007) showed that *L. monocytogenes* prevalence varies greatly over time with both outbreak and sporadic cases in dairy cattle; the presence of the pathogen was associated with feed contamination with the organism. *L. monocytogenes* have also been isolated from the hides and carcasses of cattle at slaughter (Guerini *et al.*, 2007; Rivera-Betancourt *et al.*, 2004). There are no published reports evaluating the prevalence of *Listeria* in the feces of organic cattle. A study evaluating the presence of *L. monocytogenes* in beefsteaks were compared using samples from certified organic and conventionally produced meat (Miranda *et al.*, 2009). The percent of samples positive was not different between production systems and was 29.3% in conventional and 36% in organic beefsteaks. Antimicrobial susceptibility patterns to nine antimicrobials were not different in the *L. monocytogenes* isolates obtained from conventional and organic samples in this study (Miranda *et al.*, 2009). More research is needed to understand the epidemiology of *Listeria* in cattle and beef products, especially in establishing the prevalence and antimicrobial susceptibility of these organisms in organic cattle.

17.6 CONCLUSIONS

Consumers of organic products have cited safer food as one of the reasons for selecting organic or naturally labeled products over food produced conventionally (Harper & Makatouni, 2002; O'Donovan & McCarthy, 2002). Traditionally, "food safety" has been left open to interpretation and undefined (Hughner *et al.*, 2007). Two facets of food safety likely to influence consumers' perception of a safe beef product are the presence of food-borne pathogens (e.g., *E. coli* O157, non-O157 STEC, *Salmonella, Campylobacter,* or *Listeria*), and the presence of antimicrobial resistance in these and other organisms. As previous reviews have suggested, there does not appear to be enough existing research to support differences in the prevalence of food-borne pathogens or antimicrobial resistance in food-borne pathogens between isolates obtained from organic or conventional cattle production systems (Wilhelm *et al.*, 2009). Still, the impact of these systems on food-borne pathogen epidemiology within the cattle populations has only been evaluated during the past 10 years, and there are relatively few studies available. In addition, we know little of the influence of these production systems by the contamination of feed or other products, which may be a source of the organisms to cattle. Previous studies have shown that *E. coli* O157, *Salmonella*, and other organisms can be cultured from cattle feed (Dargatz *et al.*, 2005; Dodd *et al.*, 2003). Others have speculated that organic farms would have an increased *Salmonella* prevalence because of the use of manure fertilizers for crops (Fossler *et al.*, 2005a). Unfortunately, at this time, there is not enough scientific evidence to support or refute this hypothesis; however, it seems likely the presence of these organisms in the feed would perpetuate their transmission throughout the production system.

No report has found a difference in the presence or prevalence of specific food-borne pathogens in cattle or beef products between organic and conventional production systems. Perhaps this is not surprising given that studies of single management factors (in-feed antimicrobial use, growth hormone implants, etc.) that represent the major differences between organic and conventional production systems rarely, and inconsistently, report finding any differences in prevalence (Fox *et al.*, 2008a). Often, the expected prevalence of these organisms within cattle production environments is highly variable and the reports of prevalence in organic cattle generally fall well within the expected range. In addition, contamination of the animal's carcass at slaughter is usually highly associated with the contamination by the animal's hide. The processes that occur once the animal is harvested are more likely to impact the contamination of the final beef product, and are most likely differences in processing plant practices rather than organic or conventional production systems.

The role or impact of food animals in the reservoir of antimicrobial resistance has been highly debated. One regulation that distinguishes organic and natural food animal production from conventional production is the use of antibiotics. Antimicrobial use is not permitted in organic food animal production systems, and animals that require therapy must be treated and moved to a conventional production system (USDA-AMS, 2008). In conventional food animal production, antimicrobials are given for both disease therapy and prevention (USDA-APHIS, 1999).

The use of any antimicrobial is thought to be a selection pressure for the development of antimicrobial resistance; therefore, the antimicrobial susceptibility of isolates from these two different production systems is of interest. The susceptibility of STEC, *Salmonella*, *Campylobacter*, and *Listeria* isolates has been examined in several of the studies mentioned previously. No clear relationship between production system and resistance can be observed from these data, but occasionally, increased resistance is seen in isolates from conventional systems. Much more work is needed to clarify this possible relationship. Antimicrobial susceptibility of food-borne pathogens is important to human and animal health; risk factors for increased resistance should be identified. In addition, as with food-borne pathogen prevalence, the impact of production practices throughout the entire system (e.g., feed production, etc.) need to be evaluated before large, far reaching conclusions can be made.

REFERENCES

Alam, M. J., D. G. Renter, S. E. Ives, D. U. Thomson, M. W. Sanderson, L. C. Hollis, and T. G. Nagaraja. 2009. Potential associations between fecal shedding of *Salmonella* in feedlot cattle treated for apparent respiratory disease and subsequent adverse health outcomes. *Vet. Res.* 40:02.

Alcaine, S. D., L. D. Warnick, and M. Wiedmann. 2007. Antimicrobial resistance in nontyphoidal *Salmonella*. *J. Food Prot.* 70:780–790.

Armstrong, G. L., J. Hollingsworth, and J. G. Morris Jr. 1996. Emerging foodborne pathogens: *Escherichia coli* O157:H7 as a model of entry of a new pathogen into the food supply of the developed world. *Epidemiol. Rev.* 18:29–51.

Arthur, T. M., G. A. Barkocy-Gallagher, M. Rivera-Betancourt, and M. Koohmaraie. 2002. Prevalence and characterization of non-O157 Shiga toxin-producing *Escherichia coli* on carcasses in commercial beef cattle processing plants. *Appl. Environ. Microbiol.* 68:4847–4852.

Berge, A. C., D. D. Hancock, W. M. Sischo, and T. E. Besser. 2010. Geographic, farm, and animal factors associated with multiple antimicrobial resistance in fecal *Escherichia coli* isolates from cattle in the western United States. *J. Am. Vet. Med. Assoc.* 236:1338–1344.

Berger, C. N., S. V. Sodha, R. K. Shaw, P. M. Griffin, D. Pink, P. Hand, and G. Frankel. 2010. Fresh fruit and vegetables as vehicles for the transmission of human pathogens. *Environ. Microbiol.* 12:2385–2397.

Besser, T. E., J.T. LeJeune, D. H. Rice, J. Berg, R. P. Stilborn, K. Kaya, W. Bae, and D. D. Hancock. 2005. Increasing prevalence of *Campylobacter jejuni* in feedlot cattle throughout the feeding period. *Appl. Environ. Microbiol.* 71:5752–5758.

Bettelheim, K. A. 2007. The non-O157 Shiga-toxigenic (verocytotoxigenic) *Escherichia coli*; under-rated pathogens. *Crit. Rev. Microbiol.* 33:67–87.

Bosilevac, J. M., T. M. Arthur, J. L. Bono, D. M. Brichta-Harhay, N. Kalchayanand, D. A. King, S. D. Shackelford, T. M. Wheeler, and M. Koohmaraie. 2009. Prevalence and enumeration of *Escherichia coli* O157:H7 and *Salmonella* in U.S. abattoirs that process fewer than 1000 head of cattle per day. *J. Food Prot.* 72:1272–1278.

Brennan, C., K. Gallagher, and M. McEachern. 2003. A review of the 'consumer interest' in organic meat. *Int. J. Consumer Stud.* 27:381–394.

Callaway, T. R., T. S. Edrington, A. D. Brabban, J. E. Keen, R. C. Anderson, M. L. Rossman, M. J. Engler, K. J. Genovese, B. L. Gwartney, J. O. Reagan, T. L. Poole, R. B. Harvey, E. M. Kutter, and D. J. Nisbet. 2006. Fecal prevalence of *Escherichia coli* O157, *Salmonella*, *Listeria*, and bacteriophage infecting *E. coli* O157:H7 in feedlot cattle in the southern plains region of the United States. *Foodborne Pathog. Dis.* 3:234–244.

Callaway, T. R., T. S. Edrington, R. C. Anderson, J. A. Byrd, and D. J. Nisbet. 2008. Gastrointestinal microbial ecology and the safety of our food supply as related to *Salmonella*. *J. Anim. Sci.* 86:E163–E172.

Centers for Disease Control and Prevention (CDC). 2010. Preliminary FoodNet data on the incidence of infection with pathogens transmitted commonly through food – 10 States, 2009. *Morb. Mortal. Wkly. Rep.* 59:418–422.

Cho, S., C. P. Fossler, F. Diez-Gonzalez, S. J. Wells, C. W. Hedberg, J. B. Kaneene, P. L. Ruegg, L. D. Warnick, and J. B. Bender. 2007. Antimicrobial susceptibility of Shiga toxin-producing *Escherichia coli* isolated from organic dairy farms, conventional dairy farms, and county fairs in Minnesota. *Foodborne Pathog. Dis.* 4:178–186.

Cho, S., J. B. Bender, F. Diez-Gonzalez, C. P. Fossler, C. W. Hedberg, J. B. Kaneene, P. L. Ruegg, L. D. Warnick, and S. J. Wells. 2006a. Prevalence and characterization of *Escherichia coli* O157 isolates from Minnesota dairy farms and county fairs. *J. Food Prot.* 69:252–259.

Cho, S., F. Diez-Gonzalez, C. P. Fossler, S. J. Wells, C. W. Hedberg, J. B. Kaneene, P. L. Ruegg, L. D. Warnick, and J. B. Bender. 2006b. Prevalence of Shiga toxin-encoding bacteria and shiga toxin-producing *Escherichia coli* isolates from dairy farms and county fairs. *Vet. Microbiol.* 118: 289–298.

Dargatz, D. A., R. A. Strohmeyer, P. S. Morley, D. R. Hyatt, and M. D. Salman. 2005. Characterization of *Escherichia coli* and *Salmonella enterica* from cattle feed ingredients. *Foodborne Pathog. Dis.* 2: 341–347.

Dewell, G. A., C. A. Simpson, R. D. Dewell, D. R. Hyatt, K. E. Belk, J. A. Scanga, P. S. Morley, T. Grandin, G. C. Smith, D. A. Dargatz, B. A. Wagner, and M. D. Salman. 2008. Impact of transportation and lairage on hide contamination with *Escherichia coli* O157 in finished beef cattle. *J. Food Prot.* 71: 1114–1118.

Dodd, C. C., M. W. Sanderson, J. M. Sargeant, T. G. Nagaraja, R. D. Oberst, R. A. Smith, and D. D. Griffin. 2003. Prevalence of *Escherichia coli* O157 in cattle feeds in Midwestern feedlots. *Appl. Environ. Microbiol.* 69:5243–5247.

Edrington, T. S., M. L. Looper, S. E. Duke, T. R. Callaway, K. J. Genovese, R. C. Anderson, and D. J. Nisbet. 2006. Effect of ionophore supplementation on the incidence of *Escherichia coli* O157:H7 and *Salmonella* and antimicrobial susceptibility of fecal coliforms in stocker cattle. *Foodborne Pathog. Dis.* 3: 284–291.

Elder, R. O., J. E. Keen, G. R. Siragusa, G. A. Barkocy-Gallagher, M. Koohmaraie, and W. W. Laegreid. 2000. Correlation of enterohemorrhagic *Escherichia coli* O157 prevalence in feces, hides, and carcasses of beef cattle during processing. *Proc. Natl. Acad. Sci. U.S.A.* 97:2999–3003.

Englen, M. D., A. E. Hill, D. A. Dargatz, S. R. Ladely, and P. J. Fedorka-Cray. 2007. Prevalence and antimicrobial resistance of *Campylobacter* in U. S. dairy cattle. *J. Appl. Microbiol.* 102:1570–1577.

Federal Meat Inspection Act. 2010. Bill to Amend the Federal Meat Inspection Act. 111th Congress, 2nd ses. S.3435

Fossler, C. P., S. J. Wells, J. B. Kaneene, P. L. Ruegg, L. D. Warnick, J. B. Bender, L. E. Eberly, S. M. Godden, and L. W. Halbert. 2005a. Herd-level factors associated with the isolation of *Salmonella* in a multi-state study of conventional and organic dairy farms I. *Salmonella* shedding in cows. *Prev. Vet. Med.* 70:257–277.

Fossler, C. P., S. J. Wells, J. B. Kaneene, P. L. Ruegg, L. D. Warnick, J. B. Bender, L. E. Eberly, S. M. Godden, and L. W. Halbert. 2005b. Herd-level factors associated with the isolation of *Salmonella* in a multi-state study of conventional and organic dairy farms II. *Salmonella* shedding in calves. *Prev. Vet. Med.* 70:279–291.

Fox, J. T., D. G. Renter, M. W. Sanderson, A. L. Nutsch, X. Shi, and T. G. Nagaraja. 2008b. Associations between the presence and magnitude of *Escherichia coli* O157 in feces at harvest and contamination of preintervention beef carcasses. *J. Food. Prot.* 71:1761–1767.

Fox, J. T., S. Reinstein, M. E. Jacob, and T. G. Nagaraja. 2008a. Niche marketing production practices for beef cattle in the United States and prevalence of foodborne pathogens. *Foodborne Pathog. Dis.* 5:559–569.

Freitag, N. E., G. C. Port, and M. D. Miner. 2009. *Listeria monocytogenes* – from saprophyte to intracellular pathogen. *Nat. Rev. Microbiol.* 7:623–637.

Guerini, M. N., D. M. Brichta-Harhay, S. D. Shackelford, T. M. Arthur, J. M. Bosilevac, N. Kalchayanand, T. L. Wheeler, and M. Koohmaraie. 2007. *Listeria* prevalence and *Listeria monocytogenes* serovar diversity at cull cow and bull processing plants in the United States. *J. Food Prot.* 70:2578–2582.

Halbert, L. W., J. B. Kaneene, P. L. Ruegg, L. D. Warnick, S. J. Wells, L. S. Mansfield, C. P. Fossler, A. M. Campbell, A. M. Geiger-Zwald. 2006. Evaluation of antimicrobial susceptibility patterns in *Campylobacter* spp. Isolated from dairy cattle and farms managed organically and conventionally in the midwestern and northeastern United States. *J. Am. Vet. Med. Assoc.* 228:1074–1081.

Hannon, S. J., B. Allan, C. Waldner, M. L. Russell, A. Potter, L.A. Babiuk, and H. G. G. Townsned. 2009. Prevalence and risk factor investigation of *Campylobacter* species in beef cattle feces from seven large commercial feedlots in Alberta, Canada. *Can. J. Vet. Res.* 73:275–282.

Harper, G. C. and A. Makatouni. 2002. Consumer perception of organic food production and farm animal welfare. *Br. Food J.* 104:287–299.

Ho, A. J., R. Ivanek, Y. T. Grohn, K. K. Nightingale, and M. Wiedmann. 2007. *Listeria monocytogenes* fecal shedding in dairy cattle shows high levels of day-to-day variation and includes outbreaks and sporadic cases of shedding of specific *L. monocytogenes* subtypes. *Prev. Vet. Med.* 80:287–305.

Hughner, R. S., P. McDonagh, A. Prothero, C. J. Shultz II, and J. Stanton. 2007. Who are organic food consumers? A compilation and review of why people purchase organic food. *J. Consumer Behav.* 6: 94–110.

Jacob, M. E., D. G. Renter, and T. G. Nagaraja. 2010. Animal- and truckload-level associations between *E. coli* O157:H7 in feces and on hides at harvest and contamination of pre-evisceration beef carcasses. *J. Food Prot.* 73:1030–1037.

Jacob, M. E., J. T. Fox, S. L. Reinstein, and T. G. Nagaraja. 2008. Antimicrobial susceptibility of foodborne pathogens in organic or natural production systems: an overview. *Foodborne Pathog. Dis.* 5:721–730.

Kuhnert, P., C. R. Dubosson, M. Roesch, E. Homfeld, M. G. Doherr, and J. W. Blum. 2005. Prevalence and risk-factor analysis of Shiga toxigenic *Escherichia coli* in faecal samples of organically and conventionally farmed dairy cattle. *Vet. Microbiol.* 109:37–45.

Lianou, A., and J. N. Sofos. 2007. A review of the incidence and transmission of *Listeria monocytogenes* in ready-to-eat products in retail and food service environments. *J. Food Prot.* 70:2172–2198.

Looper, M. L., C. F. Rosenkrans Jr, G. E. Aiken, and T. S. Edrington. 2003. *Escherichia coli* and *Salmonella* in beef cattle grazing tall fescue. *Ark. Exp. Stat. Res. Ser.* 509:58–60.

Magnusson, M. K., A. Arvola, U. K. Hursti, L. Aberg, and P. O. Sjoden. 2003. Choice of organic foods is related to perceived consequences for human health and to environmentally friendly behaviour. *Appetite*. 40:109–117.

McAllister, T. A., S. J. Bach, K. Stanford, and T. R. Callaway. 2006. Shedding of *Escherichia coli* O157:H7 by cattle fed diets containing monensin or tylosin. *J. Food. Prot.* 69:2075–2083.

Miranda, J. M., A. Mondragon, B. I. Vazquez, C. A. Fente, A. Cepeda, and C. M. Franco. 2009. Influence of farming methods on microbiological contamination and prevalence of resistance to antimicrobial drugs in isolates from beef. *Meat Sci.* 82:284–288.

Naylor, S. W., J. C. Low, T. E. Besser, A. Mahajan, G. J. Gunn, M. C. Pearce, I. J. McKendrick, D. G. E. Smith, and D. L. Gally. 2003. Lymphoid follicle-dense mucosa at the terminal rectum in the principal site of colonization of enterohemorrahagic *Escherichia coli* O157:H7 in the bovine host. *Infect. Immun.* 71:1505–1512.

Nightingale, K. K., Y. H. Schukken, C. R. Nightingale, E. D. Fortes, A. J. Ho, Z. Her, Y. T. Grohn, P. L. McDonough, and M. Wiedmann. 2004. Ecology and transmission of *Listeria monocytogenes* infecting ruminants in the farm environment. *Appl. Environ. Microbiol.* 70:4458–4467.

O'Donovan, P. and M. McCarthy. 2002. Irish consumer preference for organic meat. *Br. Food J.* 104:353–370.

Rawls, E., L. Meyer, and K. Burdine. 2002. Niche marketing of cattle/beef. Managing for today's cattle market and beyond. pp. 1–4. Available at: http://www.agecon.ksu.edu/livestock/Master%20Web%20Page%20Folder/Extension%20Bullentins.Research/ManageForTodaysCattleMkt/Niche_Marketing.pdf (accessed September 16, 2010).

Ray, K. A., L. D. Warnick, R. M. Mitchell, J. B. Kaneene, P. L. Ruegg, S. J. Wells, C. P. Fossler, L. W. Halbert, and K. May. 2006. Antimicrobial susceptibility of *Salmonella* from organic and conventional dairy farms. *J. Dairy Sci.* 89:2038–2050.

Rees, J. H., S. E. Soudain, N. A. Gregson, and R. A. C. Hughes. 1995. *Campylobacter jejuni* infection and Guillain-Barre syndrome. *N. Engl. J. Med.* 333:1374–1379.

Reinstein, S., J. T. Fox, X. Shi, M. J. Alam, D. G. Renter, and T. G. Nagaraja. 2009. Prevalence of *Escherichia coli* O157:H7 in organically and naturally raised beef cattle. *Appl. Environ. Microbiol.* 75: 5421–5423.

Renter, D. G., D. R. Smith, R. King, R. Stilborn, J. Berg, J. Berezowski, and M. McFall. 2008. Detection and determinants of *Escherichia coli* O157:H7 in Alberta feedlot pens immediately prior to slaughter. *Can. J. Vet. Res.* 72:217–227.

Rivera-Betancourt, M., S. D. Shackelford, T. M. Arthur, K. E. Westmoreland, G. Bellinger, M. Rossman, J. O. Reagan, and M. Koohmaraie. 2004. Prevalence of *Escherichia coli* O157:H7, *Listeria monocytogenes*, and *Salmonella* in two geographically distant commercial beef processing plants in the United States. *J. Food Prot.* 67:295–302.

Sargeant, J. M., M. W. Sanderson, R. A. Smith, and D. D. Griffin. 2003. *Escherichia coli* O157 in feedlot cattle feces and water in four major feeder-cattle states in the USA. *Prev. Vet. Med.* 61:127–135.

Sato, K., P. C. Bartlett, J. B. Kaneene, and F. P. Downes. 2004. Comparison of prevalence and antimicrobial susceptibilities of *Campylobacter* spp. isolates from organic and conventional dairy herds in Wisconsin. *Appl. Environ. Microbiol.* 70:1442–1447.

Sato, K., P. C. Bartlett, and M. A. Saeed. 2005. Antimicrobial susceptibility of *Escherichia coli* isolates from dairy farms using organic versus conventional production methods. *J. Am. Vet. Med. Assoc.* 226: 589–594.

Sofos, J. N. 2008. Challenges to meat safety in the 21st century. *Meat Sci.* 78:3–13.

Swaminathan, B. and P. Gerner-Smidt. 2007. The epidemiology of human listeriosis. *Microbes Infect.* 9:1236–1243.

United States Department of Agriculture – Agriculture Marketing Service. 2008. The national organic program. Washington, DC: Agricultural Marketing Service/United States Department of Agriculture. Available at: http://www.ams.usda.gov/nop/indexIE.htm (accessed November, 2008).

United States Department of Agriculture – Animal and Plant Health Inspection Service (USDA-APHIS). 2009. *Salmonella* and *Campylobacter* on U.S. dairy operations, 1996–2007. Available at: http://www.aphis.usda.gov/animal_health/nahms/dairy/index.shtml (accessed November 12, 2010).

United States Department of Agriculture – APHIS/VS/CEAH/CEI. 1999. Antimicrobial resistance issues in animal agriculture. Available at: http://www.aphis.usda.gov/animal_health/emergingissues/index.shtml. (accessed November 20, 2010).

United States Department of Agriculture Economic Research Service. 2010a. U.S. beef and cattle industry: background statistics and information. Available at: http://www.ers.usda.gov/news/BSECoverage.htm (accessed September 1, 2010).

United States Department of Agriculture Economic Research Service. 2010b. Foodborne illness cost calculator. Available at: http://www.ers.usda.gov/data/foodborneillness (accessed September 1, 2010).

United States Department of Agriculture Food Safety and Inspection Service. 2010c. FSIS Recalls. Available at: http://www.fsis.usda.gov/FSIS_Recalls/ (accessed October 3, 2010).

United States Department of Agriculture Food Safety and Inspection Service. 2010d. Federal Meat Inspection Act. Available at: http://www.fsis.usda.gov/regulations_&_Policies/Federal_Meat_Inspection_Act/index.asp (accessed October 3, 2010).

Walk, S. T., J. M. Mladonicky, J. A. Middleton, A. J. Heidt, J. R. Cunningham, P. Bartlett, K. Sato, and T. S. Whittam. 2007. Influence of antibiotic selection on genetic composition of *Escherichia coli* populations from conventional and organic dairy farms. *Appl. Environ. Microbiol.* 73:5982–5989.

Wileman, B. W., D. U. Thomson, C. D. Reinhardt, and D. G. Renter. 2009. Analysis of modern technologies commonly used in beef cattle production: conventional versus nonconventional production using meta-analysis. *J. Anim. Sci.* 87:3418–3426.

Wilhelm, B., A. Rajic, L. Waddell, S. Parker, J. Harris, K. C. Roberts, R. Kydd, J. Greig, and A. Baynton. 2009. Prevalence of zoonotic or potentially zoonotic bacteria, antimicrobial resistance, and somatic cell counts in organic dairy production: current knowledge and research gaps. *Foodborne Pathog. Dis.* 6:525–539.

18 Incidence of Food-Borne Pathogens in Organic Swine

Marcos H. Rostagno and Paul D. Ebner

Abstract: Organic products are often assumed to be healthier and safer than conventional products. The preharvest microbial safety of conventional pork production has been extensively studied, whereas research in organic pork production is very scarce. Differences have been reported on the prevalence of bacterial food-borne pathogens in alternative and conventional pork production systems; however, findings have been inconsistent. Bacteria isolated from antimicrobial-free swine farms are often less likely to be resistant to antimicrobials. However, antimicrobial-resistant bacteria have also been isolated from organic pork. In organic pork production, animals are particularly at risk of parasitism due to outdoor rearing and ban of prophylactic medication. In summary, it is impossible to draw general conclusions regarding preharvest food safety risks in organic versus conventional pork production. From the limited number of published data, it seems that there is no clear evidence that organically produced pork is safer than conventionally produced pork, and vice versa.

Keywords: Epidemiology; food safety; organic pork; pathogens; swine

18.1 INTRODUCTION

18.1.1 Organic pork and food safety: why does it matter?

Niche markets such as those for organic foods are rapidly increasing worldwide (Box 18.1). Although consumer behavior is affected by various issues, organic products are often purchased because it is believed that they are healthier and safer than conventional products.

> **Box 18.1 Niche Pork**
>
> Niche pork is pork with attributes that appeal to a specific market segment (i.e., niche market), or consumers that want something different. Organic pork is an example of niche pork.

Organic Meat Production and Processing, First Edition. Edited by Steven C. Ricke, Ellen J. Van Loo, Michael G. Johnson, and Corliss A. O'Bryan.

In the case of products of animal origin, additional purchase reasons for consumers include the perception of reduced environmental impact and superior animal welfare (Blokhuis *et al.*, 2003; Hermansen, 2003; Honeyman *et al.*, 2006). Marketing of organic animal products usually stresses these issues, but often neglects potential food safety impacts, because they are in conflict with the consumer-held perception of the health benefits of organic products (Magkos *et al.*, 2006; Siekkinen *et al.*, 2006; Kijlstra *et al.*, 2009). There are many implications (direct and indirect) of the organic principles for food safety, and as consumer interest and demand for organic products increase, a better understanding of the microbial food safety of organic production becomes necessary.

Production systems that are considered more animal welfare friendly, such as free-range and organic systems, allow outdoor access to farm animals, and therefore, may create new or reintroduce old public health risks. A more open farming system can result in increased risk of transfer of zoonotic pathogens to livestock, negatively affecting food safety. Compared to conventional (indoor) systems, outdoor production systems are inherently less controllable from a hygiene (microbial) point of view. In current organic production systems, pigs are kept at lower stocking densities, with outdoor access and bedding, and fed organic feed. Organic feed for the animals is grown without chemical pesticides or artificial fertilizers, is not produced from genetically modified seeds, and manure used for fertilizer to reuse grain for feed should come from organically kept animals (Hovi *et al.*, 2003; von Borell & Sorensen, 2004; Kijlstra *et al.*, 2009; Merrigan *et al.*, 2010).

Public risk perception and demand for safer food are important factors shaping agricultural production practices in the world today. However, despite documented food safety concerns, little attempt has been made to elicit consumers' subjective risk judgment for a range of food safety hazards or to identify predictive factors of food safety risks. Consumers often associate consumption of organic products to reduced food safety hazards. Although health is seen as a vital part of animal welfare, there is growing evidence that animal health in organic farms is not always better than in conventional livestock production systems (Williams & Hammitt, 2001; Hovi *et al.*, 2003). Consumers of organic products are often not aware of these issues. In general, food safety problems associated with alternative animal production systems have been controversial and require serious consideration. Organic livestock production has experienced a rapid growth in the past decade, generating many challenges for the animal production industry. While emphasizing the importance of a systems approach to animal health and welfare protection, organic livestock production standards place considerable restrictions on the use of many animal health inputs that are routinely used in conventional production systems. According to the European Union regulations on organic farming, the use of antimicrobials is restricted, and the withdrawal times after treatments are doubled. In contrast to European regulations, the use of antimicrobials is prohibited in organic livestock production in the United States, and when an antimicrobial treatment is applied, the animal loses its organic status and has to be sold as a conventional product (von Borell & Sorensen, 2004; Winter & Davis, 2006; Jacob *et al.*, 2008; Merrigan *et al.*, 2010).

Obviously, high expectations and major questions about food quality and safety exist for organic pork production systems. Outdoor exposure, in combination with minimized/prohibited use of antimicrobial treatments, presents a challenge or even a risk with respect to zoonotic diseases. The consumer perception of organic livestock systems as being welfare-friendly and producing safe products must be verified by collecting unbiased data from such systems. In this review, we aim at analyzing and discussing the current published data available on organic pork safety.

18.2 INCIDENCE OF BACTERIAL FOOD-BORNE PATHOGENS

The preharvest microbial food safety of conventional pork production has been extensively studied and reviewed (Sanchez *et al.*, 2007; Fosse *et al.*, 2009; Rostagno, 2009). However, research in organic pork production is very scarce, perhaps due to the relatively recent growth in popularity of organic animal production. For instance, Young *et al.* (2009) attempted to develop a systematic review and meta-analysis study to compare the prevalence of bacterial food-borne pathogens in organic and conventional swine production systems and concluded that limited or inconsistent research was available for a comparison. Nevertheless, many believe that alternative, more "natural," pork production systems (such as organic) are associated with decreased risk of pathogens, including zoonotic pathogens. This presumed association is usually based on the perception that animals experience reduced stress when raised in alternative systems. Although stress in farm animals (Box 18.2) has been linked to increased risk of infection with food-borne pathogens (Rostagno, 2009), the increased outdoor exposure may generate favorable ecological and epidemiological conditions for the occurrence and persistence of pathogens. In fact, several studies have shown the ability of *Salmonella* to survive/persist in the outdoor pig farm environment, including a variety of environmental samples (Callaway *et al.*, 2005; Jensen *et al.*, 2006; Rodriguez *et al.*, 2006).

Box 18.2 Stress

Stress represents the reaction of the body (a biological response) to stimuli that disturb its normal physiological equilibrium or homeostasis. Stressors can be defined as conditions that endanger, or are perceived to endanger, the homeostasis of an individual and, therefore, cause a stress response or reaction.

Differences have been reported in studies investigating the prevalence of *Salmonella* in alternative and conventional pork production systems. However, findings have been inconsistent. In Europe, Van der Wolf *et al.* (2001) reported that *Salmonella* seroprevalence in free range finishers was significantly higher than in intensively housed finishers in the Netherlands, whereas Zheng *et al.* (2007) observed no difference in the proportion of *Salmonella* seropositive animals in organic, outdoor, and indoor pig farms in Denmark. In the United States, Gebreyes *et al.* (2008) reported significantly higher seroprevalence of *Salmonella* in outdoor herds than in conventional/indoor herds. Also in the United States, studies have been conducted, focused on the bacteriological prevalence of *Salmonella* (i.e., based on the isolation of the bacteria) in different pork production systems. Gebreyes *et al.* (2006) reported that *Salmonella* prevalence was significantly higher in outdoor systems than in conventional/indoor systems, with an odds ratio of 4.23. In the same study, more *Salmonella* were recovered from the carcasses of the outdoor pigs than from the conventional/indoor ones, with an odds ratio of 2.34. In contrast, Thakur *et al.* (2007) reported a significantly higher *Salmonella* prevalence in pigs from conventional/indoor production systems than in pigs from outdoor systems. However, significantly higher number of *Salmonella* was isolated at the pre- and post-evisceration stages from carcasses of outdoor than from indoor reared pigs (Table 18.1). This observation highlights the critical role played by the abattoir in the

Table 18.1 Proportion of *Salmonella* and *Campylobacter* isolates recovered from feces and carcasses of pigs, originating from indoor and outdoor production systems.

Pathogen	Location	Sample source	Indoor	Outdoor
Salmonella[a]	Farm	Feces	99.6%	0.4%
	Slaughter	Carcass pre-evisceration	10.4%	89.6%
	Slaughter	Carcass post-evisceration	14.1%	85.9%
	Slaughter	Carcass post-chill	37.5%	62.5%
Campylobacter[b]	Farm	Feces	55.7%	44.3%
	Slaughter	Carcass pre-evisceration	9.4%	90.6%
	Slaughter	Carcass post-evisceration	44.2%	55.8%
	Slaughter	Carcass post-chill	0%	100%

[a]Thakur *et al.* (2007).
[b]Gebreyes *et al.* (2005).

contamination of carcasses, particularly from pigs raised in alternative production systems. In general, alternative/outdoor pigs are processed in what the USDA classifies as "small" or "very small" abattoirs. These types of abattoirs have traditionally had difficulties establishing and maintaining efficient hazard analysis critical control point (HACCP) programs.

Thakur and Gebreyes (2005) reported significantly higher prevalence of *Campylobacter coli* in outdoor pigs (77.3%) than in conventional/indoor ones (27.6%) at the nursery stage; however, no differences were observed between the production systems in the prevalence at the finishing stage. In another study, Gebreyes *et al.* (2005) also reported no difference of *C. coli* prevalence in finishers. However, at the post-chill carcass processing stage, *C. coli* were isolated only from the outdoor reared pigs (Table 18.1), further suggesting, in agreement with Thakur *et al.* (2007), that the abattoir plays a critical role in the dissemination of pathogens and contamination of carcasses from pigs raised in alternative/outdoor systems.

Carcass contamination essentially results from the carriage of pathogens by the pig itself, but it also results from contact with other contaminated carcasses and/or surfaces in the abattoir. Overall, slaughter and processing can have substantial effects on bacterial carcass contamination, leading to quantitative and qualitative changes based on the number of contaminant bacteria and the incidence of pathogens, such as *Salmonella* and *Campylobacter*. Hygiene varies between abattoirs and has been shown to have an important impact on carcass contamination (Botteldoorn *et al.*, 2004; Eblen *et al.*, 2006; Delhalle *et al.*, 2008; Baptista *et al.*, 2010), in addition to preslaughter transportation and lairage/holding, which have been shown to increase the contamination risk of the abattoir's slaughter and processing line (Hurd *et al.*, 2002; Rostagno *et al.*, 2003).

Although there are other food-borne pathogens commonly associated with pork, they have received much less attention than *Salmonella* and *Campylobacter*. For example, pathogenic strains of *Yersinia enterocolitica* have been frequently isolated from pigs and pork products and are assumed to be important sources of human infection (Bhaduri *et al.*, 2005; Fredriksson-Ahomaa *et al.*, 2006). Nowak *et al.* (2006) studied the incidence of *Y. enterocolitica* in pigs from conventional and alternative (organic) production systems in Germany. The authors reported a prevalence of 29% in conventional versus 18% in alternative systems. However, the ecology and epidemiology of this pathogen is complex and poorly understood, requiring further investigation.

Another pathogen, *Listeria monocytogenes*, is of particular concern in ready-to-eat foods. Pork meat and processed pork products, such as deli meats, have been implicated in human outbreaks (Thevenot *et al.*, 2006). In Finland, Hellstrom *et al.* (2010) reported higher on-farm prevalence of *L. monocytogenes* in organic than in conventional systems based on a variety of animal and environmental samples. However, as for *Yersinia*, little is known about the ecology and epidemiology of *Listeria* along the pork production and processing chain.

Other bacterial pathogens with potential food safety implications are methicillin-resistant *Staphylococcus aureus* (MRSA) and *Clostridium difficile* (de Boer *et al*., 2009; Rupnik & Songer, 2010; Weese, 2010; Weese *et al*., 2010). Currently, it is unclear if these pathogens constitute significant pork safety risks, with many questions unanswered regarding their ecology and epidemiology along the pork production chain. To the best of our knowledge, there are no studies comparing the incidence of these pathogens in different swine production systems. In fact, their role as relevant food-borne pathogens along the pork production chain remains to be determined.

18.3 ANTIMICROBIAL RESISTANCE IN CONVENTIONAL VERSUS ORGANIC PORK PRODUCTION

In the United States, antimicrobials are used in pork production for two purposes: (1) to prevent diseases, and (2) to treat diseases once they become established (Box 18.3). The use of antimicrobials in any form can lead to the generation of antimicrobial-resistant bacteria, which can present a food safety risk as antimicrobial-resistant bacteria generated on the farm could be transmitted to humans via consumption or handling of contaminated pork products.

Box 18.3 Preventative Antimicrobial Use

Preventative antimicrobial use is often referred to as subtherapeutic use. Often the drugs are direct-fed or included in the water. Pigs treated with antimicrobials in this manner often grow more efficiently. As such, some antimicrobials are also referred to as antimicrobial growth promoters.

One hallmark of organic pork production is the exclusion of antimicrobials. Pigs and pork products marketed as organic cannot be treated with antimicrobials in any form, whether for disease prevention or treatment. Naturally, two questions arise: Do pigs raised organically or antimicrobial-free produce fewer antimicrobial-resistant bacteria? How about their finished products?

Several studies that have examined antimicrobial-resistant bacteria development and transmission on antimicrobial-free pork farms are now available. Although the systems examined varied across studies, a general consensus exists that bacteria isolated from antimicrobial-free pigs are often less likely to be resistant to various antimicrobials compared to similar isolates obtained from pigs treated with antimicrobials (Table 18.2). Bunner *et al*. (2007) collected over 1400 *Escherichia coli* isolates from conventional or antimicrobial-free pig fecal samples in the United States and measured resistance to a battery of antimicrobials. *E. coli* isolates obtained from conventional farms were more frequently resistant to ampicillin, sulfamethoxazole, tetracycline, and chloramphenicol compared to similar isolates obtained from antimicrobial-free fecal samples. In a similar study, Nulsen *et al*. (2008) screened *E. coli* and *Enterococcus* spp. from three conventional pork farms and one antimicrobial-free pork farm in New Zealand. Although a small study, both *E. coli* and *Enterococcus* spp.

Table 18.2 Comparison of antimicrobial resistance in bacteria isolated from conventional versus antimicrobial free pigs and pork products.

	% Resistant Isolates Conventional/Antimicrobial-free													
Bacteria	**AUX**	**AMP**	**CEP**	**ERY**	**CHL**	**CIP**	**GEN**	**KAN**	**NAL**	**STR**	**SUL**	**TET**	**NEO**	**CTR**
Gram (−)														
Fecal Sample														
E. coli[a]	0.5/0.2	24.1/13.1	3.7/1.6	nt	8.2/3.2	0/0	0.8/0.2	20.2/15.7	0/0	28.9/20.9	31.6/21.9	90.9/71.3	nt	Nt
E. coli[b]	nt	2.4/0	nt	nt	nt	0/0	0.6/0	nt	nt	28.7/2	nt	65.6/5	0.6/1	13.4/0
Campylobacter[c]	nt	nt	nt	77/34.5	1.7/1.4	2.8/0.6	.8/0	nt	7.3/8.4	nt	nt	83.4/56.2	nt	nt
Salmonella[d]	14.1/0.7	35.3/12.4	5.9/0.4	nt	25.9/10.6	nt	2.4/0	18.8/1.4	nt	85.9/36.2	64.7/30.1	93/88.3	nt	nt
Carcass or Pork														
E. coli[e]	nt	81.1/23.3	14.4/6.7	nt	2.2/5.6	nt	0/1.1	nt	nt	26.7/43.3	70/20	70/31.1[f]	nt	nt
Salmonella[d]	3.5/1.8	23.5/3.2	2.6/2.3	nt	21.7/0	nt	0/0	17.4/0	nt	54.8/30.3	50.4/25.3	97.4/55.7	nt	nt
Campylobacter[c]	nt	nt	nt	81.4/40.4	4.3/1.2	0/0	0.5/0.6	nt	1.5/0.6	nt	nt	80.8/36.6	nt	nt

	AMP	**ERY**	**GEN**	**VAN**	**STR**	**TET**	**VIR**
Gram (+)							
Enterococcus[b]	0/0	65/1	0/0	0/0	48.7/0	70.4/4.5	46.6/0

[a]Bunner *et al.*, 2007.
[b]Nulsen *et al.*, 2008.
[c]Thakur & Gebreyes, 2005.
[d]Gebreyes *et al.*, 2006.
[e]Miranda *et al.*, 2008.
[f]doxycycline.
AUX, clavulanic acid/amoxicillin; AMP, ampicillin; CEP, cephalothin; ERY, erythromycin; CHL, chloramphenicol; GEN, gentamicin; KAN, kanamycin, NAL, nalidixic acid; STR, streptomycin; SUL, sulfamethoxazole; TET, tetracycline, NEO, neomycin; CTR, co-trimoxazole; VAN, vancomycin; VIR, virginiamycin.

isolated from conventional farms were more likely to be resistant to several test antimicrobials including tetracycline, streptomycin, erythromycin, and virginiamycin, compared to *E. coli* and *Enterococcus* spp. isolated from antimicrobial-free farms.

Patterson *et al.* (2007) used a macroarray-based technique to detect tetracycline (*tet*) and erythromycin (*erm*) resistance genes in bacteria isolated from different pork farms in Europe. Both *tet* and *erm* genes were detected in higher frequency in conventional versus antimicrobial-free farms and herds. Similar results were reported by Jindal *et al.* (2006) looking at *tet* genes in conventional and organic swine herds in the United States, offering further indication that excluding antimicrobials in pork production results in fewer antimicrobial-resistant bacteria.

Several groups have also looked at antimicrobial resistance patterns in food-borne pathogens isolated from conventional and antimicrobial-free pork farms. Gebreyes *et al.* (2006) found that *Salmonella* isolated from conventional herds were more frequently resistant to most classes of antibiotics compared to *Salmonella* isolated from antimicrobial-free herds. Isolates from conventional herds were also more likely to be resistant to multiple antimicrobials. In a similar study, Thakur and Gebreyes (2005) showed that *Campylobacter* isolates obtained from conventional herds were more frequently resistant to 2 of the 12 tested antimicrobials (tetracycline and erythromycin).

It is important to note that in all of the aforementioned studies, antimicrobial-resistant bacteria were isolated from organic pork farms as well, albeit with less frequency. In many cases, isolates from organic farms also had multiple drug-resistant phenotypes. The previously mentioned study by Gebreyes *et al.* (2006) reported that the most common multiple drug resistance pattern in their sample set (streptomycin, sulfamethoxazole, and tetracycline) was detected with similar frequency in bacteria isolated from both conventional and organic pork farms. This observation serves to illustrate the complexity and widespread nature of antimicrobial resistance.

Most agree that the main risk of antimicrobial use in pork products is through the handling or consumption of pork products containing antimicrobial-resistant bacteria. Once pigs leave the farm, however, they are transported and held in lairage prior to being processed. These events average approximately 8–12 hours and present numerous opportunities for cross-contamination and rapid infection via other pigs, contaminated trailers or holding pens, or processing itself. While it is clear that excluding antibiotics in pork production results in fewer antimicrobial-resistant bacteria in the pig, there are currently very few studies that examine whether these reductions translate to fewer antimicrobial-resistant bacteria on finished products. One study from Europe examined resistance to nine different antimicrobials in *E. coli* isolated from conventional versus organic pork products. Although only a small number of isolates were examined (54 conventional isolates and 67 organic isolates), *E. coli* obtained from conventional pork products were more frequently resistant to sulfisoxazole, ampicillin, and doxycycline (Miranda *et al.*, 2008). The aforementioned study by Gebreyes *et al.* (2006) also showed that *Salmonella* isolates from conventional pig carcasses were more likely to be resistant to most test antimicrobials than similar isolates obtained from antimicrobial-free carcasses. More comprehensive studies will be needed, however, to accurately quantify the impact of antimicrobial exclusion on antimicrobial-resistant bacteria on finished pork products.

Antimicrobial resistance impacts human health when illnesses are prolonged, hospital stays are extended, or treatments fail. There still is considerable debate as to whether antimicrobial-resistant bacteria that develop on the farm impact human health and, if so, to what extent. Although thorough examination of this issue is outside of the scope of this chapter, it deserves acknowledgment. It is clear, however, that pigs raised without antimicrobials produce fewer antimicrobial-resistant bacteria in their feces and it appears that reductions in antimicrobial-resistant bacteria on the farm can result in fewer antimicrobial-resistant bacteria on finished products.

18.4 INCIDENCE OF PARASITES

The development and survival of pig parasites in the environment are dependent on a number of factors. Housing system, hygiene, and management practices constitute determinant factors for the transmission rate and the consequent risks due to parasitism (Nansen & Roepstorff, 1999). Endo- and ectoparasites appear to be the most common concern for organic pig producers. Direct contact with soil, combined with restrictions of prophylactic use of antiparasitic products, is likely to increase the risk of parasitic infections, particularly endoparasitic infections, in organic pigs.

Outdoor access has been associated with high rates of parasitic infections (particularly, helminth species) in pigs (Roepstorff & Nansen, 1994; Roepstorff *et al.*, 1998; Carstensen *et al.*, 2002). However, in intensive indoor production systems (with concrete slatted floors and no bedding), parasitic infections have become less frequent (Table 18.3; Nansen & Roepstorff, 1999). However, even though slaughter statistics show significantly more lesions from parasites in organic pig carcasses (Hansson *et al.*, 2000), the precise impact of different levels of parasitic infection on final pork quality and safety, as well as on consumer risk, remains unknown.

Pigs and undercooked pork are considered potential sources of human infection with *Toxoplasma gondii* and *Trichinella spiralis*, which are known to have historical significance

Table 18.3 Occurrence of helminths in different types of pork production systems.

Helminth	Outdoor[a]	Indoor (extensive)[b]	Indoor (intensive)[c]
Ascaris	+	+	+
Oesophagostomum	+	+	(+)
Trichuris	+	+	(+)
Strongyloides	+	+	
Hyostrongylus	+	(+)	
Metastrongylus	+		
Stephanurus	(+)	(+)	
Ascarops	(+)		
Physocephalus	(+)		
Macracanthorhynchus	(+)		
Trichinella[d]	(+)	(+)	
Taenia[d]	(+)		
Schistosoma	(+)		
Fasciola	(+)		
Dicrocoelium	(+)		

Reproduced from Nansen & Roepstorff (1999), copyright 1999, with permission of Elsevier.
[a]Outdoor: Pigs raised entirely in outdoor environments.
[b]Indoor (Extensive): Pigs raised indoor with bedding and/or access to outdoor environment.
[c]Indoor (Intensive): Pigs raised entirely in indoor environments.
[d]Potential food-borne parasites of major importance.
+, Frequent occurrence; (+), occasional occurrence.

in pork production (Hovi *et al.*, 2003; Gottstein *et al.*, 2009). As the pork industry shifted to a more intensive indoor production system, complemented with stringent biosecurity measures, the prevalence of these pathogens declined greatly, and human infections originating from indoor pork production systems have rarely been reported (Gottstein *et al.*, 2009; Takumi *et al.*, 2009). Over the last decades, conventional (indoors) production systems have been very successful in controlling *Toxoplasma* infections, maintaining very low prevalence levels or no infected animals at all. However, according to Kijlstra *et al.* (2004), alternative production systems that provide outdoor access to pigs may lead to a reemergence of *Toxoplasma* infections. Gebreyes *et al.* (2008) found significantly higher seroprevalence of *Toxoplasma* in outdoor herds (6.8%) than conventional/indoor herds (1.1%). In the same study, only pigs from outdoor herds were found to be seropositive for *Trichinella*.

The pork tapeworm, *Taenia solium*, is the causative organism of porcine cysticercosis and human neurocysticercosis (Krecek *et al.*, 2008). Cysticercosis is emerging as a serious public health and agricultural problem in many poor countries. It occurs where pigs range freely, sanitation is poor, and meat inspection is absent or inadequate, and is thus strongly associated with poverty and smallholder farming (Flisser *et al.*, 2006). According to Sikasunge *et al.* (2007), free-range rearing is a significant risk factor for porcine cysticercosis, with an odds ratio of 1.68.

As presented, in organic pork production systems, the animals are particularly at risk of parasitism due to outdoor rearing and ban of prophylactic medication. Even if these parasites do not pose a direct public health threat (because they are destroyed or removed from the final product), their simple presence is perceived negatively by consumers (Kouba, 2003; Magkos *et al.*, 2006). Moreover, according to Steenhard *et al.* (2002), interactions between intestinal parasites and bacteria play an important role in the dynamics of *Salmonella* infections in pigs. Pigs infected with *Oesophagostomum* spp. excreted significantly higher amounts of *Salmonella* in feces, and for more days, than pigs not infected with the parasite. Mansfield *et al.* (2003) also showed that *Trichuris suis* and *Campylobacter jejuni* synergise within the intestinal tract of experimentally infected pigs. Currently, very little is known about these and other potential interactions in the intestinal tract of the pigs. However, based on these initial evidences, more research is needed.

18.5 CONCLUSIONS

Consumer concern over the quality and safety of conventional food has intensified in recent years, and primarily drives the increasing demand for organically produced food, which is perceived as healthier and safer. Relevant scientific evidence, however, is scarce. Although there is an urgent need for more information related to health benefits and hazards of food products from both conventional and organic systems, generalized conclusions remain tentative in the absence of adequate comparative data. It is difficult, therefore, to determine the risks. The number of studies on this topic is very small, and therefore, a thorough analysis is not currently possible. Furthermore, the comparison of the studies described in this review is very difficult, because of a variety of potential confounders, including differing sample sizes, methodologies, times, and locations, as well as several other variables. However, what should be made clear is that "organic" does not automatically equal "safe."

In general, the management of specific public health risks associated with organic animal production is difficult, because high biosecurity levels often are not feasible. Although organic animal husbandry is well defined, the regulations still allow considerable degree of freedom, resulting in a large number of producers with their own ideas and practices. This variability should be taken into account when comparing organic with conventional livestock systems.

Consumers play a major role in the development of the organic food market. They define the needs and their interest guides the market. However, there is very low market transparency connected to organic food production (Vaarst *et al.*, 2005; Magkos *et al.*, 2006). Risk management will continue to pose a dilemma for consumer communication. The design of new animal production systems (with outdoor access) requires both a thorough analysis of possible risks and optimal communication of these risks throughout the food chain, in addition to appropriate distribution of responsibility concerning these risks. Some risks are inherent to the choice of keeping animals in a more natural and open environment, and could be seen as an inherent responsibility of the consumer, whereas other risks may be mitigated by further refinement or adjustment of the housing or farm management system applied. As shown, outdoor life is associated to some risks for the animals as well as for the humans consuming their final product. Therefore, knowledge and willingness are necessary to develop a production framework that provides outdoor life for the animals and at the same time minimizes risks.

Despite the limited number of comparative preharvest food safety data from organic and conventional pork production systems, it is clear that organic production systems are associated with lower frequency of antimicrobial-resistant bacteria. However, organic animal husbandry has been strongly criticized by veterinarians, claiming that organic livestock often are not treated properly when sick, because of longer withdrawal times or complete restrictions by the organic standards, and also, because alternative approaches are preferred, which are sometimes not recognized by science (Lund & Algers, 2003).

As seen, there are several issues related to aspects of the production systems, and ultimately linked to food safety, that need to be addressed. Although there have been many investigations into the occurrence and transmission of microbial pathogens in conventional systems, little similar relevant information is available regarding organic livestock. Absence of data, however, does not necessarily translate into absence of hazard. While the high concentration of animals in conventional indoor production systems favor pathogen dissemination, organic systems favor the exposure of animals to outdoor environment and other species, facilitating introduction and maintenance of pathogens.

Based on the current review, it is impossible to draw general conclusions regarding preharvest food safety risks in organic versus conventional pork production systems. Therefore, the question persists: Is organic pork safer? As long as there is insufficient epidemiological evidence, it remains very difficult to answer this question. The existence of opposing trends within the food market has given rise to some degree of polarization among different interest groups. From the limited number of published data, it seems that there is no clear evidence that organically produced food of animal origin is safer than conventionally produced food, and vice versa. However, it is critical to keep in mind that whatever pork production system is considered, the entire process from the pigs living in the farm to the pork product marketed and prepared in the home of the consumer should be analyzed and discussed in relation to the overall aims of pork safety (i.e., farm-to-fork approach).

REFERENCES

Baptista, F. M., J. Dahl, and L. R. Nielsen. 2010. Factors influencing *Salmonella* carcass prevalence in Danish pig abattoirs. *Prev. Vet. Med.* 95:231–238.

Bhaduri, S., I. V. Wesley, and E. J. Bush. 2005. Prevalence of pathogenic *Yersinia enterocoli*tica strains in pigs in the United States. *Appl. Environ. Microbiol.* 71:7117–7121.

Blokhuis, H. J., R. B. Jones, R. Geers, M. Miele, and I. Veissier. 2003. Measuring and monitoring animal welfare: Transparency in the food product quality chain. *Anim. Welf.* 12:445–455.

Botteldoorn, N., L. Herman, N. Rijpens, and M. Heyndrickx. 2004. Phenotypic and molecular typing of *Salmonella* strains reveals different contamination sources in two commercial pig slaughterhouses. *Appl. Environ. Microbiol.* 70:5305–5314.

Bunner, C. A., B. Norby, P. C. Bartlett, R. J. Erskine, F. P. Downes, and J. B. Kaneene. 2007. Prevalence and pattern of anti-microbial susceptibility in *Escherichia coli* isolated from pigs reared under antimicrobial-free and conventional production methods. *J. Am. Vet. Med. Assoc.* 231:275–283.

Callaway, T. R., J. L. Morrow, A. K. Johnson, J. W. Dailey, F. M. Wallace, E. A. Wagstrom, J. J. McGlone, A. R. Lewis, S. E. Dowd, T. L. Poole, T. S. Edrington, R. C. Anderson, K. J. Genovese, J. A. Byrd, R. B. Harvey, and D. J. Nisbet. 2005. Environmental prevalence and persistence of *Salmonella* spp. in outdoor swine wallows. *Foodborne Pathog. Dis.* 2:263–273.

Carstensen, L., M. Vaarst, and A. Roepstorff. 2002. Helminth infections in Danish organic swine herds. *Vet. Parasitol.* 106:253–264.

de Boer, E., J. T. Zwartkruis-Nahuis, B. Wit, X. W. Huijsdens, A. J. de Neeling, T. Bosch, R. A. van Oosterom, A. Vila, and A. E. Heuvelink. 2009. Prevalence of methicillin-resistant *Staphylococcus aureus* in meat. *Int. J. Food Microbiol.* 134:52–56.

Delhalle, L., L. De Sadeleer, K. Bollaerts, F. Farnir, C. Saegerman, N. Korsak, J. Dewulf, L. De Zutter, and G. Daube. 2008. Risk factors for *Salmonella* and hygiene indicators in the 10 largest Belgian pig slaughterhouses. *J. Food Prot.* 71:1320–1329.

Eblen, D., K. E. Barlow, and A. L. Naugle. 2006. U.S. Food Safety and Inspection Service testing for *Salmonella* in selected raw meat and poultry products in the United States, 1998 through 2003: An establishment-level analysis. *J. Food Prot.* 69:2600–2606.

Flisser, A., R. Rodriguez-Canul, and A. L. Williangham III. 2006. Control of the taeniosis/cysticercosis complex: Future developments. *Vet. Parasitol.* 139:283–292.

Fosse, J., H. Seeger, and C. Magras. 2009. Prevalence and risk factors for bacterial food-borne zoonotic hazards in slaughter pigs: A review. *Zoonoses Public Health* 56:429–454.

Fredriksson-Ahomaa, M., A. Stolle, and H. Korkeala. 2006. Molecular epidemiology of *Yersinia enterocolitica* infections. *FEMS Immunol. Med. Microbiol.* 47:315–329.

Gebreyes, W. A., P. B. Bahnson, J. A. Funk, J. McKean, and P. Patchanee. 2008. Seroprevalence of *Trichinella*, *Toxoplasma*, and *Salmonella* in antimicrobial-free and conventional swine production systems. *Foodborne Pathog. Dis.* 5:199–203.

Gebreyes, W. A., S. Thakur, and W. E. M. Morrow. 2005. *Campylobacter coli*: prevalence and antimicrobial resistance in antimicrobial-free (ABF) swine production systems. *J. Antimicrob. Chemother.* 56:765–768.

Gebreyes, W. A., S. Thakur, and W. E. M. Morrow. 2006. Comparison of prevalence, antimicrobial resistance, and occurrence of multidrug-resistant *Salmonella* in antimicrobial-free and conventional pig production. *J. Food Prot.* 69:743–748.

Gottstein, B., E. Pozio, and K. Nockler. 2009. Epidemiology, diagnosis, treatment, and control of trichinellosis. *Clin. Microbiol. Rev.* 22:127–145.

Hansson, I., C. Hamilton, T. Ekman, and K. Forslund. 2000. Carcass quality in certified organic production compared with conventional livestock production. *J. Vet. Med. B* 47:111–120.

Hellstrom, S., R. Laukkanen, K. M. Siekkinen, J. Ranta, R. Maijala, and H. Korkeala. 2010. *Listeria monocytogenes* contamination in pork can originate from farms. *J. Food Prot.* 73:641–648.

Hermansen, J. E. 2003. Organic livestock production systems and appropriate development in relation to public expectations. *Livest. Prod. Sci.* 80:3–15.

Honeyman, M. S., R. S. Pirog, G. H. Huber, P. J. Lammers, and J. R. Hermann. 2006. The United States pork niche market phenomenon. *J. Anim. Sci.* 84:2269–2275.

Hovi, M., A. Sundrum, and S. M. Thamsborg. 2003. Animal health and welfare in organic livestock production in Europe: current state and future challenges. *Livest. Prod. Sci.* 80:41–53.

Hurd, H. S., J. D. McKean, R. W. Griffith, I. V. Wesley, and M. H. Rostagno. 2002. *Salmonella enterica* infections in market swine with and without transport and holding. *Appl. Environ. Microbiol.* 68:2376–2381.

Jacob, M. E., J. T. Fox, S. L. Reinstein, and T. G. Nagaraja. 2008. Antimicrobial susceptibility of foodborne pathogens in organic or natural production systems: An overview. *Foodborne Pathog. Dis.* 5:721–730.

Jensen, A. N., A. Dalsgaard, A. Stockmarr, E. M. Nielsen, and D. L. Baggesen. 2006. Survival and transmission of *Salmonella enterica* serovar Typhimurium in an outdoor organic pig farming environment. *Appl. Environ. Microbiol.* 72:1833–1842.

Jindal, A., S. Kocherginskaya, A. Mehboob, M. Robert, R. I. Mackie, L. Raskin, and J. L. Zilles. 2006. Antimicrobial use and resistance in swine waste treatment systems. *Appl. Environ. Microbiol.* 72: 7813–20.

Kijlstra, A., B. G. Meerburg, and A. P. Bos. 2009. Food safety in free-range and organic livestock systems: Risk management and responsibility. *J. Food Prot.* 72:2629–2637.

Kijlstra, A., B. G. Meerburg, and M. F. Mul. 2004. Animal-friendly production systems may cause re-emergence of *Toxoplasma gondii*. *NJAS-Wageningen J. Life Sci.* 52:119–132.

Kouba, M. 2003. Quality of organic animal products. *Livest. Prod. Sci.* 80:33–40.

Krecek, R. C., L. M. Michael, P. M. Schantz, L. Ntanjana, M. F. Smith, P. Dorny, L. J. S. Harrison, F. Grimm, N. Praet, and A. L. Willingham III. 2008. Prevalence of *Taenia solium* cysticercosis in swine from a community-based study in 21 villages of the Eastern Cape Province, South Africa. *Vet. Parasitol.* 154:38–47.

Lund, V. and B. Algers. 2003. Research on animal health and welfare in organic farming – a literature review. *Livest. Prod. Sci.* 80:55–68.

Magkos, F., F. Arvaniti, and A. Zampelas. 2006. Organic food: Buying more safety or just peace of mind? A critical review of the literature. *Crit. Rev. Food Sci. Nut.* 46:23–56.

Mansfield, L. S., D. T. Gauthier, S. R. Abner, K. M. Jones, S. R. Wilder, and J. F. Urban. 2003. Enhancement of disease and pathology by synergy of *Trichuris suis* and *Campylobacter jejuni* in the colon of immunologically naïve swine. *Am. J. Trop. Med. Hyg.* 68:70–80.

Merrigan, K. A., M. R. Bailey, and W. Lockeretz. 2010. Strengthening US organic standards on animal health and welfare. *Anim. Welf.* 19:45–54.

Miranda, J. M., B. I. Vázquez, C. A. Fente, J. Barros-Velázquez, A. Cepeda, and C. M. Franco Abuín. 2008. Antimicrobial resistance in *Escherichia coli* strains isolated from organic and conventional pork meat: a comparative survey. *Eur. Food Res. Technol.* 226:371–375.

Nansen, P. and A. Roepstorff. 1999. Parasitic helminthes of the pig: factors influencing transmission and infection levels. *Int. J. Parasitol.* 29:877–891.

Nowak, B., T. V. Mueffling, K. Caspari, and J. Hartung. 2006. Validation of a method for the detection of virulent *Yersinia enterocolitica* and their distribution in slaughter pigs from conventional and alternative housing systems. *Vet. Microbiol.* 117:219–228.

Nulsen, M. F., M. B. Mor, and D. E. Lawton. 2008. Antibiotic resistance among indicator bacteria isolated from healthy pigs in New Zealand. *N. Z. Vet. J.* 56:29–35.

Patterson, A. J., R. Colangeli, P. Spigaglia, and K. P. Scott. 2007. Distribution of specific tetracycline and erythromycin resistance genes in environmental samples assessed by macroarray detection. *Environ. Microbiol.* 9:703–715.

Rodriguez, A., P. Pangloli, H. A. Richards, J. R. Mount, and F. A. Draughon. 2006. Prevalence of *Salmonella* in diverse environmental farm samples. *J. Food Prot.* 69:2576–2580.

Roepstorff, A. and P. Nansen. 1994. Epidemiology and control of helminth infections in pigs under intensive and non-intensive production systems. *Vet. Parasitol.* 54:69–85.

Roepstorff, A., O. Nilsson, A. Oksanen, B. Gjerde, S. H. Richter, E. Ortenberg, D. Christensson, K. B. Martinsson, P. C. Bartlett, P. Nansen, L. Eriksen, O. Helle, S. Nikander, and K. Larsen. 1998. Intestinal parasites in swine in the Nordic countries: prevalence and geographical distribution. *Vet. Parasitol.* 76:305–319.

Rostagno, M. H. 2009. Can stress in farm animals increase food safety risk? *Foodborne Pathog. Dis.* 6:767–776.

Rostagno, M. H., H. S. Hurd, J. D. McKean, C. J. Ziemer, J. K. Gailey, and R. C. Leite. 2003. Preslaughter holding environment in pork plants is highly contaminated with *Salmonella enterica*. *Appl. Environ. Microbiol.* 69:4489–4494.

Rupnik, M. and J. G. Songer. 2010. *Clostridium difficile*: Its potential as a source of foodborne disease. *Adv. Food Nut. Res.* 60:53–66.

Sanchez, J., I. R. Dohoo, J. Christensen, and A. Rajic. 2007. Factors influencing the prevalence of *Salmonella* spp. in swine farms: A meta-analysis approach. *Prev. Vet. Med.* 81:148–177.

Siekkinen, K. M., L. Nuotio, J. Ranta, R. Laukkanen, S. Hellstrom, H. Korkeala, and R. Maijala. 2006. Assessing hygiene proficiency on organic and conventional pig farms regarding pork safety: A pilot study in Finland. *Livest. Sci.* 104:193–202.

Sikasunge, C. S., I. K. Phiri, A. M. Phiri, P. Dorny, S. Siziya, and A. L. Williangham III. 2007. Risk factors associated with porcine cysticercosis in selected districts of Eastern and Southern provinces of Zambia. *Vet. Parasitol.* 143:59–66.

Steenhard, N. R., T. K. Jensen, D. L. Baggesen, A. Roepstorff, and K. Moller. 2002. Excretion in feces and mucosal persistence of *Salmonella* ser. Typhimurium in pigs subclinically infected with *Oesophagostomum* spp. *Am. J. Vet. Res.* 63:130–136.

Takumi, K., P. Teunis, M. Fonville, I. Vallee, P. Boireau, K. Nockler, and J. van der Giessen. 2009. Transmission risk of human trichinellosis. *Vet. Parasitol.* 159:324–327.

Thakur, S. and W. A. Gebreyes. 2005. Prevalence and antimicrobial resistance of *Campylobacter* in antimicrobial-free and conventional pig production systems. *J. Food Prot.* 68:2402–2410.

Thakur, S., D. A. Tadesse, M. Morrow, and W. A. Gebreyes. 2007. Occurrence of multidrug resistant *Salmonella* in antimicrobial-free (ABF) swine production systems. *Vet. Microbiol.* 125:362–367.

Thevenot, D., A. Dernburg, and C. Vernozy-Rozand. 2006. An updated review of *Listeria monocytogenes* in the pork meat industry and its products. *J. Appl. Microbiol.* 101:7–17.

Vaarst, M., S. Padel, M. Hovi, D. Younie, and A. Sundrum. 2005. Sustaining animal health and food safety in European organic livestock farming. *Livest. Prod. Sci.* 94:61–69.

Van der Wolf, P. J., A. R. W. Elbers, H. M. J. F. van der Heijden, F. W. van Schie, W. A. Hunneman, and M. J. M. Tielen. 2001. *Salmonella* seroprevalence at the population and herd level in pigs in The Netherlands. *Vet. Microbiol.* 80:171–184.

Von Borell E. and J. T. Sorensen. 2004. Organic livestock production in Europe: aims, rules and trends with special emphasis on animal health and welfare. *Livest. Prod. Sci.* 90:3–9.

Weese, J. S. 2010. *Clostridium difficile* in food. Innocent bystander or serious threat? *Clin. Microbiol. Infect.* 16:3–10.

Weese, J. S., B. P. Avery, and R. J. Reid-Smith. 2010. Detection and quantification of methicillin-resistant *Staphylococcus aureus* (MRSA) clones in retail meat products. *Lett. Appl. Microbiol.* 51:338–342.

Williams, P. R. and J. K. Hammitt. 2001. Preceived risks of conventional and organic produce: pesticides, pathogens, and natural toxins. *Risk Anal.* 21:319–330.

Winter, C. K. and S. F. Davis. 2006. Organic foods. *J. Food Sci.* 71:R117–R124.

Young, I., A. Rajic, B. J. Wilhelm, L. Waddell, S. Parker, and S. A. McEwen. 2009. Comparison of the prevalence of bacterial enteropathogens, potentially zoonotic bacteria and bacterial resistance to antimicrobials in organic and conventional poultry, swine and beef production: A systematic review and meta-analysis. *Epidemiol. Infect.* 137:1217–1232.

Zheng, D. M., M. Bonde, and J. T. Sorensen. 2007. Associations between the proportion of *Salmonella* seropositive slaughter pigs and the presence of herd level risk factors for introduction and transmission of *Salmonella* in 34 Danish organic, outdoor (non-organic) and indoor finishing pig farms. *Livest. Sci.* 106:189–199.

19 Food-borne Pathogen Occurrence in Organically and Naturally Raised Poultry

Ellen J. Van Loo, Sherry N. Melendez, Irene B. Hanning, and Steven C. Ricke

Abstract: The organic meat sector is currently the fastest growing segment of the US organic food industry, with poultry accounting for nearly two-thirds of this sector. In addition, the popularity for other alternatively raised poultry, such as pasture raised, has grown. However, little is known about the prevalence and characteristics of food-borne pathogens in nonconventionally produced poultry products. This chapter gives an overview of the studies on prevalence of food-borne pathogens and antibiotic resistance in pastured and organically raised poultry.

Keywords: Poultry; organic; free-range; pasture; resistance

19.1 INTRODUCTION

The organic meat sector is currently the fastest growing segment of the US organic food industry, with poultry accounting for nearly two-thirds of this sector (Oberholtzer *et al.*, 2006). The majority of the consumers perceive organic and natural products as safer and healthier than the conventionally produced alternatives (Harper & Makatouni, 2002; Magnusson *et al.*, 2003). However, little is known about the prevalence and characteristics of food-borne pathogens in nonconventionally produced meat products. The growing interest in organic and natural foods warrants a greater need for information on the food safety of these products.

While certified organic products are required to meet all United States Department of Agriculture (USDA) organic certification standards, non-certified organic, free-range, and pastured poultry rearing systems may represent a fairly wide range of rearing conditions (Bailey & Cosby, 2005). Non conventional systems are not required to adhere to any USDA standards. These systems allow flocks access to the outdoors and often limit or avoid the use of antibiotics. The lack of biosecurity poses an increased food safety risk due to increased potential contact with vectors of pathogenic microorganisms (Bailey & Cosby, 2005; Cui *et al.*, 2005; Miranda *et al.*, 2008).

Little is known about the frequency of *Salmonella* and other food-borne pathogens in organic and natural poultry. *Salmonella* is a leading cause of food-borne illness in the United States and worldwide (Wegener *et al.*, 2003; Zhao *et al.*, 2006). Approximately 800,000 to 4 million salmonellosis infections occur annually in the United States alone (Jain & Chen,

Organic Meat Production and Processing, First Edition. Edited by Steven C. Ricke, Ellen J. Van Loo, Michael G. Johnson, and Corliss A. O'Bryan.

2006). Poultry and poultry products have been linked to many of the salmonellosis cases, accounting for 29% of all *Salmonella* infections (Braden, 2006; Jain & Chen, 2006). Much less is known about other food-borne pathogens and their prevalence in organically raised flocks.

19.2 BROILER PRODUCTION IN THE UNITED STATES

19.2.1 Commercial (conventional) system

The conventional or commercial poultry industry has been one of the fastest growing sectors of the animal agricultural industry (Herren, 2000). In 2006, commercial poultry systems represented 95% of the total poultry production in the United States (MacDonald, 2008). The majority of broilers are reared in large housing operations consisting of 6–40 thousand birds per housing unit (Herren, 2000). Broilers in the conventional poultry industry are generally raised cage-free in barns on litter, the stocking density requirements range from 6.5 to 8.5 lbs/ft^2 (31.7–41.5 kg/m^2) depending on the size of the birds (National Chicken Council, 2010). This space allotment allows less than 1 ft^2 (0.093 m^2) of room for each bird (Fanatico, 2006). The main genotype chosen for commercial broiler production is a Cornish and White Rock chicken cross, due to its high feed conversion and ability to finish in as few as 6 weeks (Fanatico, 2006). This breed, however, is prone to health problems such as leg tendon slippage (Henry, 2002).

Bird health is an important factor within commercial poultry production, as lowering mortality results in greater economic returns. Antibiotics are currently approved to be utilized only for the treatment or prevention of diseases in chickens, some of which include bacitracin, chlortetracycline, erythromycin, lincomycin, and oxytetracycline (Bell & Weaver, 2002). Anticoccidials are also often included in poultry feed to prevent coccidiosis, an amoebic disease that can result in death (Bell & Weaver, 2002; Mortier *et al.*, 2005).

Another aspect often practiced in the commercial poultry industry to promote overall bird health and welfare is beak trimming. Beak trimming is routinely done in commercial poultry operations to avoid injury, feather plucking, and cannibalism among birds (Henderson *et al.*, 2009). While it is currently not prohibited in certified organic poultry, beak trimming remains controversial in commercial and noncommercial systems with regard to animal welfare due to potential for acute and chronic pain (Hughes & Gentle, 1995; Cheng, 2005; Kuenzel, 2007; Henderson *et al.*, 2009).

19.2.2 Nonconventional systems

Nonconventional poultry systems, such as free-range, pasture, and organic poultry systems generally use slow-growing breeds that have a longer growout period than the traditional conventional breed (Henry, 2002; Fanatico *et al.*, 2009). Larger organic broiler operations have the potential for increased bird mortality (5%–10%) due to the restriction of medications (Fanatico *et al.*, 2009).

A survey conducted by the USDA in 2006, reported that 1.7% of broiler operations were certified organic and 0.44% were free-range (MacDonald, 2008). The poultry package claim of "Certified Organic" is a highly regulated and specific animal production claim in the United States (Bailey & Cosby, 2005). The use of the claim "natural" on poultry packaging has recently come under scrutiny and is now defined as products that are "minimally processed," "do not contain artificial additives or chemical preservatives," and "have been raised from birth to slaughter without growth promotants" (USDA, 2005). Animal production claims on product labels that have been passed by USDA include "free-range," "raised without antibiotics," and "raised without hormones"; however, these claims are not specific and one claim can represent different animal rearing conditions (USDA, 2008).

Currently, the only regulation on free-range poultry issued by the USDA is that the poultry has had access to the outdoors during the production cycle (USDA, 2008). The term free-range, however, is more highly regulated in other countries. The European Union (EU) legislation on "free-range" states that chickens must be exposed to an outdoor run (1 m^2 per bird) for at least half their life, while "traditional free-range" birds have a minimum stocking density based on age and must have access to an outdoor run for at least 6 weeks (Commission Regulation, 2008).

Consumer attitudes towards "natural" and "organic" products have created high demand for these unique commodities (Jacob *et al.*, 2008a). Consumer-based studies have reported that the primary drive to purchase organic is based on health, since the majority of consumers view organic food products as healthier and safer than conventional (Harper & Makatouni, 2002; Magnusson *et al.*, 2003; Jacob *et al.*, 2008a; Van Loo *et al.*, 2010). Van Loo *et al.* (2010) surveyed 976 consumers about their attitudes toward organic poultry meat with an online survey. The main motivation to buy organic chicken was the perception that organic chicken has fewer residues (pesticides, hormones, and antibiotics), is safer, and is healthier than its conventionally produced option. The study from Van Loo *et al.* (2010) reported the high price and the poor availability as the key factors limiting the organic meat purchase.

Sofos (2008) expressed the opinion that organically grown products may carry a heavier microbial load than those produced by conventional methods. Several studies within the literature attribute the increased risk for bacterial contamination on organic farms to increased access to transmission agents, such as birds, rodents, insects, and wild animals (Rodenburg *et al.*, 2004; Cui *et al.*, 2005; Miranda *et al.*, 2007; Lestari *et al.*, 2009). A study conducted by Meerburg *et al.* (2006) revealed that a small proportion, 10% and 1.2%, of the house mice population on organic farms were infected with *Campylobacter* and *Salmonella*, respectively.

19.2.3 Alternative poultry housing

Because of the lack of specific regulations on animal production claims in the United States, several different definitions of nonconventional poultry production systems exist. A review published by the National Center for Appropriate Technology (NCAT) (Fanatico, 2006) extensively described the details of the use of varying housing units including fixed houses, portable houses, and pasture pens in free-range systems. These different systems are briefly discussed in this section. Siemon *et al.* (2007) defined the term "pasture raised" as poultry that have been reared outside in small, ventilated, movable pens. Many differing structural designs exist for alternative poultry housing units; however, the two main types of broiler houses are movable pens and fixed housing units (Plamondon, 2003).

Lightweight pens can be manually moved either daily or weekly; however, larger machine portable housing units built on skids require a tractor or other mechanized means for relocation (Plamondon, 2003; Fanatico, 2006). Rotation of poultry onto new pasture is important in preventing the death of vegetation (Fanatico, 2006). Machine portable housing units can be moved as often as once every 3 days to as infrequent as once a year; however, the pasture around these units must be subdivided with movable fencing to provide birds with fresh pasture (Plamondon, 2003; Fanatico, 2006). Movable pens are required to be easily moved without injury to chickens, but still remain in place during high winds and other weather extremes. These units should also shield the birds from the elements and predators (Plamondon, 2003; Fanatico *et al.*, 2009). Various predators posing a danger to outdoor poultry include skunks, dogs, coyotes, mink, weasels, owls, and foxes (Fanatico, 2006).

Fixed housing units are usually larger, with a heavier, more solid structure (Palmondon, 2003). Hoop houses, originally used as greenhouses for vegetables, are being implemented as fixed housing units by alternative poultry production farmers (Fanatico, 2006; USDA, 2009). Hoop houses are generally made of polyvinyl chloride (PVC) piping and can be covered with a clear tarp to allow solar heating during winter months or shaded tarp to block the sun during summer (Fanatico, 2006; USDA, 2009). Hoop houses, or high tunnels, are frequently used in poultry operations due to cost efficiency. In addition, they are generally considered easier to move and maintain compared to other housing systems (USDA, 2009).

19.3 PREVALENCE OF FOOD-BORNE PATHOGENS IN PASTURE AND ORGANICALLY RAISED POULTRY

Unlike conventional poultry, where birds are reared in large barns, pastured poultry are reared outside on pasture in smaller, open-air, movable pens (Siemon *et al.*, 2007). An increase in microbiological transmission routes exists on nonconventional poultry farms, including handlers, flies, rodents, wildlife, other farm animals, and birds (Rodenburg *et al.*, 2004; Meerburg *et al.*, 2005). Organic poultry are required to have outdoor access but the access may be more limited than in pastured poultry; nevertheless, similar issues apply.

19.3.1 *Salmonella*

Salmonella prevalence rates in conventionally produced poultry in the United States have been monitored continuously by the USDA-FSIS (2010a). Within the literature, greatly varying prevalence rates of 18%, 16%, 35%, and 3% (White *et al.*, 2001; Goncagual *et al.*, 2003; Jain & Chen, 2006; Zhao *et al.*, 2006, respectively) have been reported. According to the fourth-quarter progress report for 2010 by the USDA-FSIS, 9.5% of the broilers tested positive for *Salmonella*. For small and very small broiler establishments specifically, this number was higher (17.6%) (USDA-FSIS, 2010b).

Some variation exists in the reported prevalence rates of *Salmonella* in nonconventional poultry productions systems. Authors reported levels of *Salmonella* contamination to be 20.8% (Lestari *et al.*, 2009) and 16.2% (Siemon *et al.*, 2007) for organic and pasture poultry, respectively. Bailey and Cosby (2005) reported a prevalence rate for *Salmonella* of 31% and 25% for free-range and all natural chickens, respectively. Melendez *et al.* (2010) collected isolates of *Salmonella* from pasture-raised poultry (farms, retail store carcasses, and processing plants). They reported 29.5% (n = 200) of their samples to be positive for *Salmonella*. This included 50% (n = 36) of the retail carcasses and 25% (n = 164) of the farm samples. Even higher prevalence rates were reported by Cui *et al.* (2005), 61% of the tested organic chickens were contaminated with *Salmonella*. However, some studies reported very low numbers of *Salmonella* prevalence on free-range farms (2.9%) and organic (0%) (Esteban *et al.*, 2008; Hoogenboom *et al.*, 2008). This difference may be attributed to the smaller sample size used by Hoogenboom *et al.* (2008) and Esteban *et al.* (2008) or the difference in location. While most of the reported studies have been conducted in the United States, these particular studies were conducted in Europe (Spain and the Netherlands).

The reported prevalence of *Salmonella* in nonconventional (organically/free-range/pasture raised) versus conventionally raised poultry is inconsistent in the literature. While two recent comparison studies conducted by Overbeke *et al.* (2006) and by Lund *et al.* (2003) found no significant differences in the *Salmonella* prevalence between organic and conventional broilers, more studies appear to support the conclusion that differences between both production systems do exist. Prevalence within organic poultry was reported to be significantly higher than conventional by Cui *et al.* (2005); however, the isolation and enrichment techniques were not described and more than double the number of organic samples were taken for analysis as compared to conventionally raised poultry. Conversely, the findings of Siemon *et al.* (2007) indicated that the *Salmonella* prevalence in sampled conventional poultry was higher than that of pastured poultry. Siemon *et al.* (2007) attributed the findings to high-density living quarters and very short flock turnover times. Similarly, Alali *et al.* (2010) reported the prevalence of *Salmonella* in fecal samples from conventional farms was higher than from large scale USDA certified organic broiler farms (38.8% versus 5.6%). As reported by Alali *et al.* (2010), many organic chicken studies do not clearly indicate if they refer to USDA-certified bird or pasture raised birds. For example, Cui *et al.* (2005) and Lestari *et al.* (2009) reported results for organic chicken but failed to describe whether these were USDA-certified organic birds or pasture or free-range birds. As indicated earlier in this chapter, there are important differences between both systems that may influence the pathogen prevalence. USDA organically raised birds are grown according to the strict USDA National Organic Program (NOP) rules (USDA, 2008). Pasture broiler farms are usually smaller farms and are often not classified as USDA organic (Fanatico, 2006; Jacob *et al.*, 2008b).

Pathogen prevalence on poultry farms has been observed to differ by geographical location as well (Pieskus *et al.*, 2008). *Salmonella* prevalence on farms in the Netherlands has been reported to be within the range of 3.7%–13% of conventional broilers. Fifty-three percent of retail chicken samples were positive for *Salmonella* from Vietnam, using the standard isolation procedures given by the Nordic Committee on Food Analysis (Oslo, Norway) (Van *et al.*, 2007). A Danish study, following the same isolation procedure, reported 5.5% of commercial broiler flocks to be positive for *Salmonella* (Wedderkopp *et al.*, 2001). Interestingly, greater than 65% of Danish broiler flocks tested positive for *Salmonella* from 1989 to 1993, prior to the administration of the EU salmonellosis control program (Wegener *et al.*, 2003). The initial aim of the program was that less than 5% of broiler flocks would be infected with *Salmonella*. The program was successful and was gradually revised towards assurance of complete freedom from *Salmonella* in broiler production (Wegener *et al.*, 2003). Pieskus *et al.* (2008) also reported *Salmonella*, isolated by standard methods, in 29% of the Lithuanian and 9% of the Italian conventional poultry systems. Retail broilers from processing plants in Malaysia have been reported to exhibit a *Salmonella* prevalence of 50% (Rusul *et al.*, 1996) while antibiotic-free poultry farms in Spain had a prevalence rate of only 2.9% ($n = 34$) (Esteban *et al.*, 2008). The considerable variation in prevalence in these studies can be attributed to a variety of factors including differences in sampling, enrichment, and isolation techniques used, as well as various animal-rearing conditions.

In addition to overall prevalence, *Salmonella* serotype identification is an important factor in epidemiological studies. From the CDC data during the period of 1998–2006, it was reported that *Salmonella* Typhimurium and *Salmonella* Enteritidis were the most frequently isolated from human clinical cases (Finstad *et. al.*, 2011; Foley *et al.*, 2011). *Salmonella enterica* serovar Kentucky has been reported as the major serotype in several conventional poultry (Li *et al.*, 2007; Parveen *et al.*, 2007; Lestari *et al.*, 2009), as well as nonconventional poultry studies (Parveen *et al.*, 2007; Lestari *et al.*, 2009). *S.* Kentucky was also the most frequent serotype isolated from chickens in 2006 (CDC, 2006). *Salmonella* serotypes also appear to vary based on geographical location. Studies conducted by Galanis *et al.* (2006) support this observation and they suggested that certain serotypes may be geographically restricted based on ecological niches. One of the more highly examined regionally specific serotypes is *Salmonella* Weltevreden, which has been the dominant serotype associated with food-borne illness in Southeast Asia for several years (Bangtrakulnonth *et al.*, 2004). Zhao *et al.* (2006) observed that out of their *S.* Weltevreden isolates 78% originated in Southeast Asia and a global surveillance study by Herikstad *et al.* (2002) reported that *S.* Weltevreden only made the top 15 serotype lists in the Southeast Asian and Western Pacific regions. Galanis *et al.* (2006) also reported *S.* Weltevreden as the second most prevalent serotype in Asia in 2000 and 2001. In the United States, Melendez *et al.* (2010) identified *S.* Kentucky as the most prevalent serotype detected from the sampled pastured poultry sources (53%), followed by *S.* Enteritidis (24%), Bareilly (10%), Mbandaka (7%), Montevideo (5%), and Newport (2%).

19.3.2 *Campylobacter*

Food-borne Campylobacteriosis is a major public health concern and it is estimated that each year 845,024 cases of food-borne illnesses caused by *Campylobacter* exist in the United States (Scallan *et al.*, 2011). *Campylobacter* spp. are highly prevalent in poultry production systems and cause a risk for the contamination of poultry meat (Sahin *et al.*, 2002).

Similar to the reported data on *Salmonella* in organic or free-range poultry, there is some variation on the prevalence levels of *Campylobacter*. Some studies reported values between 70% and 80% of *Campylobacter* contamination in organically raised poultry carcasses (Cui *et al.*, 2005; Esteban *et al.*, 2008; Hanning *et al.*, 2010). Similarly, Griggs *et al.* (2006) reported 96% of the carcasses of chickens raised without antibiotics to be positive for *Campylobacter*. Han *et al.* (2009) detected substantially lower levels, with only 43.4% being infected with *Campylobacter*.

Most comparative studies of organic and conventional poultry indicate a similar prevalence of organic and conventional *Campylobacter* contamination, illustrating that the rearing conditions do not appear to influence the prevalence of *Campylobacter* (Cui *et al.*, 2005; Luangtongkum *et al.*, 2006; Han *et al.*, 2009). However, Heuer *et al.* (2001) found that all 22 organic broilers sampled were contaminated with *Campylobacter* compared to only one-third of the conventionally raised broilers. The incidence of *Campylobacter* spp. in conventionally raised broilers and turkeys was 66% and 83%, respectively, while the prevalence in organically raised broilers and turkey was 89% and 87% respectively (Luangtongkum *et al.*, 2006).

In addition to overall prevalence, *Campylobacter* strain identification is an important factor in epidemiological studies. *Campylobacter jejuni* is a major cause of food-borne illness (Snelling *et al.*, 2005) and has a low infective dose of nearly 500 cells. An estimated 98% of the chicken flocks in the United States are contaminated with *C. jejuni* (Stern *et al.*, 2001) and consequently, mishandled poultry products have a high risk of causing food-borne illness. Most studies on organic poultry reported *C. jejuni* as being the predominant *Campylobacter* species. Han *et al.* (2009), Hanning *et al.* (2010), Luangtongkum *et al.* (2008), and Price *et al.*

(2007) identified 67%, 62%, 100%, and 56%, respectively, of their *Campylobacter* isolates as *C. jejuni*. Hanning *et al.* (2010) characterized *Campylobacter* isolated from pasture poultry utilizing polymerase chain reaction (PCR) specific primers and reported the majority as *C. jejuni* and classified the remaining as *Campylobacter coli* (30%) or as others (8%) (Hanning *et al.*, 2010).

19.3.3 Other food-borne pathogens

There is little known about the prevalence of food-borne pathogens other than *Campylobacter* and *Salmonella* in pastured or organic poultry production systems. Miranda *et al.* (2007) reported *Enterococcus* prevalence in organic chicken at retail to be higher than in conventional chicken or turkey meat. Milillo *et al.* (2011) detected *Listeria monocytogenes* contamination of pasture reared chickens in cecal, soil, and grass samples. Approximately 1.75% of the cecal samples were positive for *Listeria*. Only 4.5% of the soil and grass samples before rearing were contaminated with *Listeria* compared to 53% after pasture rearing of the chickens. Both *L. monocytogenes* and *Listeria innocua* were recovered from cecal and environmental samples. These results indicate that pasture reared poultry, as well as their environment, may be contaminated with *Listeria*.

19.4 ANTIBIOTIC RESISTANCE

Antibiotics in livestock production can be used for different reasons including therapeutic purposes. USDA national organic livestock production regulations prohibit the use of antibiotics for any use (USDA National Organic Program, 2008). In recent years, rising antibiotic resistance trends as well as the emergence of multidrug resistant (MDR) strains have been the focus of public health officials. Antibiotic resistance was first reported in the early 1960s by Bulling *et al.* (1973) and the history of the supplementation of animal feeds with antibiotics has been described by Jones and Ricke (2003). Today, MDR pathogens are becoming a major public health concern, which is generating a growing interest in alternative approaches to replace the functional properties of antibiotics by nonantibiotic strategies (Griggs & Jacob, 2005; Sirsat *et al.*, 2009). The highly disparate systems of certified organic, nonconventional, and conventional poultry production have also been linked to variance in antibiotic resistance (Miranda *et al.*, 2007; Jacob *et al.*, 2008a).

19.4.1 *Salmonella*

Antibiotic-resistant *Salmonella* are of great concern to public health personnel who must treat salmonellosis, particularly in immunocompromised patients. Resistance to antibiotics such as ampicillin, chloramphenicol, and trimethoprim–sulfonamide has led to the use of fluoroquinolones in the treatment of invasive *Salmonella* infections. However, quinolone resistance in *Salmonella* is emerging, which may be linked to horizontal gene transfer through mobile genetic elements, such as plasmids possessing key virulence factors. Epidemiological studies are important in tracking resistance and understanding current and emerging resistance factors.

Salmonella isolates from organic chicken meat have been reported as less resistant to several antibiotics as compared to commercially produced broilers (Cui *et al.*, 2005; Miranda *et al.*, 2007; Siemon *et al.*, 2007; Schwaiger *et al.*, 2008; Lestari *et al.*, 2009; Melendez *et al.*, 2010). A study conducted by Griggs *et al.* (2006) on antibiotic resistant *Salmonella* isolates from antibiotic-free chickens reported 69.4% (n = 63) of the isolates as resistant to at least one antibiotic. Although this percentage appeared to be very high, only 17.5% (n = 63) of the isolates were resistant to five antibiotics while none were detected as being resistant to more than five. Some studies have compared the antibiotic resistance in conventional and organic poultry side by side. Lestari *et al.* (2009) reported that the antibiotic resistance profiles greatly differed among the *Salmonella* serovars, as well as the type of chicken (conventional or organic). *Salmonella* Kentucky isolated from organic chicken samples were susceptible to 11 of the 15 tested antibiotics while isolates of this organism from conventional chickens were susceptible to only four antibiotics (Lestari *et al.*, 2009). Siemon *et al.* (2007) detected multidrug resistance from *Salmonella* isolates on conventional and pasture farms in the southeast of 69% and 11%, respectively. However, none of the isolates of Wisconsin were found to be MDR (Siemon *et al.*, 2007). Cui *et al.* (2005) reported 79% of the *S. enterica* Typhimurium recovered from organic chicken samples to be resistant to all tested antibiotics while none of those recovered from the conventional samples were resistant to all antibiotics.

Melendez *et al.* (2010) reported all *Salmonella* isolates (59) from pastured poultry farms and carcasses to be resistant to at least two antibiotics, sulfisoxazole and novobiocin. However, when excluding novobiocin, an antibiotic to which *Salmonella* is naturally resistant to and therefore, is often used in pre-enrichments and selective media (Restaino *et al.*, 1977; Jain & Chen, 2006), none of the pastured poultry isolates were resistant to five or more antibiotics. Alali *et al.* (2010) observed the most common resistance phenotypes were single resistance to streptomycin and multidrug resistance to six antibiotic agents. However, there were differences among conventional and organic *Salmonella* isolates. Approximately 36% of the conventional and 25% of the organic isolates exhibited resistance to streptomycin, and multidrug resistance to the six antibiotics was only detected in conventional isolates (40%), but not in any of the organic isolates.

19.4.2 *Campylobacter*

Antibiotic resistance in *Campylobacter* has increased considerably (Gupta *et al.*, 2004). This includes various antibiotics such as fluoroquinolone, ciprofloxacin, erythromycin, clindamycin, kanamycin, and ampicillin. Several studies have evaluated *Campylobacter* antibiotic resistance in nonconventional poultry products and most studies conclude that there is a higher antibiotic resistance frequency in conventional as compared to organically raised poultry products. Avrain *et al.* (2003) reported higher resistance to tetracycline in *Campylobacter* isolates from conventional broilers than free-range broilers. Similarly, Cui *et al.* (2005) reported less than 5% of the *Campylobacter* recovered from organic chickens to be resistant to ciprofloxacin compared to 20% for conventional chickens. When comparing the resistance to fluoroquinolone, an important antibiotic used for human health, only 2% of *Campylobacter* species from organic poultry were resistant compared to 46%–67% for conventional poultry products (Luangtongkum *et al.*, 2006). Similarly, Price *et al.* (2007) reported *Campylobacter* strains from the conventional producers to be more likely to be fluoroquinolone resistant than those from the antibiotic-free producers. Isolates from both conventional (80%) and organic poultry operations (50%–60%) showed high resistance to tetracycline (Luangtongkum *et al.*, 2006). Soonthornchaikul *et al.* (2006) reported 100% of Campylobacter isolated from conventional and organic retail chicken products were resistant to erythromycin. Approximately 27% of the unpackaged and 9% of the prepackaged conventional chicken products were contaminated with *Campylobacter* isolates that were resistant to ciprofloxacin, while none of the organic chicken product contained resistant bacteria. Most studies evaluating the effect of conventional versus organic production practices on the antibiotic resistance of *Campylobacter* in poultry exhibited similar results and indicate, overall, a higher antibiotic resistance in conventional than organic poultry products. Results suggest that antibiotic use in the conventional and organic poultry production practices may correlate with the prevalence of antibiotic resistant *Campylobacter*. However, even in antibiotic-free organic poultry practices, high levels of antibiotic resistance have been reported, such as tetracycline resistance in 50% of the organic poultry *Campylobacter* isolates, indicating that other factors such as mobile genetic elements in addition to antibiotic use may influence the antibiotic resistance in *Campylobacter* (Luangtongkum *et al.*, 2006).

19.4.3 Antibiotics and dissemination of resistance

The use of antibiotics in animal production systems is the speculated culprit for bacterial resistance (McEwen & Fedorka-Cray, 2002; Martin *et al.*, 2008; Miranda *et al.*, 2008). However, as many studies of antibiotic-free and organically raised poultry have shown, the elimination of therapeutic antibiotic use does not necessarily eliminate resistance (Cui *et al.*, 2005; Luangtongkum *et al.*, 2006; Miranda *et al.*, 2007; Siemon *et al.*, 2007; Schwaiger *et al.*, 2008; Lestari *et al.*, 2009). Thus, antibiotic resistant pathogens on these antibiotic-free farms may have acquired resistance through alternative routes. The increased occurrence and spread of antibiotic resistant pathogens may be based on the exchange of genetic elements (Carattoli, 2003). MDR genes have been found to be carried on large mobile extrachromosomal plasmids (Guerra *et al.*, 2001). The dissemination of resistance genes through horizontal gene transfer is attributed to the rise in resistance as well as the appearance of similar antibiotic resistance profiles among different strains of pathogens (Krauland *et al.*, 2009).

Integrons were first described in the late 1980s as possible mobile genetic elements associated with antibiotic resistance (Stokes & Hall, 1989). It is now known that integrons do not encode for self-mobilization, although they possess the ability to excise gene cassettes from plasmids and integrate them into the integron (Rowe-Magnus *et al.*, 2001; Liebana *et al.*, 2002; Martin *et al.*, 2008; Zaneveld *et al.*, 2008; Krauland *et al.*, 2009). However, integrons are often present on mobile genetic elements such as transposons and plasmids (Barlow *et al.*, 2004; Martin *et al.*, 2008; Krauland *et al.*, 2009). The integron contains a gene encoding for an integrase (int), a promoter (P_{ant}), as well as a recombination site (attI) (Rosser & Young, 1999; Fluit & Schmitz, 2004). An integron may or may not also contain gene cassettes (Rosser & Young, 1999; Fluit & Schmitz, 2004). The 3′ end of the integron is denoted the conserved sequence (CS) and, in class I integrons, may include combinations of genes for sulfonamide resistance (sul1, sul2, or sul3), an open reading frame (orf5), and qacEΔ 1, a gene for ethidium bromide resistance (Rosser & Young, 1999; Fluit & Schmitz, 2004; Hammerum *et al.*, 2006; Martin *et al.*, 2008). For commercial farm animals, 6.9% (Liebana *et al.*, 2002) and 29% (Martin *et al.*, 2008) of the *Salmonella* isolates were reported to have class I integrons. Melendez *et al.* (2010) observed a higher frequency of integrons (68%) among *Salmonella* isolates from pasture-raised poultry. This might be caused by greater contact with environment compared to the conventional poultry production (Melendez *et al.*, 2010).

19.5 CONCLUSIONS

Alternative or nonconventional poultry production is growing in popularity due to increasing consumer interest. These products are perceived as healthier and safer than conventionally produced products; however, few studies have focused on microbiological aspects of free-range, natural, or pasture poultry production systems or compared organic and conventional operations in parallel. Additional research on these poultry production systems is necessary due to the increase in microbial transmission routes such as outdoor access, which can create potential risks for increased contamination compared to conventional production. The increased transmission routes in alternative poultry production may lead to cross-contamination among flocks at the farm. The results from previous studies show that the prevalence and antibiotic resistance may vary among different poultry production systems. There is a need for effective control measures to reduce the food-borne pathogen contamination in both conventionally and organically produced poultry products. Consequently, there is a need for development and implementation of Good Agricultural practices (GAPS) and Good Manufacturing practices (GMPS) on alternative poultry farms and processing plants.

ACKNOWLEDGMENT

Support is gratefully acknowledged from a National Integrated Food Safety Initiative (NIFSI) (2008-51110-04339) and a USDA Food Safety Consortium grant to coauthor Steven C. Ricke.

REFERENCES

Aarestrup, F. M. 1999. Association between the consumption of antimicrobial agents in animal husbandry and the occurrence of resistant bacteria among food animals. *Int. J. Antimicrob. Agents*. 12:279–285.

Alali, W. Q., S. Thakur, R. D. Berghaus, M. P. Martin, and W. A. Gebreyes. 2010. Prevalence and distribution of *Salmonella* in organic and conventional broiler poultry farms. *Foodborne Pathog. Dis*. 7:1363–1371.

Avrain, L., F. Humbert, R. L'Hospitalier, P. Sanders, C. Vernozy-Rozand, and I. Kempf. 2003. Antimicrobial resistance in *Campylobacter* from broilers: association with production type and antimicrobial use. *Vet. Microbiol*. 96:267–276.

Bailey, J. S. and D. E. Cosby. 2005. *Salmonella* prevalence in free-range and certified organic chickens. *J. Food Prot*. 68:2451–2453.

Bangtrakulnonth, A., S. Pornreongwong, C. Pulsrikarn, P. Sawanpanyalert, R. S. Hendriksen, D. M. A. Lo Fo Wong, and F. M. Aarestrup. 2004. *Salmonella* serovars from humans and other sources in Thailand, 1993-2002. *Emerging Infect. Dis*. 10:131–136.

Barlow, R. S., J. M. Pemberton, P. M. Desmarchelier, and K. S. Gobius. 2004. Isolation and characterization of integron-containing bacteria without antibiotic selection. *Antimicrob. Agents Chemother*. 48:838–842.

Bell, D. D. and W. D. Weaver. 2002. *Commercial Chicken Meat and Egg Production*. 5th edn. Kluwer Academic Publishers, Norwell, MA. p. 1416.

Braden, C. R. 2006. *Salmonella enteric* serotype Enteritidis and eggs: a national epidemic in the United States. *Clin. Infect. Dis*. 43:838–842.

Bulling, E. R., R. Stephan, and V. Sebek. 1973. The development of antibiotic resistance among *Salmonella* bacteria of animal origin in the Federal Republic of Germany and West Berlin: 1st communication: a comparison between the years of 1961 and 1970–1971. *Zentralbl Bakteriol Orig A*. 225:245–256.

Carattoli, A. 2003. Plasmid-medicated antimicrobial resistance in *Salmonella* entercia. *Curr. Issues Mol. Biol.* 5:113–122.

CDC. 2008. *Salmonella* Surveillance: Annual Summary, 2006. US Department of Health and Human Services, CDC Atlanta, GA.

CDC. 2010. *Salmonella*. Atlanta, GA. Available at: www.cdc.gov/Salmonella (accessed October 24, 2011).

Cheng, H. W. 2005. Acute and chronic pain in beak trimmed chickens. In: P. Glatz (ed.) *Poultry Welfare Issues: Beak Trimming*. Nottingham University Press, Nottingham. pp. 31–49.

Commission Regulation. 2008. Laying down detailed rules for the application of council regulation (EC) No 1234/2007 as regards the marketing standards for poultrymeat. *Off. J. Eur. Union Commission Reg.* (EC) No. 543/2008. pp. 46–87.

Cui, S., B. Ge, J. Zheng, and J. Meng. 2005. Prevalence and antimicrobial resistance of *Campylobacter* spp. and *Salmonella* serovars in organic chickens from Maryland retail stores. *Appl. Environ. Microbiol.* 71:4108–4111.

Esteban, J. I., B. Oporto, G. Aduriz, R. A. Juste, and A. Hurtado. 2008. A survey of food-borne pathogens in free-range poultry farms. *Int. J. Food Microbiol.* 123:177–182.

Fanatico, A. C. 2006. Alternative poultry production systems and outdoor access. ATTRA publication. IP300. Available at: http://attra.ncat.org/attra-pub/poultryoverview.html (accessed October 24, 2011).

Fanatico, A. C., C. M. Owens, and J. L. Emmert. 2009. Organic poultry production in the United States: Broilers. *J. Appl. Poult. Res.* 18:355–366.

Fanatico, A. C., P. B. Pillai, J. L. Emmert, and C. M. Owens. 2007. Meat quality of slow- and fast-growing chicken genotypes fed low-nutrient or standard diets and raised indoors or with outdoor access. *Poult. Sci.* 86:2245–2255.

Finstad, S., C. A. O'Bryan, J. A. Marcy, P. G. Crandall, and S. C. Ricke. 2011. *Salmonella* and broiler processing in the United States: Relationship to foodborne salmonellosis. *Food Res. Int.* (In Press). doi:10.1016/j.foodres.2011.03.057.

Fluit, A. C. and F. J. Schmitz. 2004. Resistance integrons and super-integrons. *Clin. Microbiol. Infect.* 10:272–288.

Foley, S. L., R. Nayak, I. B. Hanning, T. J. Johnson, J. Han, and S. C. Ricke. 2011. Population dynamics of *Salmonella enterica* serotypes in commercial egg and poultry production. *Appl. Environ. Microb.* 77:4273–4279.

Galanis, E., D. M. A. Lo Fo Won, M. E. Patrick, N. Binsztein, A. Cieslik, T. Charlermchaikit, A. Aidara-Kane, A. Ellis, F. J. Angulo, and H. C. Wegener. 2006. Web-based surveillance and global *Salmonella* distribution, 2000-2002. *Emerging Infect. Dis.* 12:381–388.

Goncagual, G., E. Gunaydin, and K. T. Carli. 2003. Prevalence of *Salmonella* serogroups in chicken meat. *Turkish J. Vet. Anim. Sci.* 29:103–106.

Griggs, J. P., J. B. Bender, and J. P. Jacob. 2006. Microbial safety of chickens raised without antibiotics. *J. Appl. Poult. Res.* 15:475–482.

Griggs, J. P. and J. P. Jacob. 2005. Alternatives to antibiotics for organic poultry production. *J. Appl. Poult. Res.* 14:750–756.

Grimont, P. A. D. and F. Weill. 2007. Antigenic formulae of the *Salmonella* serovars. WHO collaborating centre for reference and research on *Salmonella*. 9th edn. Institut Pasteur, Paris. Available at: http://www.pasteur.fr/ip/portal/action/WebdriveActionEvent/oid/01s-000036-089.

Guerra, B., Soto, S. M., Arguelles, J. M., and Mendoza, M. C. 2001. Multidrug resistance is mediated by large plasmids carrying a class 1 integron in the emergent *Salmonella enteric* serotype [4,5.12:i:-] Antimicrob. *Agents Chemother.* 45:1305–1308.

Gupta, A., J. M. Nelson, T. J. Barrett, R. V. Tauxe, S. P. Rossiter, C. R. Friedman, K. W. Joyce, K. E. Smith, T. F. Jones, M. A. Hawkins, B. Shiferaw, J. L. Beebe, D. J. Vugia, T. Rabatsky-Ehr, J. A. Benson, T. P. Root, F. J. Angulo, and NARMS Working Group. 2004. Antimicrobial resistance among *Campylobacter* strains, United States, 1997–2001. *Emerging Infect. Dis.* 10:1102–1109.

Hammerum, A. M., D. Sandvang, S. R. Andersen, A. M. Seyfarth, L. J. Porsbo, N. Frimodt-Moller, and O. E. Heuer. 2006. Detection of *sul1*, *sul2*, and *sul3* in sulphonamide resistant *Escherichia coli* isolates obtained from healthy humans, pork, and pigs in Denmark. *Int. J. Food Microbiol.* 106:235–237.

Han, F., S. I. Lestari, S. Pu, and B. Ge. 2009. Prevalence and antimicrobial resistance among *Campylobacter* spp. in Louisiana retail chickens after the enrofloxacin ban. *Foodborne Pathog. Dis.* 6:163–171.

Hanning, I., D. Biswas, P. Herrera, M. Roesler, and S. C. Ricke. 2010. Prevalence and characterization of *Campylobacter jejuni* isolated from pasture flock poultry. *J. Food Sci.* 75:M496–M502.

Harper, G. C. and A. Makatouni. 2002. Consumer perception of organic food production and farm animal welfare. *Br. Food J.* 104:287–299.

Herikstad, H., Y. Motarjemi, and R. V. Tauxe. 2002. *Salmonella* surveillance: a global survey of public health serotyping. *Epidemiol. Infect.* 129:1–8.

Henderson, S. N., J. T. Barton, A. D. Wolfenden, S. E. Higgins, J. P. Higgins, W. J. Kruenzel, C. A. Lester, G. Tellez, and B. M. Hargis. 2009. Comparison of beak-trimming methods on early broiler breeder performance. *Poult. Sci.* 88:57–60.

Henry, R. 2002. Organic poultry – Meat birds. Maritime certified organic growers, organic farm profiles. Available at: http://www.acornorganic.org/pdf/poultrymeatprofile.pdf (accessed April 5, 2011).

Herren, R. 2000. *The Science of Agriculture*. 2nd edn. Delmar, Albany, NY. p. 592.

Heuer, O. E., K. Pedersen, J. S. Andersen, and M. Madsen. 2001. Prevalence and antimicrobial susceptibility of thermophilic *Campylobacter* in organic and conventional broiler flocks. *Lett. Appl. Microbiol.* 33:269–274.

Hoogenboom, L. A. P., J. G. Bokhorst, M. D. Northolt, L. P. L. van de Vijver, N. J. G. Broex, D. J. Mevius, J. A. C. Meijs, and J. Van der Roest. 2008. Contaminants and microorganisms in Dutch organic food products: a comparison with conventional products. *Food Addit. Contam.* 25:1197–1209.

Hughes, B. O. and M. J. Gentle. 1995. Beak trimming of poultry: its implications for welfare. World's Poult. *Sci. J.* 51:51–61.

Jacob, J. P., J. P. Griggs, and J. B. Bender. 2008b. Characterization of small-scale antibiotic-free broiler production in Minnesota. *J. Appl. Poult. Res.* 17:412–420.

Jacob, M. E., J. T. Fox, S. L. Reinstein, and T. G. Nagaraja. 2008a. Antimicrobial susceptibility of foodborne pathogens in organic or natural production systems: An overview. *Foodborne Pathog. Dis.* 5:721–730.

Jain, S. and J. Chen. 2006. Antibiotic resistance profiles and cell surface components of *Salmonella. J. Food Protect.* 69:1017–1023.

Jones, F. T., and S. C. Ricke. 2003. Observations on the history of the development of antimicrobials and their use in poultry feeds. *Poult. Sci.* 82:613–617.

Krauland, M. G., J. W. Marsh, D. L. Paterson, and L. H. Harrison. 2009. Integron-mediated multidrug resistance in a global collection of nontyphoidal *Salmonella enterica* isolates. *Emerging Infect. Dis.* 15:388–396.

Kuenzel, W. J. 2007. Neurobiological basis of sensory perception: welfare implications of beak trimming. *Poult. Sci.* 86:1273–1282.

Lestari, S. I., F. Han, F. Wang, and B. Ge. 2009. Prevalence and antimicrobial resistance of *Salmonella* serovars in conventional and organic chickens from Louisiana retail stores. *J. Food Prot.* 72:1165–1172.

Li, X., J. B. Payne, F. B. Santos, J. F. Levine, K. E. Anderson, and B. W. Sheldon. 2007. *Salmonella* populations and prevalence in layer feces from commercial high-rise houses and characterization of the *Salmonella* isolates by serotyping, antibiotic resistance analysis, and pulsed field gel electrophoresis. *Poult. Sci.* 86:591–597.

Liebana, E., C. Clouting, C. A. Cassar, L. P. Randall, R. A. Walker, E. J. Threlfall, F. A. Clifton-Hadley, A. M. Ridley, and R. H. Davies. 2002. A comparative study of *gyr* A mutations, cyclohexane resistance and class I integron presence in *Salmonella enterica* from farm animals in England and Wales. *J. Clinical Microbiol.* 40:1481–1486.

Luangtongkum, T., T. Y. Morishita, A. J. Ison, S. Huang, P. F. McDermott, and Q. Zhang. 2006. Effect of conventional and organic production practices on the prevalence and antimicrobial resistance of *Campylobacter* spp. in poultry. *Appl. Environ. Microbiol.* 72:3600–3607.

Lund, M., T. K. Welch, K. Griswold, J. B. Endres, and B. Shepherd. 2003. Occurrence of *Campylobacter* and *Salmonella* in broiler chickens raised in different production systems and fed organic and traditional feed. *Food Prot. Trends.* 23:252–256.

MacDonald, J. M. 2008. The economic organization of U.S. broiler production. Economic Information Bulletin, No. 38. Economic Research Service, USDA, Washington, DC. p. 26.

Magnusson, M. K., A. Arvola, U. K. Hursti, L. Aberg, and P. Sjoden. 2003. Choice of organic foods is related to perceived consequences for human health and to environmentally friendly behavior. *Appetite.* 40:109–117.

Martin, B. S., L. Lapierre, J. Cornejo, and S. Bucarey. 2008. Characterization of antibiotic resistance genes linked to class 1 and 2 integrons in strains of *Salmonella* spp. isolated from swine. *Can. J. Microbiol.* 54:569–576.

McEwen, S. A., and P. Fedorka-Cray. 2002. Antimicrobial use and resistance in animals. *Clin. Infect. Dis.* 34:S93–S106.

Meerburg, B. G., W. F. Jacobs-Reitsma, J. A. Wagenaar, and A. Kijlstra. 2005. Presence of *Salmonella* and *Campylobacter* spp. in wild small mammals on organic farms. *Appl. Environ. Microbiol.* 72:960–962.

Melendez, S. N., I. Hanning, J. Han, R. Nayak, A. R. Clement, A. Wooming, P. Hererra, F. T. Jones, S. L. Foley, and S. C. Ricke. 2010. *Salmonella enterica* isolates from pasture-raised poultry exhibit antimicrobial resistance and class I integrons. *J. Appl. Microbiol.* 109:1957–1966.

Milillo, S. R., J. C. Stout, I. Hanning, E. D. Fortes, H. C. den Bakker, M. Wiedmann, and S. C. Ricke. 2011. Isolation and characterization of *Listeria* from pasture-reared chickens and their environment. 111th General Annual Meeting of American Society for Microbiology. New Orleans, LA. May 21st–24th.

Miranda J. M., B. I. Vázquez, C. A. Fente, P. Calo-Mata, A. Cepeda, and C. M. Franco. 2008. Comparison of antimicrobial resistance in *Escherichia coli, Staphylococcus aureus*, and *Listeria monocytogenes* strains isolated from organic and conventional poultry meat. *J. Food Prot.* 71:2537–2542.

Miranda, J. M., M. Guarddon, B. I. Vazquez, C. A. Fente, J. Barros-Velazquez, A. Cepeda, and C. M. Franco. 2007. Antimicrobial resistance in Enterobacteriaceae strains isolated from organic chicken, conventional chicken and conventional turkey meat: A comparative study. *Food Control.* 19:412–416.

Mortier, L., E. Daeseleire, and C. V. Peteghem. 2005. Determination of the ionophoric coccidiostats narasin, monensin, lasolocid and salinomycin in eggs by liquid chromatography/tandem mass spectrometry. *Rapid Commun. Mass Spectrom.* 19:533–539.

National Chicken Council. 2010. National chicken council animal welfare guidelines and audit checklist for broilers. Washington, DC. Available at: http://www.fao.org/ag/againfo/themes/animal-welfare/aw-docum/codes-of-practice-and-recommendations/detail/it/item/40522/icode/6/ (accessed October 24, 2011).

Oberholtzer, L., C. Greene, and E. Lopez. 2006. Organic poultry and eggs capture high price premiums and growing share of specialty markets. U.S. Department of Agriculture (USDA), Economic Research Service (ERS). Available at: http://www.ers.usda.gov/Publications/LDP/2006/12Dec/LDPM15001/ (accessed October 24, 2011).

Overbeke, I. V., L. Duchateau, L. De Zutter, G. Albers, and R. Ducatelle. 2006. A comparison survey of organic and conventional broiler chickens for infectious agents affecting health and food safety. *Avian Dis.* 50:196–200.

Parveen, S., M. Taabodi, J. G. Schwarz, T. P. Oscar, J. Harter-Dennis, and D. G. White. 2007. Prevalence and antimicrobial resistance of *Salmonella* recovered from processed poultry. *J. Food Prot.* 70:2466–2472.

Pieskus, J., M. P. Franciosini, P. C. Proietti, F. Reich, E. Kazeniauskas, C. Butrimaite-Ambrozeviciene, M. Mauricas, and N. Bolder. 2008. Preliminary investigations on *Salmonella* spp. incidence in meat chicken in Italy, Germany, Lithuania and the Netherlands. *Int. J. Poult. Sci.* 7:813–817.

Plamodon, R. 2003. Range poultry housing: Livestock production guide. ATTRA publication. CT 125. p. 16. Available at: http://attra.ncat.org/attra-pub/poulthous.html (accessed October 24, 2011).

Price, L. B., L. G. Lackey, R. Vailes, and E. Silbergeld. 2007. The persistence of fluoroquinolone-resistant *Campylobacter* in poultry production. *Environ. Health Perspect.* 115:1035–1039.

Restaino, L., G. S. Grauman, W. A. McCall, and W. M. Hill. 1977. Effects of varying concentrations of novobiocin incorporated into two *Salmonella* plating media on the recovery of four *Enterobacteriaceae*. *Appl Environ Microbiol.* 33:585–589.

Rodenburg, T. B., M. C. Van Der Hulst-Van Arkel, and R. P. Kwakkel. 2004. *Campylobacter* and *Salmonella* infections on organic broiler farms. *Nat. J. Appl. Sci.* 52:101–108.

Rosser, S. J., and H. K. Young. 1999. Identification and characterization of class 1 integrons in bacteria from an aquatic environment. *J. Antimicrob. Chemother.* 44:11–18.

Rowe-Magnus, D. A., A. M. Gueroun t, P. Ploncard, B. Dychinco, J. Davies, and D. Mazel. 2001. The evolutionary history of chromosomal super-integrons provides an ancestry for multiresistant integrons. *Proc. Natl. Acad. Sci. U.S.A.* 98:652–657.

Rusul, G., J. Khair, S. Radu, C. T. Cheah, and R. Yassin. 1996. Prevalence of *Salmonella* in broilers at retail outlets, processing plant and farms in Malaysia. *Int. J. Food Microbiol.* 33:183–194.

Sahin, O., T. Y. Morishita, and Q. Zhang. 2002. *Campylobacter* colonization in poultry: sources of infection and modes of transmission. *Anim. Health Res. Rev.* 3:95–105.

Scallan, E., R. M. Hoekstra, F. J. Angulo, R. V. Tauxe, M. Widdowson, S. L. Roy, J. L. Jones, and P. M. Griffin. 2011. Foodborne illness acquired in the United States – Major pathogens. *Emerging Infect. Dis.* 17:7–15.

Schwaiger, K., E. M. V. Schmied, and J. Bauer. 2008. Comparative analysis of resistance characteristics of gram-negative bacteria isolated from laying hens and eggs in conventional and organic keeping systems in Bavaria, Germany. *Zoonoses Public Health.* 55:331–341.

Siemon, C. E., P. B. Bahnson, and W. A. Gebreyers. 2007. Comparative investigation of prevalence and antimicrobial resistance of *Salmonella* between pasture and conventionally reared poultry. *Avian Dis.* 51:112–117.

Sirsat, S. A., A. Muthaiyan, and S. C. Ricke. 2009. Antimicrobials for pathogen reduction in organic and natural poultry production. *J. Appl. Poult. Res.* 18:379–388.

Snelling, W. J., M. Matsuda, J. E. Moore, and J. S. G. Dooley. 2005. Under the microscope: *Campylobacter jejuni*. *Lett. Appl. Microbiol.* 41:297–302.

Sofos, J. N. 2008. Challenges to meat safety in the 21st century. *Meat Sci.* 78:3–13.

Soonthornchaikul, N., H. Garelick, H. Jones, J. Jacobs, D. Ball, and M. Choudhury. 2006. Resistance to three antimicrobial agents of *Campylobacter* isolated from organically- and intensively-reared chickens purchased from retail outlets. *Int. J. Antimicrob. Agents*. 27:125–130.

Stern, N. J., P. Fedorka-Cray, J. S. Bailey, N. A. Cox, S. E. Craven, K. L. Hiett, M. T. Musgrove, S. Ladely, D. Cosby, and G. C. Mead. 2001. Distribution of *Campylobacter* spp. in selected U.S. poultry production and processing operations. *J. Food Prot.* 64:1705–1710.

Stokes, H. W. and R. M. Hall. 1989. A novel family of potentially mobile DNA elements encoding site-specific gene-integration functions: integrons. *Molec. Microbiol.* 3:1669–1683.

Swick, R. A. 1996. Growth promotants in poultry and swine feed. ASA Technical Builletin. vol. An04. pp. 1–9.

USDA. 2005. Food standards and labeling policy book. Available at: http://www.fsis.usda.gov/OPPDE/larc/Policies/Labeling_Policy_Book_082005.pdf (accessed May 2011).

USDA. 2009. News release: USDA to launch high tunnel pilot study to increase availability of locally grown foods, 3-year project to verify effectiveness of high tunnels in natural resource conservation. Release No. 0617.09.

USDA-FSIS. 2008. Product labeling: use of animal raising claims in the labeling of meat and poultry products. *Fed. Reg.* 73:60228–60230.

USDA-FSIS. 2010a. Serotypes profile of *Salmonella* isolates from meat and poultry products. January 1998 through December 2010. Available at: http://www.fsis.usda.gov/Science/Serotypes_Profile_Salmonella_Isolates/index.asp.

USDA-FSIS. 2010b. Quarterly progress report on *Salmonella* testing of selected raw meat and poultry products: Preliminary Results, Oct–Dec 2010. Available at: http://www.fsis.usda.gov/Science/Quarterly_*Salmonella*_Results/index.asp (accessed March 2011).

USDA National Organic Program. 2008. Agricultural Marketing Service 7, Code of Federal Regulations (CFR), Part 205: National Organic Program.

Van Loo, E. J., V. Caputo, R. M. Nayga, J.-F. Meullenet, P. G. Crandall, and S. C. Ricke. 2010. Effect of organic poultry purchase frequency on consumer attitudes toward organic poultry meat. *J. Food Sci.* 75:S384–S397.

Van Loo, E. J., V. Caputo, R. M. Nayga, J.-F. Meullenet, and S. C. Ricke. 2011. Consumers' willingness to pay for organic meat: Experimental evidence from chicken breast. *Food Qual. Prefer.* 22:603–613.

Van, T. T. H., G. Moutafis, T. Istivan, L. T. Tran, and P. J. Coloe. 2007. Detection of *Salmonella* spp. in retail raw food samples from Vietnam and characterization of their antibiotic resistance. *Appl. Environ. Microbiol.* 73:6885–6890.

Wedderkopp, A., K. O. Gradel, J. C. Jorgensen, and M. Madsen. 2001. Pre-harvest surveillance of *Campylobacter* and *Salmonella* in Danish broiler flocks: a 2-year study. *Int. J. Food Microbiol.* 68:53–59.

Wegener, H. C., T. Hald, D. L. F. Wong, M. Madsen, H. Korsgaard, F. Bager, P. Gerner-Smidt, and K. Molbak. 2003. *Salmonella* control program in Denmark. *Emerging Infect. Dis.* 9:774–780.

White, D. G., S. Zhao, R. Sudler, S. Ayers, S. Friedman, S. Chen, P. F. McDermott, S. McDermott, D. D. Wagner, and J. Meng. 2001. The isolation of antibiotic-resistant *Salmonella* from retail ground meats. *N. Engl. J. Med.* 345:1147–1154.

Zaneveld, J. R., D. R. Nemergut, and R. Knight. 2008. Are all horizontal gene transfers created equal? Prospects for mechanism-based studies of HGT patterns. *Microbiol.* 154:1–15.

Zhao, S., P. F. McDermott, S. L. Friedman, S. Qaiyumi, J. W. Abbott, C. Kiessling, S. Ayers, R. Singh, S. Hubert, J. Sofos, and D. G. White. 2006. Characterization of antimicrobial resistant *Salmonella* isolated from imported foods. *J. Food Prot.* 69:500–507.

Section V

Preharvest Control Measures for Assuring the Safety of Organic Meats

20 Probiotics as Pathogen Control Agents for Organic Meat Production

Gregory R. Siragusa and Steven C. Ricke

Abstract: Probiotics, also referred to as competitive exclusion cultures or direct-fed microbials, consist of either single microorganism or groups of microorganisms that when administered to a host animal are capable of eliciting an identifiable beneficial response in the recipient host. Beneficial responses include limiting food-borne or disease-causing pathogen colonization and/or infection, as well as a myriad of host performance indicators such as improved or more efficient growth and better health. Although probiotics have been examined for most food animal species, it is still unclear which mechanisms are employed by these cultures. Both host and microbial factors are involved in an interactive fashion between the commensal bacteria and the corresponding host animal. As more sophisticated molecular tools are applied to the gastrointestinal tract microbiota, a much more precise and detailed picture will emerge without doubt. Once more is known about the gut microbiome, it should be possible to develop more consistent and effective probiotic cultures.

Keywords: probiotics; competitive exclusion; antibiotic growth promoters; poultry; ruminants; swine; microbiome; direct-fed microbials; company-specific inoculant

20.1 INTRODUCTION

Drug-free and organic poultry production within the United States is a relatively small sector compared to conventionally reared production (USDA-ERS, 2010), but antibiotic-free production is increasing in volume (O'Bryan *et al.*, 2008). From 1997 to 2008, total poultry numbers (broilers, layers, and turkeys) certified in the United States Department of Agriculture (USDA) organic program increased from 798,250 to 15,518,075 (USDA-ERS, 2010). Growing pressure from the regulatory and public health sectors as well as consumer groups is providing a motive for the industry to consider alternative pathogen control interventions for production on a wider scale within the traditionally conventional production sector of the food animal industry.

Although a variety of biological agents have been suggested and in some cases experimentally applied to food animal production systems, much remains to be determined regarding practicality and commercialization (Nisbet, 2002; Joerger, 2003; Berghman *et al.*, 2005; Ricke *et al.*, 2005; Sirsat *et al.*, 2009). Among the available biologicals, probiotics and competitive exclusion cultures have been fairly extensively examined for their potential abilities

Organic Meat Production and Processing, First Edition. Edited by Steven C. Ricke, Ellen J. Van Loo, Michael G. Johnson, and Corliss A. O'Bryan.

to limit colonization of food-borne pathogens in the gastrointestinal (GI) tract. Mechanistically, probiotic or competitive exclusion cultures serve more as a GI tract microflora barrier that "competitively excludes" or prevents colonization of the respective food-borne pathogen. The purpose of this chapter is to highlight administration of probiotics as an alternative biological intervention for pathogen control in food animal production to decrease reliance on antibiotic growth promoters (AGPs) or conventional antibiotics.

20.2 ANTIBIOTICS IN FOOD ANIMAL PRODUCTION

While antimicrobial resistant (AR) bacteria and their association to both human disease and AGPs usage in agriculture is a contentiously debated issue, there is little question of the adverse impacts on poultry production efficiency arising from withdrawal of AGPs. For example, since original work published by Jukes and Williams (1953), poultry production has clearly been more consistently efficient and highly reliable with the routine usage of AGPs of various classes. While estimates for nontherapeutic antibiotic usage in the US poultry industry varies (Jones & Ricke, 2003), the impact of feeding low levels of antibiotics has been proven to select for resistant bacterial pathogens and many examples have been published on the incidence of specific antibiotic resistance expression among poultry associated zoonotic pathogens from antibiotic-fed poultry (Butaye *et al.*, 2003; Emborg *et al.*, 2003; Ashayerizadeh *et al.*, 2009; Young *et al.*, 2009). Challenges to the routine usage of AGPs in poultry production is strongly driven by the development and increased incidence of AR human pathogenic bacteria, as well as a basic understanding of the propagation of antibiotic-resistant bacteria (Aminov & Mackie, 2007). While the scientific evidence for reducing AGP usage is relatively consistent, the economic case for AGP usage has been analyzed for its high cost to producers with questionable return (Graham *et al.*, 2007). Not all interpretations of the data support a ban on AGPs (Phillips *et al.*, 2004).

Poultry welfare and disease incidence are reported to be negatively impacted from AGP withdrawal. In production terms, poultry disease levels rise, mortalities increase, feed conversion worsens, and final weights decrease significantly. From a microbiological viewpoint, AGP withdrawal presents conditions favorable to clostridial overgrowth and subsequent necrotic enteritis, avian pathogenic *Escherichia coli* (APEC) proliferation, and resulting clinical or subclinical disease. The ever-present challenge from *Eimeria* spp. coccidiosis, which is considered to precede onset of clostridial disease, becomes even more critical to bird health (Williams, 2005).

Within the European Union, AGPs are banned from usage in livestock production. The so-called "Danish Experiment," whereby a complete ban of AGPs was mandated, has led to an industry which is able to meet production demands albeit at a greater food cost to the consumer (Aarestrup *et al.*, 2008). From a worldwide poultry production viewpoint, countries that ultimately export their poultry meat must also consider and adhere to the food laws of the importing country and produce meat for final delivery within that standard.

Another consequence of subtherapeutic antibiotic withdrawal from poultry production is that there appears to be an increased rate of carriage of some food-borne bacterial pathogens including *Campylobacter* while the case for *Salmonella* is less clear (Ashayerizadeh *et al.*, 2009; Bokkers *et al.*, 2009; Young *et al.*, 2009). Increased governmental and public health sector oversight and standards for lower levels of these pathogens will elevate the rising rate of zoonotic pathogens to an even greater level of importance for the producer, whether conventional or AGP free.

The food animal sector should continue to strive for producing meat that is reduced in zoonotic pathogen carriage, efficiently produced, enhances animal welfare, and does not deposit high levels of antibiotics in the environment to select for AR bacteria. On a more practical level, veterinarians, nutritionists, and production managers would benefit by having a set of alternative interventions that offer different advantages specific for their operations over time. A case in point is the usage of alternative feed ingredients such as dried or wet distillers grains (DDGS) that is steadily rising as a result of the current bioethanol industry; however, the impact of these ingredients have been shown to adversely affect microbial profiles of livestock by predisposing cattle to higher carriage rates of enterohemorrhagic *E. coli* (Wells *et al.*, 2009). Conversely, in broilers, the impact of DDGS on the general microbial levels was not observed, but it remains to be shown whether such alternate feed ingredients impact levels of pathogenic subtypes of APEC or type A *Clostridium perfringens* (Loar *et al.*, 2010). In the future, use of new and varied feedstuffs available to both the conventional, drug-free, and organic poultry producers will require pathogen control interventions that are effective within a variety of changing feed compositions.

20.3 DEVELOPMENT OF PROBIOTICS

Probiotics or direct-fed microbials (DFM) are live bacterial cultures which when administered to the host animal impart a beneficial influence on the host in some detectable manner. The majority of the focus for probiotic and DFM application in food animals has been on their potential ability to limit food-borne pathogen colonization in the GI tract. Only recently have other benefits to the host animal become a focus of extensive research (Yan *et al.*, 2011). Development and application of probiotics in feeds and food production has been extensively reviewed. Bernardeau and Vernoux (2009) differentiated probiotic use into the era of 1950–1993 and post-1993. This chronological division was derived from a consideration of regulatory status and oversight, safety, and product efficacy and claims. Although probiotic and DFM cultures have historically garnered interest as a potential means to improve overall host animal or human health via introduction of bacteria identified as beneficial to the host, considerable uncertainty persists on how best to achieve success (Klaenhammer, 2000; Saarela *et al.*, 2000). Part of the problem rests with a lack of understanding and appreciation of just how complex the GI tract microflora truly is in most farm production animals.

The classic approach to probiotic or DFM administration to animals has been to introduce bacterial cultures to relatively young animals such as day-old chicks where the gut microflora has yet to develop and the GI tract remains relatively unoccupied. Early successes with undefined culture mixtures proved that this concept would be effective against food-borne pathogen invaders such as *Salmonella* (Nurmi & Ratala, 1973). Subsequent efforts to refine this approach proved to be more inconsistent and less predictable, primarily due to minimal attempts to identify or at least characterize the bacteria in these mixtures administered to the animal. This in turn led to efforts to use more defined cultures where the microorganisms that made up the probiotic culture were identified and administration could be more consistent for potential large-scale commercial production.

However, not knowing the fate of the probiotic culture once administered to the animal is problematic for any attempt to standardize for potential commercialization. Consequently, development of defined probiotic cultures where all bacteria were characterized became the preferred research direction for potential commercialization (Nisbet, 2002). As these cultures were developed, it became apparent that better methods were needed not only for identification of the probiotic microorganisms but elucidation of their metabolic role in the GI tract (Ricke & Pillai, 1999; Vaughan *et al*., 2000). Consequently, molecular techniques such as denatured gradient gel electrophoresis (DGGE), %G (Guanine) + C (Cytosine) profiling, and 16S rDNA sequencing were applied over the past decade to profile GI tract microflora responses and develop a better understanding of their responses to external factors such as shifts in diets (Apajalahti *et al*., 2001, 2002; Gong *et al*., 2002, 2006; Zhu *et al*., 2002; Hume *et al*., 2003; Wagner *et al*., 2003; Hanning & Ricke, 2011).

With the advent of economical next-generation sequencing technologies, it has now become possible to conduct metagenomic analyses of entire GI tract populations from extracts containing the genetic material of the microbial consortia (Kim & Mundt, 2011). Combining high throughput pyrosequencing with quantitative real time polymerase chain reaction analyses represents the possibility for assessment of specific genes and thus, the potential for tracking specific microorganisms within the background of the GI tract microbial population (Zhang *et al*., 2011). In the near future it should become likely to follow the introduction of probiotic cultures in the GI tract and not only assess their ability to become established, but any influence they might have on the GI tract microbial population. Acquiring some of these types of data may go a long way toward anticipating variability in host responses to probiotic administration and, in turn, help to design more optimal and consistent probiotic cultures. Such methods will also allow assessment and comparison of probiotic efficacy in different food animal genotypes.

20.4 PROBIOTICS AND THE GI TRACT

20.4.1 Ruminants

Developing an understanding of GI microbial ecology and the metabolic roles that individual organisms play as members of that consortia has been studied in some gut systems, such as the rumen of ruminants, for over 60 years (Hungate, 1950, 1966, 1979; Bryant, 1959; Hungate *et al*., 1964). The rumen or the forestomach of ruminants has been studied extensively, not only from a microbial ecology standpoint, but physiological functions such as rumen motility, as well as comprehensive characterization of digesta flow rates and turnover for various dietary regimes have been quantified (Church, 1976; Van Soest, 1994). However, it has always been difficult to match the dynamics of the rumen with its unique microbial consortia. In the past few years, as more sophisticated molecular analyses have been applied, some of these pieces have been put together. Consequently, the details of the metabolic complexity of the rumen function in ruminants have become clearer (Stevenson & Weimer, 2007; Uyeno *et al*., 2007; Weimer *et al*., 2009).

It is now known that not only primary microflora are present to degrade dietary components such as cellulose and hemicelluloses, but numerous individual bacterial species play important intermediate metabolic roles, such as cross feeding on hydrolyzed carbohydrate fragments and fermentation of end products, which in turn can serve as substrates for other unique microorganisms in the gut (Ricke *et al*., 1996; Mackie, 2002; Weimer *et al*., 2009). The relatively slow turnover of the rumen also supports a methanogenic population that possesses slower growth rates and produces methane as their primary end product (Karasov & Carey, 2009). In addition, the methanogens utilize hydrogen to form methane to the extent that hydrogen levels are lowered sufficiently to influence the thermodynamics of fermentation pathways toward more oxidized products such as acetate rather than ethanol (Wolin & Miller, 1982).

Despite these complexities, some studies have reported success in limiting food-borne pathogen *E. coli* O157:H7 in calves and feedlot cattle with specific probiotic *Lactobacillus* strains and nonpathogenic *E. coli* (Brashears *et al.*, 2003a, 2003b; Zhao *et al.*, 2003; Loneragan & Brashears, 2005; Younts *et al.*, 2004; Younts-Dahl *et al.*, 2005). Fungal additives such as *Aspergillus oryzae* and *Aspergillus Niger*, and the yeast *Saccharomyces cerevisiae* have also been extensively studied for potential ability to influence rumen fermentation but their biological modes of action may differ (Wallace & Newbold, 1992).

20.4.2 Swine

In pigs, the GI tract is a fairly complex microbial ecosystem with a wide range of anaerobic bacteria, many of which remain to be identified (Ghnassia, 1979; Makala *et al.*, 2000; Leser *et al.*, 2003; Chaucheyras-Durand & Durand, 2010). Methanogens are located in both the cecum and the colon (Miller *et al.*, 1986; Mao *et al.*, 2011). Leser *et al.* (2003) generated a library of bacterial 16S rDNA sequences cloned from the GI tract microbial consortia from pigs of various ages and health status when fed different diets. The resulting characterized phylotypes represented 13 major phylogenetic linkages, and of the over 300 phylotypes 81% belonged to the low $G + C$ Gram-positive group and 11.2% were associated with *Bacteroides* and *Prevotella* related microorganisms (Leser *et al.*, 2003).

Probiotics are administrated in swine either not long after birth, during times of anticipated disease outbreaks, or combined with feed as an ongoing continuous supplementation (Jonsson & Conway, 1992). In general, probiotic bacteria used in swine either originate from the indigenous microflora or they do not (Jonsson & Conway, 1992). Several bacteria including *Lactobacillus* spp., *Enterococcus* spp., *Bacillus* spp., *Bifidobacterium* spp., and *Streptococcus* spp., along with yeast and fungi such as *S. cerevisiae* and *A. oryzae* have been administered as probiotics to swine (Collado *et al.*, 2007; Scharek *et al.*, 2007a, 2007b; Choi *et al.*, 2011). In summarizing a series of studies, Mulder *et al.* (1997) concluded that lactobacilli could be administrated orally such that they were sufficiently stable to resist the weaning of the host pigs and still elicit activity against enteropathogens. However, inconsistent pig performance responses to lactobacilli probiotic administration have also been noted and a more targeted approach of selecting epithelial cell associated lactobacilli isolated from weanling pigs has been suggested (Krause *et al.*, 1997).

20.4.3 Poultry

Chickens may not have the same GI complexity as the rumen or even the pig GI tract, but many of the same ecological metabolic relationships still appear to be relevant. For example, chickens have been shown to harbor methanogens and produce methane (Ricke *et al.*, 2004). More recently, it has been shown that although the methanogen species involved are not nearly as diverse as those found in the rumen, they do become detectable in fairly young chicks (Saengkerdsub *et al.*, 2007a, 2007b). Likewise, microbial populations associated with the mucosal layer that lines the GI tract wall of the chicken may be an important component. Drolesky *et al.* (1995) used scanning electron microscopy to demonstrate that visual increases in mucosal epithelial colonization by a 29 microorganism competitive exclusion culture paralleled increased resistance to *Salmonella* Typhimurium colonization and increased concentrations of short chain fatty acids (SCFA).

How extensively and consistently probiotic establishment influences overall fermentation in the chicken GI tract remains to be determined but alterations in fermentation could certainly have impacts on the bird's metabolism as well. Evidence of these host changes in response to probiotics is reflected in some key bird performance parameters. Luo *et al.* (2010) demonstrated that layer hens fed a probiotic *Propionibacterium jensenii* supplement exhibited increased average egg weights and altered fatty acid profiles in eggs. Performance may also be enhanced by poultry disease prevention properties associated with probiotic administration, which would be an attractive alternative for organic poultry production. McReynolds *et al.* (2009) demonstrated that combinations of a DFM multistrain product and phytogenic products could lower intestinal lesions, mortalities, and levels of *C. perfringens*, the causative microorganism for necrotic enteritis. Combining plant-derived phytogenic feed additives with probiotics to replace AGPs to improve growth performance in poultry is becoming more prominent in commercial markets (Applegate *et al.*, 2010). Addition of probiotics may also enhance performance in chickens from an improved digestive function standpoint as well. Sieo *et al.* (2005) reported that supplementing diets with β-glucanase-producing *Lactobacillus* increased jejunum villus height, decreased the length of duodenum, jejunum, ileum, and ceca, and reduced fecal passage rate in broiler chickens.

Based on these results, it would appear that even in a presumably fairly simplistic GI ecosystem such as the chicken there still remains a complex interrelated and metabolically connected microbial population. This does influence strategy for development of commercially successful probiotics. For a probiotic to be able to ward off food-borne pathogens in the chicken ceca not only requires establishment of the culture in the GI tract, but retention of the metabolic relationship among the members of that probiotic consortia (Nisbet *et al.*, 1994, 1996a, 1996b). Consequently, there is considerable opportunity for such cultures to be variable in their ability to uniformly function under all conditions. This has led to the rethinking of the application of probiotics and how best to use them effectively. For example, combinations of competitive exclusion cultures with lytic bacteriophage, such as *Salmonella*-specific phage, have been used in chickens to take advantage of the ability to use the phage to eliminate *Salmonella* in already infected chickens in conjunction with administration of a probiotic culture (Toro *et al.*, 2005).

More recently, it appears that the original concept to use probiotics to accelerate colonization by the organisms, considered to be indigenous but not necessarily members of the traditional probiotic cultures, may be the better application. Along with that comes the philosophy of designing cultures to improve overall health and efficiency with more focus on broad-spectrum limitation of subclinical problems associated with a variety of pathogens. Newer generation products such as spore-forming bacteria as probiotics also offer specific advantages (Hong *et al.*, 2005). Significant evidence exists to support the probiotic enhancement of poultry production as documented in more recently conducted research (Tellez *et al.*, 2011).

20.5 PROBIOTICS AND MECHANISMS OF PROTECTION

A wide range of studies has identified several potential mechanisms elicited by probiotic bacteria against food-borne pathogens. These mechanisms fall into two broad categories, namely those directly associated with metabolic activities of the probiotic culture and those indirectly associated via stimulation of a host system, such as an altered immune response.

20.5.1 Host system–GI tract responses

Early development of specific components of the immune system in both swine and poultry has been attributed to probiotic supplementation (Scharek *et al.*, 2005, 2007a, 2007b; Farnell *et al.*, 2006; Haghighi *et al.*, 2006; Schierack *et al.*, 2007, 2009; Yoshimura *et al.*, 2010). Recently, it has been demonstrated that different *Bifidobacterium* strains will induce different *in vitro* immune responses (López *et al.*, 2009). This suggests that not only the presence of probiotic bacteria, but their specific composition also, could be an important modulator of specific host immune responses in the GI tract of young animals.

Interactions between GI tract microflora and the host immune system may not always be expressed as a general stimulation of the immune response but in some cases may actually repress certain components to aid establishment. For example, Peterson *et al.* (2007) using immunodeficient mice demonstrated that when the host was exposed to *B. thetaiotamicron*, a bacterial species that can be isolated from the GI tract, specific IgA was detected that reduced proinflammatory host response and epitope expression by *B. thetaiotamicron*. They suggested that this immunoselection of bacterial epitope expression is part of the mechanism for adaptive immune response that contributes to a sustained host–GI tract microbiota relationship. Hapfelmeier *et al.* (2010), using germ-free mice, demonstrated that although upon introduction of commensal GI tract bacteria induced concomitant IgA response in these mice, there was no evidence of an immune memory similar to what is seen with classical pathogen colonization and IgG responses (Cerutti, 2010). This is supported by recent evidence that peptidoglycan from bacteria can activate the innate immune system (Clarke *et al.*, 2010). More recently, Round *et al.* (2011) demonstrated that the human commensal *Bacteroides fragilis* could activate Toll-like receptors on the mucosal surface of mice by secreting a symbiosis factor to promote immunotolerance and facilitate colonization by this microorganism. Screening probiotic cultures for such factors may be a means to optimize early establishment of niche-specific microorganisms.

20.5.2 Fermentation acid production

From a metabolic standpoint, generation of metabolites such as fermentation acids produced by probiotic cultures and GI tract microflora have historically been identified as a potential inhibitory mechanism toward food-borne pathogens for a number of years (Russell, 1992; Ricke, 2003a; Van Immerseel *et al.*, 2006). Traditionally, it was believed that GI tract microorganisms were more resistant to their own end products than were the pathogens that they were able to outcompete, thereby preventing pathogen establishment (Wallace *et al.*, 1989; Russell, 1992; Ricke, 2003a). However, as more work has been done to characterize metabolic products of food-borne pathogens, such as *Salmonella* and their ecology in the GI tract, it is less clear how important this proposed mechanism is. For example, *Salmonella* produce fermentation acids when grown anaerobically in continuous culture (Dunkley *et al.*, 2009), suggesting that these cells at least when grown under these conditions, may be capable of protecting themselves from similar end products produced by indigenous GI tract organisms. This is supported by a series of studies that demonstrated that *S.* Typhimurium could elicit acid tolerance mechanisms when exposed to SCFA at neutral pH and in concentrations similar to those found in the GI tract of poultry (Kwon & Ricke, 1998a, 1998b, 1999; Kwon *et al.*, 1998, 2000).

The SCFA profile produced by GI tract microflora may exert considerable influence on the colonization and pathogenesis of food-borne pathogens entering the GI tract. Durant *et al.* (1999b, 2000c, 2000d) demonstrated with HEp2 tissue culture cell studies that growth phase and type of SCFA could influence the extent of both, attachment to and invasion of, these cells by *S.* Typhimurium. Follow-up studies with *S.* Typhimurium gene fusion assays of key virulence genes *hilA* and *invF* indicated that individual SCFAs could modulate expression of these genes, depending on the pH level (Durant *et al.*, 2000a). Based on a range of approaches, it appears that the type of SCFA produced is important in *Salmonella* GI tract ecology with some SCFA tending to enhance *Salmonella* virulence gene expression while other SCFA can repress expression (Durant *et al.*, 2000a; Lawhon *et al.*, 2002; Gantois *et al.*, 2006; Huang *et al.*, 2008). However, *Salmonella* responses could be influenced by the growth conditions, such as anaerobic versus aerobic atmosphere, growth physiology of the organism, and serotype. For example, there is some indication that *Salmonella* Enteritidis can adapt to propionate (Calhoun & Kwon, 2010) while *Salmonella* Kentucky appears to differ in its acid sensitivity as compared to other *Salmonella* serotypes (Joerger *et al.*, 2009). The implications for these varied responses for *Salmonella* ecology in the GI tract remain to be sorted out via more comprehensive genomic responses applied to *Salmonella* cultures directly recovered from GI tracts. Likewise, presumably there are differences among non-*Salmonella* food-borne pathogens and their responses to pH and SCFA, which may explain why probiotics may be either more or less effective in limiting establishment and persistence of individual pathogen strains in the GI tract.

20.5.3 Nutrient utilization

In theory, it is believed that GI tract bacteria can selectively consume certain nutrients that could be utilized by the food-borne pathogens, hence limiting the latter's ability to sustain themselves in the GI tract. Indeed, competition between probiotic bacteria and pathogens for specific nutrients such as amino acids and particular carbon sources has been documented with pure culture *in vitro* studies as a possible mechanism for successful probiotic exclusion of pathogens (Ha *et al.*, 1994, 1995; Durant *et al.*, 2000b; Herrera *et al.*, 2011). However, demonstrating similar outcomes with *in vivo* tests has yielded less clear results and thus may be highly dependent on the particular probiotic strain used. The nature of the specific nutrient involved is also probably critical as general decreases in nutrient availability in the GI tract can influence food-borne pathogen ecology and pathogenesis. The best example of this is the historical practice of feed withdrawal used to induce molt in laying hens (Ricke, 2003b; Norberg *et al.*, 2010). When feed is removed even for a few days, *S.* Enteritidis not only rapidly colonizes the hen's GI tract but infects the bird's organs including the spleen, liver, and ovaries (Holt *et al.*, 1995; Corrier *et al.*, 1997; Durant *et al.*, 1999a; Dunkley *et al.*, 2007a).

Molecular studies, traditional enumeration methods, and GI tract lumen metabolite characterization studies have revealed that feed removal from the GI tract contents appears to decrease fermentation activities and either shift microbial populations or in some cases actually decrease numbers and diversity of the indigenous GI tract microflora (Durant *et al.*, 1999a; Hume *et al.*, 2003; Woodward *et al.*, 2005; Dunkley *et al.*, 2007b; Callaway *et al.*, 2009). Under the extreme nutrient limiting conditions, invasiveness of *S.* Enteritidis has been directly linked with increased expression of a key virulence regulatory gene *hilA* (Durant *et al.*, 1999a; Dunkley *et al.*, 2007a). Providing diets with fermentable substrates such as lactose or dietary fiber or cereal grain by-products sources has been demonstrated to alleviate and/or prevent some of the more dramatic shifts in GI tract populations and fermentation as well as limit *S.* Enteritidis and its ability to colonize and invade the GI tract (Corrier *et al.*, 1997; Seo *et al.*, 2001; Woodward *et al.*, 2005; Dunkley *et al.*, 2007a, 2007b).

The contribution of feed and dietary composition to the makeup of the GI tract microbial population has always been somewhat of an enigma. Assessing and comparing mammalian fecal populations by shotgun sequencing and targeted sequencing of bacterial 16S ribosomal RNA genes indicates that there is some evidence suggesting a similar shift on adaptation of these microbial populations to changes in diet among the different foregut and hindgut fermenting herbivores as well as carnivores and omnivores (Muegge *et al.*, 2011). This has implications for development of probiotic cultures, as successful administration and retention of these cultures in the GI tract for maximum effect on the host may well depend on their ability to match their responses with the indigenous microbial population to changes in dietary composition.

An additional confounding factor may be in play as well. Sonnenburg *et al.* (2005) demonstrated that *B. thetaiotamicron* could induce outer membrane polysaccharide-binding proteins and glycoside hydrolases to utilize host mucus glycans in the absence of dietary polysaccharides. The ability of GI tract microflora to switch to host endogenous carbon and energy sources in times of nutrient limitation may explain some of the shifts in microbial populations observed under exogenous nutrient limit stress conditions. In addition, both virulent and avirulent *S.* Typhimurium strains have been shown to bind to specific glycoproteins in the rat mucus layer with the virulent strain exhibiting sixfold more binding versus the avirulent strain (Vimal *et al.*, 2000). Such localization of attachment may play a role in pathogen nutrition as well. Krivan *et al.* (1992) reported that *Salmonella* possessed the ability to utilize intestinal mucus phosphatidylserine as their sole carbon and nitrogen source. Thus, for probiotic cultures, a determining competition factor may be not so much as how they compete for dietary nutrient sources but how well they compete with pathogens for host endogenous nutrient sources.

As advances have been made in studying the interaction between the GI tract microbiome and the host, more mechanistically defined functions have been identified that are specifically associated with probiotic bacteria nutrient acquisition and symbiotic-like activities at the molecular level. Using germ-free mice and metabolome approaches, Fukuda *et al.* (2011) demonstrated that only specific probiotic *Bifidobacterium* possessing certain carbohydrate transporters could produce acetate from nonglucose sources to prevent death in mice from lethal *E. coli* O157:H7. As Fukuda *et al.* (2011) points out, while many probiotic bacteria can produce sufficient quantities of acetate, it is the ability to produce acetate from nonglucose sources that may enable this subset of probiotics strains to play a more significant role in GI tract niches, where exogenous glucose is limited, and may explain why such strains perform more consistently as successful preventive probiotic cultures. Using functional genomic analysis in combination with a mouse model, Motherway *et al.* (2011) demonstrated that *Bifidobacterium breve* expressed a tight adherence pili that imparted host-specific colonization characteristics to this probiotic bacteria that, based on genomic sequences of other *Bifidobacterium* spp., could be a common property for this group of bacteria.

As more detailed studies are conducted, there will no doubt be further elucidation of the highly intricate and specific mechanisms of symbiotic-like associations between GI tract microflora and their respective host. As these factors become known, this will certainly lead to more directed selection approaches for isolating probiotic bacteria that exhibit more targeted benefits to the host. An example of how this might be done is discussed in the forthcoming section.

20.6 COMPANY-SPECIFIC INOCULANT

A system of formulating a probiotic mixture that is specific to production site pathogen loads has been developed. Termed company-specific inoculants (CSI), this system has been used to successfully produce efficacious probiotic mixtures of *Bacillus* spp. for use in poultry operations.

CSI began as a research process to identify a set of probiotic strains or DFMs. There are several stages to the process, which will be outlined here. A starting premise is that two of the main drivers of poultry production inefficiency are subclinical diseases from high levels of intestinal *C. perfringens* and APEC. Given the high diversity of pathogenic bacteria impacting poultry production efficiency, it is unlikely that a single DFM will be inhibitory to those entire groups of pathogens and microorganisms. Therefore, one approach is the combination of multiple strains.

Stepwise, the process begins by isolation of *C. perfringens* and APEC from a sampling of GI tracts from broiler, layer, or turkey operations. This library of bacterial isolates is confirmed for identity by a gene marker. Confirmed isolates are then genotyped by a standard pattern-based technique, randomly amplified polymorphic DNA (RAPD), and resolved into clades using dendrogram generation and analysis software. Next, members of representative clades are selected based on their percentage association. Bacterial isolate representatives of these clades are subjected to assays determining their susceptibility to cell-free supernatants of *Bacillus* spp. DFMs. Based on the greatest coverage of inhibition of the selected pathogens, a mixture of probiotics is then made to scale up production, into feed in feed mills, and finally onto animal feeding to operations.

The impact of a CSI-derived DFM *Bacillus* spp. mixture on intestinal levels of type A *C. perfringens* and virulence gene positive APEC in a broiler operation pre-CSI-DFM (AGP in usage) and CSI-DFM fed (AGP-free) is illustrated in a typical example (Table 20.1). Analysis of this same data (Table 20.2) illustrated that CSI-DFM usage resulted in an increase in the lower *C. perfringens* categories (<3 $\log_{10}$ CFU/g mucosal homogenate)

Table 20.1 *Clostridium perfringens* Type A levels (top) and avian pathogenic *Escherichia coli* levels (bottom) in broiler chicken intestinal samples fed a CSI-DFM *Bacillus* product. Pre-CSI DFM represents the time period with AGP usage.

	Pre-CSI *C. perf.*	**CSI-treated *C. perf.***
Mean ($\log_{10}$ cfu/g count)	2.85	0.88
Standard deviation (SD)	2.43	1.59
Sample size (*N*)	52	16
Statistical test	Mann-Whitney test (nonpara.)	
P value	$P = 0.0003$	

C. perf., *Clostridium perfringens*; CSI, company-specific inoculants; DFM, direct-fed microbials.

	Pre-CSI APEC	**CSI-treated APEC**
Mean ($\log_{10}$ cfu/g count)	5.38	4.41
Standard deviation (SD)	1.51	1.11
Sample size (*N*)	52	16
Statistical test	Unpaired *t* test	
P value	$P = 0.0197$	

APEC, avian pathogenic *Escherichia coli*.

Table 20.2 Categorical tabulation of *Clostridium perfringens* Type A levels (top) and avian pathogenic *Escherichia coli* levels in broiler chicken intestinal samples fed a CSI-DFM *Bacillus* product. Pre-CSI DFM represents the time period with AGP usage, refer to Table 20.1.

***C. perfringens* type A**			**Proportion (%)**	
$\log_{10}$ CFU/g	**Pre-CSI**	**CSI**	**Pre-CSI**	**CSI**
0–1	18	11	34.62	68.75
1–2	3	2	5.77	12.50
2–3	5	1	9.62	6.25
3–4	7	1	13.46	6.25
4–5	9	0	17.31	0.00
5–6	4	1	7.69	6.25
6–7	3	0	5.77	0.00
>7	3	0	5.77	0.00
n	52	16	100	100

from 48% (no CSI-DFM usage) frequency to 87% (while fed CSI-DFM). Levels for cpa + C. *perfringens* of greater than 5 $\log_{10}$ CFU/g of broiler gut mucosal homogenate has routinely been associated with overt necrotic disease in broilers.

Another important aspect of DFM usage is not only the potential for disease and zoonotic pathogen mitigation but nutritional performance enhancement. Through successive use and practice in one field example presented in Table 20.3, broiler live weight, feed conversion ratio, age-to-market, and the daily weight gain were all improved from a CSI-DFM feeding regimen. Another benefit this same operation experienced was a more rapid growout time leading to 1–2 days of additional time between new flocks (data not shown). This additional time allows for houses to dry out more efficiently and thereby creating less favorable environments for bacterial growth prior to the introduction of a new flock.

20.7 CONCLUSIONS

Biocontrol of poultry disease bacterial pathogens as well as human zoonotic pathogens within an AGP-free or an organic setting is achievable with alternative interventions. Alternative biocontrol technologies were presented that are either in usage (CSI-derived DFMs) or have undergone some proof of concept that set the stage for further scientific and technical development. This chapter, while not an exhaustive or complete review of all probiotic-based

Table 20.3 Broiler chicken nutritional performance values from a commercial operation fed CSI-DFM formulated *Bacillus* direct-fed microbial.

Mean performance parameters	**Growout period coverage fed CSI-DFM**		
	60%	**80%**	**100%**
Live weight	3.98	4.12	4.16
FCR[a]	1.628	1.635	1.605
Age (days)	35.18	35.11	34.80
Daily gain	11.32	11.72	11.95

[a]Feed conversion ratio (lb feed/lb gain).

CSI–DFM, company-specific inoculants—direct-fed microbials.

alternatives for antibiotic-free operations, does present several perspectives on probiotics and DFM.

The mechanisms of action of feeding antibiotic growth promotants has long been considered to be the result of both intestinal microbial control, and reduction in sequelae of that infectious process, resulting in enhanced nutritional performance. Specific mechanisms purported include immune enhancement effects. Niewold (2007) presented a hypothetical case for the direct interaction of antibiotics with host cells resulting in an undefined immune enhancement effect and subsequent overall performance improvement. The use of *Bacillus*-based DFMs and purified bacteriocins were both demonstrated to elicit measurable immune enhancement impacts on broilers and turkeys, respectively (Cole *et al.*, 2006; Lee *et al.*, 2010, 2011).

It is worth noting here that, although a major result of AGP-withdrawal can be the overgrowth of pathogenic clostridia, the even greater challenge is the persistence of disease-causing Gram-negative *E. coli* pathotypes that still represents a formidable challenge to the large-scale poultry producer. This is especially evident in the hatchery and early life stages of production. Thus, effective Gram-negative bacterial inhibitors are apparently less frequently encountered within the realm of natural products such as the DFM most often studied as feed agents.

A major issue facing the usage of many bio-derived antimicrobial interventions is delivery and administration of the agents to the bird. *Bacillus* based probiotics have the advantage of natural encapsulation through the *Bacillus* endospore. While encapsulation of enzymes, organic molecules, and nonsporulating probiotic bacteria to achieve some level of heat stability is possible, it is costly. Research in this area is vital for the practical application of economically feasible alternative interventions for poultry (Gibbs *et al.*, 1999; Kuang *et al.*, 2010).

One other approach for antibiotic-free poultry production not frequently discussed is the potential for genetic selection and breeding of food animals such as poultry genetic lines based on the criteria for strains of birds that are better able to resist or eliminate gut colonization of pathogenic bacteria or at least maintain performance. To date, breeding for rapid growth and muscle deposition has been successfully achieved, but the additional desired trait of subclinical disease resistance remains to be accomplished for avian as well as other food animal species breeding and selection programs.

The makeup of the intestinal microbiome as an overall driver in bird development and yield is perhaps the central target of study for the future of organic production in the poultry sector. The ability to control the status of the gut microbiome development as well as mitigate pathogenic forces, whether protozoal, viral, or bacterial, will become even more critical as AGPs are either reduced or discontinued from the poultry production sector. Not discussed extensively herein was the concept of combining alternative interventions to achieve synergistic effects to enhance production efficiency and health of animals. In the future, it is likely that antimicrobial interventions and performance-improving feeds will possibly be the result of combined biocontrol strategies with synergistic results from the additivity of different mechanisms of inhibition conferred by differing classes of antimicrobials.

ACKNOWLEDGMENT

This review was partially supported by a USDA Food Safety Consortium grant and USDA National Integrated Food Safety Grant #2008-51110-04339 to Steven C. Ricke.

REFERENCES

Aarestrup, F. M., H. C. Wegener, and P. Collignon. 2008. Resistance in bacteria of the food chain: epidemiology and control strategies. *Expert Rev. Anti Infect. Ther.* 6:733–750.

Aminov, R. I. and R. I. Mackie. 2007. Evolution and ecology of antibiotic resistance genes. *FEMS Microbiol. Lett.* 271:147–161.

Apajalahti, J. H. A., H. Kettunen, M. R. Bedford, and W. E. Holben. 2001. Percent G+C profiling accurately reveals diet-related differences in the gastrointestinal microbial community of broiler chickens. *Appl. Environ. Microbiol.* 67:5656–5667.

Apajalahti, J. H., H. Kettunen, A. Kettunen, W. E. Holben, P. H Nurminen, N. Rautonen, and M. Mutanen. 2002. Culture-independent microbial community analysis reveals that inulin in the diet primarily affects previously unknown bacteria in the mouse cecum. *Appl. Environ. Microbiol.* 68:4986–4995.

Applegate, T. J., V. Klose, T. Steiner, A. Ganner, and G. Schatzmayr. 2010. Probiotics and phytogenics for poultry: Myth or reality? *J. Appl. Poultry Res.* 19:194–210.

Ashayerizadeh, A., N. Dabiri, O. Ashayerizadeh, K. H. Mirzadeh, H. Roshanfekr, and M. Mamooee. 2009. Effect of dietary antibiotic, probiotic and prebiotic as growth promoters, on growth performance, carcass characteristics and hematological indices of broiler chickens. *Pak. J. Biol. Sci.* 12:52–57.

Berghman, L. R., D. Abi-Ghanem, S. D. Waghela, and S. C. Ricke. 2005. Antibodies: An alternative for antibiotics. *Poult. Sci.* 84:660–666.

Bernardeau, M. and J. P. Vernoux. 2009. Overview of the use of probiotics in the feed/food chain. In: M. Bernardeau and J. P. Vernoux (eds) *Probiotics: Production, Evaluation, and Uses in Animal Feed.* Research Signpost, Kerala. pp. 15–45.

Bokkers, E. A. M. and I. J. M. de Boer. 2009. Economic, ecological, and social performance of conventional and organic broiler production in the Netherlands. *Br. Poult. Sci.* 50:546–557.

Brashears, M. M., D. Jaroni, and J. Trimble. 2003b. Isolation, selection, and characterization of lactic acid bacteria for a competitive exclusion product to reduce shedding of *Escherichia coli* O157:H7 in cattle. *J. Food Prot.* 66:355–363.

Brashears, M. M., M. L. Galyean, G. H. Loneragan, J. E. Mann,and K. Killinger-Mann. 2003a. Prevalence of *Escherichia coli* O157:H7 and performance by beef feedlot cattle given *Lactobacillus* direct-fed microbials. *J. Food Prot.* 66:748–754.

Bryant, M. P. 1959. Bacterial species of the rumen. *Bacteriol. Rev.* 23:125–153.

Butaye, P., L. A. Devriese, and F. Haesebrouck. 2003. Antimicrobial growth promoters used in animal feed: effects of less well known antibiotics on gram-positive bacteria. *Clin. Microbiol. Rev.* 16:175–188.

Calhoun, L. N. and Y. M. Kwon. 2010. The effect of long term propionate adaptation on the stress resistance of *Salmonella* Enteritidis. *J. Appl. Microbiol.* 109:1294–1300.

Callaway, T. R., S. E. Dowd, R. D. Wolcott, Y. Sun, J. L. McReynolds, T. S. Edrington, J. A. Byrd, R. C. Anderson, N. Krueger, and D. J. Nisbet. 2009. Evaluation of the bacterial diversity in cecal contents of laying hens fed various molting diets by using bacterial tag-encoded FLX amplicon pyrosequencing. *Poult. Sci.* 88:298–302.

Cerutti, A. 2010. IgA changes the rules of memory. *Science.* 328:1646–1647.

Chaucheyras-Durand, F. and H. Durand. 2010. Probiotics in animal nutrition and health. *Beneficial Microbes.* 1:3–9.

Choi, J. Y., P. L. Shinde, S. L. Ingale, J. S. Kim, Y. W. Kim, I. K. Kwon, and B. J. Chae. 2011. Evaluation of multi-microbe probiotics prepared by submerged liquid or solid substrate fermentation and antibiotics in weaning pigs. *Livestock Sci.* 138:144–151.

Church, D. C. 1976. Digestive physiology and nutrition of ruminants. 2nd edn. O&B Books, Corvallis, OR. p. 350.

Clarke, T. B., K. M. Davis, E. S. Lysenko, A. Y. Zhou, Y. Yu, and J. N. Weiser. 2010. Recognition of peptidoglycan from the microbiota by Nod1 enhances systemic innate immunity. *Nat Med.* 16:228–231.

Cole, K., M. B. Farnell, A. M. Donoghue, N. J. Stern, E. A. Svetoch, B. N. Eruslanov, L. I. Volodina, Y. N. Kovalev, V. V. Perelygin, E. V. Mitsevich, I. P. Mitsevich, V. P. Levchuk, V. D. Pokhilenko, V. N. Borzenkov, O. E. Svetoch, T. Y. Kudryavtseva, I. Reyes-Herrera, P. J. Blore, F. Solis de los Santos, and D. J. Donoghue. 2006. Bacteriocins reduce *Campylobacter* colonization and alter gut morphology in turkey poults. *Poult. Sci.* 85:1570–1575.

Collado, M. C., L. Grzéskowiak, and S. Salminen. 2007. Probiotic strains and their combination inhibit adhesion of pathogens to pig intestinal mucosa. *Curr. Microbiol.* 55:260–265.

Corrier, D. E., D. J. Nisbet, B. M. Hargis, P. S. Holt, and J. R. DeLoach. 1997. Provision of lactose to molting hens enhances resistance to *Salmonella enteritidis* colonization. *J. Food Prot.* 60:10–15.

Drolesky, R. E., D. E. Corrier, D. J. Nisbet, and J. R. DeLoach. 1995. Colonization of cecal mucosal epithelium in chicks treated with a continuous flow culture of 29 characterized bacteria: Confirmation by scanning electron microscopy. *J. Food Prot.* 58:837–842.

Dunkley, K. D., J. L. McReynolds, M. E. Hume, C. S. Dunkley, T. R. Callaway, L. F. Kubena, D. J. Nisbet, and S. C. Ricke. 2007a. Molting in *Salmonella* Enteritidis challenged laying hens fed alfalfa crumbles I. *Salmonella* Enteritidis colonization and virulence gene *hilA* response. *Poult. Sci.* 86:1633–1639.

Dunkley, K. D., J. L. McReynolds, M. E. Hume, C. S. Dunkley, T. R. Callaway, L. F. Kubena, D. J. Nisbet, and S. C. Ricke. 2007b. Molting in *Salmonella* Enteritidis challenged laying hens fed alfalfa crumbles II. Fermentation and microbial ecology response. *Poult. Sci.* 86:2101–2109.

Dunkley, K. D., T. R. Callaway, C. O'Bryan, M. M. Kundinger, C. S. Dunkley, R. C. Anderson, D. J. Nisbet, P. G. Crandall, and S. C. Ricke. 2009. Cell yields and fermentation responses of a *Salmonella* Typhimurium poultry isolate at different dilution rates in an anaerobic steady state continuous culture (CC). *Antonie van Leeuwenhoek J. Gen. Mol. Microbiol.* 96:537–544.

Durant, J. A., D. E. Corrier, J. A. Byrd, L. H. Stanker, and S. C. Ricke. 1999a. Feed deprivation affects crop environment and modulates *Salmonella enteritidis* colonization and invasion of Leghorn hens. *Appl. Environ. Microbiol.* 65:1919–1923.

Durant, J. A., D. E. Corrier, and S. C. Ricke. 2000a. Short-chain volatile fatty acids modulate the expression of the *hilA* and *invF* genes of *Salmonella* Typhimurium. *J. Food Prot.* 63:573–578.

Durant, J. A., D. E. Corrier, L. H. Stanker, and S. C. Ricke. 2000b. *Salmonella enteritidis hilA* gene fusion response after incubation in a spent media from either *S. enteritidis* or a probiotic *Lactobacillus* strain. *J. Environ. Sci. Health* B35:599–610.

Durant, J. A., V. K. Lowry, D. J. Nisbet, L. H. Stanker, D. E. Corrier, and S. C. Ricke. 1999b. Short-chain volatile fatty acids affect the adherence and invasion of HEp-2 cells by *Salmonella typhimurium*. *J. Environ. Sci. Health* B34:1083–1099.

Durant, J. A., V. K. Lowry, D. J. Nisbet, L. H. Stanker, D. E. Corrier, and S. C. Ricke. 2000c. Late logarithmic *Salmonella typhimurium* HEp-2 cell-association and invasion response to short chain volatile fatty acid addition. *J. Food Safety.* 20:1–11.

Durant, J. A., V. K. Lowry, D. J. Nisbet, L. H. Stanker, D. E. Corrier, and S. C. Ricke. 2000d. Short-chain fatty acids alter HEp-2 cell association and invasion by stationary growth phase *Salmonella typhimurium*. *J. Food Sci.* 65:1206–1209.

Emborg, H. D., J. S. Andersen, A. M. Seyfarth, S. R. Andersen, J. Boel, and H. C. Wegener. 2003. Relations between the occurrence of resistance to antimicrobial growth promoters among *Enterococcus faecium* isolated from broilers and broiler meat. *Int. J. Food Microbiol.* 84:273–284.

Farnell, M. B., A. M. Donoghue, F. S. de Los Santos, P. J. Blore, B. M. Hargis, G. Tellez, and D. J. Donoghue. 2006. Upregulation of oxidative burst and degranulation in chicken heterophils stimulated with probiotic bacteria. *Poult. Sci.* 85:1900–1906.

Fukuda, S., H. Toh, K. hase, K. Oshima, Y. Nakanishi, K. Yoshimura, T. Tobe, J. M. Clarke, D. L. Topping, T. Suzuki, T. D. Taylor, K. Itoh, J. Kikuchi, H. Morita, M. Hattori, and H. Ohno. 2011. Bifidobacteria can protect from enteropathogenic infection through production of acetate. *Nature* 469:543–549.

Gantois, I., R. Ducatelle, F. Pasmans, F. Haesebrouck, I. Hautefort, A. Thompson, J. C. Hinton, and F. Van Immerseel. 2006. Butyrate specifically down-regulates *Salmonella* pathogenicity island 1 gene expression. *Appl. Environ. Microbiol.* 72:946–949.

Ghnassia, J. C. 1979. Quantitative study of gastrointestinal bacterial-flora. *Medecine et Maladies Infectieuses* 9:507–510.

Gibbs, B. F., S. Kermasha, I. Alli, and C. N. Mulligan. 1999. Encapsulation in the food industry: a review. *Int. J. Food Sci. Nutr.* 50:213–224.

Gong, J., R. J. Forster, H. Yu, J. R. Chambers, P. M. Sabour, R. Wheatcroft, and S. Chen. 2002. Diversity and phylogenetic analysis of bacteria in the mucosa of chicken ceca and comparison with bacteria in the cecal lumen. *FEMS Microbiol. Lett.* 208:1–7.

Gong, J., W. Si, R. J. Forster, R. Huang, H. Yu, Y. Yin, C. Yang, and Y. Han. 2006. 16S rRNA gene-based analysis of mucosa-associated bacterial community and phylogeny in the chicken gastrointestinal tracts: from crops to ceca. *FEMS Microbiol. Ecology* 59:147–157.

Graham, J. P., J. J. Boland, and E. Silbergeld. 2007. Growth promoting antibiotics in food animal production: an economic analysis. *Public Health Rep.* 122:79–87.

Ha, S. D., D. J. Nisbet, D. E. Corrier, J. R. DeLoach, and S. C. Ricke. 1995. Comparison of *Salmonella typhimurium* and selected facultative cecal bacteria survivability after specific amino acid-limited batch growth. *J. Food Prot.* 58:1335–1339.

Ha, S. D., S. C. Ricke, D. J. Nisbet, D. E. Corrier, and J. R. DeLoach. 1994. Serine utilization as a potential competition mechanism between *Salmonella* and a chicken cecal bacterium. *J. Food Prot.* 57:1074–1079.

Haghighi, H. R., J. H. Gong, C. L. Gyles, and M. A. Hayes, H. Zhou, B. Sanei, J. R. Chambers, and S. Sharif. 2006. Probiotics stimulate production of natural antibodies in chickens. *Clin. Vaccine Immunol.* 13:975–980.

Hanning, I. and S. C. Ricke. 2011. Prescreening methods of microbial populations for the assessment of sequencing potential. In: Y. M. Kwon, and S. C. Ricke (eds) *Methods in Molecular Microbiology 733 – High-Throughput Next Generation Sequencing: Methods and Applications*. Springer Protocols, Humana Press, New York. pp. 159–170.

Hapfelmeier, S., M. A. E. Lawson, E. Slack, J. K. Kirundi, M. Stoel, M. Heikenwalder, J. Cahenzli, Y. Velykoredko, M. L. Balmer, K. Endt, M. B. Geuking, R. Curtiss 3rd, K. D. McCoy, and A. J. Macpherson. 2010. Reversible microbial colonization of germ-free mice reveals the dynamics of IgA immune responses. *Science* 328:1705–1709.

Herrera, P., C. A. O'Bryan, P. G. Crandall, and S. C. Ricke. 2011. Growth response of *Salmonella enterica* Typhimurium in co-culture with ruminal bacterium *Streptococcus bovis* is influenced by time of inoculation and carbohydrate substrate. *Food Res. Int.* (In press).

Holt, P. S., N. P. Macri, and R. E. Porter Jr. 1995. Microbiological analysis of the early *Salmonella* enteritidis infection in molted and unmolted hens. *Avian Dis.* 39:55–63.

Hong, H. A., L. H. Duc, and S. M. Cutting. 2005. The use of bacterial spore formers as probiotics. *FEMS Microbiol. Rev.* 29:813–835.

Huang, Y., M. Suyemoto, C. D. Garner, K. M. Cicconi, and C. Altier. 2008. Formate acts as a diffusible signal to induce *Salmonella* invasion. *J. Bacteriol.* 190:4233–4241.

Hume, M. E., L. F. Kubena, T. S. Edrington, C. J. Donskey, R. W. Moore, S. C. Ricke, and D. J. Nisbet. 2003. Poultry digestive microflora diversity as indicated by denaturing gradient gel electrophoresis. *Poult. Sci.* 82:1100–1107.

Hungate, R. E. 1950. The anaerobic mesophilic cellulolytic bacteria. *Bacteriol. Rev.* 14:1–49.

Hungate, R. E. 1966. The rumen and its microbes. Academic Press, Inc., New York. p. 533.

Hungate, R. E. 1979. Evolution of a microbial ecologist. *Ann. Rev. Microbiol.* 33:1–20.

Hungate, R. E., M. P. Bryant, and R. A. Mah. 1964. The rumen bacteria and protozoa. *Annu. Rev. Microbiol.* 18:131–166.

Joerger, R. D. 2003. Alternatives to anitobiotics: Bacteriocins, antimicrobial peptides and bacteriophages. *Poult. Sci.* 82:640–647.

Joerger, R. D., C. A. Sartori, and K. E. Kniel. 2009. Comparison of genetic and physiological properties of *Salmonella enterica* isolates from chickens reveals one major difference between serovar Kentucky and other serovars: response to acid. *Foodborne Pathog. Dis.* 6:503–512.

Jones, F. T. and S. C. Ricke. 2003. Observations on the history of the development of antimicrobials and their use in poultry feeds. *Poult. Sci.* 82:613–617.

Jonsson, E. and P. Conway. 1992. Probiotics for pigs. In: R. Fuller (ed.) *Probiotics: The Scientific Basis*. Chapman and Hall, London. pp. 259–316.

Jukes, T. H. and W. L. Williams. 1953. Nutritional effects of antibiotics. *Pharmacol. Rev.* 5:381–420.

Karasov, W. H., and H. V. Carey. 2009. Metabolic teamwork between gut microbes and hosts. *Microbe* 4:323–328.

Kim, T. and E. Mundt. 2011. Metagenomic analysis of intestinal microbiomes in chickens. In: Y. M. Kwon and S. C. Ricke (eds) *Methods in Molecular Microbiology 733. High-Throughput Next Generation Sequencing: Methods and Applications*. Springer Protocols, Humana Press, New York. pp. 185–194.

Klaenhammer, T. R. 2000. Probiotic bacteria: today and tomorrow. *J. Nutr.* 130:415s–416s.

Krause, D. O., B. A. White, and R. I. Mackie. 1997. Ribotyping of adherent *Lactobacillus* from weaning pigs: A basis for probiotic selection based on diet and gut department. *Anaerobe* 3:317–325.

Krivan, H. C., D. P. Franklin, W. Wang, D. C. Laux and P. S. Cohen. 1992. Phosphatidylserine found in intestinal mucus serves as a sole source of carbon and nitrogen for salmonellae and *Escherichia coli*. *Infect Immun.* 60:3943–3946.

Kuang, S. S., J. C. Oliveira, and A. M. Crean. 2010. Microencapsulation as a tool for incorporating bioactive ingredients into food. *Crit. Rev. Food Sci. Nutr.* 50:951–968.

Kwon, Y. M. and S. C. Ricke. 1998a. Induction of acid resistance of *Salmonella typhimurium* by exposure to short-chain fatty acids. *Appl. Environ. Microbiol.* 64:3458–3463.
Kwon, Y. M. and S. C. Ricke. 1998b. Survival of a *Salmonella typhimurium* poultry isolate in the presence of propionic acid under aerobic and anaerobic conditions. *Anaerobe* 4:251–256.
Kwon, Y. M. and S. C. Ricke. 1999. *Salmonella typhimurium* poultry isolate growth response to propionic acid and sodium propionate under aerobic and anaerobic conditions. *Int. Biodeterioration and Biodegradation* 43:161–165.
Kwon, Y. M., S. D. Ha, and S. C. Ricke. 1998. Growth response of a *Salmonella typhimurium* poultry isolate to propionic acid in aerobic and anaerobic growth conditions. *J. Food Safety* 18:139–149.
Kwon, Y. M., S. Y. Park, S. G. Birkhold, and S. C. Ricke. 2000. Induction of resistance of *Salmonella typhimurium* to environmental stresses by exposure to short-chain fatty acids. *J. Food Sci.* 65:1037–1040.
Lawhon, S. D., R. Maurer, M. Suyemoto, and C. Altier. 2002. Intestinal short-chain fatty acids alter *Salmonella* Typhimurium invasion gene expression and virulence through BarA/SirA. *Mol. Microbiol.* 46:1451–1464.
Lee, K.-W., G. Li, H. S. Lillehoj, S.-H. Lee, S. I. Jang, U. S. Babu, E. P. Lillehoj, A. P. Neumann, and G. R. Siragusa. 2011. *Bacillus subtilis*-based direct-fed microbials augment macrophage function in broiler chickens. *Res Vet Sci.* (In Press).
Lee, K. W., S. H. Lee, H. S. Lillehoj, G. X. Li, S. I. Jang, U. S. Babu, M. S. Park, D. K. Kim, E. P. Lillehoj, A. P. Neumann, T. G. Rehberger, and G. R. Siragusa. 2010. Effects of direct-fed microbials on growth performance, gut morphometry, and immune characteristics in broiler chickens. *Poult. Sci.* 89:203–216.
Leser, T. D., J. Z. Amenuvor, T. K. Jensen, R. H. Lindecrona, M. Boye, and K. Møller. 2003. Culture-independent analysis of gut bacteria: The pig gastrointestinal tract microbiota revisited. *Appl. Environ. Microbiol.* 68:673–690.
Loar, R. E., J. S. Moritz, J. R. Donaldson, and A. Corzo. 2010. Effects of feeding distillers dried grains with solubles to broilers from 0 to 28 days posthatch on broiler performance, feed manufacturing efficiency, and selected intestinal characteristics. *Poult. Sci.* 89:2242–2250.
Loneragan, G. H. and M. M. Brashears. 2005. Pre-harvest interventions to reduce carriage of *E. coli* O157 by harvest-ready feedlot cattle. *Meat Sci.* 71:72–78.
López, P., M. Gueimonde, A. Margolles, and A. Suárez. 2009. Distinct *Bifidobacterium* strains drive different immune responses *in vitro*. *Int. J. Food Microbiol.* 138:157–165.
Luo, J., S. King, and M. C. Adams. 2010. Effect of probiotic *Propionibacterium jensenii* 702 supplementation on layer chicken performance. *Beneficial Microbes* 1:53–60.
Mackie, R. I. 2002. Mutualistic fermentative digestion in the gastrointestinal tract: Diversity and evolution. *Integ. Comp. Biol.* 42:319–326
Makala, L. H. C., T. Kamada, Y. Nishikawa, H. Nagasawa, I. Igarashi, K. Fujisaki, N. Suzuki, T. Mikami, K. Haverson, M. Bailey, C. R. Stokes, and P. W. Bland. 2000. Ontogeny of pig discrete Peyer's patches: distribution and morphometric analysis. *Pathobiology* 68:275–282.
Mao, S. Y., C. F. Yang, and W. Y. Zhu. 2011. Phylogenetic analysis of methanogens in the pig feces. *Curr. Microbiol.* 62:1386–1389.
McReynolds, J., C. Wanek, J. Byrd, K. Genovese, S. Duke, and D. Nisbet. 2009. Efficacy of multistrain direct-fed microbial and phytogenic products in reducing necrotic enteritis in commercial broilers. *Poult. Sci.* 88:2075–2080.
Miller, T. L., M. J. Wolin, and E. A. Kusel. 1986. Isolation and characterization of methanogens from animal feces. *System. Appl. Microbiol.* 8:234–238.
Motherway, M. O., A. Zomer, S. C. Leahy, J. Reunanen, F. Bottacini, M. J. Claesson, F. O'Brien, K. Flynn, P. G. Casey, J. A. M. Munoz, B. Kearney, A. M. Houston, C. O'Mahony, D. G. Higgins, F. Shanahan, A. Palva, W. M. de Vos, G. F. Fitzgerald, M. Ventura, P. W. O'Toole, and D. van Sinderen. 2011. Functional genome analysis of *Bifidobacterium breve* UCC2003 reveals type IVb tight adherence (Tad) pili as an essential and conserved host-colonization factor. *Proc. Natl. Acad. Sci.* Early edition pp. 1–6.
Muegge, B. D., J. Kuczynski, D. Knights, J. C. Clemente, A. González, L. Fontana, B. Henrissat, R. Knight, and J. I. Gordon. 2011. Diet drives convergence in gut microbiome functions across mammalian phylogeny and within humans. *Science* 332:970–974.
Mulder, R. W. A. W., R. Havenaar, and J. H. J. Huis in't Veld. 1997. Intervention strategies: the use of probiotics and competitive exclusion microfloras against contamination with pathogens in pigs and poultry. In: R. Fuller (ed.) *Probiotics 2: Applications and Practical Aspects*. Chapman and Hall, London. pp. 187–207.
Niewold, T. A. 2007. The nonantibiotic anti-inflammatory effect of antimicrobial growth promoters, the real mode of action? A hypothesis. *Poult. Sci.* 86:605–609.

Nisbet, D. 2002. Defined competitive exclusion cultures in the prevention of enteropathogen colonisation in poultry and swine. *Antonie Van Leeuwenhoek* 81:481–486.

Nisbet, D. J., D. E. Corrier, S. C. Ricke, M. E. Hume, J. A. Byrd II, and J. R. DeLoach. 1996a. Maintenance of the biological efficacy in chicks of a cecal competitive-exclusion culture against *Salmonella* by continuous-flow fermentation. *J. Food Prot.* 59:1279–1283.

Nisbet, D. J., D. E. Corrier, S. C. Ricke, M. E. Hume, J. A. Byrd II, and J. R. DeLoach. 1996b. Cecal propionic acid as a biological indicator of the early establishment of a microbial ecosystem inhibitory to *Salmonella* in chicks. *Anaerobe* 2:345–350.

Nisbet, D. J., S. C. Ricke, C. M. Scanlan, D. E. Corrier, A. G. Hollister, and J. R. DeLoach. 1994. Inoculation of broiler chicks with a continuous-flow derived bacterial culture facilitates early native cecal bacterial colonization and increases resistance to *Salmonella typhimurium*. *J. Food Prot.* 57:12–15.

Norberg, L. M., J. L. McReynolds, W.-K. Kim, V. I. Chalova, D. J. Nisbet, and S. C. Ricke. 2010. Colonization and pathogenesis of foodborne *Salmonella* in egg – laying hens. In: S. C. Ricke, and F. T. Jones (eds) *Perspectives on Food Safety Issues of Food Animal Derived Foods*. University of Arkansas Press, Fayetteville, AR. pp. 63–84.

Nurmi, E. and M. Rantala. 1973. New aspects of *Salmonella* infection in broiler production. *Nature* 241:210–211.

O'Bryan, C. A., P. G. Crandall, and S. C. Ricke. 2008. Organic poultry pathogen control from farm to fork. *Foodborne Pathog. Dis.* 5:709–720.

Peterson, D. A., N. P. McNulty, G. L. Guruge, and J. I. Gordon. 2007. IgA response to symbiotic bacteria as a mediator of gut homeostasis. *Cell Host and Microbe* 2:328–339.

Phillips, I., M. Casewell, T. Cox, B. De Groot, C. Friis, R. Jones, C. Nightingale, R. Preston, and J. Waddell. 2004. Does the use of antibiotics in food animals pose a risk to human health? A critical review of published data. *J. Antimicrob. Chemother.* 53:28–52.

Ricke, S. C. 2003a. Perspectives on the use of organic acids and short chain fatty acids as antimicrobials. *Poultry Sci.* 82:632–639.

Ricke, S. C. 2003b. The gastrointestinal tract ecology of *Salmonella* Enteritidis colonization in molting hens. *Poultry Sci.* 82:1003–1007.

Ricke, S. C., C. L. Woodward, Y. M. Kwon, L. F. Kubena, and D. J. Nisbet. 2004. Limiting avian gastrointestinal tract *Salmonella* colonization by cecal anaerobic bacteria and a potential role for methanogens. In: R.C. Beier, S. D. Pillai, T. D. Phillips, and R. L. Ziprin (eds) *Pre-Harvest and Post-Harvest Food Safety: Contemporary Issues and Future Directions*. Blackwell Publishing Professional, Ames, IA. pp. 141–150.

Ricke, S. C., M. M. Kundinger, D. R. Miller, and J. T. Keeton. 2005. Alternatives to antibiotics: Chemical and physical antimicrobial interventions and foodborne pathogen response. *Poultry Sci.* 84:667–675.

Ricke, S. C., S. A. Martin, and D. J. Nisbet. 1996. Ecology, metabolism, and genetics of ruminal selenomonads. *Crit. Rev. Microbiol.* 22:27–65.

Ricke, S. C. and S. D. Pillai. 1999. Conventional and molecular methods for understanding probiotic bacteria functionality in gastrointestinal tracts. *Crit. Rev. Microbiol.* 25:19–38.

Round, J. L., M. Lee, J. Li, G. Tran, B. Jabri, T. A. Chatila, and S. K. Mazmanian. 2011. The toll-like receptor 2 pathway establishes colonization by a commensal of the human microbiota. *Science* 332:974–977.

Russell, J. B. 1992. Another explanation for the toxicity of fermentation acids at low pH – anion accumulation versus uncoupling. *J. Appl. Bacteriol.* 73:363–370.

Saarela, M., G. Mogensen, R. Fonden, J. Matoo, and T. Mattila-Sandholm. 2000. Probiotic bacteria: safety functional and technological properties. *J. Biotechnol.* 84:197–215.

Saengkerdsub, S., P. Herrera, C. L. Woodward, R. C. Anderson, D. J. Nisbet, and S. C. Ricke. 2007b. Detection of methane and quantification of methanogenic archaea in faeces from young broiler chickens using real-time PCR. *Lett. Appl. Microbiol.* 45:629–634.

Saengkerdsub, S., R. C. Anderson, H. H. Wilkinson, W.-K. Kim, D. J. Nisbet, and S. C. Ricke. 2007a. Identification and quantification of methanogenic archaea in adult chicken ceca. *Appl. Environ. Microbiol.* 73:353–356.

Scharek, L., B. J. Altherr, C. Tölke, and M. F. G. Schmidt. 2007a. Influence of the probiotic *Bacillus cereus* var. *toyoi* on the intestinal immunity of piglets. *Vet. Immun. Immunopath.* 120:136–147.

Scharek, L., J. Guth, K. Reiter, K. D. Weyrauch, D. Taras, P. Schwerk, P. Schierack, M. F. Schmidt, L. H. Wieler, and K. Tedin. 2005. Influence of a probiotic *Enterococcus faecium* strain on development of the immune system of sows and piglets. *Vet. Immun. Immunopath.* 105:151–161.

Scharek, L., J. Guth, M. Filter, and M. F. G. Schmidt. 2007b. Impact of the probiotic bacteria *Enterococcus faecium* NCIMB 10415 (SF68) and *Bacillus cereus* var. *toyoi* NCIMB 40112 on the development of serum IgG and faecal IgA of sows and their piglets. *Arch. Anim. Nutr.* 61:223–234.

Schierack, P., L. H. Wieler, D. Taras, V. Herwig, B. Tachu, A. Hlinak, M. F. G. Schmidt, and L. Scharek. 2007. *Bacillus cereus* var. *toyoi* enhanced systemic immune response in piglets. *Vet. Immun. Immunopath.* 118:1–11.

Schierack, P., M. Filter, L. Scharek, C. Toelke, D. Taras, K. Tedin, K. Haverson, A. Lübke-Becker, and L. H. Wieler. 2009. Effect of *Bacillus cereus* var. *toyoi* on immune parameters of pregnant sows. *Vet. Immun. Immunopath.* 27:26–37.

Seo, K.H., P. S. Holt, and R. K. Gast. 2001. Comparison of *Salmonella* Enteritidis infection in hens molted via longterm withdrawal versus full-fed wheat middling. *J. Food Prot.* 64:1917–1921.

Sieo, C. C., N. Abdullah, W. S. Tan, and Y. W. Ho. 2005. Influence of β-glucanase-producing *Lactobacillus* strains on intestinal characteristics and feed passage rate of broiler chickens. *Poult. Sci.* 84:734–741.

Sirsat, S. A., A. Muthaiyan, and S. C. Ricke. 2009. Antimicrobials for pathogen reduction in organic and natural poultry production. *J. Appl. Poultry Res.* 18:379–388.

Sonnenburg, J. L., J. Xu, D. D. Leip, C.-H. Chen, B. P. Westover, J. Weatherford, J. D. Buhler, and J. I. Gordon. 2005. Glycan foraging in vivo by an intestine-adapted bacterial symbiot. *Science* 307:1955–1959.

Stevenson, D. M., and P. J. Weimer. 2007. Dominance of *Prevotella* and low abundance of classical ruminal bacteria species in the bovine rumen revealed by relative quantification real-time PCR. *Appl. Microbiol. Biotechnol.* 75:165–174.

Tellez, G., S. L. Layton, B. M. Hargis. 2011. Probiotics/direct fed microbials for *Salmonella* control in poultry. *Food Res. Int.* (In Press).

Toro, H., S. B. Price, S. McGee, F. J. Hoerr, J. Krehling, M. Perdue, and L. Baurmeister. 2005. Use of bacteriophages in combination with competitive exclusion to reduce *Salmonella* from infected chickens. *Avian Dis.* 49:118–124.

USDA-ERS. 2010. Organic Production: Data sets. Table 5: Certified organic livestock. Available at: http://www.ers.usda.gov/Data/Organic/#national.

Uyeno, Y., Y. Sekiguchi, K. Tajima, A. Takenaka, M. Karihara, and Y. Kamagata. 2007. Evaluation of group-specific, 16S rRNA-targeted scissor probes for quantitative detection of predominant bacterial populations in dairy cattle rumen. *J. Appl. Microbiol.* 103:1995–2005.

Van Immerseel, F., J. B. Russell, M. D. Flythe, I. Gantois, L. Timbermont, F. Pasmans, F. Haesebrouck, and R. Ducatelle. 2006. The use of organic acids to combat *Salmonella* in poultry: a mechanistic explanation of the efficacy. *Avian Pathol.* 35:182–188.

Van Soest, P. J. 1994. *Nutritional Ecology of the Ruminant.* 2nd edn. Cornell University Press, Ithaca, NY. p. 476.

Vaughan, E. E., H. G. G. J. Heilig, E. G. Zoetendal, R. Satokari, J. K. Collins, A. D. L. Akkermans, and W. M. de Vos. 2000. Molecular approaches to study probiotic bacteria. *Trends Food Sci. Technol.* 10:400–404.

Vimal, D. B., M. Khullar, S. Gupta, and N. K. Ganguly. 2000. Intestinal mucins: The binding sites for *Salmonella typhimurium*. *Mol. Cell. Biochem.* 204:107–117.

Wagner, R. D., D. D. Paine, and C. E. Cerniglia. 2003. Phenotypic and genotypic characterization of competitive exclusion products for use in poultry. *J. Appl. Microbiol.* 94:1098–1107.

Wallace, R. J. and C. J. Newbold. 1992. Probiotics for ruminants. In: R. Fuller (ed.) *Probiotics: The Scientific Basis.* Chapman and Hall, London. pp. 317–353.

Wallace, R. J., M. L. Falconer and P. K. Bhargava. 1989. Toxicity of volatile fatty acids at rumen pH prevents enrichment of *Escherichia coli* by sorbitol in rumen contents. *Curr. Microbiol.* 19:277–281.

Weimer, P. J., J. B. Russell, and R. E. Muck. 2009. Lessons from the cow: what the ruminant animal can teach us about consolidated bioprocessing of cellulosic biomass. *Biores. Technol.* 100:5323–5331.

Wells, J. E., S. D. Shackelford, E. D. Berry, N. Kalchayanand, M. N. Guerini, V. H. Varel, T. M. Arthur, J. M. Bosilevac, H. C. Freetly, T. L. Wheeler, C. L. Ferrell, and M. Koohmaraie. 2009. Prevalence and level of *Escherichia coli* O157:H7 in feces and on hides of feedlot steers fed diets with or without wet distillers grains with solubles. *J. Food Prot.* 72:1624–1633.

Williams, R. B. 2005. Intercurrent coccidiosis and necrotic enteritis of chickens: rational, integrated disease management by maintenance of gut integrity. *Avian Pathol.* 34:159–180.

Wolin, M. J. and T. L. Miller. 1982. Interspecies hydrogen transfer: 15 years later. *ASM News* 48:561–565.

Woodward, C. L., Y. M. Kwon, L. F. Kubena, J. A. Byrd, R. W. Moore, D. J. Nisbet, and S. C. Ricke. 2005. Reduction of *Salmonella enterica* serovar Enteritidis colonization and invasion by an alfalfa diet during molt in Leghorn hens. *Poultry Sci.* 84:185–193.

Yan, F., H. Cao, T. L. Cover, M. K. Washington, Y. Shi, L. Liu, R. Chaturvedi, R. M. Peek, K. T. Wilson, and D. B. Polk. 2011. Colon-specific delivery of a probiotic-derived soluble protein ameliorates intestinal inflammation in mice through an EGFR-dependent mechanism. *J. Clin. Invest.* 121:2242–2253.

Yoshimura, Y., M. Oda and N. Isobe. 2010. Effects of feeding probiotics on the localization of cells containing immunoreactive interleukin-6 in the intestine of broiler chicks. *J. Poult. Sci.* 47:250–255.

Young, I., A. Rajić, B. J. Wilhelm, L. Waddell, S. Parker, and S. A. McEwen. 2009. Comparison of the prevalence of bacterial enteropathogens, potentially zoonotic bacteria and bacterial resistance to antimicrobials in organic and conventional poultry, swine and beef production: a systematic review and meta-analysis. *Epidemiol. Infect.* 137:1217–1232.

Younts-Dahl, S. M., G. D. Osborn, M. L. Galyean, J. D. Rivera, G. H. Loneragan and M. M. Brashears. 2005. Reduction of *Escherichia coli* O157 in finishing beef cattle by various doses of *Lactobacillus acidophilus* in direct-fed microbials. *J. Food Prot.* 68:6–10.

Younts, S. M., M. L. Galyean, G. H. Loneragan, N. A. Elam, and M. M. Brashears. 2004. Dietary supplementation with *Lactobacillus*- and *Propionibacterium*-based direct-fed microbials and prevalence of *Escherichia coli* O157 in beef feedlot cattle and on hides at harvest. *J. Food Prot.* 67:889–893.

Zhang, H., P. Parameswaran, J. Badalamenti, B.E. Rittman, and R. Krajmalnik-Brown. 2011. Integrating high-throughput pyrosequencing and quantititative real-time PCR to analyze complex microbial communities. In: Y. M. Kwon, and S. C. Ricke (eds) *Methods in Molecular Microbiology 733. High-Throughput Next Generation Sequencing: Methods and Applications*. Springer Protocols, Humana Press, New York. pp. 107–128.

Zhao, T., S. Tkalcic, M. P. Doyle, B. G. Harmon, C. A. Brown, and P. Zhao. 2003. Pathogenicity of enterohemorrhagic *Escherichia coli* in neonatal calves and evaluation of fecal shedding by treatment with probiotic *Echerichia coli*. *J. Food Prot.* 66:924–930.

Zhu, X. Y., T. Zhong, Y. Pandya, and R. D. Joerger. 2002. 16s rRNA-based analysis of microbiota from the caecum of broiler chickens. *Appl. Environ. Microbiol.* 68:124–137.

21 Gut Health and Organic Acids, Antimicrobial Peptides, and Botanicals as Natural Feed Additives

Jacqueline Jacob and Anthony Pescatore

Abstract: The organic food animal industry is looking for alternatives to antibiotics to maintain intestinal health, reduce animal production's impact on the environment, and allow for the economical production of animal food products. This includes prevention of necrotic enteritis in poultry, postweaning diarrhea in pigs, digestive upsets of rabbits, and reduction in methane production in ruminants. Alternatives discussed include organic acids, antimicrobial peptides, and botanicals.

Keywords: gut health; organic acids; antimicrobial peptides (AMPs); phytogenic compounds; botanicals

21.1 INTRODUCTION

Antibiotics have had a dominant role in animal production for many years. When antibiotics are administered at the recommended dose to control disease, it is referred to as a "therapeutic" dose. Antibiotics can also be administered at a lower dose as a way to prevent disease or to stimulate growth. This lower level is referred to as "subtherapeutic." With interest in reducing or eliminating the subtherapeutic use of all antibiotics in food animal production, producers are looking for "natural" feed additives that can substitute for previously used antimicrobials. While antibiotics can be used to treat disease, in animal agriculture they are more commonly used to prevent disease. Antibiotics are frequently given at subtherapeutic levels to animals at high risk of developing a disease. When used to prevent subclinical infections, antibiotics also improve growth rates and feed efficiency. For these reasons they are often referred to as "antibiotic growth promoters" (AGP), but they are actually achieving this improved growth performance through better gut health. For example, necrotic enteritis (NE) is a common intestinal disease of meat chickens (broilers) and antibiotics have been routinely added to broiler feeds at subtherapeutic levels to prevent this problem (Gillespie, 1997). Similarly, postweaning diarrhea caused by *Escherichia coli* can be a recurring problem in pork production and antibiotics are added to the feed to reduce or eliminate this problem, especially during and after weaning (Hedegaard, 2000). Antibiotics play a similar role in rabbit production. Because of their very complex and peculiar digestion, rabbits are also susceptible to digestive problems. Subtherapeutic levels of antibiotics are used to prevent enteric diseases, particularly after weaning (Cs *et al.*, 2008).

Organic Meat Production and Processing, First Edition. Edited by Steven C. Ricke, Ellen J. Van Loo, Michael G. Johnson, and Corliss A. O'Bryan.

Suckling ruminants such as cows and sheep are considered preruminant. It is not until they have been weaned and placed on solid food that the rumen begins to play a major role in digestion. The early weaning of veal calves, together with the stress of transportation and regrouping of individuals, is a major stress on the animals and frequently results in digestive disorders. Antibiotics are typically used at this time in conventional animal production (Verdonk *et al.*, 2005). Subtherapeutic antibiotics are also used with a number of ruminants. Antibiotics are added to feed for beef cattle to improve rate of gain and feed efficiency, especially when cattle are under stress or just starting on finishing feed in the feedlot (Gillespie, 1997). Antibiotics are added to lamb feed to promote gut health and control pneumonia, and in doing so, increase growth rates (Lupton, 2008).

Ruminant animals, such as cattle, sheep, and goats, are able to use dietary fiber and nonprotein nitrogen sources thanks to the symbiotic relationship that they have with the microflora in the rumen. The microorganisms in the rumen break down these ingredients and produce short-chain fatty acids (SCFA; e.g., acetic, butyric, propionic, and lactic acids) and microbial protein, which the animal is then able to use as energy and protein sources, respectively. Under the anaerobic conditions of the rumen, the oxidation reactions that are taking place during fermentation release hydrogen. The amount of hydrogen that is produced will depend on the type of microorganisms in the rumen. The formation of propionic acid consumes hydrogen, while the formation of acetic and butyric acids releases hydrogen. A build up of hydrogen inhibits dehydrogenases, which are involved in hydrogen transfers, affecting the fermentation process. Methanogenic bacteria in the rumen use the hydrogen and carbon dioxide during metabolism, with methane and acetate being the by-products. When these bacteria are inhibited, methane production is reduced and fermentation is shifted to propionate, with lactate and butyrate as by-products. Ruminants cannot make use of the methane, heat, and ammonia that are produced as by-products of rumen fermentation. A common class of antimicrobials used in beef production is the ionophores, with monensin as a common example. Although ionophores were originally developed as coccidiostats, they can also alter fermentation in the rumen (Callaway *et al.*, 1997; Russell & Mantovani, 2002). By reducing methane production, ionophores increase energy retention, thus improving feed efficiency. This shift in fermentation also reduces methane emissions (Martin *et al.*, 2009).

Ruminants produce 80 million metric tons of methane each year (Johnson & Johnson, 1995). The methane is a concern for two reasons. First, methane is considered a greenhouse gas associated with environmental problems. Second, ruminants typically lose 6% of their dietary energy as methane (Johnson & Johnson, 1995) reducing feed efficiency. Reducing methane production, therefore, would be good for the environment as well as for the farmer.

The current European ban on subtherapeutic antibiotic use in animal production can be used to identify potential problems that need to be addressed. In Denmark, the ban on subtherapeutic antibiotic use in pig production started with the finisher phase and only minimal health problems were encountered (Hayes & Jenson, 2003). The antibiotic ban was then extended to the weaning stages of pig production and several health problems occurred. The piglets were weaker and more vulnerable to disease when they were moved to the finishing barns. The main problem was identified as postweaning diarrhea that had significant effects on the health of the pigs throughout the rest of the production system (Wierup, 2001). Similarly, the European ban of subtherapeutic antibiotic use in poultry production resulted in an increase in the incidences of NE, as well as leg and skin problems (Casewell *et al.*, 2003; Dahiya *et al.*, 2006; Flint & Garner, 2009). These developments highlight the fact that antibiotics acted by preventing disease in addition to improving growth and feed efficiency.

Ionophores are a group of antimicrobials used for the control of coccidiosis (caused by a protozoan parasite) and are routinely added to poultry feeds for this purpose. Ionophores have also been shown to improve animal performance of both, dairy and beef cattle (Jouany & Morgavi, 2007), as well as meat goats and sheep (Lupton, 2008). While ionophores were developed as coccidiostats, they have also been found to alter fermentation in the rumen. Supplementation of ruminant diets with ionophores has been shown to inhibit methanogenic bacteria, thus reducing the amount of methane produced. The use of ionophore coccidiostats is not allowed in organic animal production. In poultry production, it is being replaced by vaccinations. In many cases, it resulted in an increased incidence of NE (Dahiya *et al.*, 2006). In addition to protecting poultry from coccidiosis, ionophores were also acting to control several intestinal bacteria, including *Clostridium perfringens* that can cause NE.

21.2 GUT HEALTH AND MICROBIAL POPULATION

Any antibiotic alternative needs to be able to maintain animal health and well-being through promotion of gut health, reduce animal production's impact on the environment by maximizing nutrient utilization, and allow for the economical production of animal food items. Gut health is a reflection of the status of the microbiota ("*micro*" meaning extremely small and "*biota*" meaning life) present. The intestinal microbiota of all animals is made up of a vast array of microorganisms, with bacteria being the most common, Gram-positive bacteria in particular (Richards *et al.*, 2005). A symbiotic relationship has evolved over time between this community of microorganisms and the host animal. In general, intestinal bacteria are classified as being beneficial to the host or as exerting harm (pathogenic). *Bifidobacteria* and *Lactobacilli* species are considered to be beneficial bacteria since they act as important barriers to the establishment of pathogens. Both types of bacteria produce a range of antimicrobial agents that are active against both Gram-positive and Gram-negative bacteria (Rastall, 2004).

There are important differences in the digestive tracts of pigs, chickens, and rabbits. As a result, there are differences in the "normal" intestinal microbiota of these animals (Flickinger *et al.*, 2003). In the pig, the highest concentration of microorganisms is found in the cecum and colon. Swine microflora contain bacteria common in ruminants as well as the *Bacteriodes*, *Lactobacilli*, *Enterobacteria*, and *Bifidobacteria* (Varel & Yen, 1997). In chickens, the two ceca are the main site of bacterial fermentation, with the colon playing only a minor role. *Lactobacilli* and *Bifidobacteria* are among the common species of bacteria present (Amit-Romach *et al.*, 2004). Rabbits have extensive hindgut fermentation. The most common bacteria in rabbit intestines include *Bacteriodes* and *Enterobacteria*. A few *Bifidobacteria* are occasionally present but *Lactobacilli* are not normally found in rabbit feces (Kovács *et al.*, 2006).

When an animal is born (or hatched), its digestive tract is basically sterile but they are exposed to bacteria from the moment they come into the world (Bauer *et al.*, 2006). The gut then begins to be colonized over time until the "mature" microbiota is established. The gastrointestinal (GI) tract of mammals typically receives its first exposure to bacteria from the feces and breast milk of their mothers (Abecia *et al.*, 2007). For newly hatched chicks, the first bacteria are obtained from the surface of the egg shells (Apajalahti, 2005). If raised by their mothers, chicks also get bacteria inocula from the feces of their mothers, but this source is not available if the chicks are raised artificially.

The progression of different bacterial species colonizing the GI tract is very similar in many animals, including chicks, piglets, calves, and humans (Richards *et al.*, 2005) with maximum bacterial densities being established around 5 days of age (Apajalahti, 2005). For many animal species, *Lactobacilli* are the predominant bacteria in the young, while *Bifidobacteria* dominate in older individuals (Tellez *et al.*, 2006). The role of the beneficial bacteria in excluding pathogens is well documented (Patterson & Burkholder, 2003). The microbiota is also important in the synthesis of vitamins, minerals, and other biologically active compounds. They have also been reported to stimulate the immune system (Oviedo-Rondón *et al.*, 2006).

The development of the normal microbiota of an animal is affected by the environment. This has been demonstrated by the difference in bacteria that inhabit the gut of breast-fed versus formula-fed human infants (Bauer *et al.*, 2006). During the first week of life, the microbiota of both is relatively simple with the dominant species being enterobacteria, which metabolizes oxygen. The reduction in oxygen in the environment encourages the growth of anaerobic bacteria such as *Lactobacilli*, *Bifidobacteria*, *Bacteriodes*, and *Clostridia*. The proportions of the different bacterial types depend on the diet. In breast-fed infants, *Bifidobacteria* and *Lactobacilli* dominate, while in formula-fed infants, there are more *Bacteriodes*, *Clostridium*, and *Enterobacteria* (Bauer *et al.*, 2006). The difference is believed to be due to the lower diversity of bacteria in breast milk.

In suckling pigs, however, levels of *Bifidobacteria* are low or undetectable (Bauer *et al.*, 2006). Instead, *Lactobacilli* are the main bacterial species to become established and typically remain the predominant bacteria in the intestinal microbiota of healthy pigs. As the suckling pig transitions from milk to a solid diet during weaning, the environment of the GI tract undergoes a dramatic change (Bauer *et al.*, 2006). During this time, the number of beneficial bacteria decreases making it easier for potential pathogens such as coliforms (e.g., *E. coli*) to take over. It is for this reason that many piglets suffer from postweaning diarrhea (Gillespie, 1997). In the past, the subtherapeutic use of antibiotics prevented postweaning GI stress (Hedegaard, 2000; Hayes & Jensen, 2003).

In nonruminants, the main area of microbial fermentation is in the hindgut. For most animals, this is the cecum. Ceca are blind pouches located at the junction of the small and large intestines. The number and size of cecum vary depending on the species. For example, most mammals, including pigs and rabbits, have a single cecum. Poultry typically have two ceca and fish may have several (McBee, 1971). In pigs, the main location for microbial fermentation occurs most frequently in the colon (Apajalahti, 2005), while in poultry it is the two ceca. In rabbits, the site of microbial fermentation is their large cecum. As a result of hindgut fermentation in nonruminants, the materials available for microbial fermentation are only those feed components that escape digestion in the earlier sections of the GI tract. In contrast, in ruminant animals, the microbiota in their GI tracts are located early in the digestive system and are the main source of dietary energy. It is the nutritional diversity of these microorganisms that allows ruminants to obtain energy from the cellulose and hemicellulose components of the diet. Ruminants typically receive 70% of their dietary energy from the products of rumen fermentation, mainly the SCFA, acetic, propionic, and butyric acid (Bergman, 1990).

The bacteria in the GI tract interact with each other as well as the host animal. The animal must be able to recognize and tolerate the beneficial bacteria while protecting themselves from potential pathogens. Several strategies are involved (Apajalahti, 2005) and include modifying the environment with acid and bile that favors the beneficial bacteria; lots of surface area to compete with the bacteria for nutrient absorption; relatively fast rate of passage through the gut so that bacteria have less time to colonize; continuous sloughing of the intestinal lining and mucus, along with any adhering bacteria; and the animal's innate immune system (Forchielli & Walker, 2005).

Bacteroides and *Bifidobacterium* spp. have been shown to inhibit *Salmonella* growth *in vitro* (Barnes *et al.*, 1979). This has been attributed to the production of SCFA coupled with a low pH. The researchers were not able to repeat this effect *in vivo* with broiler chicks and hypothesized that it was insufficient to establish a single SCFA producing bacterium without also establishing those organisms that contribute to lowering the pH of the cecum. This is the role of *Lactobacilli* that do not usually inhabit the ceca until the second or third day post hatch.

Intestinal bacteria, while having benefits for the health of the animal, compete with the host for nutrients. In most nonruminant animals, the intestinal bacteria metabolize 10%–20% of the nutrients in feed, making them unavailable to the host (Apajalahti *et al.*, 2004). By keeping total numbers of bacteria in the gut low, subtherapeutic levels of antibiotics reduce this competition and increase the level of nutrients available for absorption. This results in improved feed conversion for the animal.

Another important role that beneficial bacteria play in animal production is controlling food-borne pathogens to reduce or eliminate carcass contamination. Based on the incidences of food-borne disease in the United States, the main bacteria of concern are *E. coli* O157:H7, *Listeria monocytogenes*, *Salmonella* species and *Campylobacter jejuni* (Sofos, 2008). Control of these food-borne pathogens begins on the farm with a reduction or elimination of these bacteria from the GI tract.

Effects on intestinal morphology should also be taken into consideration when evaluating an alternative feed supplement since it is another measure of gut health. A large crypt indicates a fast tissue turnover rate (Xu *et al.*, 2003). Conversely, a low villus height to crypt depth ratio indicates the presence of damaging substances. Such conditions are typically associated with poor nutrient absorption, diarrhea, reduced disease resistance, and an overall lower performance.

Several feed additives have shown promise in achieving good gut health, which ultimately will result in better nutrient utilization and economical animal production. They include probiotics and prebiotics (discussed in other chapters), organic acids, antimicrobial peptides (AMPs), and various phytogenic compounds.

21.3 ORGANIC ACIDS

By definition, organic acids are any organic compounds with acidic properties. The most common are the carboxylic acids that have a carbon chain ending with a carboxyl group (i.e., R–COOH). The group of organic acids identified as having antimicrobial activity is the SCFA (one to six carbons), which include formic, acetic, propionic, butyric, valeric, caproic, and lactic acids. Organic acids have a long history as food additives and preservatives (Ricke, 2003). The antimicrobial effect of organic acids tends to increase as environmental pH decreases. Salts of some organic acids have also been shown to have some performance benefits.

Most of the organic acids are pH dependent. Most pathogens grow at a pH close to 7 or slightly higher. Beneficial bacteria, on the other hand, are able to grow in an acidic environment (pH 5.8–6.2) and thus compete with the pathogens in the GI tract (Ozduven *et al.*, 2009). Feeding of organic acids is particularly beneficial in controlling acid-intolerant bacteria such as *E. coli*, *Salmonella*, and *Campylobacter* (Dibner & Buttin, 2002). Each

SCFA has its own spectrum of antimicrobial activity (Partanen & Mroz, 1999). For example, lactic acid is most effective against bacteria while sorbic acid is more of an antifungal. Formic and propionic acids have a broader activity including bacteria, fungi, and yeast. This diversity has led to the use of acid blends as feed additives rather than the use of any single organic acid.

Unlike antibiotics, organic acids have a common mode of action despite their different chemical structures. They act primarily by altering the pH (Sterzo *et al.*, 2007). By decreasing the intestinal pH, they inhibit bacterial growth; by decreasing cytoplasmic pH, they interfere with bacterial metabolism, and inhibit enzymatic action and DNA synthesis. Organic acids are weak acids and typically do not kill microorganisms, but rather inhibit their growth (Lambert & Statford, 1999). They are more effective at low pH values (Chaveerach *et al.*, 2002) where there are increased concentrations of undissociated acids (R-COOH). These undissociated acid molecules pass through the cell's plasma membrane and dissociate (release H^+ ions) within the cell, acidifying the contents (Lambert & Statford, 1999; Ghosh *et al.*, 2008). The invaded organism expends energy in an effort to export protons to restore pH balance. In addition, the $RCOO^-$ anions produced by the dissociation of the acid molecules disrupt DNA synthesis, which hampers protein synthesis, making it difficult for the organism to replicate, and thus no growth occurs.

21.3.1 Using organic acids with nonruminants

There is a considerable body of research looking at organic acids and their salts in swine production (Partanen & Mroz, 1999). Early weaning is stressful for piglets as their diet changes from milk to solid feed. The passive immunity acquired from the mother's milk is lost when the immune system of the piglet is not fully developed, making them more vulnerable to the introduction of pathogens. The use of SCFA such as citric, formic, fumaric, lactic, and propionic acid has been shown to be effective in overcoming the problems associated with early weaning (Freitag *et al.*, 1998; Dibner & Buttin, 2002). SCFA have also been shown to be beneficial in increasing feed efficiency. For example, the addition of fumaric acids to starter diets during the first 3–4 weeks after weaning increased the ileal digestibilities of energy, protein, and amino acids (Blank *et al.*, 1999).

In addition to helping alleviate problems with weaned piglets, SCFA have also been shown to be beneficial for fattening pigs. Organic acid supplementation has been shown to improve protein and amino acid digestibility and increase absorption of minerals. While this often leads to improved growth performance, it also reduces the amount of nitrogen and phosphorus excreted, reducing the impact of swine production on the environment.

Organic acids also have benefits beyond their antimicrobial properties with effects in the feed as well as in the digestive tract (Schöner, 2001; Dibner & Buttin, 2002). Organic acids add protection to stored feed preventing bacterial and fungal growth (Partanen & Mroz, 1999). In the digestive tract, organic acids reduce the pH, which is beneficial for the weanling pig. Gastric acid secretion in piglets does not reach appreciable levels until 3–4 weeks postweaning. Pepsinogen activation occurs rapidly at pH 2 but occurs very slowly at pH 4. As a result, at higher pH values, protein digestion is reduced. The addition of organic acids to the diet induces a rapid reduction in the pH value in the stomach, aiding digestion (Schöner, 2001).

While organic acids are widely used in the swine industry, their effects in poultry have not been clearly shown (Mateos *et al.*, 2001). In poultry production, organic acids have been used primarily to sanitize feed or water where problems are occurring with bacterial infection

(Ghosh *et al.*, 2008). *In vitro*, formic, acetic, and propionic acids have been shown to have bactericidal effects on *Campylobacter*. The combination of organic acids showed synergistic bactericidal activity at pH 4.5 (Chaveerach *et al.*, 2002). The researchers concluded that routine application of organic acids to the water supply on poultry farms could prevent or diminish *Campylobacter* transmission.

There has been an increased interest in using organic acids as replacements for AGP in poultry production. The body of research on feeding organic acids and/or their salts to poultry is smaller than that for feeding pigs. Several studies have looked at fumaric acid supplementation of broiler diets (Dibner & Buttin, 2002). Supplementation with 3.5%–4.0% fumaric acid has been shown to improve feed efficiency of broilers (Vogt & Matthes, 1981). Skinner *et al.* (1991) reported that supplementing the diet of male broilers with 0.5% fumaric acid resulted in improved bodyweight gain and feed efficiency. Similarly, Runho *et al.* (1997) reported that the growth performance of broilers receiving 1.0% fumaric supplementation was comparable to that of broilers receiving an AGP (Nitrovin, a nitrofuran feed additive). In contrast, Patten and Waldroup (1988) reported that supplementation of broiler diets with 0.5%–1.0% fumaric acid improved bodyweight gain but did not affect feed intake or feed efficiency. Including fumaric acid at high levels, however, resulted in reduced feed consumption as well as reduced bodyweight gain.

Butyric acid is another possible alternative to the use of AGP and coccidiostats in broiler diets. Leeson *et al.* (2005) noted that 0.2% butyric acid, supplemented as butyrate glyceride, can be added to diets to maintain the performance and carcass quality of broilers, especially the previously vaccinated broilers challenged with coccidiosis. Previously vaccinated broilers supplemented with 0.2% butyric acid and challenged with a mixed culture of coccidia oocysts achieved better growth rates than those broilers receiving the unsupplemented control diet.

Other organic acids that have been shown to improve broiler performance include malic, sorbic, and tartaric acids (Vogt *et al.*, 1982; Izat *et al.*, 1990) as well as lactic and formic acids (Vogt *et al.*, 1982; Vertseegh & Jongbloed, 1999). Malic acid supplementation of broiler diets has also been shown to reduce intestinal *E. coli* populations (Moharrery & Mahzonieh, 2005). In contrast, Pirgozliev *et al.* (2008) found that fumaric and sorbic acid supplementation of broiler diets resulted in reduced feed consumption with no effect on bodyweight gain or feed efficiency. Biggs and Parsons (2008) evaluated citric, fumaric, and malic acids on the growth and nutrient digestibility in New Hampshire × Columbian male chicks. The results on 21-day live weight and nutrient digestibility were not consistent for the organic acids. Some organic acids depressed growth and the cause was not clear. Palatability of the feed did not play a factor. Similarly, Talebi *et al.* (2010) found no effect of citric, benzoic, or tartaric acids on broiler growth performance.

In addition to SCFA, the medium-chain caproic, caprylic, and capric acids have been shown to have bactericidal affects *in vitro* (Hermans *et al.*, 2010). Caprylic acid is a medium-chain (eight carbons) fatty acid found naturally in breast milk, bovine milk (Jensen, 2002), and coconut oil (Sprong *et al.*, 2001). It is a food-grade compound classified as GRAS (Generally Recognized As Safe) and has been shown to be effective in killing a variety of bacterial pathogens, including *C. jejuni*. Caprylic acid supplementation of broiler diets has been shown to be effective in decreasing *Campylobacter* contamination of market-aged chickens (Solis de los Santos *et al.*, 2009). In contrast, Hermans *et al.* (2010) reported that, under the conditions of their research, caproic, caprylic, and capric acids were not able to reduce cecal *Campylobacter* colonization in 27-day-old broiler chickens. The authors proposed three possible explanations for the contradictory results. First, there may have been differences in the formulation of the acids in the feed so that higher concentrations of the acids reached the ceca in the study of Solis de los Santos *et al.* (2009). The ceca are the predominant site for *Campylobacter* colonization. A second possible explanation may be differences in the properties of the *C. jejuni* strains used in the studies. Hermans *et al.* (2010) used a highly invasive strain that might be less susceptible to the effects of medium-chained fatty acids. And finally, there may have been differences in the genetics of the strains of chickens used in the two studies with different dietary influences on cecal biochemistry (Hermans *et al.*, 2010).

Most of the studies with organic acids have been with broilers reared in a healthy environment, but most of the positive effects have been shown when the health status of the animal is impaired. One model to study the effect of antibiotic alternatives on growth performance of broilers chickens suffering from compromised intestinal integrity is the malabsorption syndrome model. This condition is achieved by dosing day-old chicks with a chyme homogenate from birds suffering from malabsorption syndrome (den Hartog *et al.*, 2005). The addition of short- and medium-chained fatty acids has been shown to result in significantly higher growth performance in broilers whose health was impaired by malabsorption syndrome (Gutierrez del Alamo *et al.*, 2007). At 14 days of age, the addition of the short- and medium-chained fatty acids to broiler diets resulted in significantly higher bodyweights compared to the unsupplemented controls, with the combination of short and medium-chained fatty acids having a synergistic effect. By 42 days, however, the difference was no longer statistically significant.

It has been hypothesized that SCFA are antibacterial in the crop but have no additional activity further down the digestive tract (Thompson & Hinton, 1997). A new range of products has been developed by encapsulating SCFA (acetic acid, formic acid, propionic acid, and butyric acid) in mineral carriers so that they can be released slowly during transport down through the GI tract (Van Immerseel *et al.*, 2004). A challenge study looking at the effects of supplementing broiler diets with encapsulated SCFA showed that acetic and formic acids increased *Salmonella* Enteritidis colonization in the ceca and internal organs. Broilers receiving propionic acid were colonized to the same extent as the unsupplemented controls. Butyric acid supplementation, however, resulted in a significant decrease in *Salmonella* in the ceca but not in the liver and spleen.

Supplementation of turkey diets with propionic acid was found to help in controlling Poult Enteritis and Mortality Syndrome (PEMS) (Roy *et al.*, 2002). The occurrence of PEMS has been associated with two viruses and at least three *E. coli* isolates. Turkeys supplemented with a commercial product containing short-chain organic acids had reduced mortality associated with PEMS. The reduction in mortality was attributed to a reduced bacterial level in the gut. The commercial product acts primarily via propionic acid (Roy *et al.*, 2002).

Studies of organic acids in rabbit diets are few, and their results far from consistent (Maertens *et al.*, 2006). Supplementation of growing rabbits with commercial acidifiers did not affect any performance characteristics (Scapinello *et al.*, 2001). Fumaric acid had no effect on weight gain but improved feed efficiency when added to rabbit diets at 20 g/kg. Caprylic acid at 5 g/kg had no effect on rate of growth but reduced mortality in the postweaning period (Scapinello *et al.*, 2001).

21.3.2 Using organic acids with ruminants

Organic acids have a different mode of action in ruminants than in nonruminants. Fumaric and malic acids have been shown, *in vitro*, to reduce methane production in rumen fluid

(Beauchemin & McGinn, 2006). The organic acids act as hydrogen sinks (i.e., consume more H^+ than they generate) and interfere with methane production (Martin *et al.*, 2009). Instead, there is an increased production of propionate. The addition of the salts, fumarate and malate, has been reported to reduce rumen methane production and improve the feed efficiency (Jouany & Morgavi, 2007). Unfortunately, based on *in vitro* tests, a supplementation with 2.9 kg of sodium fumerate would be needed to decrease methane production by 10% (Newbold *et al.*, 2005). Although free acids are more effective at reducing methane production than their salts, their use is not recommended for use *in vivo* because of the unwanted drop in rumen pH. Fumaric acid can be encapsulated to protect it in the rumen (Wallace *et al.*, 2006). *In vitro* encapsulated fumaric acid did not cause a drop in pH but retained its ability to suppress methane production. When encapsulated fumaric acid was fed to growing sheep, there were no adverse effects on growth performance, while methane production was reduced by 75%. Lambs receiving the supplemented diets had a higher rate of gain during the first 22 days, but there was no effect on final bodyweights. Feed intake was reduced with supplementation so that feed efficiency was higher (Wallace *et al.*, 2006). The use of encapsulated SCFA is currently not economical (Jouany & Morgavi, 2007). The malate concentration in alfalfa and Bermudagrass hay has been shown to range from 1.9% to 4.5% of the dry matter suggesting that the use of forages rich in organic acids may provide a more economical option (Martin, 1998).

Organic acids can play a role in animal production as alternatives to the subtherapeutic use of antibiotics. More research is required, however, on the type and form of acid to be used as well as the level of inclusion. Ruminants have different mechanisms of action resulting in different requirements than that of nonruminants.

21.4 ANTIMICROBIAL PEPTIDES

AMPs are part of the innate immunity genetically coded in all living cells. These peptides are produced by all living organisms, including all animals, plants, invertebrates, and microorganisms (Boman, 2003; Parisien *et al.*, 2008). In humans, AMPs are present in white blood cells and are secreted by the outer layers of the skin and the mucosal surfaces. AMPs kill a broad spectrum of organisms, including bacteria, fungi, parasites, and some viruses. They are believed to act by disrupting the cell membrane (Gordon *et al.*, 2005; Parisien *et al.*, 2008). AMPs typically have a positive charge at physiological pH allowing them to interact with the negatively charged lipids in the cytoplasmic membranes of Gram-positive bacteria. The AMPs also have portions that are hydrophobic and portions that are hydrophilic (Moll *et al.*, 1999). This allows them to form pores in the cell membrane of the target organism, resulting in the loss of the cell content and death.

Sang and Blecha (2008) reviewed the antibiotic properties of AMPs, highlighting their potential as alternatives to conventional antibiotics. While conventional antibiotics primarily affect bacteria and/or fungi, AMPs have been shown to act on bacteria, fungi, parasites, enveloped viruses, and even cancer cells. Antibiotic use in animal agriculture has been criticized because of the selection for resistant microorganisms. Because AMPs act on the cell membrane, the development of resistance is believed to be difficult and unlikely to occur (van 't Hof *et al.*, 2001; Boman, 2003; Hancock & Sahl, 2006). This has been found to be not completely true. There have been some incidences of resistance development to AMPs through modification of the cell membrane. Certain genes can confer increased resistance to AMPs. An example is a gene in *Legionella pneumophila* that is involved in AMP resistance. To date, however, this resistance has only been generated in the laboratory (Robey *et al.*, 2001). Resistance to some AMPs has also been observed in *Staphylococcus* (Peschel & Collins, 2001), but it is not clear if this resistance can be transferred between different bacteria.

There has been a considerable amount of research done with AMPs from eukaryotic cells. They have been shown to have a broad spectrum of activity (antibacterial, antiviral, and antifungal) and kill their targets quickly. As previously mentioned, the probability of a target organism developing resistance to AMPs is low. Although there have been several clinical trials, no eukaryotic AMP has thus far obtained FDA approval (Gordon *et al.*, 2005).

Broadly speaking, bacteriocins are AMPs produced by bacteria and have a killing action on bacteria other than the producing strain (Daw & Falkiner, 1996). It is assumed that production of bacteriocins is the mechanism used by intestinal bacteria to achieve a competitive advantage (Joerger, 2003). Bacteriocins were first reported by Gratia in 1925 when he noticed an antagonism between two strains of *E. coli*. This antibacterial activity was then found to be produced by various enterobacteria species and the generic name "colicine" was given (Daw & Falkiner, 1996). When it was discovered that this activity was produced by other types of bacteria, Jacob *et al.* (1952) proposed the generic name "bacteriocine," which was later adopted but spelled without the final "e". Since then at least 99% of bacteria are believed to produce at least one bacteriocin (Parisien *et al.*, 2008).

Previous reviews by Daw and Falkiner (1996) and Cotter *et al.* (2005) have described in detail the history, biology, classification, biochemical nature, and mode of action for bacteriocins. It is beyond the scope of this chapter to cover these topics in detail. Instead, this subsection will focus more on the use of bacteriocins as alternatives to antibiotics in meat production.

When bacteriocins recovered from bacteria commonly inhabiting the intestinal tracts of broiler chickens (*Paenibacillus polymyxa* and *Lactobacillus salivarius*) were microencapsulated and added to turkey feed, *Campylobacter* colonization was reduced to undetectable levels (Cole *et al.*, 2006). Changes to the digestive tract were also noted. The duodenum crypt depth and the number of goblet cell were reduced. Joerger (2003) commented that it may be more cost-effective to administer the bacteriocin-producing organisms themselves rather than isolate the bacteriocins. There are cases, however, where feeding the bacteriocins is effective, while feeding the bacteria that produce them is not. For example, Stern *et al.* (2008) identified two bacteria in the chicken GI tract that have been shown to inhibit *C. jejuni* and they had hoped to develop a probiotic treatment for poultry. When these organisms failed to control a *Campylobacter* challenge in turkeys, the researchers looked instead at the bacteriocins these two bacteria produce. Feeding the bacteriocins reduced *C. jejuni* colonization in turkeys significantly. In a subsequent study, Svetoch and Stern (2010) added bacteriocins to the drinking water of broiler chickens. They found that, when given for the last 3 days of the growout, the bacteriocins achieved a five- to sixfold reduction in cecal colonization of *Campylobacter* compared to untreated chickens.

The previous studies did not use one specific bacteriocin, but used instead a collection of bacteriocins produced by the bacteria in the digestive tract of chickens. Piva and Headon (1994) purified pediocin A, a bacteriocin produced by a strain of *Pediococcus pentosaceus*. Pediocin A was found to inhibit, *in vitro*, the growth of *Lactobacillus*, *Lactococcus*, *Leuconostoc*, *Pediococcus*, *Staphylococcus*, *Enterococcus*, *Listeria*, and *Clostridium*. Using a model swine intestinal fermentation system, pediocin A was shown to be antagonistic to a broad spectrum of Gram-positive bacteria and, *in vitro*, was able to directly decrease clostridia and indirectly decrease coliform counts (Casadei *et al.*, 2009). The researchers recommended pediocin A as a potential substitute for AGP. Grilli *et al.* (2009) also isolated pediocin A from a strain of *P. pentosaceus* and fed it to broilers during a 42-day growout period. Their results indicate that feeding pediocin A can reverse the adverse affects of a *C. perfringens* challenge. The occurrence of NE is associated with the presence of *C. perfringens*. Pediocin A appears to be, therefore, an alternative for subtherapeutic antibiotics that are normally used for the control of NE in chicken production. One of the constraints to its use, however, has been the cost involved in its production and purification (Dickson *et al.*, 1998).

Infections caused by *E. coli*, *Staphylococcus aureus*, *Clostridium* spp., and other pathogens can be a serious problem in commercial rabbit production and antibiotics are often used to control this problem. Simonová and Lauková (2007) isolated bacteriocin-producing enterococcal species from the feces of rabbit and studied their biochemical properties. One of the bacteriocins isolated from a strain of *E. faecium* had a broad spectrum of inhibitory activity against both Gram-positive and Gram-negative bacteria. It controlled, *in vitro*, *Listeria* contamination as well as infections caused by *E. coli*, *Staphylococcus* sp., *Clostridium* sp., or *Pseudomonas* sp.

Commercial applications of bacteriocins in veterinary and human medicine are currently being explored. Since bacteriocins are proteins, they are vulnerable to breakdown in the gastric juices of an animal's digestive system. This does not, however, prevent them from being used to control microbial growth on surfaces (i.e., act as a disinfectant) or in preventing food spoilage or controlling food-borne pathogens during storage. They also have potential in biological materials such as vaccines. In these applications enzymatic degradation is less of a concern. Research has also been conducted with feeding bacteriocins to food animals, both ruminant and nonruminant.

The first bacteriocin approved for commercial use was nisin (Vessoni Penna *et al.*, 2006). Nisin is a bacteriocin produced during the exponential growth phase of certain strains of *Lactococcus lactis* (Vessoni Penna *et al.*, 2006). It has a broad spectrum of activity against Gram-positive bacteria but shows little or no activity against Gram-negative bacteria, yeasts, or molds (Hurst, 1981). The outer membrane of Gram-negative bacteria prevents nisin from reaching the site of action. The permeability of this outer membrane can be altered by the addition of chelators such as EDTA (disodium ethylenediaminetetraacetate) (Vessoni Penna *et al.*, 2006). Nisin has been shown to inhibit *E. coli*, but only in the presence of EDTA.

Nisin is currently the only bacteriocin approved for limited food uses by the US Food and Drug Administration (FDA). In 1998, the FDA-approved nisin as an additive in canned products to inhibit the growth of *Clostridium botulinum*. Nisin is now classified as GRAS and used in a wide variety of food and beverage products (Delves-Broughton *et al.*, 1996).

Nisin has been successful in reducing methane production in ruminants. Although ionophores were originally developed as coccidiostats, they have also been used to alter fermentation in the rumen (Callaway *et al.*, 1997; Russell & Mantovani, 2002). The ionophores are believed to act by inhibiting Gram-positive bacteria in the rumen and bacteriocins have been shown to do the same. Bacteriocins, therefore, may serve as a replacement for ionophores in beef production. Using *in vitro* studies, it was shown that monensin and nisin decrease methane production by 43% and 36%, respectively (Callaway *et al.*, 1997). Jalc and Lauková (2002) also found that nisin was not as effective as monensin for reducing methane production in an artificial rumen.

Kišdayová *et al.* (2003) looked at the effect, *in vitro*, of nisin on two species of ciliate protozoa from sheep rumen fluid, the bacterial co-population, and the levels of different SCFA produced. Long-term exposure to nisin concentrations had no effect on one of the protozoa species (*Entodinium caudatum*), while the other (*Epidinium ecaudatum*) was more sensitive. The authors were unable to explain the difference in sensitivity. They did speculate that the cell surfaces of the two protozoa are different enough to show different permeabilities to nisin. The rumen bacteria appeared to be more sensitive than the protozoa. In the same cultures, the growth of *Lactobacilli*, *Enterococci*, *Staphylococci*, and amylolytic *Streptococci* was reduced. No resistance was observed in the 30-day trial period. There was also a modification in the levels of SCFA produced with a decrease in acetate and an increase in propionate levels.

A similar study was done by Kišdayová *et al.* (2009) but they compared, *in vitro*, the effects of nisin and monensin alone or in combination on ovine rumen ciliates. The levels of two species of ciliate protozoa (*Entodinium* spp. and *Dasytricha ruminantium*) were increased with nisin supplementation and decreased by monensin. The combined supplementation was similar to that of the effects of monensin alone. In contrast to previous research, neither nisin nor monensin affected the levels of amylolytic streptococci and enterococci.

The addition of nitrates to a rumen's diet has been shown to reduce methane production (Takahashi *et al.*, 1997) but nitrates can be toxic to ruminants if fed at high levels. Simultaneous supplementation with nisin has been shown to counter the toxic effects of the nitrates while keeping the production of methane low (Sar *et al.*, 2004).

The potential for developing resistance to an antimicrobial is always a concern, as was seen with the prolonged use of antibiotics. Until recently, resistance to bacteriocins was considered extremely unlikely because of their structure and mode of action (Hancock, 1997; van 't Hof *et al.*, 2001; Joerger, 2003). However, the development of nisin resistance has been achieved in the laboratory. Strains of *Streptococcus bovis* in the rumen of cattle have been shown, *in vitro*, to initially be nisin sensitive but later developed nisin resistance, which persisted even in the absence of nisin (Mantovani & Russell, 2001). Resistant strains of *L. monocytogenes* can be created in the laboratory by exposing the bacteria to high concentrations of nisin (Crandall & Montville, 1998). At least some of this resistance was hypothesized as being due to changes to the cell wall. In nisin-susceptible bacterial strains, nisin passes through the cell wall and acts on the cytoplasmic membrane. In the suggested model, nisin-resistant strains developed changes to the cell wall that prevent nisin from passing through to reach the cytoplasmic membrane (Crandall & Montville, 1998).

Until recently, development of resistance to bacteriocins was not considered a threat to the development of antibiotic resistance in bacteria. However, Carlson *et al.* (2001) reported that exposing *Salmonella* to the bacteriocin microcin-24 resulted in microsin-resistant cells, exhibiting resistance to multiple common antibiotics. Mantovani and Russell (2001) reported that ampicillin resistance in nisin-resistant mutants of *S. bovis* is 1000 times that seen with nisin-sensitive isolates. It is unclear if these situations are unique or if certain other bacteriocins can produce similar effects.

AMPs show potential as alternatives to antibiotics. Many studies comparing AMPs to traditional antibiotics have had comparable results. However, the potential of organisms to develop resistance to the AMP or to antibiotics is a factor that must be considered.

21.5 PHYTOGENIC COMPOUNDS/BOTANICALS

Phytogenic compounds, also referred to as phytobiotics or botanicals, are simply the materials derived from plants. They can be derived from different parts of the plant including flowers, buds, leaves, twigs, bark, fruit, and roots. Herbs are "flowering, nonwoody and nonpersistent plants" that are well known for providing flavor to foods. Herbs are also valued for their smell and are often used as air fresheners. For generations, herbs have been used in human medicine and only recently are they being investigated for use in animal production. This includes herbs such as oregano, rosemary, sage, thyme, etc. Essential oils are a group of volatile oils extracted from plants using "cold expression" or by steam or alcohol distillation. They are often referred to as "ethereal" oils. When solvents are used to extract the oils, they are typically referred to as "oleoresins."

Phytogenics in animal production is still a relatively new area of research but various plant extracts have received increased attention as potential alternatives for AGP. Bodyweight gain and feed efficiency are common measures of effectiveness. Animal health is another important factor. The quality and oxidation status of the meat product also need to be considered since they determine shelf-life. Shelf-life is the length of time a perishable item can be stored before it is unsuitable for human consumption.

There are a wide variety of phytogenic products being investigated for use in animal production and they vary considerably with respect to botanical origin, processing method used, and composition. They also show a wide range of antimicrobial activities (Windisch *et al.*, 2008). The active ingredients of a phytogenic feed additive will vary depending on the part of the plant used, the geographical location where it was grown, and when the material was harvested. A variety of phytogenic feed additives have been used and the effects vary from dramatic enhancement in production performance, to none or even negative effects.

Consumption of fruits and vegetables has long been associated with good health (Sun *et al.*, 2002). Phenolics in these foods are suggested to be the major bioactive compounds involved. In the past, the phenolic content and their antioxidant activity have been underestimated in the literature because bound phenolics were not included. Sun *et al.* (2002) determined the total phenolics, including the bound forms, in common fruits. Cranberry had the highest total phenolic content and the highest total antioxidant activity. Leusink *et al.* (2010) looked at the effect of a cranberry extract on growth performance and intestinal health of broilers and found no significant effects. The authors suggested that the high hygienic and biosecurity practices used in the study masked any effects of the cranberry extract and that perhaps a more stressful environment is required to observe any benefit.

Grape seed proanthocyanidin extract is widely used as a human food supplement for health promotion and disease prevention (Bagchi *et al.*, 2000). Proanthocyanidins are a class of condensed tannins known to have antioxidant properties. Adding as little as 10–20 mg/kg of grape seed proanthocyanidin extract to broiler diets enhanced growth performance and reduced mortality after a coccidiosis infection with *Eimeria tenella* (Wang *et al.*, 2008). The extract was believed to counter the oxidative stress caused by the coccidiosis infection.

Grape pomace, a by-product of the wine industry, is a rich source of polyphenols that have the ability to act as antioxidants (Goñi *et al.*, 2007; Brenes *et al.*, 2008). Supplementing broiler diets with grape pomace increased the antioxidant activity in the diet and ileal contents but did not adversely affect growth performance or protein and amino acid digestibilities (Brenes *et al.*, 2008). In addition, grape pomace in broiler diets reduced the lipid oxidation of the meat during refrigerated storage (Goñi *et al.*, 2007; Brenes *et al.*, 2008), thus increasing its shelf-life.

The inclusion of dried tomato pulp, a by-product of the tomato processing industry, in the diet of Japanese quail had both anti- and pro-oxidant effects depending on the level of inclusion. At 5% inclusion, there was an antioxidant effect, while the effect was pro-oxidant at 10% inclusion (Botsoglou *et al.*, 2004a). The active components of tomatoes are carotenoids. Several researchers have reported on the delicate balance between the anti- and pro-oxidant effects of dietary carotenoids (Botsoglou *et al.*, 2004a). Inclusion of 10% dried tomato pulp in the diet of Japanese quail also resulted in higher total polyunsaturated fatty acids and greater unsaturated to saturated fatty acids ratios in the meat compared to the unsupplemented controls.

Echinacea is a genus of flowering plants in the daisy family. They are commonly called purple coneflowers and are native to North America. Some species of echinacea are used in herbal medicines. Maass *et al.* (2005) looked at the use of dried echinacea *(Echinacea purpurea)* leaves as a replacement for growth promoting antibiotics in three stages of swine production. When included in the diets of gestating sows, there was a significant reduction in bodyweight gain for the sow, but higher bodyweights of the piglets at birth. There was no significant effect of echinacea supplementation on the bodyweights of lactating sows. The piglets born to sows that had been supplemented with echinacea were then raised without supplementation and there were no significant differences in bodyweights 4 weeks after weaning (Maass *et al.*, 2005). Those piglets from supplemented sows, however, had improved feed efficiency. In a second trial, piglets from unsupplemented sows were raised for 6 weeks and no effect was noted on bodyweight of echinacea- or antibiotic-supplemented piglets as compared to the unsupplemented controls. Feed efficiency, however, was improved with the echinacea supplementation (Maass *et al.*, 2005). In a third study with grower/finisher pigs, there was no significant effect of echinacea supplementation on feed intake or growth performance.

Echinacea extract has been shown to boost the immune response of pigs (Maass *et al.*, 2005). Pigs were vaccinated against swine erysipelas and the antibody response was measured. The immune response in the pigs fed echinacea-supplemented diets was greater than in the unsupplemented control group. A similar response was reported in broilers. The antibody titers of broilers vaccinated against infectious bursal disease (IBD) (Ma *et al.*, 2009) or Newcastle disease (Zhang, 2005) were greater than the levels in the unsupplemented control group. Repeated short-term supplementation with echinacea juice has been reported to have immune stimulating effects in laying hens and fattening pigs (Böhmer *et al.*, 2009). Repeating for 2 days with supplementation, followed by 12 days without, were sufficient to enhance the immune reactions. In contrast, other researchers have reported that echinacea supplementation of pig diets did not enhance growth, exhibit protection against Porcine Reproductive and Respiratory Syndrome (PRRS) infection, and did not show any evidence of immune enhancing properties (Hermann *et al.*, 2003).

A number of herbs have shown antimicrobial activity. For example, supplementing rabbit diets with dried Siberian ginseng improved weight gain and feed efficiency, and decreased mortality. There were also reduced numbers of parasitic *Eimeria* oocysts in the intestines (Simonová *et al.*, 2008). Feeding dried peppermint leaves to broilers has been reported to affect growth performance (Ocak *et al.*, 2008). Bodyweight gain early in the trial (35 days) was higher in the broilers supplemented with peppermint leaves as compared to the unsupplemented controls, but the differences disappeared by the end of the 42-day trial. Feed intake, feed efficiency, carcass weight, carcass yield, and the relative weights of the edible inner organs and whole gut were not affected by peppermint or thyme supplementation. However, both supplements did increase the size of the abdominal fat pad. Supplementing broiler diets with a mixture of anise, cinnamon, and peppermint leaves was reported to improve bodyweight gain and feed efficiency compared to the unsupplemented controls (Al-Kassie & Witwit, 2010). Similarly, supplementing the broiler diets with dried dandelion also resulted in improved growth performance, though the improvement was not as great as that observed with the mixture of herbs (Al-Kassie & Witwit, 2010).

The addition of dried oregano leaves (up to 20 g/kg inclusion) to broiler diets had no effect on growth performance or carcass traits (Halle *et al.*, 2004; Karimi *et al.*, 2010). The authors reported no effect on bodyweight gain, feed efficiency, or mortality rate but suggested that higher levels of inclusion in a more challenging environment may be needed to elicit an effect. In contrast, Bampidis *et al.* (2005) supplemented the diets of female early maturing turkeys with dried oregano, at a maximum inclusion rate of 3.75 g/kg, and noted improved feed efficiency.

Tulbaghia violacea (wild garlic) has been shown to have potential as a therapeutic or prophylactic anticoccidial agent in broilers (Naidoo *et al.*, 2008). Supplementation of broiler diets with garlic bulb or husks resulted in increased protein and lower fat content in the thigh muscles as compared to those from broilers fed unsupplemented control diets (Kim *et al.*, 2009). Sensory panelists scored the samples high for texture and flavor. Similar results were obtained by Choi *et al.* (2010) who reported that supplementing broiler diets with garlic powder enhanced lipid and color stability of the meat.

Broilers fed diets supplemented with dried ginger root powder were reported to have higher live bodyweights and carcass yields compared to unsupplemented controls (Zhang *et al.*, 2009). There was no effect on feed intake or feed efficiency. The antioxidant status of broilers was improved with ginger root supplementation.

Lipid oxidation is one of the primary mechanisms causing deterioration of meats during storage (Kanner, 1994). This is especially important with the trend toward increasing amounts of omega-3 fatty acids in meat. Smet *et al.* (2008) compared the antioxidant potential of rosemary, green tea, grape seed, and tomato extracts. All of the extracts were less effective than synthetic antioxidants but there were marked differences between them. The oxidative protection was greatest with the rosemary extract. O'Grady *et al.* (2006) also reported antioxidant properties of rosemary during meat storage. Rosemary extract supplemented to the diets of beef cattle for 103 days before slaughter increased the lipid stability of the meat during storage. Green tea extracts were found to have a similar effect (O'Grady *et al.*, 2006). A similar response to green tea supplementation has been reported for pork production (Mason *et al.*, 2005). Additional herb extracts are being investigated for antioxidant properties. Jang *et al.* (2008) reported that the breast meat of broilers fed a mixture of herb extracts (mulberry leaf, Japanese honeysuckle, and goldthread) had increased anti-oxidant potential during storage. In a sensory taste test, the breast meat from the supplemented broilers was preferred over that from the unsupplemented controls.

The olive tree shows strong resistance against attack by microorganisms and insects (Kubo *et al.*, 1995). Olive leaf extracts are rich in phenolic compounds (Kubo *et al.*, 1995; De Nino *et al.*, 1997), most of which have been shown by *in vitro* studies to possess antimicrobial and antioxidant activity (Benavente-Garcia *et al.*, 2000; Bisignano *et al.*, 2001). Antioxidant-rich plant extracts have been shown to be beneficial in treating coccidial infections in broiler chickens (Naidoo *et al.*, 2008). Nonphenolic compounds found in olive leaves, such as aldehydes, have also been studied for their antimicrobial properties (Kubo *et al.*, 1995). Incorporation of olive leaves in the diet of turkeys delayed lipid oxidation in the raw breast fillets stored for 12 days at 4°C (Botsoglou *et al.*, 2010). There was no effect on bodyweight gain.

Of 13 plant extracts screened *in vitro* for their action on fermentation and the protozoa population of rumen fluid, four showed promise as feed additives for ruminants (Broudiscou *et al.*, 2000). Extracts from lavender (*Laricifomes officinalis*) and goldenrod (*Solidago virgaurea*) flowers were found to increase the level of fermentation. Horsetail (*Equisetum arvense*) and sage (*Salvia officinalis*) leaves were found to inhibit methane production. None of the extracts affected the protozoa numbers. It is unclear whether the effects of feed additives would have continued long-term. Cardozo *et al.* (2004) evaluated the effects extracts from garlic (*Allium sativa*), cinnamon (*Cinnamomum cassia*), yucca (*Yucca schidigera*), anise (*Pimpinella anisum*), oregano (*Origanum vulgare*), and pepper (*Capsicum annuum*) during a 10-day *in vitro* study with ruminal fluid. Between days 2 and 6 of fermentation, the proportions of acetate, propionate, and butyrate were altered by the cinnamon, garlic, oregano, and anise extracts. All these effects, however, disappeared by day 6. The results of this *in vitro* study bring into question the reliability of short-term studies (Cardozo *et al.*, 2004). In the previous study, the ruminal fluid was obtained from two fistulated dairy cows fed a total mixed diet. In a follow up study, the ruminal fluid was obtained from beef heifers fed a high-concentrate finishing diet and the same six plant extracts were evaluated *in vitro* (Cardozo *et al.*, 2005). All the effects of the plant extracts were pH dependent. At high pH, all plant extracts except cinnamon maintained or decreased the level of total SCFA and maintained or increased the acetate to propionate ratio. At pH 7.0, none of the extracts would be beneficial for beef production. At pH 5.5, however, total SCFA concentration was greater with garlic, pepper, and yucca extracts and the acetate to propionate ratio was less with oregano, garlic, pepper, and yucca extracts. The researchers concluded that at the low ruminal pH values, expected when feeding high concentrate feeds, oregano, garlic, pepper, and yucca extracts are potentially useful in beef diets.

Essential oils are a class of volatile oils obtained from plants via hydro-distillation and have been shown to have antibacterial activity (Dorman & Deans, 2000). Essential oils have been studied as a feed additive to improve feed characteristics, improve digestion and performance of the animal being fed, or to improve characteristics of the meat produced. Essential oils have also been used to reduce methane (Chaves *et al.*, 2008) and ammonia production (Ando *et al.*, 2003) of ruminants.

Essential oils have been shown to work primarily against Gram-negative bacteria (de Lange *et al.*, 2010). *In vitro* studies using pig cecal digesta, essential oils were reported to be effective against *Salmonella* Typhimurium, *E. coli* O157:H7, and *E. coli* K88 but had little effect on *Lactobacillus* and *Bifidobacterium*. Essential oils also retain the characteristic odor and flavor of the plant they were extracted from. It is possible that improved performances reported with essential oil supplementation may be due to increased palatability of the diet (de Lange *et al.*, 2010).

Cross *et al.* (2007) looked at the effects of five culinary herbs: (1) thyme (*Thymus vulgaris* L.), (2) oregano (*O. vulgare* subsp. Hirtum), (3) marjoram (*Origanum majorana* L.), (4) rosemary (*Rosmarinus officinalis* L.), and (5) yarrow (*A. millefolium* L.), or their essential oils on the performance of female broiler chicks. Only those broilers fed the diet supplemented with thyme oil were significantly heavier than the unsupplemented controls. There were no significant differences in overall feed efficiency. None of the dietary treatments had any effect on the intestinal microflora populations or nutrient digestibility. Based on this study, supplementing diets with a single herb or its extracted essential oil does not always have similar results.

Several essential oils have been studied for use in animal production but the reported effects on growth performance and meat quality have been inconsistent. Oregano is a traditional Mediterranean spice. The essential oil from oregano contains several compounds, with thymol and carvacrol contributing the most to its activity (Lambert *et al.*, 2001). Oregano supplementation has been reported to have no effect on growth performance of poultry including medium-growing broilers (Symeon *et al.*, 2009), fast-growing broilers (Basmacioğlu *et al.*, 2004), turkeys (Botsoglou *et al.*, 2003a; Papageorgiou *et al.*, 2003; Florou-Paneri *et al.*,

2005), and Japanese quail (Sarica *et al.*, 2009). There have also been reports of no effect of oregano essential oil supplementation with pigs (Simitzis *et al.*, 2010), rabbits (Botsoglou *et al.*, 2004b), or lamb (Simitzis *et al.*, 2008).

In addition to not having beneficial effects on growth performance, oregano essential oil supplementation has been reported to negatively affect the feeding and drinking behavior, as well as the overall activity, of broilers (Symeon *et al.*, 2010). Chicks supplemented with oregano essential oil made fewer trips to the feeders than the unsupplemented controls. The level of feed intake for the two groups, however, was not reported. The decrease in activity was correlated to the increase in oregano essential oil inclusion. Oregano essential oil supplementation also decreased the trips to the drinkers. This is expected since there is a direct relationship between drinking and eating. It is not known whether the reduced number of trips to the drinkers reduced overall water intake since this parameter was not measured in this study.

Thymol is a major component of thyme essential oils (Lee *et al.*, 2003). Carvacrol is an isomer of thymol and is found in essential oils isolated from oregano, thyme, marjoram, and summer savory. Although thymol and carvacrol are isomers, they have very different effects in animal feeding. While thymol supplementation of broiler diets had no effect on growth performance, carvacrol supplementation resulted in reduced feed intake, bodyweight gain, and feed efficiency (Lee *et al.*, 2003). It may be that carvacrol is negatively affecting the flavor of the feed, resulting in reduced feed consumption.

While oregano essential oil may give conflicting results on growth performance, it may play an important role in meat quality. There is a trend in the food industry to shift from synthetic antioxidants to more "natural" additives that can inhibit the development of oxidative rancidity in meat (Diplock *et al.*, 1998). Oregano essential oil is obtained by steam distillation of the leaves and flowers and is known for its antioxidative activity (Economou *et al.*, 1991). Cavacrol and thymol are the two main phenols in oregano essential oil and are primarily responsible for its antioxidative activity. Feeding turkeys diets supplemented with oregano herb or oregano essential oil was found to increase the oxidative stability of the raw turkey meat during storage at 4°C, thus increasing its shelf-life (Botsoglou *et al.*, 2003a; Papageorgiou *et al.*, 2003; Govaris *et al.*, 2004; Florou-Paneri *et al.*, 2005). There was a similar effect on the oxidative stability of cooked breast and thigh turkey meat during storage (Botsoglou *et al.*, 2003b). Dietary oregano essential oil has also been reported to increase the oxidative stability of chicken (Botsoglou *et al.*, 2002), lamb (Simitzis *et al.*, 2008), and rabbit (Botsoglou *et al.*, 2004b).

Basmacioğlu *et al.* (2004) found that feeding essential oils from oregano and rosemary to broilers reduced lipid oxidation during meat storage. The combination of the oils was more effective than either individually, suggesting a possible synergism between the two essential oils. In contrast, Lopez-Bote *et al.* (1998) reported that rosemary and sage essential oil supplementation had no effect on lipid stability of chicken; Janz *et al.* (2007) reported that supplementation with rosemary, garlic, oregano, and ginger essential oils had no effect on lipid stability of pork; Galobart *et al.* (2001) reported that supplementation with rosemary essential oil had no effect on the lipid stability of eggs; and Simitzis *et al.* (2010) reported that supplementation of pig diets with oregano oil had no effect on lipid stability of pork.

Rosemary (*R. officinalis*) is another typical Mediterranean plant. It contains natural polyphenols that significantly influence the oxidative processes (Karpinská *et al.*, 2000). Basmacioğlu *et al.* (2004) reported that supplementing the diets of fast-growing broilers with the essential oils from rosemary had no effect on growth performance over the 42-day trial. In contrast, Šperňáková *et al.* (2007) reported that broilers supplemented with rosemary powder had higher bodyweights than those broilers receiving the unsupplemented control diet. Rosemary supplementation was also effective in delaying lipid oxidation of the meat and improved the sensory properties of the meat.

Essential oils may also have a role in postharvest food safety. Soultos *et al.* (2009) reported that feeding oregano essential oil to rabbits reduced microbial growth during meat storage. Rosemary essential oil has been reported to positively affect the intestinal microflora of poultry and other monogastric animals (Hernández *et al.*, 2004).

Thyme (*T. vulgaris*) is yet another Mediterranean spice. The essential oil in thyme has been shown to vary depending on the climatic conditions in different geographical locations (Abu-Darwish & Abu-Dieyeh, 2009). For example, thyme from northern Jordan typically contains more essential oil than plants from the south and middle regions. Sarica *et al.* (2005) reported that thyme and garlic supplementation of broiler diets had no effect on growth performance or the intestinal microbiota. The authors noted that the trial was conducted using clean cages and suggested that the beneficial effects of the supplementations may only be observed under less hygienic housing conditions.

Supplementation of broiler diets with clove essential oil has been shown to have no effect on the lipid content and fatty acid profile of breast and thigh meat (Hernández *et al.*, 2009). Lipid oxidation was also not affected. The addition of savoury (*Nigella sativa* L.) essential oil to the diets of slow-growing broilers improved daily feed intake over the 84 days of the trial (Halle *et al.*, 2004). While early bodyweight gain was improved by the supplementation, there were no differences in final bodyweights.

Anise (*P. anisum* L.) is an aromatic plant that has been used in a number of treatments including stimulation of digestion and control of parasites. Simsek *et al.* (2007) reported that anise essential oil supplementation of broiler diets resulted in improved weight gain and carcass yield as compared to broilers fed diets supplemented with an AGP (avilamycin) or the unsupplemented control diets. The chicken from the group supplemented with anise essential oil also had better organoleptic properties as determined by a taste panel.

While individual essential oils may have no effect on growth performance and feed efficiency of meat animals, the use of blends of essential oils have shown potential. A blend of essential oils (from oregano, clove, and anise) added to broiler diets improved growth performance (Ertas *et al.*, 2005). The addition of essential oil blends has also been shown to increase immunity to coccidiosis in broiler chicks receiving vaccinations on the day of hatch (Oviedo-Rondón *et al.*, 2006). When broilers are fed essential oils post vaccination, they are better able to withstand a cocci challenge. In addition, supplementing broiler diets with blends of essential oils are reported to reduce *C. perfringens* in the intestines compared to those fed the unsupplemented control diet (Mitsch *et al.*, 2004). Blends of essential oils, therefore, may be an alternative to antibiotics in the control of NE in broilers.

Oleoresins differ from essential oils with regards to the manner of extraction. Essential oils are volatile lipophilic compounds derived by cold expression or by steam or alcohol distillation, while oleoresins are extracts derived by nonaqueous solvents. Oleoresins are used in a wide variety of products including beverages, soup powders, curry powders, noodles, canned meats, sauces, and poultry products. Oleoresins from cloves have been reported to control bacterial growth in marinated chicken products (Carlos & Harrison, 1999). Addition of rosemary oleoresin to ground chicken had an overall positive effect on raw meat appearance during storage and on cooked meat flavors (Keokamnerd *et al.*, 2008). The oxidation was slowed and the color (redness) was more stable. Rosemary oleoresin was also effective in maintaining quality of precooked roast beef slices during frozen storage (Murphy *et al.*, 1998).

Janz *et al.* (2007) compared the effects of dietary essential oils (rosemary and garlic) and oleoresins (oregano and ginger) on pork quality. Feed consumption and average daily gains were increased by supplementation with garlic products compared to the other diets but there was no difference in feed efficiency. The dietary supplementations did not affect carcass and meat quality attributes and sensory panels were unable to detect flavor or aroma differences between pork from the pigs fed the supplemented diets versus the unsupplemented controls. The researchers speculated that a higher level of supplementation may be necessary before any statistically significant differences are noted.

The feeding of a commercial mixture of oregano, cinnamon, thyme, and capsicum essential oil blends to broilers have led to varied results with some researchers reporting improved growth performance (Mathis & Scicutella, 2007; Tollba *et al.*, 2010), while others have reported no overall effect (Zhang *et al.*, 2005). There are also commercial products combining both essential oils and oleoresins. Again, there were varied results reported from the use of a commercial product containing carvacrol, cinamaldehyde, and capsicum oleoresins when included in animal feeds (Manzanilla *et al.*, 2004, 2006, 2009; Jamroz *et al.*, 2005, 2006; Muhl & Liebert, 2007).

A variety of phytogenic compounds, or botanicals as they are often referred to, have been studied for use as alternative feed additives with varying degrees of success. The use of botanicals is a relatively new area of research but several products have shown potential and are already commercially available.

21.6 CONCLUSIONS

Several "natural" feed additives are available that may be able to substitute for AGP in the diets of food animals. These include organic acids, AMPs, and various phytogenic compounds. The reported effectiveness of these products varies and further research is required to determine which experimental test factors are involved in the success or failure of a particular feed additive with a particular species of food animal.

It has been suggested by many researchers that the absence of significant differences from the various feed additives tested were the result of the experimental conditions and that more benefits would be observed in cases of stress. For example, in an effort to reduce feed costs and environmental impacts, researchers have been looking at using low protein (18%) broiler diets. Combining herbs (thyme and turmeric) and organic acids (citric and lactic acids) in such low-protein diets had no effect on final bodyweight, feed conversion, carcass quality, fecal nitrogen, or nitrogen retention (Abd El-Hakim *et al.*, 2009).

Any evaluation of a potential feed additive must look at the effect on animal growth performance as measured by bodyweight gain and feed efficiency; animal welfare, through improved gut health and reduced mortality; food safety, through reduced carcass contamination by food-borne pathogens; and environmental impact by increasing nutrient digestibility. For some feed additives, the beneficial effect on shelf-life for the meat produced is also an important factor to consider.

Organic acids have been successfully used as feed additives in pork production but in poultry production, organic acids have been used primarily to sanitize water lines. Research has shown that organic acids do not have consistent benefits when added to poultry diets. Most of the research with poultry, however, was with birds raised in a clean environment. More benefits may be seen with a subclinical disease challenge. The use of organic acids in rabbit diets has also shown inconsistent results. In ruminants, supplementation with encapsulated organic acids may have a role in reducing methane emissions but their use is not currently economical.

Bacteriocins are a group of AMPs with potential as alternative feed additives for both ruminants and nonruminants. Nisin is a bacteriocin that is commercially available. Nisin supplementation has been shown to reduce methane production in ruminants making it a possible alternative to the use of ionophores for this purpose. In nonruminants, bacteriocins have been shown to control colonization of food-borne pathogens.

Several phytogenic compounds have been investigated for their potential to improve gut health, animal performance, or characteristics of the meat produced. A variety of parts or extracts from several different plants have been studied with varying degrees of success. No single phytogenic compound has been found to be successful in achieving all three goals but some have been identified as potential substitutes for subtherapeutic antibiotics. The essential oils appear to have the greatest potential and some commercial products are already available. Supplementing the diets of beef cattle with essential oils has also been shown to reduce methane production.

REFERENCES

Abd El-Hakim, A.S., G. Cherian, and M. N. Ali. 2009. Use of organic acid, herbs and their combination to improve the utilization of commercial low protein broiler diets. *Int. J. Poult. Sci.* 8:14–20.

Abecia, L., M. Fondevilla, J. Balcells, and N. R. McEwan. 2007. The effect of lactating rabbit does on the development of the caecal microbial community in the pups they nurture. *J. Appl. Microbiol.* 103: 557–564.

Abu-Darwish, M. S., and Z. H. M. Abu-Dieyeh. 2009. Essential oil content and heavy metals composition of *Thymus vulgaris* cultivated in various climatic regions of Jordan. *Int. J. Agric. Biol.* 11:59–63.

Al-Kassie, G. A. M. and N. M. Witwit. 2010. A comparative study of diet supplementation with a mixture of herbal plants and dandelion as a source of prebiotics on the performance of broilers. *Pakistan J. Nutr.* 9:67–71.

Amit-Romach, E., D. Sklan, and Z. Uni. 2004. Microflora ecology of the chicken intestine using 16S ribosomal DNA primers. *Poult. Sci.* 83:1093–1098.

Ando, S., T. Nishida, M. Ishida, K. Hospda, and E. Bayaru. 2003. Effect of peppermint feeding on the digestibility, ruminal fermentation and protozoa. *Livest. Prod. Sci.* 82:245–248.

Apajalahti, J. 2005. Comparative gut microflora, metabolic challenges, and potential opportunities. *J. Appl. Poult. Res.* 14:444–453.

Apajalahti, J., A. Kettunen, and H. Graham. 2004. Characteristics of the gastrointestinal microbial communities, with special reference to the chicken. *World's Poult. Sci. J.* 60:223–232.

Bagchi, D., M. Bagchi, J. S. Stohs, K. D. Das, S. D. Ray, C. A. Kuszynski, S. S. Joshi, and H. G. Pruess. 2000. Free radical and grape seed proanthocyanidin extract: importance in human health and disease prevention. *Toxicology* 148:187–197.

Bampidis, V. A., V. Christodoulou, P. Florou-Paneri, E. Christaki, P. S. Chatzopoulou, T. Tsiligianni, and A. B. Spais. 2005. Effect of dietary dried oregano leaves on growth performance, carcass characteristics and serum cholesterol of female early maturing turkeys. *Br. Poult. Sci.* 46:595–601.

Barnes, E. M., C. S. Impey, and B. J. H. Stevens. 1979. Factors affecting the incidence and anti-salmonella activity of the anaerobic caecal flora of the young chick. *J. Hyg.* 82:263–283.

Basmacioğlu, H., Ö. Tokuşoğlu, and M. Ergül. 2004. The effect of oregano and rosemary essential oils or α-tocopheryl acetate on performance and lipid oxidation of meat enriched with n-3 PUFAs in broilers. *S. Afr. J. Anim. Sci.* 34:197–210.

Bauer, E., B. A. Williams, H. Smidt, R. Mosenthin, and M. W. A. Verstegen. 2006. Influence of dietary components on development of the microbiota in single-stomached species. *Nutr. Res. Rev.* 19:63–78.

Beauchemin, K. A. and S. M. McGinn. 2006. Methane emissions from beef cattle: effects of fumaric acid, essential oil, and canola oil. *J. Anim. Sci.* 84:1489–1496.

Benavente-Garcia, O., J. Castillo, J. Lorente, A. Ortuno, and J. A. Del Rio. 2000. Antioxidant activity of phenolics extracted from *Olea europea* L. leaves. *Food Chem.* 68:457–462.

Bergman, E. N. 1990. Energy contributions of volatile fatty acids from the gastrointestinal tract in various species. *Physiol. Rev.* 70:567–590.

Biggs, P. and C. M. Parsons. 2008. The effects of several organic acids on growth performance, nutrient digestibilities, and cecal, microbial populations in young chicks. *Poult. Sci.* 87:2581–2589.

Bisignano, G., M. G. Lagana, D. Trombetta, S. Arena, A. Nostro, N. Uccella, G. Mazzani, and A. Saija. 2001. *In vitro* antibacterial activity of some aliphatic aldehydes from *Olea europea* L. *FEMS Microbiol. Lett.* 198:9–13.

Blank, R., R. Mosenthin, W. C. Sauer, and S. Huang. 1999. Effect of fumaric acid and dietary buffering capacity on ileal and fecal amino acid digestibilities in early-weaned pigs. *J. Anim. Sci.* 77:2974–2984.

Böhmer, B. M., H. Salisch, B. R. Paulicks, and F. X. Roth. 2009. *Echinacea purpurea* as a potential immunostimulatory feed additive in laying hens and fattening pigs by intermittent application. *Livest. Sci.* 122:81–85.

Boman H. G. 2003. Antibacterial peptides: basic facts and emerging concepts. *J. Intern. Med.* 254:197–215.

Botsoglou, E., A. Govaris, E. Christaki, and N. Botsoglou. 2010. Effect of dietary olive leaves and/or α-tocopheryl acetate supplementation on microbial growth and lipid oxidation of turkey breast fillets during refrigerated storage. *Food Chem.* 121:17–22.

Botsoglou, N. A., P. Florou-Paneri, E. Christaki, D. J. Fletouris, and A. B. Spais. 2002. Effect of dietary oregano essential oil on performance of chickens and on iron-induced lipid oxidation of breast, thigh and abdominal fat tissues. *Br. Poult. Sci.* 43:223–230.

Botsoglou, N. A., A. Govaris, E. N. Botsoglou, S. H. Grigoropoulou, and G. Papageorgiou. 2003a. Antioxidant activity of dietary oregano essential oil and α-tocopheryl acetate supplementation in long-term frozen stored turkey meat. *J. Agric. Food Chem.* 51:2930–2936.

Botsoglou, N. A., S. H. Grigoropoulou, E. Botsoglou, A. Govaris, and G. Papageorgiou. 2003b. The effects of dietary oregano essential oil and α-tocopheryl acetate on lipid oxidation in raw and cooked turkey during refrigerated storage. *Meat Sci.* 65:1193–1200.

Botsoglou, N. A., G. Papageorgiou, I. Nikolakakis, P. Florou-Paneri, I. Giannenas, V. Dotas, and E. Sinapis. 2004a. Effect of dietary dried tomato pulp on oxidative stability of Japanese quail meat. *J. Agric. Food Chem.* 52:2982–2988.

Botsoglou, N. A., P. Florou-Paneri, E. Christaki, I. Giannenas, and A. B. Spais. 2004b. Performance of rabbits and oxidative stability of muscle tissues as affected by dietary supplementation with oregano essential oil. *Arch. Anim. Nutr.* 58:209–218.

Brenes, A., A. Viveros, I. Goñi, C. Centeno, S. G. Sáyago-Ayerdy, I. Arija, and F. Saura-Calixto. 2008. Effect of grape pomace concentrate and vitamin E on digestibility of polyphenols and antioxidant activity in chickens. *Poult. Sci.* 87:307–316.

Broudiscou, L. P., Y. Papon, and A. F. Broudiscou. 2000. Effects of dry plant extracts on fermentation and methanogenesis in continuous culture of rumen microbes. *Anim. Feed Sci. Technol.* 87:263–277.

Callaway, T. R., A. M. S. Carneiro De Melo, and J. B. Russell. 1997. The effect of nisin and monensin on ruminal fermentations *in vitro*. *Curr. Microbiol.* 35:90–96.

Cardozo, P. W., S. Calsamiglia, A. Ferret, and C. Kamel. 2004. Effects of natural plant extracts on ruminal protein degradation and fermentation profiles in continuous culture. *J. Anim. Sci.* 82:3230–3236.

Cardozo, P. W., S. Calsamiglia, A. Ferret, and C. Kamel. 2005. Screening for the effects of natural plant extracts at different pH on *in vitro* rumen microbial fermentation of a high-concentrate diet for beef cattle. *J. Anim. Sci.* 83:2572–2579.

Carlos, A. M. A. and M. A. Harrison. 1999. Inhibition of selected microorganisms in marinated chicken by pimento leaf oil and clove oleoresin. *J. Appl. Poult. Res.* 8:100–109.

Carlson, S. A., T. S. Frana, and R. W. Griffith. 2001. Antibiotic resistance in *Salmonella enterica* serovar Typhimurium exposed to microcin-producing *Escherichia coli*. *Appl. Environ. Microbiol.* 67:3763–3766.

Casadei, G., E. Grilli, and A. Piva. 2009. Pediocin A modulates intestinal microflora metabolism in swine in vitro intestinal fermentations. *J. Anim. Sci.* 87:2020–2028.

Casewell, M., C. Friis, E. Marco, P. McMullin, and I. Phillips. 2003. The European ban on growth-promoting antibiotics and emerging consequences for human and animal health. *J. Antimicrob. Chemother.* 52:159–161.

Chaveerach, P., D. A. Keuzenkamp, H. A. P. Urlings, L. J. A. Lipman, and F. van Knapen. 2002. *In vitro* study on the effect of organic acids on *Campylobacter jejuni/coli* populations in mixtures of water and feed. *Poult. Sci.* 81:621–628.

Chaves, A. V., M. L. He, W. Z. Yang, A. N. Hristov, T. A. McAllister, and C. Benchaar. 2008. Effects of essential oils on proteolytic, deaminative and methanogenic activities of mixed ruminal bacteria. *Can. J. Anim. Sci.* 88:117–122.

Choi, I. H., W. Y. Park, and Y. J. Kim. 2010. Effects of dietary garlic powder and α-tocopherol supplementation on performance, serum cholesterol levels, and meat quality of chicken. *Poult. Sci.* 89: 1724–1731.

Cole, K., M. B. Farnell, A. M. Donoghue, N. J. Stern, E. A. Svetoch, B. N. Eruslanov, L. I. Volodina, Y. N. Kovalev, V. V. Perelygin, E. V. Mitsevich, I. P. Mitsevich, V. P. Levchuk, V. D. Pokhilenko, V. N. Borzenkov, O. E. Svetoch, T. Y. Kudryavtseva, I. Reyes-Herrera, P. J. Blore, F. Solis de los Santos, and D. J. Donoghue. 2006. Bacteriocins reduce campylobacter colonization and alter gut morphology in turkey poults. *Poult. Sci.* 85:1570–1575.

Cotter, P. D., C. Hill, and R. P. Ross. 2005. Bacteriocins: developing innate immunity for food. *Nature Rev. Microbiol.* 3:777–788.

Crandall, A. D. and T. J. Montville. 1998. Nisin resistance in *Listeria monocytogenes* ATCC 700302 is a complex phenotype. *Appl. Environ. Microbiol.* 64:231–237.

Cross, D. E., R. M. McDevitt, K. Hillman, and T. Acamovic. 2007. The effect of herbs and their associated essential oils on performance, dietary digestibility and gut microflora in chickens from 7 to 28 days of age. *Br. Poult. Sci.* 48:496–506.

Cs, E., T. Gippert, K. Gódor-Surmann, and K. Kustos. 2008. Feed additives as they affect the fattening performance of rabbits. *9th World Rabbit Congress*, Verona, Italy, June 10th–13th. pp. 625–630. Available at: http://world-rabbit-science.com/WRSA-Proceedings/Congress-2008-Verona/Papers/N-Eiben1.pdf.

Dahiya, J. P., D. C. Wilkie, A. G. Van Kessel, and M. D. Drew. 2006. Potential strategies for controlling necrotic enteritis in broiler chickens in post-antibiotic era. *Anim. Feed Sci. Technol.* 129:60–88.

Daw, M. D. and F. R. Falkiner. 1996. Bacteriocins: nature, function and structure. *Micron.* 27:467–479.

de Lange, C. F. M., J. Pluske, J. Gong, and C. M. Nyachoti. 2010. Strategic use of feed ingredients and feed additives to stimulate gut health and development in young pigs. *Livest. Sci.* 134:124–134.

Delves-Broughton, J., P. Blackburn, R. J. Evans, and J. Hugenholtz. 1996. Applications of the bacteriocin, nisin. *Antonie van Leewenhoek* 69:193–202.

den Hartog, L. A., A. Gutiérrez del Álamo, J. Doorenbos, and A. Flores. 2005. The effect of natural alternatives for anti-microbial growth promoters in broiler diets. *15th European Symposium on Poultry Nutrition*, Balatonfüred, Hungary. pp. 224–232.

De Nino, A., N. Lombardo, E. Perri, A. Procopio, A. Raffaelli, and G. Sindona. 1997. Direct identification of phenolic glucosides from olive leaf extracts by atmospheric pressure ionization tandem mass spectrometry. *J. Mass Spectrom.* 32:533–541.

Dibner, J. J. and P. Buttin. 2002. Use of organic acids as a model to study the impact of gut microflora on nutrition and metabolism. *J. Appl. Poult. Res.* 11:453–463.

Dickson, J. S., P. Sundaram, and R. J. Hubert. 1998. Use of pediocin AcH in meat preservation. Iowa State University Swine Report. Available at: http://www.extension.iastate.edu/Pages/ansci/swinereports/asl-1610.pdf.

Diplock, A. T., J. L. Charleux, G. Crozier-Willi, F. J. Kok, C. Rice-Evans, M. Roberfroid, W. Stahl, and J. Vina-Ribes. 1998. Functional food science and defense against reactive oxidative species. *Br. J. Nutr.* 80:S77–S112.

Dorman, H. J. D. and Deans, S. G. 2000. Antimicrobial agents from plants: antibacterial activity of plant oils. *J. Appl. Microbiol.* 88:308–316.

Economou, K. D., V. Oreopoulou, and C. Thomopoulos, 1991. Antioxidant properties of some plant extracts of the family labiatae. *J. Am. Oil Chem. Soc.* 68:109–113.

Ertas, O. N., T. Güler, M. Çiftçi, B. DalkIlIç, and Ü. G. Simsek. 2005. The effect of an essential oil mix derived from oregano, clove and anise on broiler performance. *Int. J. Poult. Sci.* 4:879–884.

Flickinger, E. A., J. Van Loo, and G. C. Fahey. 2003. Nutritional responses to the presence of inulin and oligofructose in the diets of domesticated animals: a review. *Crit. Rev. Food Sci. Nutr.* 43:19–60.

Flint, J. F. and M. R. Garner. 2009. Feeding beneficial bacteria: a natural solution for increasing efficiency and decreasing pathogens in animal agriculture. *J. Appl. Poult. Res.* 18:367–378.

Florou-Paneri, P., G. Palatos, A. Govaris, D. Botsoglou, I. Giannenas, and I. Ambrosiadis. 2005. Oregano herb versus oregano essential oil as feed supplements to increase the oxidative stability of turkey meat. *Int. J. Poult. Sci.* 4:866–871.

Forchielli, M. L. and W. A. Walker. 2005. The role of gut-associated lymphoid tissues and mucosal defence. *Br. J. Nutr.* 93:S41–S48.

Freitag, M., H. U. Hensche, H. Schulte-Sienbeck, and B. Reichelt. 1998. Kritische Betrachtung des Einsatzes von Leistungsförderern in der Tierernährung. Forschungsberichte des Fachbereichs Agrarwirtschaft Soest, Universität-Gesamthochschule Paderborn, Nr. 8. Cited by Schöner F.J. (2001). Nutritional effects of organic acids. In: Brufau J. (ed.) *Feed Manufacturing in the Mediterranean Region. Improving Safety: From feed to Food.* Zaragoza: CIHEAM-IAMZ, 2001. pp. 55–61 *Conference of Feed Manufacturers of the Mediterranean*, March 22nd–24th, 2000, Reus (Spain). Available at: http://ressources.ciheam.org/om/pdf/c54/01600011.pdf.

Galobart, J., A. C. Barroeta, M. D. Baucells, R. Codony, and W. Ternes. 2001. Effect of dietary supplementation with rosemary extract and α-tocopheryl acetate on lipid oxidation in eggs enriched with ω3-fatty acids. *Poult. Sci.* 80:460–467.

Ghosh, H. K., G. Halder, G. Samanta, and S. Koley. 2008. Effect of dietary supplementation of organic acid and mannan oligosaccharide on the plasma minerals and carcass traits of Japanese quail (*Coturnix coturnix japonica*). *Res. J. Vet. Sci.* 1:44–49.

Gillespie, J. R. 1997. *Modern Livestock and Poultry Production.* 6th edn. Delmar Publishing, Albany, NY. p. 1060.

Goñi, I., A. Brenes, C. Centeno, A. Viveros, F. Saura-Calixto, A. Rebolé, I. Arija, and R. Estevez. 2007. Effect of dietary grape pomace and vitamin E on growth performance, nutrient digestibility, and susceptibility to meat lipid oxidation in chickens. *Poult. Sci.* 86:508–516.

Gordon, Y. J., E. G. Romanowski, and A. M. McDermott. 2005. A review of antimicrobial peptides and their therapeutic potential as anti-infective drugs. *Curr. Eye Res.* 30:505–515.

Govaris, A., N. Botsoglou, G. Papageorgiou, E. Botsoglou, and I. Amvrosiadis. 2004. Dietary versus postmortem use of oregano oil and/or a-tocopherol in turkeys to inhibit development of lipid oxidation in meat during refrigerated storage. *Int. J. Food Sci. Nutr.* 55:115–123.

Gratia, A. 1925. Sur un remarquable example d'antagonisme entre deux souches de colibacille. *Compt. Rend. Soc. Biol.* 93:1040–1042.

Grilli, E., M. R. Messina, E. Catelli, M. Morlacchini, and A. Piva. 2009. Pediocin A improves growth performance of broilers challenged with *Clostridium perfringens. Poult. Sci.* 88:2152–2158.

Gutierrez del Alamo, A., J. De los Mozos, J. T. P. Van Dam, and P. Perez de Ayala. 2007. The use of short and medium chain fatty acids as an alternative to antibiotic growth promoters in broilers infected with malabsorption syndrome. *16th European Symposium on Poultry Nutrition*, Strasbourg, France, August 26–30. Available at: http://www.cabi.org/animalscience/Uploads/File/AnimalScience/additionalFiles/WPSAStrasbourgAug2007/47.pdf.

Halle, I., R. Thomann, U. Bauermann, M. Henning, and P. Köhler. 2004. Effects of a graded supplementation of herbs and essential oils in broiler feed on growth and carcass traits. *Landbauforsch. Volk.* 54: 219–229.

Hancock, R. E. W. 1997. Peptide antibiotics. *Lancet* 349:418–422.

Hancock, R. E. W. and H.-G. Sahl. 2006. Antimicrobial and host-defense peptides as new anti-infective therapeutic strategies. *Nat. Biotechnol.* 24:1551–1557.

Hayes, D. J. and H. H. Jensen. 2003. Lessons from the Danish ban on feed-grade antibiotics. *Choices (A publication of the American Agricultural Economics Association).* Available at: http://www.farmdoc.illinois.edu/policy/choices/20033/20033.pdf.

Hedegaard, A. 2000. Danish approach on swine production with no antibiotics. Available at: http://ressources.ciheam.org/om/pdf/c54/01600007.pdf.

Hermann, J. R., M. S. Honeyman, J. J. Zimmerman, B. J. Thacker, P. J. Holden and C. C. Chang. 2003. Effect of dietary Echinacea purpurea on viremia and performance in porcine reproductive and respiratory syndrome virus-infected nursery pigs. *J. Anim. Sci.* 81:2139–2144.

Hermans, D., A. Martel, K. Van Deun, M. Verlinden, F. Van Immerseel, A. Garmyn, W. Messens, M. Heyndrickx, F. Haesebrouck, and F. Pasmans. 2010. Intestinal mucus protects *Campylobacter jejuni* in the ceca of colonized broiler chickens against the bactericidal effects of medium-chain fatty acids. *Poult. Sci.* 89:1144–1155.

Hernández F., J. Madrid, V. García, J. Orengo, and M. D. Megías. 2004. Influence of two plant extracts on broilers performance, digestibility, and digestive organ size. *Poult. Sci.* 83:169–174.

Hernández, P., V. Juste, C. Zomeño, J. R. Moreno, and P. Peñalver. 2009. Effect of dietary clove essential oil on poultry meat quality. Lidervet technical publication. Available at: http://www.lidervet.com/Pdf/ICOMST2009-Hernandez_texto_ingl%C3%A9s_.pdf.

Hurst, A. 1981. Nisin. *Adv. Appl. Microbiol.* 27:85–123.

Izat, A. L., N. M. Tidwell, R. A. Thomas, M. A. Reiber, M. H. Adams, M. Colberg, and P. W. Waldroup. 1990. Effects of a buffered propionic acid in diets on the performance of broiler chickens and on microflora of the intestine and carcass. *Poult. Sci.* 69:818–826.

Jacob, F., L. Siminovitch, and E. Wollman. 1952. Sur la biosynthese d'une colicine et sur son mode d'action [On the biosynthesis of a colicine and its mode of action]. *Ann. Inst. Pasteur* 83:295–315.

Jalc D. and A. Lauková. 2002. Effect of nisin and monensin on rumen fermentation in the artificial rumen. *Berl. Munch. Tierarztl. Wochenschr.* 115:6–10.

Jamroz D., A. Wiliczkiewicz, T. Wertelecki, J. Orda, and J. Skorupinska. 2005. Use of active substances of plant origin in chicken diets based on maize and locally cereals. *Br. Poult. Sci.* 46:485–493.

Jamroz D., T. Wertelecki, M. Houszka, and C. Kamel. 2006. Influence of diet type on the inclusion of plant origin active substances on morphological and histochemical characteristics of the stomach and jejunum walls in chickens. *J. Anim. Physiol. Anim. Nutr.* 90:255–268.

Jang, A., X. D. Liu, M. H. Shin, B. D. Lee, S. K. Lee, J. H. Lee, and C. Jo. 2008. Antioxidative potential of raw breast meat from broiler chicks fed a dietary medicinal herb extract mix. *Poult. Sci.* 87:2382–2389.

Janz, J. A. M., P. C. H. Morel, B. H. P. Wilkinson, and R. W. Purchas. 2007. Preliminary investigation of the effects of low-level dietary inclusion of fragrant essential oils and oleoresins on pig performance and pork quality. *Meat Sci.* 75:350–355.

Jensen, R. G. 2002. The composition of bovine milk lipids: January 1995 to December 2000. *J. Dairy Sci.* 85:295–350.

Joerger, R. D. 2003. Alternatives to antibiotics: bacteriocins, antimicrobial peptides and bacteriophages. *Poult. Sci.* 82:640–647.

Johnson, K. A. and D. E. Johnson. 1995. Methane emissions from cattle. *J. Anim. Sci.* 73:2483–2492.

Jouany, J. P. and D. P. Morgavi. 2007. Use of 'natural' products as alternatives to antibiotic feed additives in ruminant production. *Animal* 1:1443–1466.

Kanner, J. 1994. Oxidative processes in meat and meat products: quality implications. *Meat Sci.* 36:169–189.

Karimi, A., F. Yan, C. Coto, J. H. Park, Y. Min, C. Lu, J. A. Gidden, J. O. Lay Jr., and P. W. Waldroup. 2010. Effects of level and source of oregano leaf in starter diets for broiler chicks. *J. Appl. Poult. Res.* 19:137–145.

Karpinská M., J. Borowski, and M. Danowska-Oziewicz. 2000. Antioxidative activity of rosemary extract in lipid fraction of minced meatballs during storage in a freezer. *Nahrung.* 44:38–41.

Keokamnerd, T., J. C. Acton, I. Y. Han, and P. L. Dawson. 2008. Effect of commercial rosemary oleoresin preparations on ground chicken thigh meat quality packaged in a high-oxygen atmosphere. *Poult. Sci.* 87:170–179.

Kim, Y. J., S. K. Jin, and H. S. Yang. 2009. Effect of dietary garlic bulb and husk on the physicochemical properties of chicken meat. *Poult. Sci.* 88:398–405.

Kišdayová, S., P. Siroka, and A. Lauková. 2003. Effect of nisin on two cultures of rumen ciliates. *Folia Microbiol.* 48:408–412.

Kišdayová, S., A. Lauková, and D. Jalč. 2009. Comparison of nisin and monensin effects of ciliate and selected populations of artificial rumen. *Folia Microbiol.* 54:527–532.

Kovács, M., Z. Szendrö, G. Milisitis, B. Bóta, E. Bíró-Németh, I. Radnai, R. Pósa, A. Bónai, F. Kovács, and P. Horn. 2006. Effects of nursing methods and faeces consumption on the development of bacteriodes, lactobacillus and coliform flora in the caecum of the newborn rabbits. *Reprod. Nutr. Dev.* 46:205–210.

Kubo, A., C. S. Lunde, and I. Kuboi. 1995. Antimicrobial activity of the olive oil flavor compounds. *J. Agric. Food Chem.* 43:1629–1633.

Lambert, R. J. and M. Statford. 1999. Weak acid preservatives: modeling microbial inhibition and response. *J. Appl. Microbiol.* 86:157–164.

Lambert, R. J., P. N. Skandamis, P. J. Coote, and G. J. Nychas. 2001. A study of the minimum inhibitory concentration and mode of action of oregano essential oil, thymol and carvacrol. *J. Appl. Microbiol.* 91:453–462.

Lee, K. W., H. Everts, H. J. Kappert, K. H. Yeom, and A. C. Beynen. 2003. Dietary carvacrol lowers body weight gain but improves feed conversion in female broiler chickens. *J. Appl. Poult. Res.* 12:394–399.

Leeson, S., H. Namkung, M. Antongiovanni, and E. H. Lee. 2005. Effect of butyric acid on the performance and carcass yield of broiler chickens. *Poult. Sci.* 84:1418–1422.

Leusink, G., H. Rempel, B. Skura, M. Berkyto, W. White, Y. Yang, J. Y. Rhee, S. Y. Xuan, S. Chiu, F. Silversides, S. Fitzpatrick, and M. S. Diarra. 2010. Growth performance, meat quality, and gut microflora of broiler chickens fed with cranberry extract. *Poult. Sci.* 89:1514–1523.

Lopez-Bote, C. J., J. I. Gray, E. A. Gomaa, and C. J. Flegal. 1998. Effect of dietary administration of oil extracts from rosemary and sage on lipid oxidation in broiler meat. *Br. Poult. Sci.* 39:235–240.
Lupton, C. J. 2008. ASAS centennial paper: impacts of animal science research on United States sheep production and predictions for the future. *J. Anim. Sci.* 86:3252–3274.
Ma, A., W. Shi, X. Niu, M. Wang, and X. Zhong. 2009. Effects of *Echinacea purpurea* extract on the immunological response to infectious bursal disease vaccine in broilers. *Front. Agric. China* 3: 452–456.
Maass, N., J. Bauer, B. R. Paulicks, B. M. Böhmer, and D. A. Roth-Maier. 2005. Efficiency of *Echinacea purpurea* on performance and immune status in pigs. *J. Anim. Physiol. Anim. Nutr.* 89:244–252.
Maertens L., L. Falcão-e-Cunha, and M. Marounek. 2006. Feed additives to reduce the use of antibiotics. In: L. Maertens and P. Coudert (eds) *Recent advances in Rabbit Science*. ILVO, Melle. pp. 259–265.
Mantovani, H. C. and J. B. Russell. 2001. Nisin resistance of *Streptococcus bovis*. *Appl. Environ. Microbiol.* 67:808–813.
Manzanilla, E. G., J. F. Perez, M. Martin, C. Kamel, F. Baucells, and J. Gasa. 2004. Effect of plant extracts and formic acid on the intestinal equilibrium of early-weaned pigs. *J. Anim. Sci.* 82:3210–3218.
Manzanilla, E. G., M. Nofrarías, M. Anguita, M. Castillo, J. F. Perez, S. M. Martín-Orúe, C. Kamel, and J. Gasa. 2006. Effects of butyrate, avilamycin, and a plant extract combination on the intestinal equilibrium of early-weaned pigs. *J. Anim. Sci.* 84:2743–2751.
Manzanilla, E. G., J. F. Pérez, M. Martín, J. C. Blandón, F. Baucells, C. Kamel, and J. Gasa. 2009. Dietary protein modifies effect of plant extracts in the intestinal ecosystem of the pig at weaning. *J. Anim. Sci.* 87:2029–2037.
Martin, S. A. 1998. Manipulation of ruminal fermentation with organic acids: a review. *J. Anim. Sci.* 76:3123–3132.
Martin, C., M. Doreau, and D. P. Morgavi. 2009. Methane mitigation in ruminants: from rumen microbes to the animal. INRA Herbivores Research Unit, France. Available at: http://www.animalbytes.org/wp-content/uploads/2009/06/edition_2_lgccmartin.pdf.
Mason, L. M., S. A. Hogan, A. Lynch, K. O'Sullivan, P. G. Lawlor, and J. P. Kerry. 2005. Effect of restricted feeding and antioxidant supplementation on pig performance and quality characteristics of *longissimus dorsi* from Landrace and Duroc pigs. *Meat Sci.* 70:307–317.
Mateos, G. G., R. Lázaro, and P. Medel. 2001. Feeding strategies for intensive livestock production without in feed antibiotic growth promoters. In: Brufau J. (ed.) *Feed Manufacturing in the Mediterranean Region. Improving Safety: From feed to Food*, vol. 54. Cahiers Options Méditerranéennes, CIHEAM, Zaragoza, Spain. pp. 11–16
Mathis, G. and N. Scicutella. 2007. A comparison of performance of coccidiosis vaccinated broilers fed a coated blend of essential oils, a coated blend of organic and inorganic acids with essential oils, or bacitracin methylen disalicylate. *16th European Symposium on Poultry Nutrition*. pp. 573–575.
McBee, R. H. 1971. Significance of intestinal microflora in herbivory. *Annu. Rev. Ecol. Syst.* 2:165–176.
Mitsch, P., K. Zitterl-Eglseer, B. Köhler, C. Gabler, R. Losa, and I. Zimpernik. 2004. The effect of two different blends of essential oil components on the proliferation of *Clostridium perfringens* in the intestines of broiler chickens. *Poult. Sci.* 83:669–675.
Moharrery, A. and M. Mahzonieh. 2005. Effect of malic acid on visceral characteristics and coliform counts in small intestine in the broiler and layer chickens. *Int. J. Poult. Sci.* 4:761–764.
Moll, G. N., W. N. Konings, and A. J. M. Driessen. 1999. Bacteriocins: mechanism of membrane insertion and pore formation. *Antonie van Leeuwenhoek* 76:185–198.
Muhl, A. and F. Liebert. 2007. Growth, nutrient utilization and threonine requirement of growing chicken fed threonine limiting diet with commercial blends of phytogenic feed additives. *J. Poult. Sci.* 44: 297–304.
Murphy, A., J. P. Kerry, J. Buckley, and I. Gray. 1998. The antioxidative properties of rosemary oleoresin and inhibition of off-flavours in precooked roast beef slices. *J. Sci. Food Agric.* 77:235–243.
Naidoo, V., L. J. McGaw, S. P. Bisschop, N. Duncan, and J. N. Eloff. 2008. The value of plant extracts with antioxidant activity in attenuating coccidiosis in broiler chickens. *Vet. Parasitol.* 153:214–219.
Newbold, C. J., S. Lopez, N. Nelson, J. O. Ouda, R. J. Wallace, and A. R. Moss. 2005. Propionate precursors and other metabolic intermediates as possible alternative electron acceptors to methanogenesis in ruminal fermentation in vitro. *Br. J. Nutr.* 94:27–35.
Ocak, N., G. Erener, F. B. Ak, M. Sungu, A. Altop, and A. Ozmen. 2008. Performance of broilers fed diets supplemented with dry peppermint (*Mentha piperita* L.) or thyme (*Thymus vulgaris* L.) leaves as growth promoters. *Czech J. Anim. Sci.* 53:169–175.

O'Grady, M. N., M. Maher, D. J. Troy, A. P. Moloney, and J. P. Kerry. 2006. An assessment of dietary supplementation with tea catechins and rosemary extract on the quality of fresh beef. *Meat Sci.* 73: 132–146.

Oviedo-Rondón, E. O., M. E. Hume, C. Hernández, and S. Clemente-Hernández. 2006. Intestinal microbial ecology of broilers vaccinated and challenged with mixed *Eimeria* species, and supplemented with essential oil blends. *Poult. Sci.* 85:854–860.

Ozduven, M. L., H. E. Samli, A. A. Okur, F. Koc, H. Akyurek, and N. Senkoylu. 2009. Effects of mannanoligosaccharide and/or organic acid mixture on performance, blood parameters and intestinal microbiota of broiler chicks. *Ital. J. Anim. Sci.* 8:595–602.

Papageorgiou, G., N. Botsoglou, A. Govaris, I. Giannenas, S. Iliadis, and E. Botsoglou. 2003. Effect of dietary oregano oil and α-tocopheryl acetate supplementation on iron-induced lipid oxidation of turkey breast, thigh, liver and heart tissues. *J. Anim. Physiol. Anim. Nutr.* 87:324–335.

Parisien, A., B. Allain, J. Zhang, R. Mandeville, and C. Q. Lan. 2008. Novel alternatives to antibiotics: bacteriophages, bacterial cell wall hydroloases and antimicrobial peptides. *J. Appl. Microbiol.* 104:1–13.

Partanen, K. H. and Z. Mroz. 1999. Organic acids for performance enhancement in pig diets. *Nutr. Res. Rev.* 12:117–145.

Patten, J. D. and P. W. Waldroup. 1988. Use of organic acids in broiler diets. *Poult. Sci.* 67:1178–1182.

Patterson, J. A. and K. M. Burkholder. 2003. Application of prebiotics and probiotics in poultry production. *Poult. Sci.* 82:627–631.

Peschel, A. and L. V. Collins. 2001. Staphylococcal resistance to antimicrobial peptides of mammalian and bacterial origin. *Peptides* 22:1651–1659.

Pirgozliev, V., T. C. Murphy, B. Owens, J. George, and M. E. E. McCann. 2008. Fumaric and sorbic acid as additives in broiler feed. *Res. Vet. Sci.* 84:387–394.

Piva, A. and D. R. Headon. 1994. Pediocin A, a bacteriocin produced by *Pediococcus pentasaceus* FBB61. *Microbiology* 140:697–702.

Rastall, R. A. 2004. Bacteria in the gut: friends and foes and how to alter the balance. *J. Nutr.* 134:2022S–2026S.

Richards, J. D., J. Gong, and C. F. M. de Lange. 2005. The gastrointestinal microbiota and its role in monogastric nutrition and health with an emphasis on pigs: current understanding, possible modulations, and new technologies for ecological studies. *Can. J. Anim. Sci.* 85:421–435.

Ricke, S. C. 2003. Perspectives on the use of organic acids and short chain fatty acids as antimicrobials. *Poult. Sci.* 82:632–639.

Robey, M., W. O'Connell, and N. P. Cianciotto. 2001. Identification of *Legionella pneumophila rcp*, a *pagP*-like gene that confers resistance to cationic antimicrobial peptides and promotes intracellular infection. *Infect. Immun.* 69:4276–4286.

Roy, R. D., F. W. Edens, C. R. Parkhurst, M. A. Qureshi, and G. B. Havenstein. 2002. Influence of a propionic acid feed additive on performance of turkey poults with experimentally induced poult enteritis and mortality syndrome. *Poult. Sci.* 81:951–957.

Runho, R. C., N. K. Sakomura, S. Kuana, D. Banzatto, O. M. Junqueira, and J. H. Stringhini. 1997. Use of an organic acid (fumaric acid) in broiler rations. *Rev. Bras. Zootec.* 26:1183–1191.

Russell, J. B. and H. C. Mantovani. 2002. The bacteriocins of ruminal bacterial and their potential as alternative to antibiotics. *J. Mol. Microbiol. Biotechnol.* 4:347–355.

Sang, Y. and F. Blecha. 2008. Antimicrobial peptides and bacteriocins: alternatives to traditional antibiotics. *Anim. Health Res. Rev.* 9:227–235.

Sar, C., B. Santoso, B. Mwenya, Y. Gamoa, T. Kobayashi, R. Morikawa, K. Kimura, H. Mizukoshi, and J. Takahashi. 2004. Manipulation of rumen methanogenesis by the combination of nitrate with β1-4 galacto-oligosaccharides or nisin in sheep. *Anim. Feed Sci. Technol.* 115:129–142.

Sarica, S., A. Citci, E. Demir, K. Kiline, and Y. Yidirum. 2005. Use of an antibiotic growth promoter and two herbal natural feed additives with and without exogenous enzymes in wheat based broiler diets. *S. Afr. J. Anim. Sci.* 35:61–72.

Sarica, S., M. Corduk, G. F. Yarim, G. Yenisehirli, and U. Karatas. 2009. Effects of novel feed additives in wheat based diets on performance, carcass and intestinal tract characteristics of quail. *S. Afr. J. Anim. Sci.* 39:144–157.

Scapinello, C., H. Garcia de Faria, A. C. Furlan, and A. C. Michelan. 2001. Efeito da utilização de oligossacarídeo manose e acidificantes sobre o desempenho de coelhos em crescimento (Effect of using acidifying agents and mannose oligosaccharide on the growth performance of rabbits). *Rev. Bras. Zootec.* 30:1272–1277.

Schöner, F. J. 2001. Nutritional effects of organic acids. In: J. Brufau (ed.) *Feed manufacturing in the Mediterranean region. Improving safety: From feed to food*. Zaragoza: CIHEAM-IAMZ. *Conference of Feed Manufacturers of the Mediterranean*, Reus, Spain, March 22th–24th, 2000. pp. 55–61. Available at: http://ressources.ciheam.org/om/pdf/c54/01600011.pdf.
Simitzis, P. E., S. G. Deligeorgis, J. A. Bizelis, A. Dardamani, I. Theodosiou, and K. Fegeros. 2008. Effect of dietary oregano oil supplementation on lamb meat characteristics. *Meat Sci*. 79:217–223.
Simitzis, P. E., G. K. Symeon, M. A. Charismiadou, J. A. Bizelis, and S. G. Deligeorgis. 2010. The effects of dietary oregano oil supplementation on pig meat characteristics. *Meat Sci*. 84:670–676.
Simonová, M. and A. Lauková. 2007. Bacteriocin activity of enterococci from rabbits. *Vet. Res. Commun*. 31:143–152.
Simonová M., R. Szabóová, L. Chrastinová, A. Lauková, M. Haviarová, V. Strompfová, I. Plachá, Š. Faix, Z. Vasilková, J. Mojto, and J. Rafay. 2008. The use of ginseng extract in rabbits. *9th World Rabbit Congress*, Verona, Italy, June 10–13. pp. 809–813.
Simsek, U. G., M. Ciftci, B. Dalkilic, T. Guler, and O. N. Ertas. 2007. The effects of dietary antibiotic and anise oil supplementation on body weight, carcass characteristics and organoleptic analysis of meat in broilers. *Rev. Med. Vet*. 158:514–518.
Skinner, J. T., A. L. Izat, and P. W. Waldroup. 1991. Research note: fumaric acid enhances performance of broiler chickens. *Poult. Sci*. 70:1444–1447.
Smet, K., K. Raes, G. Huyghebaert, L. Haak, S. Arnouts, and S. De Smet. 2008. Lipid and protein oxidation of broiler meat as influenced by dietary natural antioxidant supplementation. *Poult. Sci*. 87:1682–1688.
Sofos, J. N. 2008. Challenges to meat safety in the 21st century. *Meat Sci*. 78:3–13.
Solis de los Santos, F., A. M. Donoghue, K. Venkitanarayanan, J. H. Metcalf, I. Reyes-Herrera, M. L. Dirain, V. F. Aguiar, P. J. Blore, and D. J. Donoghue. 2009. The natural feed additive caprylic acid decreases *Campylobacter jejuni* colonization in market-aged broiler chickens. *Poult. Sci*. 88:61–64.
Soultos, N., Z. Tzikas, E. Christaki, K. Papageorgiou, and V. Steris. 2009. The effect of dietary oregano essential oil on microbial growth of rabbit carcasses during refrigerated storage. *Meat Sci*. 81: 474–478.
Šperňáková, D., D. Máté, H. Róžaňa, and G. Kováč. 2007. Effects of dietary rosemary extract and α-tocopherol on the performance of chickens, meat quality and lipid oxidation in meat storage under chilling conditions. *Bull. Vet. Inst. Pulawy*. 51:585–589.
Sprong, R. C., M. F. Hulstein, and R. van Der Meer. 2001. Bactericidal activities of milk lipids. *Antimicrob. Agents Chemother*. 45:1298–1301.
Stern, N. J., B. V. Eruslanov, V. D. Pokhilenko, Y. N. Kovalev, L. L. Volodina, V. V. Perelygin, E. V. Mitsevich, I. P. Mitsevich, V. N. Borzenkov, V. P. Levchuk, O. E. Svetoch, Y. G. Stepanshin, and E. A. Svetoch. 2008. Bacteriocins reduce *Campylobacter jejuni* colonization while bacteria producing bacteriocins are ineffective. *Microb. Ecol. Health Dis*. 20:74–79.
Sterzo, E. V., J. B. Paiva, A. L. Mesquita, O. C. Freitas Neto, and A. Berchieri Jr. 2007. Organic acids and/or compounds with defined microorganisms to control *Salmonella enterica* serovar Enteritidis experimental infection in chickens. *Braz. J. Poult. Sci*. 9:69–73.
Sun, J., Y. F. Chu, X. Wu, and R. H. Liu. 2002. Antioxidant and antiproliferative activities of common fruits. *J. Agric. Food Chem*. 50:7449–7454.
Svetoch, E. A. and N. J. Stern. 2010. Bacteriocins to control *Campylobacter spp*. in poultry—a review. *Poult. Sci*. 89:1763–1768.
Symeon, G. K., C. Zintilas, A. Ayoutanti, J. A. Bizelis, and S. G. Deligeorgis. 2009. Effect of dietary oregano essential oil supplementation for an extensive fattening period on growth performance and breast meat quality of female medium-growing broilers. *Can. J. Anim. Sci*. 89:331–334.
Symeon, G. K., C. Zintilas, N. Demiris, I. A. Bizelis, and S. G. Deligeorgis. 2010. Effects of oregano essential oil dietary supplementation on the feeding and drinking behaviour as well as the activity of broilers. *Int. J. Poult. Sci*. 9:401–405.
Takahashi, J., A. S. Chaudhry, R. G. Beneke, Suhubdy, and B. A. Young. 1997. Modification of methane emission in sheep by cysteine and a microbial preparation. *Sci. Total Environ*. 204:117–123.
Talebi, E., A. Zarei, and M. E. Abolfath. 2010. Influence of three different organic acids on broiler performance. *Asian J. Poult. Sci*. 4:7–11.
Tellez, G., S. E. Higgins, A. M. Donoghue, and B. M. Hargis. 2006. Digestive physiology and the role of microorganisms. *J. Appl. Poult. Res*. 15:136–144.
Thompson, J. L. and M. Hinton, 1997. Antibacterial activity of formic and propionic acids in the diet of hens on *Salmonella* in the crop. *Br. Poult. Sci*. 38:59–65.

Tollba, A. A. H., S. A. M. Shabaan, and M. A. A. Abdel-Mageed. 2010. Effects of using aromatic herbal extract and blended with organic acids on productive and physiological performance of poultry. 2—The growth during cold winter stress. *Egypt. Poult. Sci.* 30:229–248.

Van Immerseel, F., V. Fievez, J. de Buck, F. Pasmans, A. Martel, F. Haesebrouck, and R. Ducatelle. 2004. Microencapsulated short-chain fatty acids in feed modify colonization and invasion early after infection with *Salmonella enteritidis* in young chickens. *Poult. Sci.* 83:69–74.

van 't Hof, W., E. C. I. Veernan, E. J. Helmerhorst, and A. V. N. Amerongen. 2001. Antimicrobial peptides: properties and applications. *Biol. Chem.* 382:597–619.

Varel, V. H. and J. T. Yen. 1997. Microbial perspective on fiber utilization by swine. *J. Anim. Sci.* 75:2715–2722.

Verdonk, J. M. A. J., S. B. Shim, P. van Leeuwen, and M. W. A. Verstegen. 2005. Application of inulin-type fructans in animal feed and pet food. *Br. J. Nutr.* 93:S125–S138.

Versteegh, H. A. J. and A. W. Jongbloed. 1999. *The effect of supplementary lactic acid in diets on the performance of broilers*. Institute for Animal Science and Health, Lelystad.

Vessoni Penna, T. C., A. F. Jozala, T. R. Gentille, A. Pessoa, Jr., and O. Cholewa. 2006. Detection of nisin expression by *Lactococcus lactis* using two susceptible bacteria to associate the effects of nisin with EDTA. *Appl. Biochem. Biotechnol.* 121-124:334–346.

Vogt, H. and S. Matthes. 1981. Effect of organic acids in rations on the performances of broilers and laying hens. *Arch. Gefluegelkd.* 45:221–232.

Vogt, H., S. Matthes, and S. Harnisch. 1982. Effect of organic acids in rations on the performances of broilers and laying hens. *Arch. Gefluegelkd.* 46:223–227.

Wallace, R. J., T. A. Wood, A. Rowe, J. Price, D. R. Yanez, S. P. Williams, and C. J. Newbold. 2006. Encapsulated fumaric acid as a means of decreasing ruminal methane emissions. International Congress Series Greenhouse Gases and Animal Agriculture: An Update. C. R. Soliva, J. Takahashi, and M. Kreuzer. (eds) *Second International Conference on Greenhouse Gases and Animal Agriculture, Zurich.* International Congress Series 1293, Elsevier, Amsterdam. pp. 148–151.

Wang, M. L., X. Suo, J. H. Gu, W. W. Zhang, Q. Fang, and X. Wang. 2008. Influence of grape seed proanthocyanidin extract in broiler chickens: effect on chicken coccidiosis and antioxidant status. *Poult. Sci.* 87:2273–2280.

Wierup, M. 2001. The experience of reducing antibiotics used in animal production in the Nordic countries. *Int. J. Antimicrob. Agents* 18:287–290.

Windisch, W., K. Schedle, C. Plitzner, and A. Kroismayr. 2008. Use of phytogenic products as feed additives for swine and poultry. *J. Anim. Sci.* 86:E140–E148.

Xu, Z. R., C. H. Hu, M. S. Xia, X. A. Zhan, and M. Q. Wang. 2003. Effects of dietary fructooligosaccharide on digestive enzyme activities, intestinal microflora and morphology of male broilers. *Poult. Sci.* 82:1030–1036.

Zhang, C. J. 2005. Influence of *Echinacea purpurea* extract on antibody production to Newcastle Disease and infectious bursal disease vaccination. *J. Tradition Chin. Vet. Med.* 24:26–27.

Zhang, G. F., Z. B. Yang, Y. Wang, W. R. Yang, S. Z. Jiang, and G. S. Gai. 2009. Effects of ginger root (*Zingiber officinale*) processed to different particle sizes on growth performance, antioxidant status, and serum metabolites of broiler chickens. *Poult. Sci.* 88:2159–2166.

Zhang, K. Y., F. Yan, C. A. Keen, and P. W. Waldroup. 2005. Evaluation of microencapsulated essential oils and organic acids in diets for broiler chickens. *Int. J. Poult. Sci.* 4:612–619.

22 Prebiotics

Jacqueline Jacob and Anthony Pescatore

Abstract: Prebiotics are a possible alternative to antibiotics in organic food animal production, primarily with nonruminants. The effects of several prebiotics on the composition of the gut microflora, gut morphology, nutrient availability, and animal performance are discussed. The prebiotics addressed include fructo-oligosaccharides, mannanoligosaccharides, and other oligosaccharides; and inulin, a naturally occurring polysaccharide.

Keywords: prebiotics; oligosaccharides; polysaccharides; fructo-oligosaccharides (FOS); mannanoligosaccharides (MOS)

22.1 INTRODUCTION

The concept of prebiotics was first introduced in 1995 (Gibson & Roberfroid, 1995). A prebiotic was defined as being "nondigestible food that beneficially affects the host by selectively stimulating the growth and/or activity of one or a limited number of bacteria in the colon." While probiotics are meant to bring beneficial microorganisms to the gut, prebiotics selectively stimulate the beneficial microorganisms that already live there. Prebiotics are natural feed components that typically survive the thermal processing of pelleted feed as well as the acid conditions of the stomach, giving them yet another advantage over probiotics. However, while the effects of probiotics continue long after administration, prebiotics must be consistently consumed to have a long-lasting effect (Flickinger *et al.*, 2003).

A "natural" component of feed that would fit the definition for a prebiotic is the nondigestible carbohydrate. This would include the oligosaccharides and polysaccharides (Cummings *et al.*, 2001). The term oligosaccharide refers to short chains of sugar molecules. *Oligo* means few and *saccharide* means sugar. The sugar molecules are connected via what are known as β-linkages. These linkages cannot be broken by the endogenous enzymes of animals. As a result, they are not digested in the stomach or small intestines of nonruminants but pass unaffected to the hindgut (cecum or colon). In the hindgut, they are fermented by microorganisms that can cleave the β-linkages. One end product of microbial fermentation of oligosaccharides is short-chain fatty acids (SCFA). The feeding of oligosaccharides, therefore, favors the development of bacteria able to metabolize this feed ingredient to SCFA. The SCFA reduce the intestinal pH, making the environment more favorable for *Bifidobacteria*

Organic Meat Production and Processing, First Edition. Edited by Steven C. Ricke, Ellen J. Van Loo, Michael G. Johnson, and Corliss A. O'Bryan.

and *Lactobacillus*, which, in addition to helping to exclude pathogenic bacteria, stimulate the innate immune system of the host (Forchielli & Walker, 2005).

There are several types of oligosaccharides and they are named for the type of sugar they are made up of. For example, fructo-oligosaccharides (FOS) contain the sugar fructose; galacto-oligosaccharides (GOS) contain galactose; mannanoligosaccharides (MOS) contain mannose, etc. Oligosaccharides are natural components of different raw materials. For example, FOS are common in edible parts of a variety of plants including onion, Jerusalem artichoke, chicory, leek, garlic, wheat, rye, and barley. GOS can be extracted from soybeans.

While oligosaccharides are short chains of sugar molecules, polysaccharides such as inulin are long chains of sugars. Starch and glycogen are examples of glucose-containing polysaccharides that can be digested by the endogenous enzymes in the digestive tract of animals. While cellulose is also a long chain of glucose molecules, the β-linkages between the sugar molecules makes them indigestible in nonruminants.

While there is considerable research on the use of prebiotics as feed additives for pigs and poultry, the use of prebiotics has not proven as successful in ruminant animals. This is primarily because the oligo- and polysaccharides are fermented early in the foregut part of the digestive tract and do not make it to the hindgut. Oligosaccharides have shown promise, however, as a nonantibiotic method for preventing or treating enterotoxigenic *Escherichia coli* (ETEC) in neonatal calves (Constable, 2009). Giving oligosaccharides in the water has been shown to decrease intestinal *E. coli* levels in calves inoculated with ETEC. Similarly, supplementation with 2% inulin to the milk replacer of male Holstein-Friesian calves resulted in higher daily weight gains compared to calves receiving unsupplemented control diets (Masanetz *et al.*, 2010). While improved weight gain is normally associated with better nutrient absorption because of longer villi, in this study villi in the small intestine of the inulin-supplemented calves were shorter. Prebiotics may also have a role in controlling methane production in ruminants. The addition of nitrates to the diet of a ruminant has been shown to reduce methane production (Takahashi *et al.*, 1997), but nitrates can be toxic to ruminants if fed at high levels. Simultaneous supplementation with a prebiotic has been shown to counter the toxic effects of the nitrates while keeping the production of methane low (Sar *et al.*, 2004).

The largest potential impact of prebiotics is as a feed additive for nonruminants, primarily pigs and poultry. Prebiotics have been shown to alter gut microbiota and gut morphology. They have also been shown to stimulate the innate immune system.

22.1.1 Effect of prebiotics on gut microbiota

SCFA is often used as an indirect indication of the types of microbiota present in the gut. Acetic acid is produced from fermentation of cellulose. Butyric and propionate acids are produced from fermentation of nonstructural carbohydrates. Isobutyrate, isovaleric, and valeric acids are produced from degradation of the amino acids valine, leucine, and lysine (Bovera *et al.*, 2010).

Different high-fiber feedstuffs have been shown, *in vitro*, to modify the microbial diversity in the ceca of laying hens (Dunkley *et al.*, 2007). Cecal microorganisms were able to digest these high-fiber feedstuffs with acetate, propionate, and butyrate as the major fermentation products. These SCFA have been shown to limit *Salmonella* growth *in vitro* (McHan & Shotts, 1993; Ricke, 2003). Based on polymerase chain reaction denaturing gradient gel electrophoresis (PCR-DGGE) similarity coefficients, the different dietary sources were shown *in vitro* to alter microbial diversity in the ceca (Richards *et al.*, 2005).

The metabolic activity of the microbiota of pigs *in vitro* differs considerably between adult and weanling pigs (Bauer *et al.*, 2001). Fermentability of various feed ingredients was evaluated in terms of kinetics and extent of fermentation, substrate disappearance, as well as the amount of gas and SCFA produced. The inocula from the unweaned pigs fermented inulin better (more gas, which was produced faster) than those from the adult pigs although this was not reflected in higher levels of SCFA. Fermentation of starch was more rapid and complete by the inocula from the adult pigs. The differences are most likely related to both the age and diet of the pigs.

In an *in vivo* study, Awati *et al.* (2006) monitored the changes in fermentation end products in the feces of weanling pigs receiving diets with different fermentable carbohydrates. There were no differences in concentrations of SCFA in the feces, but ammonia levels were lower in piglets receiving the fermentable carbohydrates (unmolassed sugar beet pulp, native wheat starch, lactulose, and inulin) compared to the semipurified control diet. The reduced ammonia levels indicate a reduction in protein fermentation in favor of the more desired carbohydrate fermentation.

In vivo soybean meal oligosaccharides have been shown to promote the growth of lactic acid bacteria in broilers challenged with coccidia (*Eimeria tenella*) (Lan *et al.*, 2004). The supplementation also lowered the number of oocysts per gram of fecal material when compared to the unsupplemented controls. Lactic acid bacteria are known to play a role in the competitive exclusion of pathogenic bacteria. Based on the results of Lan *et al.* (2004), it appears that they may play a similar role with coccidia. Stanley *et al.* (2004a) reported similar anticoccidial effects of yeast culture residue.

The addition of arabinoxylooligosaccharide (AXOS) from wheat bran has been reported to reduce *Salmonella* Enteritidis shedding in broilers (Eeckhaut *et al.*, 2008). However, there have been conflicting reports on the effect of oligosaccharide supplementation on *Salmonella* colonization. Oligofructose from Jerusalem artichokes was reported to promote colonization of *Salmonella* Typhimurium in the ceca of chickens, while oligofructose from chicory roots had the opposite effect (Chambers *et al.*, 1997). Complicating matters, the effects also appear to be species-specific, even between related species. Inulin and oligofructose, for example, have been reported to reduce *S.* Typhimurium and *Listeria monocytogenes* in mice (Buddington *et al.*, 2002) but increase the levels of *Salmonella* in the gut of rats (Ten Bruggencate *et al.*, 2004).

Shen *et al.* (2009) evaluated a commercial fully fermented yeast culture product as a potential substitute for antibiotic growth promoters (AGP) in weanling pigs. The researchers used SCFA levels as an indicator of changes in microbial fermentation and no differences were observed. The number of *E. coli* in the cecum was reduced in both supplemented groups. However, no differences were observed in *E. coli* levels of the colon or rectum and there was no effect on *Lactobacilli* counts.

Energy and amino acid digestibility has been shown to be reduced with supplementation of high levels of various oligosaccharides (Biggs *et al.*, 2007). Although feed efficiency was negatively impacted, these high levels of inclusion reduced cecal populations of *Clostridium perfringens* in 21-day-old chicks. Feeding the lower inclusion levels had no effect on cecal *Bifidobacterium*, *Lactobacillus*, *C. perfringens*, or *E. coli* populations.

22.1.2 Effect of prebiotics on gut morphology

The structure of the intestinal mucosa is another indicator of gut health (Xu *et al.*, 2003). Any stressors present in the digesta can result in changes in intestinal morphology due to the close proximity of the intestinal lining and contents. Shorter intestinal villi and deeper crypts have been associated with the presence of toxins (Xu *et al.*, 2003). Shortened villi result in less surface area for nutrient absorption and adversely affect feed efficiency. A large crypt indicates a high demand for new tissue. The energy and protein demand for gut maintenance is higher than that for any other organ. Any additional tissue turnover will increase maintenance requirements and reduce feed efficiency. Shorter villi and large crypts typically lead to poor nutrient absorption, diarrhea, reduced disease resistance, and lower overall performance (Xu *et al.*, 2003). A healthy gut, therefore, has a high villus height to crypt depth ratio.

The effects of prebiotics on gut morphology vary. Supplementation of broiler diets with yeast cell walls (*Saccharomyces cerevisiae*), which are a source of MOS, have been shown to increase villi height during the first week with no differences later (Santin *et al.*, 2001). In contrast, supplementation of pig diets with a yeast culture (*S. cerevisiae*) or yeast cell wall products had no effect on gut morphology (van der Peet-Schwering *et al.*, 2007). An increase in villi length in the ileum has been reported for rabbits fed MOS-supplemented diets as compared to nonsupplemented diets (Pinheiro *et al.*, 2004). A similar change in villi length was also noted in those rabbits supplemented with an antibiotic.

Varied responses to FOS supplementation have also been reported. While supplementation with FOS had no effect on the morphology of the duodenum of broiler chickens, there was an increase in villi height in the ileum and reduced crypt depths in both the jejunum and ileum (Xu *et al.*, 2003). Dietary FOS supplementation of weaned pigs had no effect on villi height in the small intestine (Shim *et al.*, 2005). Similarly, FOS supplementation of diets for growing rabbits had no effect on ileal villi measurements (Mourão *et al.*, 2004).

22.1.3 Effect of prebiotics on immune function

As previously indicated, the by-products of prebiotic fermentation are SCFA. There has been some indication, based on animal models, that SCFA have a stimulatory effect on the gut-associated lymphoid tissue (Bornet & Brouns, 2002). It is unclear, however, if this is a direct effect of the fatty acids themselves or if it is indirect via enhanced growth of lactic bacteria such as *Bifidobacteria* and *Lactobacilli*. Ten Bruggencate *et al.* (2004), however, reported that, although dietary supplementation with FOS or inulin increased fecal *Lactobacilli* and *Enterobacteria*, both prebiotics reduced the rat's resistance to *Salmonella* colonization.

Silva *et al.* (2009) reported that supplementation of broiler diets with a commercial yeast extract had no effect on the humoral immune response of broilers vaccinated with Newcastle disease and infectious bursal disease. The prebiotics were only included in the prestarter diets from 1 to 7 days of age and the chicks were vaccinated at day 8. Flickinger *et al.* (2003) demonstrated that prebiotics must be fed continuously to have any long-lasting effect.

22.1.4 Effect of prebiotics on nutrient availability

Since nonruminants are not able to digest oligosaccharides which pass directly through to the hindgut, there was concern that overall nutrient digestibility may be negatively impacted. Using cecectomized roosters to remove the cecal contribution to digestion, Biggs and Parsons (2007) showed that oligosaccharide supplementation had little overall effect on amino acid digestibility and the true metabolizable energy (TME) content of the feed. In some instances, amino acid digestibility was increased. They concluded that the oligosaccharides tested can be added to the diets of adult chickens, up to 8 g/kg, with no adverse effects on nutrient availability. Biggs *et al.* (2007), however, found that while feeding 4 or 8 g/kg of various oligosaccharides had no effect on the growth performance of broiler chicks, there was a reduction in energy and amino acid digestibility at the higher rates of inclusion. Janssens *et al.* (2004) found no effect of FOS supplementation on nutrient availability in pigeons. Houdijk *et al.* (1999) reported no effects of FOS or *trans*-galactooligosaccharides (TOS) on nutrient digestibility in weanling and growing pigs. The pigs in each group were only fed the supplemented diets for 2 weeks and the authors suggested that this may have been too short a time to see any effect on digestibility.

It has been demonstrated that, through changes in gut microbiota and intestinal mucosa as well as changes in gut-associated immunity and nutrient availability, there is a role for prebiotics as antibiotic alternatives in pigs and poultry production. However, the effectiveness of each type of carbohydrate is dependent on numerous factors, including the type of product, the species and age of the animal fed, and the duration of the treatment. The effectiveness of each substance will dictate its ultimate use.

22.2 FRUCTO-OLIGOSACCHARIDES

FOS are short chains of fructose molecules. In an *in vitro* study, Donalson *et al.* (2008) noted that addition of FOS to a layer diet increased concentrations of propionic, butyric, and lactic acid. This is an indication that fermentation could be enhanced. An increase in lactic acid production has been related to decreases in gut pH, which inhibits *Salmonella* colonization (Donalson *et al.*, 2008). *In vivo*, FOS supplementation of broiler diets has been shown to alter the makeup of the intestinal microflora of the large intestine, with increases in *Bifidobacterium* and *Lactobacillus* and decreases in *E. coli* and *Salmonella* (Xu *et al.*, 2003). Jerusalem artichokes are a source of dietary FOS. Chambers *et al.* (1997) compared the effects of Jerusalem artichoke flour (crude FOS) and refined FOS after a challenge with *S.* Typhimurium. At the end of the 6-week trial, broilers fed the crude FOS diets (5% FOS) had significantly higher levels of *Salmonella* than those fed the diets containing refined FOS. Broilers fed the refined FOS had levels lower than those broilers fed the control diets. In a separate treatment, a group of broilers originally fed the refined FOS were fed the unsupplemented control diet for the last week of the trial. The level of *Salmonella* infection returned to the same levels as the control broilers, indicating the need for continuous supplementation throughout the rearing period (Chambers *et al.*, 1997).

There have also been reports of changes in the gut morphology of broilers receiving FOS-supplemented diets (Xu *et al.*, 2003). The height of the villi in the ileum was increased, while the crypt depth in the jejunum and ileum decreased. As a result, the villus height to crypt depth ratio increased in both the jejunum and ileum indicating improved gut health. A low villus height to crypt depth ratio typically indicates a problem in the digestive tract (Xu *et al.*, 2003).

When a commercial FOS product extracted from sugar beets was fed to broilers, there was an improvement in feed efficiency but these could not be explained by changes to the intestinal morphology (Catalá-Gregori *et al.*, 2008). When a commercial FOS product made from chicory inulin was fed to turkey poults for 8 weeks, the only change in intestinal morphology noted was an increase in the relative cecal weight for the turkeys receiving the highest level of supplementation (2 g/kg). Ileal and cecal pH declined, indicating a probable change in cecal metabolism (Juśkiewicz *et al.*, 2006). There were no effects on feed intake, bodyweight, or feed efficiency.

The results with FOS supplementation in swine production vary. Dietary FOS supplementation of weaned pigs has been reported to increase levels of the SCFA, acetic, butyric, and isobutyric, indicating a shift in the microbiota in the hindgut (Shim *et al.*, 2005). Mikkelsen *et al.* (2004) reported a similar shift in the microbiota as reflected by an increase in butyric acid and a decrease in acetic acid. While Gebbink *et al.* (1999) reported an increase in the level of *Bifidobacteria* in the GI tract, with a reduction in *E. coli*, others observed no effects on bacterial populations (Mikkelsen *et al.*, 2003, 2004). FOS supplementation has also been reported to increase yeast levels in the GI tract or the feces (Mikkelsen *et al.*, 2003, 2004).

The effects of FOS supplement on pig growth performance also vary. While some researchers have reported a depression in feed intake and weight gain early in the growout (Houdijk *et al.*, 1998), others have reported no effect on growth, feed intake, or feed efficiency (Mikkelsen *et al.*, 2003; Shim *et al.*, 2005) throughout the growout period. Some researchers have reported improved growth right after weaning but not by the end of the growout period (Eugeniusz *et al.*, 2006), while others have reported improved growth throughout the entire growout (Grela *et al.*, 2006).

Couch grass (*Agropyron repens*) is a good source of FOS. Supplementing the diet of weanling pigs with couch grass reduced both the incidence and severity of diarrhea post weaning (Grela *et al.*, 2006). The health benefits carried through to the end of the 84-day trial, so the livability of the FOS-supplemented pigs was higher than that of the pigs receiving the unsupplemented controls.

FOS supplementation may have an unexpected benefit in pork production. Many countries have, or are considering, a ban on castrating pigs for pork production. Male piglets are typically castrated to eliminate boar taint of the meat. In an *in vitro* study, FOS was shown to decrease the level of L-tryptophan conversion to skatole (Xu *et al.*, 2002). The researchers speculated that the decrease in skatole levels was due to a change in the microbiota of the GI tract resulting in a shift in microbial metabolism of tryptophan toward indole at the expense of skatole. This may be a way of decreasing skatole concentration in the backfat of intact male pigs, thus reducing the incidence of boar taint, although it needs to be proven *in vivo*. Supplementing pig diets with lupines, which are high in oligosaccharides, resulted in lower skatole levels in the backfat of the pigs compared to the backfat of pigs receiving the unsupplemented control diets (Hansen & Claudi-Magnussen, 2004).

Falcão-e-Cunha *et al.* (2007) reviewed the effects of supplementing rabbit diets with FOS. The results were very inconsistent. While Aguilar *et al.* (1996) reported a positive effect on the growth rate but no effect on feed efficiency, Mourão *et al.* (2004) indicated no differences in live weights, feed intake, or feed efficiency when rabbit diets were supplemented with FOS. There were also no differences in mortality or the incidence of diarrhea of growing rabbits with or without FOS supplementation. There were also no differences in gut morphology (Mourão *et al.*, 2004). FOS appears to have negligible effects on growth performance and microbial activity in rabbits (Falcão-e-Cunha *et al.*, 2007).

22.3 MANNANOLIGOSACCHARIDES (MOS)

Santin *et al.* (2001) studied the performance of broilers fed diets supplemented with yeast cell walls (*S. cerevisiae*), which are a source of MOS. An initial study was conducted in their experimental poultry house facility and at the end of the 42-day trial period the broilers receiving the supplemented diets had significantly higher bodyweights than those receiving the unsupplemented control diet. Throughout a 49-day follow up field test with 44,000 broilers, those broilers receiving the supplemented diets had higher bodyweights and better feed efficiency than those receiving the unsupplemented control diet.

Supplementation of pig diets with a yeast culture (*S. cerevisiae*) or a yeast cell wall product, both of which contain MOS, improved growth performance and feed efficiency compared to pigs on unsupplemented control diets. The performance of the pigs on the supplemented diets was comparable to those fed diets with AGP (van der Peet-Schwering *et al.*, 2007; Shen *et al.*, 2009). Grela *et al.* (2006) reported reduced mortality when supplementing the diets of weanling pigs with a yeast cell wall extract. Losses in the unsupplemented control group were primarily due to diarrhea. Supplementation with the yeast wall extract reduced both the incidence and severity of diarrhea post weaning, and this carried through to the end of the 84-day trial. Final bodyweights of the piglets on the supplemented feed were higher than those on the unsupplemented control.

There are several commercial MOS products available. Hooge (2004) completed a detailed meta-analysis of studies from around the world that looked at supplementing broiler diets with the same commercial MOS product. His analysis of 29 pen trials concluded that MOS supplementation resulted in bodyweights and feed efficiencies comparable to those achieved with antibiotic supplementation, with both supplementations being higher than the unsupplemented controls. MOS supplementation also reduced mortality when compared to broilers fed antibiotic-supplemented or unsupplemented control diets.

Research by others, however, has not yielded such clear cut results. While there are reports of improved performance with MOS supplementation (Flemming *et al.*, 2004; Bozkurt *et al.*, 2009a), others have reported that broilers fed MOS-supplemented diets performed as well as those receiving antibiotic-supplemented diets but both groups failed to perform as well as the unsupplemented controls (Jamroz *et al.*, 2004; Baurhoo *et al.*, 2007a; Yang *et al.*, 2007). Other researchers have reported no positive effects for MOS supplementation of broiler diets (Pelicano *et al.*, 2004; Yalçinkaya *et al.*, 2008).

Flemming *et al.* (2004) reported that MOS supplementation of broiler diets improved feed intake, daily weight gain, feed efficiency, and mortality when compared to unsupplemented controls. The growth performance with MOS supplementation was similar to that observed with antibiotic supplementation. In a study by Baurhoo *et al.* (2007a), broilers receiving MOS supplementation performed as well as those receiving antibiotics, but the broilers on the unsupplemented diets grew faster than both of the supplemented groups. The broilers on MOS, however, had improved gut morphology as indicated by increased villi height in the jejunum and more goblet cells per villus than those broilers receiving antibiotic supplementation. Additionally, MOS supplementation resulted in increased levels of *Bifidobacteria* in the ceca and a major reduction in *E. coli* load in the litter. While Yang *et al.* (2007) found that MOS supplementation improved bodyweight gain of broilers in the starter phase, there were no differences in bodyweights at the end of the 42-day trial. Supplementation with MOS did not affect nutrient digestibility or intestinal morphology. Jamroz *et al.* (2004) observed similar growth performance with broilers supplemented with MOS or an antibiotic, but growth rates for both groups were reduced in the first 3 weeks, with no differences at the end of the 42-day trial. MOS supplementation, however, reduced levels of the intestinal pathogens *E. coli* and *C. perfringens*.

Positive effects of yeast supplementation in Japanese quail diets have been reported (Ghally & Abd El-Latif, 2007). There was a reduction in feed consumption with no significant effect on bodyweight gain resulting in improvement of overall feed efficiency. An economic analysis of the performance data concluded that inclusion of yeast culture in Japanese quail diets had a higher economic efficiency compared with the control diet. In contrast, while supplementing turkey diets was also shown to improve overall feed efficiency, there was an increase in bodyweights with no effect on feed intake (Zduńczyk *et al.*, 2005).

It is unclear why such widely different results have been found but possible factors include the type and quality of ingredients in the diet, the age of the litter that the broilers are raised on, the density of the broilers, and the levels of inclusion. The effect of MOS supplementation of broiler diets was shown to be affected by the cereal component of the feed. Broilers fed wheat-based diets for 21 days showed better response, in terms of growth performance and nutrient digestibility, to MOS supplementation than those broilers fed corn-based diets (Yang *et al.*, 2008). The mucosal morphology and the specific activities of brush border enzymes were improved by MOS, which varied with the type of cereal and/or the age of the broilers. Iji *et al.* (2001) looked at inclusion of a MOS in sorghum-based diets with high levels of lupin seed meal. The bodyweights and feed efficiencies observed in the study were lower than those typically attained in commercial operations. The high levels of lupin seed meal, with high levels of non-starch polysaccharides (NSP), were thought to be the cause of the reduced performance. There was no effect of MOS supplementation on broiler performance. The authors did note, however, that MOS supplementation increased the specific activities of most of the digestive enzymes in the jejunum. The jejunum villi were also longer in chicks raised on diets with the highest level of MOS inclusion. Supplementing wheat-based diets with MOS has been reported to have no effect on bodyweight gain, feed intake, or feed efficiency of Japanese quail (Sarica *et al.*, 2009). Wheat-based poultry diets are known to increase intestinal viscosity and alter the microflora. Supplementation with MOS decreased intestinal viscosity. There was no effect of MOS supplementation on intestinal *E. coli* levels or on intestinal morphology.

In a study by Waldroup *et al.* (2003), antibiotic supplementation improved broiler performance but MOS supplementation did not. All the diets were supplemented with copper sulfate and the authors suggested that there could be an interaction between the MOS product and copper sulfate. This was supported by the research with pigs where high levels of copper in the diet of weanling barrows interfered with the ability of MOS to enhance growth and feed conversion (Davis *et al.*, 2000).

Supplementation with MOS has been shown to reduce the negative effects of aflatoxins in the feed. A yeast culture residue, comprised primarily of MOS, enhanced the performance of broiler breeder hens fed diets contaminated with aflatoxin (Stanley *et al.*, 2004b). Similar benefits of MOS supplementation in aflatoxin-contaminated feed were reported with Japanese quail (Parlat *et al.*, 2001). Supplementing the diets of Japanese quail helped to overcome the deleterious effects of dietary aflatoxin on bodyweight gain, feed consumption, and feed efficiency. Quail receiving aflatoxin-contaminated diets had a 37% reduction in bodyweight. With addition of the live yeast, a source of MOS, to the aflatoxin-contaminated feed, there was a 15% gain in bodyweight compared to the aflatoxin-free controls. The researchers also noted that yeast supplementation without aflatoxin contamination resulted in a 40% increase in bodyweight gain.

More recently, MOS supplementation has been shown to alleviate some of the detrimental effects of heat stress in broilers (Sohail *et al.*, 2010). Supplementation with MOS has also been shown to reduce the negative effects associated with high-density production systems (Hooge *et al.*, 2003).

With the increased competition for wood shavings, broiler producers are recycling litter for several growout cycles. In many cases, this represents a coccidial challenge to the incoming chickens. Inclusion of a yeast culture residue in broiler diets has been shown to be successful in controlling intestinal coliforms and, to some extent, coccidial oocysts (Stanley *et al.*, 2004a). The bodyweights of broilers reared to 42 days of age on recycled litter (pine wood shavings) containing three strains of coccidia were comparable to those of broilers supplemented with a coccidiostat (lasalocid) or an antibiotic (bacitracin). The coliform counts taken from litter samples increased significantly within each of three consecutive trials. Intestinal coliform counts were consistently lower in the broilers supplemented with the yeast culture residue. Except for the litter from the lasalocid-fed broilers, oocyst counts increased at the end of the third trial. The oocyst counts in the litter from the broilers fed diets with the yeast cell residue were greater than those of the controls but higher than the litter from the lasalocid-fed broilers. The authors concluded that yeast culture residue, mainly made up of MOS, is a viable alternative to bacitracin and lasalocid medication for broilers reared on recycled litter.

Based on the meta-analysis performed by Hooge (2004), the optimal levels of inclusion of MOS in broiler chicken feeds are 2 g/kg during the first week, 1 g/kg during weeks 2 and 3, and 0.5 g/kg to market weight. The studies previously referred to have inclusion rates varying from 0.5 to 5 g/kg, which may also account for some of the differences observed.

Supplementing broiler diets with MOS has been shown to reduce intestinal levels of *E. coli* while increasing levels of *Lactobacilli* and *Bifidobacteria* (Stanley *et al.*, 2004a; Baurhoo *et al.*, 2007a, 2007b). The exact mode of action of MOS is unclear. In order for bacteria to colonize the intestinal tract of an animal, they need to attach themselves to the intestinal mucosa (Beachy, 1981). Naturally occurring mannan compounds in the mucosa surface of the intestines are required for the pathogenic bacteria to attach themselves to the mucosa (Iji *et al.*, 2001). These same pathogenic bacteria can also attach to the mannan component of MOS and pass through the GI tract without colonizing and causing disease. This may explain how MOS supplementation is able to reduce the level of pathogenic bacteria in intestinal tract of animals (Miguel *et al.*, 2004). Spring *et al.* (2000) demonstrated *in vitro* that five of seven strains of *E. coli*, four of five strains of *S.* Enteritidis and three of five strains of *S.* Typhimurium were shown to agglutinate MOS. Five strains of *Campylobacter jejuni* and five strains of *Campylobacter coli*, however, failed to agglutinate MOS.

Zduńczyk *et al.* (2005) reported that while MOS supplementation reduced *E. coli* levels in the ceca of 18-week-old turkeys, there was no effect on the *Bifidobacteria* and *Lactobacilli* populations. They also noted reduced concentrations of SCFA in the ceca, especially in the case of acetic acid, and an increase in ammonia levels with MOS supplementation. Similar changes in SCFA levels were noted by Stanczuk *et al.* (2005) during an 8-week turkey trial. While Sims *et al.* (2004) reported a reduction in the large intestinal concentrations of *C. perfringens* in 6-week MOS-supplemented turkeys, this difference was gone by the end of the 18-week trial. They did not detect any changes in *Lactobacilli* concentrations.

The intestines of newly hatched chicks and poults must adapt to the change in nutrient source as the yolk is replaced by consumed feed. When feed consumption starts, there are rapid changes in the digestive tract and associated organs (Uni *et al.*, 1998). Supplementing turkey diets with MOS was found to accelerate intestinal development in turkeys (Solis de los Santos *et al.*, 2007). This may account for some of the improvements noted early in turkey trials.

While research results have not been consistent, MOS appears to be a possible alternative to AGP in poultry production systems. Mohamed *et al.* (2008) compared MOS supplementation of broiler chickens with an AGP (enramycin). Supplementing the broiler diets with MOS, the antibiotic, or both only slightly increased bodyweight gain compared to those broilers on the unsupplemented controls. Although the differences were not statistically significantly different, the increase was sufficient to result in statistically significant differences in feed efficiency. Feed efficiency early in growth (14 and 28 days) was improved by the addition of MOS, the antibiotic, or both. However, there was no synergistic effect of combining the products. There was a significant reduction in percent abdominal fat in the final carcasses, but no differences in yield, liver, heart, gizzard, or bursa weight. The results indicate that MOS can serve as an alternative to the antibiotic enramycin in broiler diets. Similarly, Baurhoo *et al.* (2007b) compared the microbiota in the GI tract of broilers fed diets supplemented with MOS or the antibiotic virginiamycin. In this study, the virginiamycin supplementation had no effect on growth performance. The cecal populations of *Lactobacilli* and *Bifidobacteria* were higher in the broilers receiving the diets supplemented with MOS. When the broilers were challenged with *E. coli*, those receiving the MOS had reduced cecal populations of total *E. coli*.

Most commercial swine operations comprise early age weanling pigs. The transition from a liquid milk diet to solid food is often accompanied by an increase in diarrhea and poor growth rates. Miguel *et al.* (2002) completed a meta-analysis of 49 studies comparing MOS supplementation of diets for growing pigs versus unsupplemented control diets. They concluded that the addition of MOS to starter diets will enhance pig performance. The occurrence of diarrhea at postweaning was decreased, increasing the overall survival rate of the piglets. The optimal inclusion rates were 2 g/kg at an early weaning age (17–21 days) and during the first 2 weeks after weaning.

Again, the results of other studies looking at MOS supplementation of postweaning pig diets have not been as clear-cut. While some reports indicate that MOS supplementation had no affect on weight gain (Castillo *et al.*, 2008), others have reported improved growth performance (Hancock *et al.*, 2003). There have been reports of decreased levels on *Enterobacteria* with no effect on the *Lactobacilli* population (Castillo *et al.*, 2008) as well as reports of no effect on fecal concentrations of coliforms and *E. coli* (Hancock *et al.*, 2003).

The level of "Ferric Reducing Ability of Plasma" (FRAP) is a measure of the antioxidant potential. Low levels of FRAP and its components indicate that the animal has used the antioxidants for antioxidant protection during stress. In sows, this stress most often occurs during pregnancy, parturition, and postparturition lactation. In piglets, it is primarily the transition from milk to solid food during birth and weaning. MOS supplementation has been shown to increase FRAP levels in sows at 110 days of gestation, at farrowing, and after 21 days of lactation. Piglets from sows receiving MOS supplementation also increased FRAP levels at birth and at 21 days of age (Czech *et al.*, 2009).

An *in vitro* study with cecal inoculum from rabbits fed diets supplemented with 0.5, 1.0, or 1.5 g/kg of MOS demonstrated differences in fermentation activity (Bovera *et al.*, 2010). The inoculum from the rabbits receiving 1.0 g/kg of MOS had the highest organic matter digestion as well as a significantly higher level of gas production. This indicates that the cecal microbiota from these rabbits have a higher fermentative activity. The same inoculum had high levels of acetic acid indicating a more intense fermentation of the structural carbohydrates such as cellulose. There were no significant differences in the production of ammonia, which is the end product of protein fermentation.

Kocher *et al.* (2004) reviewed the data from 20 studies on feeding MOS to caged rabbits and concluded that MOS supplementation improved bodyweight gain and feed conversion ratio by 4.6% and 5.1%, respectively, compared to those rabbits on unsupplemented control diets. The results were similar to those achieved with antibiotic supplementation but MOS supplementation reduced mortality to a greater extent. The authors recommended supplementing the diets with 1–2 g/kg MOS starting at $5^1/_2$ weeks of age. In contrast, there have been reports indicating no effect of MOS supplementation on weight gain, feed intake, or feed efficiency of rabbits (Guedes *et al.*, 2009; Pinheiro *et al.*, 2009). In a study comparing the growth performance of rabbits fed MOS- or antibiotic- (oxytetracyclin) supplemented diets, bodyweights were similar but the rabbits fed the antibiotic-supplemented diets ate more feed throughout the trial (Fonseca *et al.*, 2004). As a result, the rabbits on the MOS-supplemented diets had better feed efficiency. Mortality was also lowest for the rabbits receiving the MOS supplementation. Unfortunately, the study did not compare the performance of the supplemented rabbits with an unsupplemented control.

When the intestinal morphologies of rabbits fed diets supplemented with MOS or an antibiotic (bacitracin) were compared, rabbits receiving the MOS-supplemented diets had significantly longer villi as well as a more regular villi structure (Pinheiro *et al.*, 2004). Cecal pH in the rabbits receiving the MOS-supplemented diets was also lower compared to the unsupplemented controls. There was also an increase in total SCFA concentration with a higher proportion of butyric acid. The addition of MOS to the diet of growing rabbits has been reported to result in an increase in the concentrations of acetic, propionic, and butyric acids in the GI tract (Guedes *et al.*, 2009; Pinheiro *et al.*, 2004). Increases in intestinal SCFA have been associated with improved gut health.

Both MOS and FOS appear to be possible substitutes for AGP. Comparing the performance of pigs fed FOS- or MOS-supplemented diets, Grela *et al.* (2006) reported that both decreased the incidence and severity of diarrhea and increased bodyweight gain compared to pigs on the unsupplemented control diets. Final bodyweights were highest for the pigs receiving MOS-supplemented diets. The bodyweights of the pigs receiving the FOS-supplemented diets were intermediate between those fed the control and MOS-supplemented diets.

22.4 OTHER OLIGOSACCHARIDES

While FOS and MOS have received the most interest as alternatives to AGP, a number of lesser known oligosaccharides have also been investigated. Chitooligosaccharide (COS) is made from the breakdown of chitin. Chitin is a long chain of molecules with glucose backbones, with acetyl side groups bound to the glucose units through an amino group. It is the main cell wall component of fungi as well as the exoskeletons of arthropods and insects. Low levels of COS inclusion (0.05 or 0.10 g/kg) have been reported to improve the growth rate of broilers by increasing feed intake and nutrient digestibility (Huang *et al.*, 2005; Li *et al.*, 2007). The increase in performance was comparable to that achieved with an AGP (flavomycin) (Huang *et al.*, 2005). Supplementation with COS was also found to increase the serum total protein in broilers compared with broilers fed antibiotic-supplemented or unsupplemented control diets. The serum protein level, which includes albumin and globulin, is typically a reflection of the status of protein metabolism and immune function. The increased serum protein levels, therefore, imply that dietary COS supplementation of broiler diets improved whole protein anabolism in growing broilers (Li *et al.*, 2007).

Similarly, Liu *et al.* (2008) found that including COS at low levels (0.2 g/kg) in the diet of weanling pigs improved apparent dry matter, gross energy, crude protein and crude fat digestibility, and the availability of calcium and phosphorus compared to unsupplemented controls. The incidence of diarrhea was reduced. In terms of intestinal morphology, there was an increase in villi height and villi height to crypt depth ratio in the ileum and jejunum. Pigs that had received the COS-supplemented diets also had higher *Lactobacillus* counts than those on the unsupplemented control diets. The improved growth performance obtained with early-weaned pigs may be due to increased serum growth hormone and insulin-like growth factor 1 (IGF-1) (Tang *et al.*, 2005) or improved immune function (Liu *et al.*, 2010).

AXOS are obtained by partial hydrolysis of arabinoxylan. Arabinoxylan are polysaccharides found in the cell walls of most monocotyledonous plant species and are therefore, present in cereal-based broiler diets. AXOS supplementation has been shown to improve feed efficiency in broilers (Courtin *et al.*, 2008a) and stimulate the growth of beneficial bacteria (Courtin *et al.*, 2008b). AXOS supplementation has also been shown to provide dose-dependent protection against *S*. Enteritidis infection of broilers (Eeckhaut *et al.*, 2008).

GOS are produced from lactose, which is a disaccharide of glucose and galactose. Supplementation of diets with GOS has been shown to stimulate the growth of beneficial bacteria in pigs (Smiricky-Tjardes *et al.*, 2003) and broilers (Jung *et al.*, 2008). GOS has also been shown to increase SCFA concentrations both *in vivo* and *in vitro* (Smiricky-Tjardes *et al.*, 2003).

Isomalto-oligosaccharides (IMO) are sugar substitutes with 40% of the sweetness of sucrose and have been widely used in foods and beverages (Zhang *et al.*, 2003). IMO can be produced from glucose by transgalactosylation. Since IMO can be cleaved at the brush border of the small intestine, it cannot be accurately classified as a prebiotic. IMO supplementation of broiler diets enhanced growth performance during the first 3 weeks, but had no further effects during the remainder of the growout period (Zhang *et al.*, 2003). Butyric and isobutyric acid levels increased in the jejunum of broilers fed the supplemented diets. IMO supplementation of broiler diets has also been found to reduce the levels of *S*. Typhimurium in the ceca (Thitaram *et al.*, 2005).

TOS is a synthetic compound produced from lactose by enzymatic transgalactosylation. TOS consists of lactose and several galactose molecules that are resistant to digestion by intestinal β-galactosidase and enter the lower gut intact (Alles *et al.*, 1999). A commercial TOS product, however, was reported to have no effect on nutrient digestibility in weanling and growing pigs (Houdijk *et al.*, 1999). The pigs in each age group were only fed the supplemented diets for 2 weeks and the authors suggested that this may have been too short to see any effect on digestibility. TOS was reported to increase feed intake and weight gain growing barrows early in the study but by the end of the 6-week trial there was no effect on overall performance (Houdijk *et al.*, 1998). Mikkelsen *et al.* (2003) reported that there were no effects of TOS supplementation on pig growth performance, intestinal morphology, or bacterial populations in the feces. They did observe a strong effect of TOS supplementation on the level of yeast in the feces.

Solis de los Santos *et al.* (2005) looked at the effect of adding *Aspergillus* (a fungus) meal as a prebiotic in broiler diets. *Aspergillus* meal supplementation was found to reduce the incidence of ascites in a simulated high-altitude environment. While the broilers receiving the unsupplemented control diets showed decreased gut development in the simulated high-altitude environment, those broilers receiving the diets supplemented with *Aspergillus* meal had gut development that was similar to the controls reared at local altitude. The enhanced gut development would allow the chickens to use more oxygen and maintain their metabolism in high-altitude environments.

22.5 INULIN

Inulin is a naturally occurring polysaccharide containing fructose molecules and is used by some plants to store energy. They are typically found in the plant roots. Most plants that store energy as inulin do not store energy as starch. Most inulin commercially available is extracted from chicory roots (*Cichorium intybus*) or synthesized from sucrose (Niness, 1999). As a feed additive, the shorter chained FOS is often made from inulin.

Inulin and oligofructose are β-fructans with different chain lengths. Because of the β-linkages, they cannot be digested by nonruminants, and pass to the hindgut for fermentation. Flickinger *et al.* (2003) and Verdonk *et al.* (2005) completed extensive reviews of the literature related to the supplementation of animal diets with inulin-type fructans. Overall, the studies indicated that such supplementations improve growth performance and host health as well as stimulate growth of beneficial bacteria. There was, however, a wide variation in the results reviewed.

Supplementing broiler diets with inulin has been reported to decrease the cecal concentrations of *E. coli*, *Salmonella* spp., and *Campylobacter* spp. and to increase *Bifidobacterium* spp. (Rada *et al.*, 2001; Yusrizal & Chen, 2003; Velasco *et al.*, 2010). In addition, inulin supplementation increased the concentrations of lactic acid in the jejunum and butyric acid in the ceca (Rehman *et al.*, 2008; Velasco *et al.*, 2010). An increased villus height and crypt depth in the jejunum of broilers has also been reported (Rehman *et al.*, 2007).

Traditional culture techniques used to monitor the intestinal microflora are incomplete because of the difficulty in cultivating some of the bacteria present (Rehman *et al.*, 2008). Using a PCR technique to monitor the intestinal microbiota of broilers with and without supplementation with inulin, Rehman *et al.* (2008) reported that there were no major differences in microbiota composition between the two groups. The authors concluded that inulin supplementation did not modify the cecal microbiota composition but did affect microbial activity. Rebolé *et al.* (2010), however, reported increased *Bifidobacteria* and *Lactobacilli* counts in both the ileum and cecum of broiler chickens receiving diets supplemented with inulin, indicating an alteration in the fermentation pattern in the GI tract. This was confirmed by increased concentrations of butyric and lactic acids and the butyric to acetic acid ratio. Inulin supplementation of broiler diets had no effect on villus height and crypt depth (Rebolé *et al.*, 2010). Awad *et al.* (2011), however, reported that dietary chicory, as a source of inulin, increased villus height and width as well as the villus height to crypt depth ratio of broilers, but had no effect on the duodenal crypt depth. In the jejunum, the villus height, crypt depth, and villus height to crypt depth were decreased by dietary chicory. In the ileum, villus height to villus crypt depth ratio was also increased.

The addition of different fat types in broiler diets is known to influence abdominal fat deposition. An increased consumption of polyunsaturated fatty acids (PUFAs) has been associated with improved heart health in humans. Supplementation with inulin has been shown to increase the capacity of sunflower oil to increase the ratio of PUFAs to saturated fatty acids (SFAs) in the fat in broilers (Velasco *et al.*, 2010).

Broilers appear to have gender differences in response to inulin supplementation (Yusrizal & Chen, 2003). In a study comparing two commercial products made from chicory, one containing inulin and the other oligofructose, Yusrizal and Chen (2003) reported that the body weight gain, ready-to-cook carcass weight, feed efficiency, and length of the gut intestines of the males fed either supplement did not differ from those of the negative controls. However, oligofructose supplementation of the female diets resulted in increased body weights, carcass weights, carcass yields, and gut length. The majority of the increase in the body weights of the females appears to have happened between weeks 5 and 6 of the study.

Inulin-supplementation of pig diets has been reported to have no effect on intestinal *Lactobacilli* or *Bifidobacteria* levels (Loh *et al.*, 2006). In addition, the SCFA levels were lower in the inulin-fed groups due to reduced acetic acid production. Similarly, inulin supplementation of rabbit diets did not improve growth performance or enhance the cecal environment (Bónai *et al.*, 2008). Juśkiewicz *et al.* (2007), however, reported positive effects of inulin supplementation on rabbit GI tract physiology when replacing dried sugar beet pulp with chicory flour. There was a decrease in the pH of the ileum and lower colon; an increase in the viscosity of the digesta in the ileum; and an increase in the production of SCFA. While Maertens *et al.* (2004) also reported reduced pH in the cecum of inulin-fed rabbits, total SCFA levels were not affected. There was, however, a significant increase in butyrate at the expense of acetate in the inulin-fed rabbits.

Supplementing the diets of growing rabbits with fresh chicory, a source of inulin, has been reported to improve the biochemical traits of cecum content (Castellini *et al.*, 2007). Before weaning, the level of ammonia was reduced, as were the pH values, and there was an increase in SCFA production indicating a better balanced fermentation by the gut microbiota. After weaning, there was a small effect on growth performance with the rabbits fed chicory having higher feed intakes and daily body weight gains.

Prebiotics are another possible alternative to the subtherapeutic use of antibiotics. The results, however, vary between the type of prebiotic and the species being fed (Table 22.1). Factors to consider before supplementing with prebiotics include the type of diet, especially the initial level of oligosaccharides present; the type and dose of the supplementation; the species, age, sex, and stage of production of the animals; and the condition of the farm where the animals are raised.

22.6 COMBINATIONS OF FEED ADDITIVES

Combining probiotics with prebiotics into a single dietary supplementation is often referred to as "synbiotics" (Schrezenmeir & Vreese, 2001). The term refers to a prebiotic that specifically targets the probiotic organisms so that they are acting synergistically. Sahin *et al.* (2008) coined the term "combiotics" to refer to any combination supplement of a probiotic and a prebiotic.

Commercial synbiotic products are already available on the market. One commercial product is a combination of the prebiotic FOS derived from chicory, the probiotic bacteria *Enterococcus faecium*, and an immune modulating substance from sea algae. Broiler diets supplemented with this mixture resulted in higher bodyweight gains, as well as improved feed efficiencies and carcass yields, compared to broilers fed the unsupplemented control diets as well as those broilers receiving a diet supplemented with a *Lactobacillus* sp. probiotic (Awad *et al.*, 2009). Supplementation with the synbiotic also affected intestinal morphology with an increase in villi height and villi height to crypt depth ratio in the ileum. The authors concluded that the synbiotic product is a promising alternative to antibiotics in broiler production. In contrast, no effect was found when the product was included in the diets of Japanese quail (Çakir *et al.*, 2008).

Various combinations of commercially available alternative feed additives have been studied. While supplementation of broiler diets with a commercial prebiotic or probiotic individually had no effect on bodyweight gain, feed efficiency, or livability, the combination led to improved feed efficiency, suggesting a synergism between the two products (Bozkurt *et al.*, 2009b). In contrast, a competitive exclusion probiotic was reported to reduce *S.* Typhimurium in the cecum and crop of chickens, but supplementing with FOS had no additional effects on *Salmonella* colonization and did not affect bodyweight gain, feed efficiency, or livability (Telg & Caldwell, 2009).

Table 22.1 Supplement effects of prebiotics in different animal species.

	Poultry		Pig		Rabbit		Cattle	
FOS	Layers/turkeys: Enhanced cecal fermentation	Juśkiewicz *et al.*, 2006; Donalson *et al.*, 2008	Weaned pigs: Enhanced cecal fermentation	Mikkelsen & Jensen, 2004; Shim *et al.*, 2005	Improved growth rate but no effect on feed efficiency	Aguilar *et al.*, 1996		
	Broilers: Changed intestinal microbiota	Chambers *et al.*, 1997; Xu *et al.*, 2003	Growing pigs: Changed intestinal microbiota	Gebbink *et al.*, 1999; Xu *et al.*, 2002	No effect on growth rate, feed intake, or feed efficiency	Mourão *et al.*, 2004		
	Broilers: Improved gut morphology	Xu *et al.*, 2003	Growing pigs: No effect on intestial bacteria but increased yeast	Mikkelsen *et al.*, 2003, Mikkelsen & Jensen, 2004	No differences in gut morphology	Mourão *et al.*, 2005		
	Broilers: Improved feed efficiency	Catalá-Gregori *et al.*, 2008	Growing pigs: Reduced performance	Houdijk *et al.*, 1998	No effects on mortality or incidence of diarrhea	Mourão *et al.*, 2006		
	Turkeys: No effect on performance	Juśkiewicz *et al.*, 2006	Growing pigs: No effect on overall performance	Mikkelsen *et al.*, 2003; Shim *et al.*, 2005; Eugeniusz *et al.*, 2006				
	Pigeons: No effect on nutrient digestibility	Janssens *et al.*, 2004	Growing pigs: Improved overall performance	Grela *et al.*, 2006				
			Weanling pigs: Reduced diarrhea	Grela *et al.*, 2006				
			Growing pigs: Possible reduction in skatole levels	Xu *et al.*, 2002				

(*continued*)

Table 22.1 (*Continued*)

	Poultry		Pig		Rabbit		Cattle	
MOS	Broilers: Improved performance	Santin *et al.*, 2001; Flemming *et al.*, 2004; Bozkurt *et al.*, 2009a	Growing pigs: No effect on performance	Davis *et al.*, 2000; Hancock *et al.*, 2003; Castillo *et al.*, 2008	Growing rabbits: Improved performance	Kocher *et al.*, 2004	Neonatal calves: Decreased intestinal *E. coli*	Constable, 2009
	Broilers: No effect on overall performance	Waldroup *et al.*, 2003; Jamroz *et al.*, 2004; Pelicano *et al.*, 2004; Yang *et al.*, 2007; Yalçinkaya *et al.*, 2008	Growing pigs: Improved overall performance	Hancock *et al.*, 2003; Grela *et al.*, 2006; van der Peet-Schwering *et al.*, 2007; Shen *et al.*, 2009	Growing rabbits: No effects on overall performance	Guedes *et al.*, 2009; Pinheiro *et al.*, 2009		
	Broilers: Reduced overall performance	Baurhoo *et al.*, 2007a;	Growing pigs: Changed gut microbiota	Castillo *et al.*, 2008	Growing rabbits: Improved overall performance	Kocher *et al.*, 2004		
	Broilers/turkeys: Reduced levels of intestinal SCFA and increased ammonia	Zduńczyk *et al.*, 2005/Stanczuk *et al.*, 2005	Growing pigs: No effect on gut microbiota	Hancock *et al.*, 2003	Growing rabbits: Improved gut health	Pinheiro *et al.*, 2004; Guedes *et al.*, 2009; Pinheiro *et al.*, 2009		
	Broilers: Improved gut morphology	Baurhoo *et al.*, 2007a	Weanling pigs: Reduced diarrhea	Grela *et al.*, 2006	Growing rabbits: Enhanced gut fermentation	Pinheiro *et al.*, 2004; Guedes *et al.*, 2009		
	Broilers/turkeys: Changed intestinal microbiota	Stanley *et al.*, 2004a; Jamroz *et al.*, 2004; Sims *et al.*, 2004; Zduńczyk *et al.*, 2005; Baurhoo *et al.*, 2007a, 2007b						

Broilers: No change in gut morphology	Yang *et al.*, 2007; Yang *et al.*, 2008
Japanese quail: Reduced feed intake with no effect on bodyweight gain	Ghally & Abd El-Latif, 2007
Japanese quail: Improved bodyweight gain with no effect on feed intake	Zduńczyk *et al.*, 2005
Japanese quail: No effect on feed intake or bodyweight gain	Sarica *et al.*, 2009
Japanese quail and broiler breeders: Counter negative effects of mycotoxins	Stanley *et al.*, 2004b; Parlat *et al.*, 2001
Broilers: Countered the negative effects of heat stress	Sohail *et al.*, 2010
Broilers: Countered the negative effects of high-density housing	Hooge *et al.*, 2003

(*continued*)

Table 22.1 (*Continued*)

	Poultry		Pig		Rabbit		Cattle	
MOS (*cont.*)	Broilers: Controlled coccidia	Stanley *et al.*, 2004a						
	Turkeys: Accelerated intestinal development	Solis de los Santos *et al.*, 2007						
COS	Broilers: Increased performance	Huang *et al.*, 2005; Li *et al.*, 2007	Early weaned pigs: Improved performance	Tang *et al.*, 2005				
	Broilers: Increased nutrient digestibility	Huang *et al.*, 2005; Li *et al.*, 2007	Weanling pigs: Improved nutrient digestibility	Liu *et al.*, 2008				
	Broilers: Increased immune function	Li *et al.*, 2007	Weanling pigs: Reduced diarrhea	Liu *et al.*, 2008				
	Broilers: Improved protein metabolism	Li *et al.*, 2007	Weanling pigs: Improved gut morphology	Liu *et al.*, 2008				
			Weanling pigs: Improved immune function	Liu *et al.*, 2010				
			Weanling pigs: Changed intestinal microbiotia	Liu *et al.*, 2008				
			Increased serum growth hormone and insulin-like growth factor 1	Tang *et al.*, 2005				

	Broilers: Improved feed efficiency	Courtin *et al*., 2008a		
AXOS	Broilers: Reduced *Salmonella* shedding	Eeckhaut *et al*., 2008		
	Broilers: Changed intestinal microbiota	Courtin *et al*., 2008b; Eeckaut *et al*., 2008		
GOS	Broilers: Changed intestinal microbiota	Jung *et al*., 2008	Pigs: Changed intestinal microbiota	Smiricky-Tjardes *et al*., 2003
			Pigs: Enhanced gut fermentation	Smiricky-Tjardes *et al*., 2003
	Broilers: No overall effect on performance	Zang *et al*., 2003		
IMO	Broilers: Changed intestinal microbiota	Thitaram *et al*., 2005		
	Broilers: Enhanced gut fermentation	Thitaram *et al*., 2005		
			No effect on overall growth performance	Houdijk *et al*., 1998; Mikkelson *et al*., 2003
TOS			No effect on gut morphology	Mikkelson *et al*., 2003

(*continued*)

Table 22.1 (*Continued*)

	Poultry		Pig		Rabbit		Cattle	
TOS (*cont.*)			No effect on gut bacteria but increased yeast levels	Mikkelson *et al.*, 2003				
			Weanling and growing pigs: No effect on nutrient digestibility	Houdijk *et al.*, 1999				
AOS	Broilers: Reduced *Salmonella* shedding	Eeckhaut *et al.*, 2008						
Inulin	Broilers: Enhanced gut fermentation	Rehman *et al.*, 2008; Velasco *et al.*, 2010; Rebolé *et al.*, 2010	Growing pigs: No effect on intestinal microbiota	Loh *et al.*, 2006	Improved growth performance	Castellini *et al.*, 2007	Calves: Increased weight gain	Masanetz *et al.*, 2010
	Broilers: Changed intestinal microbiota	Rada *et al.*, 2001; Yusrizal & Chen, 2003; Velasco *et al.*, 2010; Rebolé *et al.*, 2010	Growing pigs: Reduced intestinal acetic acid levels	Loh *et al.*, 2006	No effect on growth performance	Bónai *et al.*, 2008	Counter the toxic effects of nitrates in the feed	Sar *et al.*, 2004
	Broilers: No effect on intestinal microbiota	Rehman *et al.*, 2008			Improved gut physiology	Maertens *et al.*, 2004; Juśkiewick *et al.*, 2007		
	Broilers: Improved intestinal morphology	Rehman *et al.*, 2007; Awad *et al.*, 2011			Modification of cecal fermentation	Maertens *et al.*, 2004; Castellini *et al.*, 2007		
					No effect on cecal environment	Bónai *et al.*, 2008		

FOS, fructo-oligosaccharides; MOS, mannanoligosaccharides; COS, chito-oligosaccharide; AXOS, arabinoxylooligosaccharides; GOS, galacto-oligosaccharides; IMO, isomalto-oligosaccharides; TOS, transgalacto-oligosaccharide; AOS, arabinoxylooligosaccharide.

Commercial prebiotic (MOS), organic acid (formic acid), and probiotic products have individually been shown to improve growth performance of broilers. There was no additive effect of combining the prebiotic with the organic acid or probiotic on growth rate but combining the pre- and probiotics resulted in improved feed efficiency suggesting a synergism between them (Bozkurt *et al.*, 2009b).

The roots of *Astragalus membranaceus* (Huangqi), an herb used in Chinese medicine, contain a polysaccharide that has been shown to positively affect the immune system (Li *et al.*, 2009). The supplementation of broiler diets with the *Astragalus* polypeptide and a probiotic has been shown to have synergistic effects on both immunity and the intestinal microbiota (Li *et al.*, 2009). There was an enhancement in the development of immune organs as shown by increases in the thymus, bursa of fabricius, and spleen relative weights. The lymph follicles in the bursa of fabricius and the white pulp of the spleen are both major sites for lymphocyte production. There were also increased intestinal levels of *Lactobacilli* and *Bifidobacteria* and decreases in the levels of *E. coli*.

A combination of two essential oils and FOS supplemented to wheat–corn broiler diets resulted in reduced levels of *C. perfringens* in broilers compared with broilers fed the unsupplemented control diets (McReynolds *et al.*, 2009). Intestinal lesions and mortality were also reduced. Supplementing diets with a combination of MOS and essential oils has been reported to maintain the level of performance of weaned piglets equal to that obtained with an AGP (colistin) (Lipiński *et al.*, 2010).

22.7 CONCLUSIONS

Prebiotics have potential as substitutes for subtherapeutic antibiotic use, primarily with nonruminants such as pigs and poultry. Prebiotics also appear to have some coccidiostatic properties. Research related to the use of prebiotics such as MOS and FOS as feed additives in swine, poultry, and rabbit production has demonstrated the potential of prebiotics, but the results have been inconsistent. A number of factors may be involved including the composition and quality of the feed being supplemented; the housing conditions and management of the animals; the type and level of the prebiotic used; the species, age, and condition of the animals being fed; and the duration of the treatment.

REFERENCES

Aguilar J. C., T. Roca T, and E. Sanz. 1996. Fructo-oligosaccharides in rabbit diet. Study of efficiency in suckling and fattening periods. *6th World Rabbit Congress*, Toulouse, France. pp. 73–77.

Alles, M. S., R. Hartemink, S. Meyboom, J. L. Harryvan, K. M. Van Laere, F. M. Nagengast, and J. G. Hautvast. 1999. Effect of transgalactooligosaccharides on the composition of the human intestinal microflora and on putative risk markers for colon cancer. *Am. J. Clin. Nutr.* 69:980–991.

Awad, W. A., K. Ghareeb, and J. Bohm. 2011. Evaluation of the chicory inulin efficacy on ameliorating the intestinal morphology and modulating the intestinal electrophysiological properties in broiler chickens. *J. Anim. Physiol. Anim. Nutr.* 95:65–72.

Awad, W. A., K. Ghareeb, S. Abdel-Raheem, and J. Bohm. 2009. Effects of dietary inclusion of probiotic and synbiotic on growth performance, organ weights, and intestinal histomorphology of broiler chickens. *Poult. Sci.* 88:49–55.

Awati, A., B. A. Williams, M. W. Bosch, W. J. J. Gerrits, and M. W. A. Verstegen. 2006. Effect of inclusion of fermentable carbohydrates in the diet on fermentation end-product profile in feces of weanling piglets. *J. Anim. Sci.* 84:2133–2140.

Bauer, E., B. A. Williams, C. Voigt, R. Mosenthin, and M. W. A. Verstegen. 2001. Microbial activities of faeces from unweaned and adult pigs, in relation to selected fermentable carbohydrates. *Anim. Sci.* 73:313–322.

Baurhoo, B., L. Phillip, and C. A. Ruiz-Feria. 2007a. Effects of purified lignin and mannan oligosaccharides on intestinal integrity and microbial populations in the ceca and litter of broiler chickens. *Poult. Sci.* 86:1070–1078.

Baurhoo, B., A. Letellier, X. Zhao, and C. A. Ruiz-Feria. 2007b. Cecal populations of Lactobacilli and Bifidobacteria and *Escherichia coli* populations after in vivo *Escherichia coli* challenge in birds fed diets with purified lignin or mannanoligosaccharides. *Poult. Sci.* 86:2509–2516.

Beachy, E. H. 1981. Bacterial adherence: Adhesion-receptor interactions mediating the attachment of bacteria to mucosal surfaces. *J. Infect. Dis.* 143:325–345.

Biggs, P. and C. M. Parsons. 2007. The effects of several oligosaccharides on true amino acid digestibility and true metabolizable energy in cecectomized and conventional roosters. *Poult. Sci.* 86:1161–1165.

Biggs, P., C. M. Parsons, and G. C. Fahey. 2007. The effects of several oligosaccharides on growth performance, nutrient digestibilities, and cecal microbial populations in young chicks. *Poult. Sci.* 86:2327–2336.

Bónai, A., Zs. Szendrıo, L. Maertens, Zs. Matics, H. Fébel, L. Kametler, G. Tornyos, P. Horn, F. Kovács, and M. Kovács. 2008. Effect of inulin supplementation on caecal microflora and fermentation in rabbits. *9th World Rabbit Congress*, Verona, Italy, June 10–13. pp. 555–559.

Bornet, F. R. J. and F. Brouns. 2002. Immune-stimulating and gut health-promoting properties of short-chain fructo-oligosaccharides. *Nutr. Rev.* 60:326–334.

Bovera, F., S. Marono, C. Di Meo, G. Piccolo, F. Iannaccone, and A. Nizza. 2010. Effect of mannanoligosaccharides supplementation on caecal microbial activity of rabbits. *Animal* 4:1522–1527.

Bozkurt, M., K. Küçükyılmaz, A. U. Çatlı, and M. Çınar. 2009a. Effect of dietary mannan oligosaccharide with or without oregano essential oil and hop extract supplementation on the performance and slaughter characteristics of male broilers. *S. Afr. J. Anim. Sci.* 39:223–232.

Bozkurt, M., K. Küçükyılmaz, A. U. Çatlı, and M. Çınar. 2009b. The effect of single or combined dietary supplementation of prebiotics, organic acid and probiotics on performance and slaughter characteristics of broilers. *S. Afr. J. Anim. Sci.* 39:197–205.

Buddington, K. K., J. B. Donahoo, and R. K. Buddington. 2002. Dietary oligofructose and inulin protect mice from enteric and systemic pathogens and tumor inducers. *J. Nutr.* 132:472–477.

Çakir, S., M. Midilli, H. Erol, N. Şimsek, M. Çinar, A. Altintas, H. Alp, L. Altintas, Ö. Cengiz, and A. Antalyali. 2008. Use of combined probiotic-prebiotic, organic acid and avilamycin in diets of Japanese quails. *Rev. Med. Vet.* 159:565–569.

Castellini, C., R. Cardinali, P. G. Rebollar, A. Dal Bosco, V. Jimeno, and M. E. Cossu. 2007. Feeding fresh chicory (*Chicoria intybus*) to young rabbits: performance, development of gastro-intestinal tract and immune functions of appendix and Peyer's patch. *Anim. Feed Sci. Technol.* 134:56–65.

Castillo, M., S. M. Martín-Orúe, J. A. Taylor-Pickard, J. F. Pérez, and J. Gasa. 2008. Use of mannanoligosaccharides and zinc chelate as growth promoters and diarrhea preventative in weaning pigs: Effects on microbiota and gut function. *J. Anim. Sci.* 86:94–101.

Catalá-Gregori, P., S. Mallet, A. Travel, J. Orengo, and M. Lessire. 2008. Efficiency of a prebiotic and a plant extract alone or in combination on broiler performance and intestinal physiology. *Can. J. Anim. Sci.* 88:623–629.

Chambers, J. R., J. L. Spencer, and H. W. Modler. 1997. The influence of complex carbohydrates on *Salmonella typhimurium* colonization, pH, and density of broiler ceca. *Poult. Sci.* 76:445–451.

Constable, P. D. 2009. Treatment of calf diarrhea: Antimicrobial and ancillary treatments. *Vet. Clin. North Am. Food Anim. Pract.* 25:101–120.

Courtin, C. M., W. F. Broekaert, K. Swennen, O. Lescroart, O. Onagbesan, J. Buyse, E. Decuypere, T. Van de Wiele, M. Marzorati, W. Verstraete, G. Huyghebaert, and J. A. Delcour. 2008a. Dietary inclusion of wheat bran arabinoxylooligosaccharides induces beneficial nutritional effects in chickens. *Cereal Chem.* 85:607–613.

Courtin, C. M., K. Swennen, W. F. Broekaert, Q. Swennen, J. Buyse, E. Decuypere, C. W. Michiels, B. De Ketelaere, and J. A. Delcour. 2008b. Effects of dietary inclusion of xylooligosaccharides, arabinoxylooligosaccharides and soluble arabinoxylan on the microbial composition of cecal contents of chickens. *J. Sci. Food Agric.* 88:2517–2522.

Cummings, J. H., G. T. MacFarlane, and H. N. Englyst. 2001. Prebiotic digestion and fermentation. *Am. J. Clin. Nutr.* 73:415–420.

Czech A., A. Mokrzycka, E. R. Grela, and Z. Pejsak. 2009. Influence of mannanoligosaccharides additive to sow diets on blood parameters of sows and their piglets. *Bull. Vet. Inst. Pulawy* 53:89–95.

Davis, E., C. Maxwell, B. Kegley, B. de Rodas, K. Friesen, D. Hellwig, Z. B. Johnson, and D. W. Kellogg. 2000. Efficacy of mannan oligosaccharide (Bio-Mos®) addition at two levels of supplemental copper on performance and immunocompetence of early weaned pigs. *Ark. Agric. Exp. Stn. Res. Series* 470: 15–18.

Donalson, L. M., W. K. Kim, V. I. Chalova, P. Herrera, J. L. McReynolds, V. G. Gotcheva, D. Vidanović, C. L. Woodward, L. F. Kubena, D. J. Nisbet, and S. C. Ricke. 2008. In vitro fermentation response of laying hen cecal bacteria to combinations of fructooligosaccharide prebiotics with alfalfa or a layer ration. *Poult. Sci.* 87:1263–1275.

Dunkley, K. D., C. S. Dunkley, N. L. Njongmeta, T. R. Callaway, M. E. Hume, L. F. Kubena, D. J. Nisbet, and S. C. Ricke. 2007. Comparison of in vitro fermentation and molecular microbial profiles of high-fiber feed substrates incubated with chicken cecal inocula. *Poult. Sci.* 86:801–810.

Eeckhaut, V., F. Van Immerseel, J. Dewulf, F. Pasmans, F. Haesebrouck, R. Ducatelle, C. M. Courtin, J. A. Delcour, and W. F. Broekaert. 2008. Arabinoxylooligosaccharides from wheat bran inhibit *Salmonella* colonization in broiler chickens. *Poult. Sci.* 87:2329–2334.

Eugeniusz, R. G., V. Semeniuk, and A. Czech. 2006. Efficacy of fructooligosaccharides and mannanoligosaccharides in piglet diets. *Medycyna Wet.* 62:762–765.

Falcão-e-Cunha L., L. Castro-Solla, L. Maertens, M. Marounek, V. Pinheiro, J. Freire, and J. L. Mourão. 2007. Alternatives to antibiotic growth promoters in rabbit feed: A review. *World Rabbit Sci.* 15:127–140.

Flemming, J. S., J. R. S. Freitas, P. Fontoura, R. Montanhini Neto, and J. S. Arruda. 2004. Use of mannanoligosaccharides in broiler feeding. *Braz. J. Poult. Sci.* 6:159–161.

Flickinger, E. A., J. Van Loo, and G. C. Fahey. 2003. Nutritional responses to the presence of inulin and oligofructose in the diets of domesticated animals: a review. *Crit. Rev. Food Sci. Nutr.* 43:19–60.

Fonseca A. P., L. Falcão-e-Cunha, A. Kocher A, and P. Spring. 2004. Effects of dietary mannan oligosaccharide in comparison to oxytetracyclin on performance of growing rabbits. *8th World Rabbit Congress*, Puebla, México. pp. 829–833.

Forchielli, M. L. and W. A. Walker. 2005. The role of gut-associated lymphoid tissues and mucosal defence. *Br. J. Nutr.* 93:S41–S48.

Gebbink, G. A. R., A. L. Sutton, B. T. Richert, J. A. Patterson, J. Nielsen, D. T. Kelly, M. W. A. Verstegen, B. A. Williams, M. Bosch, M. Cobb, D. C. Kendall, S. DeCamp, and K. Bowers. 1999. Effects of addition of fructooligosaccharide (FOS) and sugar beet pulp to weanling pig diets on performance, microflora and intestinal health. Available at: http://www.ansc.purdue.edu/swine/swineday/sday99/9.pdf (accessed August 31, 1999).

Ghally, K. A. and S. A. Abd El-Latif. 2007. Effect of dietary yeast on some productive and physiological aspects of growing Japanese quails. African crop science conference proceedings 8:2147–2151. Available at: http://www.acss.ws/Upload/XML/Research/476.pdf.

Gibson, G. R. and M. B. Roberfroid. 1995. Dietary modulation of the human colonic microbiota: Introducing the concept of prebiotics. *J. Nutr.* 125:1401–1412.

Grela E. R., V. Semeniuk, and A. Czech. 2006. Efficacy of fructooligosaccharides and mannanoligosaccharides in piglet diets. *Med. Wet.* 62:762–765.

Guedes, C. M., J. L. Mourão, S. R. Silva, M. J. Gomes, M. A. M. Rodrigues, and V. Pinheiro. 2009. Effects of age and mannanoligosaccharides supplementation on production of volatile fatty acids in the caecum of rabbits. *Anim. Feed Sci. Technol.* 150:330–336.

Hancock, J. D., C. L. Jones, and C. W. Starkey. 2003. Mannanoligosaccharides in diets for nursery pigs. Available at: http://krex.k-state.edu/dspace/handle/2097/2116.

Hansen, L. F. and C. Claudi-Magnussen. 2004. Feeding with lupines reduces the amount of skatole in organic pigs. DARCOF enews, December 2004, Number 4. Newsletter from Danish Research Centre for Organic Farming. Available at: http://www.darcof.dk/enews/dec04/skatole.html.

Hooge, D. M. 2004. Meta-analysis of broiler chicken pen trials evaluating dietary mannan oligosaccharides, 1993-2003. *Int. J. Poult. Sci.* 3:163–174.

Hooge, D. M., M. D. Sims, A. E. Sefton, A. Connolly, and P. Spring. 2003. Effect of dietary mannan oligosaccharide, with or without bacitracin or virginiamycin, on live performance of broiler chickens at relatively high stocking density on new litter. *J. Appl. Poult. Res.* 12:461–467.

Houdijk, J. G. M., M. W. Bosch, M. W. A. Verstegen, and H. J. Berenpas. 1998. Effects of dietary oligosaccharides on the growth performance and faecal characteristics of young growing pigs. *Anim. Feed Sci. Technol.* 71:35–48.

Houdijk J. G. M., M. W. Bosch, S. Tamminga, M. W. A. Verstegen, E. B. Berenpas, and H. Knoop. 1999. Apparent ileal and total-tract nutrient digestion by pigs as affected by dietary nondigestible oligosaccharides. *J. Anim. Sci.* 77:148–158.

Huang, R. L., Y. L. Yin, G. Y. Wu, Y. G. Zhang, T. J. Li, L. L. Li, M. X. Li, Z. R. Tang, J. Zhang, B. Wang, J. H. He, and X. Z. Nie. 2005. Effect of dietary oligochitosan supplementation on ileal digestibility of nutrients and performance in broilers. *Poult. Sci.* 84:1383–1388.

Iji, P. A., A. A. Saki, and D. R. Tivey. 2001. Intestinal structure and function of broiler chickens on diets supplemented with a mannan oligosaccharide. *J. Sci. Food Agric.* 18:1186–1192.

Jamroz, D., A. Wiliczkiewicz, J. Orda, T. Wertelecki, and J. Skorupińska. 2004. Response of broiler chickens to the diets supplemented with feeding antibiotic or mannanoligosaccharides. *Electron. J. Pol. Agric. Univ.* 7:#06. Available at: http://www.ejpau.media.pl/volume7/issue2/animal/art-06.html.

Janssens, G. P., S. Millet, F. Van Immerseel, J. De Buck, and M. Hesta. 2004. The impact of prebiotics and salmonellosis on apparent nutrient digestibility and *Salmonella typhimurium* var. Copenhagen excretion in adult pigeons (*Columba livia domestica*). *Poult. Sci.* 83:1884–1890.

Jung, S. J., R. Houde, B. Baurhoo, X. Zhao, and B. H. Lee. 2008. Effects of galacto-oligosaccharides and a *Bifidobacteria lactis*-based probiotic strain on the growth performance and fecal microflora of broiler chickens. *Poult. Sci.* 87:1694–1699.

Juśkiewicz, J., J. Jankowski, Z. Zduńczyk, and D. Mikulski. 2006. Performance and gastrointestinal tract metabolism of turkeys fed diets with different contents of fructooligosaccharides. *Poult. Sci.* 85:886–891.

Juśkiewicz, J., L. Ašmanskaite, Z. Zduńczyk, P. Matusevičius, A. Wróblewska, and A. Žilinskiene. 2007. Metabolic response of the gastrointestinal tract and serum parameters of rabbits to diets containing chicory flour rich in inulin. *J. Anim. Physiol. Anim. Nutr.* 92:113–120.

Kocher, A., P. Spring, and D. M. Hooge. 2004. Summary analysis of post-weaned rabbit trials with dietary mannan oligosaccharide. *International Society for Animal Hygiene meeting*, Saint-Malo, France, October 11th–14th, 2004. pp. 261–262. Available at: http://www.isah-soc.org/documents/2004/Kocher_2.pdf.

Lan, Y., S. Xun, S. Tamminga, B. A. Williams, M. W. A. Verstegen, and G. Erdi. 2004. Real-time PCR detection of lactic acid bacteria in cecal contents of *Eimeria tenella*-infected broilers fed soybean oligosaccharides and soluble soybean polysaccharides. *Poult. Sci.* 83:1696–1702.

Li, S. P., X. J. Zhao, and J. Y. Wang. 2009. Synergy of *Astragalus* polysaccharides and probiotics (*Lactobacillus* and *Bacillus cereus*) on immunity and intestinal microbiota in chicks. *Poult. Sci.* 88:519–525.

Li, X. J., X. S. Piao, S. W. Kim, P. Liu, L. Wang, Y. B. Shen, S. C. Jung, and H. S. Lee. 2007. Effects of chito-oligosaccharide supplementation on performance, nutrient digestibility, and serum composition in broiler chickens. *Poult. Sci.* 86:1107–1114.

Lipiński, K., J. Tywończuk, C. Purwin, S. Petkevičius, P. Matusevičius, and B. Pysera. 2010. Effect of dietary mannan-oligosaccharides and essential oils on growth performance of piglets. *Vet. Med. Zoot.* 50:54–58.

Liu, P., X. S. Piao, P. A. Thacker, Z. K. Zeng, P. F. Li, D. Wang, and S. W. Kim. 2010. Chito-oligosaccharide reduces diarrhea incidence and attenuates the immune response of weaned pigs challenged with *E. coli* K88. *J. Anim. Sci.* 88:3871–3879.

Liu, P., X. S. Piao, S. W. Kim, L. Wang, Y. B. Shen, H. S. Lee, and S. Y. Li. 2008. Effects of chito-oligosaccharide supplementation on the growth performance, nutrient digestibility, intestinal morphology, and fecal shedding of *Escherichia coli* and *Lactobacillus* in weaning pigs. *J. Anim. Sci.* 86:2609–2618.

Loh, G., M. Eberhard, R. M. Brunner, U. Hennig, S. Kuhla, B. Kleessen, and C. C. Metges. 2006. Inulin alters the intestinal microbiota and short-chain fatty acid concentrations in growing pigs regardless of their basal diet. *J. Nutr.* 136:1198–1202.

Maertens L., J. Aerts, and J. De Boever. 2004. Degradation of dietary oligofructose and inulin in the gastro intestinal tract and the effects on pH and volatile fatty acids. *World Rabbit Sci.* 12:235–246.

Masanetz, S., N. Wimmer, C. Plitzner, E. Limbeck, W. Preißinger, and M. W. Pfaffl. 2010. Effects of inulin and lactulose on the intestinal morphology of calves. *Animal* 4–5:739–744.

McHan, F. and E. B. Shotts. 1993. Effect of short-chain fatty acids on the growth of Salmonella typhimurium in an in vitro system. *Avian Dis.* 37:396–398.

McReynolds, J., C. Waneck, J. Byrd, K. Genovese, S. Duke, and D. Nisbet. 2009. Efficacy of multistrain direct-fed microbial and phytogenetic products in reducing necrotic enteritis in commercial broilers. *Poult. Sci.* 88:2075–2080.

Miguel J. C., S. L. Rodriguez-Zas, and J. E. Pettigrew. 2002. Practical response to Bio-Mos in nursery pigs: a meta-analysis. Nutritional biotechnology in the feed and food industries. *Alltech's 16th international feed industry symposium.* T. P. Lyons and K. A. Jacques (eds). Nottingham University Press, Nottingham. pp. 425–433.

Miguel, J. C., S. L. Rodriguez-Zas, and J. E. Pettiegrew. 2004. Efficacy of a mannan oligosaccharide (Bio-Mos®) for improving nursery pig performance. *J. Swine Health Prod.* 12:296–307.

Mikkelsen, L. L. and B. B. Jensen. 2004. Effect of fructo-oligosaccharides and transgalacto-oligosaccharides on microbial populations and microbial activity in the gastrointestinal tract of piglets post-weaning. *Anim. Feed Sci. Technol.* 117:107–119.

Mikkelsen, L. L., M. Jakobsen, B. B. Jensen. 2003. Effects of dietary oligosaccharides on microbial diversity and fructo-oligosaccharide degrading bacteria in faeces of piglets post-weaning. *Anim. Feed Sci. Technol.* 109:133–150.

Mohamed, M. A., H. M. A. Hassan, and E. M. A. El-Barkouky. 2008. Effect of mannan oligosaccharide on performance and carcass characteristics of broiler chicks. *J. Agric. Social Sci.* 4:13–17.

Mourão J. L., A. Alves, and V. Pinheiro. 2004. Effects of fructo-oligosaccharides on performances of growing rabbits. *8th World Rabbit Congress*, Puebla, México. pp. 915–921.

Mourão J. L., V. Pinheiro, A. Alves, C. M. Guedes, L. Pinto, M. J. Saavedra, P. Spring, and A. Kocher. 2006. Effect of mannan oligosaccharides on the performance, intestinal morphology and cecal fermentation of fattening rabbits. *Anim. Feed Sci. Technol.* 126:107–120.

Mourão, J., V. Pinheiro, and L. Falcão e Cunha. 2005. Alternativas ao uso de antibióticos nas dietas para coelhos em crescimento. In: *Proc. III Jornadas Internacionais de cunicultura*, Vila Real, 27–45.

Niness, K. R. 1999. Inulin and oligofructose: What are they? *J. Nutr.* 129:1402S–1406S.

Parlat, S. S., M. Ozcan, and H. Oguz. 2001. Biological suppression of aflatoxicosis in Japanese quail (*Coturnix coturnix japonica*) by dietary addition of yeast (*Saccharomyces cerevisiae*). *Res. Vet. Sci.* 71: 207–211.

Pelicano, E. R. L., P. A. de Souza, H. B. A. de Souza, F. R. Leonel, N. M. B. L. Zeola, and M. M. Boiago. 2004. Productive traits of broiler chickens fed diets containing different growth promoters. *Braz. J. Poult. Sci.* 6:177–182.

Pinheiro V., A. Alves, J. L. Mourão, C. M. Guedes, L. Pinto, P. Spring, and A. Kocher. 2004. Effect of mannanoligosaccharides on the ileal morphometry and cecal fermentation of growing rabbits. *8th World Rabbit Congress*, Puebla. pp. 936–941.

Pinheiro, V. C. M. Guedes, D. Outor-Monteiro, and J. L. Mourao. 2009. Effects of fibre level and dietary mannanoligosaccharides on digestibility, caecal volatile fatty acids and performances of growing rabbits. *Anim. Feed Sci. Technol.* 148:288–300.

Rada, V., D. Duskova, M. Marounek, and J. Petr. 2001. Enrichment of Bifidobacteria in the hen caeca by dietary inulin. *Folia Microbiol.* 46:73–75.

Rebolé, A., L. T. Ortiz, M. L. Rodríguez, C. Alzueta, J. Treviño, and S. Velasco. 2010. Effects of inulin and enzyme complex, individually or in combination, on growth performance, intestinal microflora, cecal fermentation characteristics, and jejunal histomorphology in broiler chickens fed a wheat- and barley-based diet. *Poult. Sci.* 89:276–286.

Rehman, H., C. Rosenkranz, J. Bohm, and J. Zentek. 2007. Dietary inulin affects the morphology but not the sodium-dependent glucose and glutamine transport in the jejunum of broilers. *Poult. Sci.* 86:118–122.

Rehman, H., P. Hellweg, D. Taras, and J. Zentek. 2008. Effects of dietary inulin on the intestinal short chain fatty acids and microbial ecology in broiler chickens as revealed by denaturing gradient gel electrophoresis. *Poult. Sci.* 87:783–789.

Richards, J. D., J. Gong, and C. F. M. de Lange. 2005. The gastrointestinal microbiota and its role in monogastric nutrition and health with an emphasis on pigs: Current understanding, possible modulations, and new technologies for ecological studies. *Can. J. Anim. Sci.* 85:421–435.

Ricke, S. C. 2003. Perspectives on the use of organic acids and short chain fatty acids as antimicrobials. *Poult. Sci.* 82:632–639.

Sahin, T., I. Kaya, Y. Unal, and D. A. Elmali. 2008. Dietary supplementation on probiotic and prebiotic combination on performance, carcass quality and blood parameters in growing quail. *J. Anim. Vet. Adv.* 7:1370–1373.

Santin, E., A. Maiorka, and M. Macari. 2001. Performance and intestinal mucosa development on broiler chickens fed diets containing *Saccharomyces cerevisiae* cell wall. *J. Appl. Poult. Res.* 10:236–244.

Sar, C., B. Santoso, B. Mwenya, Y. Gamoa, T. Kobayashi, R. Morikawa, K. Kimura, H. Mizukoshi, and J. Takahashi. 2004. Manipulation of rumen methanogenesis by the combination of nitrate with β1-4 galacto-oligosaccharides or nisin in sheep. *Anim. Feed Sci. Technol.* 115:129–142.

Sarica, S., M. Corduk, G. F. Yarim, G. Yenisehirli, and U. Karatas. 2009. Effects of novel feed additives in wheat based diets on performance, carcass and intestinal tract characteristics of quail. *S. Afr. J. Anim. Sci.* 39:144–157.

Schrezenmeir, J. and M. de Vreese. 2001. Probiotics, prebiotics, and synbiotics – Approaching a definition. *Am. J. Clin. Nutr.* 73:361S–364S.

Shen, Y. B., X. S. Piao, S. W. Kim, L. Wang, P. Liu, I. Yoon, and Y. G. Zhen. 2009. Effects of yeast culture supplementation on growth performance, intestinal health, and immune response of nursery pigs. *J. Anim. Sci.* 87:2614–2624.

Shim, S. B., I. H. Williams, and M. W. A. Verstegen. 2005. Effects of dietary fructo-oligosaccharide on villous height and disaccharidase activity of the small intestine, pH, VFA and ammonia concentrations in the large intestine of weaned pigs. *Acta Agric. Scand. A Anim. Sc.* 55:91–97.

Silva, V. K., J. Della Torre da Silva, K. A. A. Torres, D. E. de Faria Filho, F. Hirota Hada, and V. M. Barbosa de Moraes. 2009. Humoral immune response of broilers fed diets containing yeast extract and prebiotics in the prestarter phase and raised at different temperatures. *J. Appl. Poult. Res.* 18:530–540.

Sims, M. D., K. A. Dawson, K. E. Newman, P. Spring, and D. M. Hooge. 2004. Effects of dietary mannan oligosaccharide, bacitracin methylene disalicylate, or both on the live performance and intestinal microbiology of turkeys. *Poult. Sci.* 83:1148–1154.

Smiricky-Tjardes, M. R., C. M. Grieshop, E. A. Flickinger, L. L. Bauer, and G.C. Fahey Jr. 2003. Dietary galactooligosaccharides affect ileal and total-tract nutrient digestibility, ileal and fecal bacterial concentrations, and ileal fermentative characteristics of growing pigs. *J. Anim. Sci.* 81:2535–2545.

Sohail, M. U., A. Ijaz, M. S. Yousaf, K. Ashraf, H. Zaneb, M. Aleem, and H. Rehman. 2010. Alleviation of cyclic heat stress in broilers by dietary supplementation of mannan-oligosaccharide and Lactobacillus-based probiotic: Dynamics of cortisol, thyroid hormones, cholesterol, C-reactive protein, and humoral immunity. *Poult. Sci.* 89:1934–1938.

Solis de los Santos, F., A. M. Donoghue, M. B. Farnell, G. R. Huff, W. E. Huff, and D. J. Donoghue. 2007. Gastrointestinal maturation is accelerated in turkey poults supplemented with a mannan-oligosaccharide yeast extract (Alphamune). *Poult. Sci.* 86:921–930.

Solis de los Santos, F., M. B. Farnell, G. Téllez, J. M. Balog, N. B. Anthony, A. Torres-Rodriguez, S. Higgins, B. M. Hargis, and A. M. Donoghue. 2005. Effect of prebiotic on gut development and ascites incidence of broilers reared in a hypoxic environment. *Poult. Sci.* 84:1092–1100.

Spring, P., C. Wenk, K. A. Dawson, and K. E. Newman. 2000. The effects of dietary mannanoligosaccharides on cecal parameters and the concentrations of enteric bacteria in the ceca of *Salmonella*-challenged broiler chicks. *Poult. Sci.* 79:205–211.

Stanczuk, J., Z. Zdunczyk, J. Juskiewicz, and J. Jankowski. 2005. Indices of response of young turkeys to diets containing mannanoligosaccharide or inulin. *Vet. Med. Zootech.* 31:98–101.

Stanley, V. G., C. Gray, M. Daley, W. F. Krueger, and A. E. Sefton. 2004a. An alternative to antibiotic-based drugs in feed for enhancing performance of broilers grown on *Eimeria* spp.-Infected litter. *Poult. Sci.* 83:39–44.

Stanley, V. G., M. Winsman, C. Dunkley, and T. Ogunleye, M. Daley, W. F. Krueger, A. E. Sefton, and A. Hinton Jr. 2004b. The impact of yeast culture residue on the suppression of dietary aflatoxin on the performance of broiler breeder hens. *J. Appl. Poult. Res.* 13:533–539.

Takahashi, J., A. S. Chaudhry, R. G. Beneke, Suhubdy, and B. A. Young. 1997. Modification of methane emission in sheep by cysteine and a microbial preparation. *Sci. Total Environ.* 204:117–123.

Tang, Z. R., Y. L. Yin, C. M. Nyachoti, R. L. Huang, T. J. Li, C. Yang, X. J. Yang, J. Gong, J. Peng, D. S. Qi, J. J. Xing, Z. H. Sun, and M. Z. Fan. 2005. Effect of dietary supplementation of chitosan and galatco-mannan-oligosaccharide on serum parameters and insulin-like growth factor-I mRNA expression in early-weaned pigs. *Domest. Anim. Endocrinol.* 28:430–441.

Telg, B. E. and D. J. Caldwell. 2009. Efficacy testing of a defined competitive exclusion product in combination with fructooligosaccharide for protection against *Salmonella* Typhimurium challenge in broiler chicks. *J. Appl. Poult. Res.* 18:521–529.

Ten Bruggencate, S. J. M., I. M. J. Bovee-Oudenhoven, M. L. G. Lettink-Wissink, M. B. Katan, and R. Van der Meer. 2004. Dietary fructo-oligosaccharides and inulin decrease resistance of rats to salmonella: protective role of calcium. *Gut.* 53:530–535.

Thitaram, S. N., C. H. Chung, D. F. Day, A. Hinton Jr., J. S. Bailey, and G. R. Siragusa. 2005. Isomaltooligosaccharide increases cecal *Bifidobacterium* population in young broiler chickens. *Poult. Sci.* 84:998–1003.

Uni, Z., S. Ganot, and D. Sklan. 1998. Posthatch development of mucosal function in the broiler small intestine. *Poult. Sci.* 77:75–82.

van der Peet-Schwering, C. M. C., A. J. M. Jansman, H. Smidt, and I. Yoon. 2007. Effects of yeast culture on performance, gut integrity, and blood cell composition of weanling pigs. *J. Anim. Sci.* 85:3099–3109.

Velasco, S., L. T. Ortiz, C. Alzueta, A. Rebolé, J. Treviño, and M. L. Rodríguez. 2010. Effect of inulin supplementation and dietary fat source on performance, blood serum metabolites, liver lipids, abdominal fat deposition, and tissue fatty acid composition in broiler chickens. *Poult. Sci.* 89:1651–1662.

Verdonk, J. M. A. J., S. B. Shim, P. van Leeuwen, and M. W. A. Verstegen. 2005. Application of inulin-type fructans in animal feed and pet food. *Br. J. Nutr.* 93:S125–S138.

Waldroup P. W., E. O. Oviedo-Rondon, and C. A. Fritts. 2003. Comparison of Bio-Mos and antibiotic feeding programs in broiler diets containing copper sulfate. *Int. J. Poult. Sci.* 2:28–31.

Xu, Z. R., C. H. Hu, and M. Q. Wang. 2002. Effects of frugtooligosaccharides on conversion of L-tryptophan to skatole and indole by mixed populations of pig fecal bacteria. *J. Gen. Appl. Microbiol.* 48:83–90.

Xu, Z. R., C. H. Hu, M. S. Xia, X. A. Zhan, and M. Q. Wang. 2003. Effects of dietary fructooligosaccharide on digestive enzyme activities, intestinal microflora and morphology of male broilers. *Poult. Sci.* 82:1030–1036.

Yalçinkaya, L., T. Güngör, M. Başalan, and E. Erdem. 2008. Mannan oligosaccharides (MOS) from *Saccharomyces cerevisiae* in broilers: Effects on performance and blood biochemistry. *Turk. J. Vet. Anim. Sci.* 32:43–48.

Yang, Y., P. A. Iji, A. Kocher, L. L. Mikkelsen, and M. Choct. 2007. Effects of mannanoligosaccharide on growth performance, the development of gut microflora, and gut function of broiler chickens raised on new litter. *J. Appl. Poult. Res.* 16:280–288.

Yang, Y., P. A. Iji, A. Kocher, L. L. Mikkelsen, and M. Choct. 2008. Effects of dietary mannanoligosaccharide on growth performance, nutrient digestibility and gut development of broilers given different cereal-based diets. *J. Anim. Physiol. Anim. Nutr.* 92:650–659.

Yusrizal and Chen T. C. 2003. Effect of adding chicory fructans in feed on broiler growth performance, serum cholesterol and intestinal length. *Int. J. Poult. Sci.* 2:214–219.

Zduńczyk, Z., J. Juskiewicz, J. Jankowski, E. Biedrzycka, and A. Koncicki. 2005. Metabolic response of the gastrointestinal tract of turkeys to diets with different levels of mannan-oligosaccharide. *Poult. Sci.* 84:903–909.

Zhang, W. F., D. F. Li, W. Q. Lu, and G. F. Yi. 2003. Effects of isomalto-oligosaccharids on broiler performance and intestinal microflora. *Poult. Sci.* 82:657–663.

23 Bacteriophages for Potential Food Safety Applications in Organic Meat Production

Steven C. Ricke, Paul Hererra, and Debabrata Biswas

Abstract: Bacteriophages are essentially bacterial viruses. They are an attractive therapeutic biological agent not only because of their specificity, but their ability to self-replicate. Most phage therapy research and applications have been done with conventional food animal production and processing. Postharvest applications on meat surfaces and poultry carcasses appear to be fairly straightforward and some commercial phage products have been approved. Preharvest gastrointestinal (GI) tract phage applications remain somewhat of an enigma with wide variations in results and interpretations. A better understanding of the bacterial tract–phage interface in these highly diverse and dense microbial populations is needed. Although more is now understood about optimizing therapeutic applications in conventional food production, research still needs to be done in organic systems.

Keywords: bacteriophage; food-borne pathogens; lytic; preharvest food safety; postharvest food safety; gastric compartment; rumen; *Salmonella*; *Escherichia coli*; rumen; virulent; temperate; lysogenic; therapeutic

23.1 INTRODUCTION

In conventional food animal production, antibiotics historically have been used to not only treat disease but also to elicit growth-promoting capabilities in animals (Jones & Ricke, 2003). However, there are limited organic counterparts to antibiotics that could serve a similar function of actually being used as treatments to eliminate already present organisms. Some botanicals and plant sources such as essential oils and citrus by-products have shown the ability to be effective in the gastrointestinal (GI) tract environment to modify fermentation, by reducing *in vitro* rumen methane production, and decreasing *in vivo* and *in vitro* ruminal levels of food-borne *Salmonella* and pathogenic *Escherichia coli*, but these are still relatively nonspecific antimicrobial activities (Callaway *et al.*, 2008a; Chaves *et al.*, 2008; Nannapaneni *et al.*, 2008; Callaway *et al.*, 2011a, 2011b, 2011c). Likewise, several of these compounds have shown the ability to limit food-borne pathogens on meat surfaces and *in vitro* laboratory incubations but unfortunately can inhibit nonpathogenic microorganisms as well (O'Bryan *et al.*, 2008a, 2008b; Friedly *et al.*, 2009; Nannapaneni *et al.*, 2009a, 2009b; Chalova *et al.*, 2010; Shannon *et al.*, 2011). In short, these compounds and their derivatives as well as other potential antimicrobials such as organic acids certainly have potential antimicrobial use in

Organic Meat Production and Processing, First Edition. Edited by Steven C. Ricke, Ellen J. Van Loo, Michael G. Johnson, and Corliss A. O'Bryan.

organic meat animal production systems but their broad-spectrum activities may limit their use for more specifically targeted applications (Ricke, 2003a; Ricke *et al.*, 2005; Sirsat *et al.*, 2009; Callaway *et al.*, 2011c).

While broad-spectrum antimicrobials such as organic acids tend to have generalized inhibitory mechanisms against microorganisms, such as membrane destabilization or energy uncoupling, bacteriophage therapy represents a completely different mechanistic type of antimicrobial (Russell, 1992; Joerger, 2003; Ricke, 2003a; Sirsat *et al.*, 2009). Bacteriophages, also commonly referred to as phages (these terms will be used interchangeably throughout this review), are biological agents that have very precisely defined targets with a genetic specificity that precludes any resemblance to broad-spectrum applications. Bacteriophages are viruses that infect bacteria, reproduce in the bacteria, and lyse the bacteria upon release of the phage particles (Ackermann & Dubow, 1987). They are essentially a bacterial parasite requiring the host bacterial cell to replicate (Freifelder, 1983).

Phage therapy for treating bacterial pathogen caused diseases is not a new concept. Bacteriophage were discovered nearly a century ago by F. W. Twort and further characterized by F. D'Hérelle who suggested their potential clinical applications for infectious pathogens (Hayes, 1968; Stent & Calendar, 1978; Summers, 2001; Sulakvelidze, 2011). It took focus on phage as a model for studying genetics beginning in the late 1930s by M. Delbrück *et al.* to uncover much of what is known currently about their life cycle, infection routes, and interaction with bacteria (Hayes, 1968; Stent & Calendar, 1978). Prior to the emergence of antibiotics in the 1940s, phage received considerable attention as potential antibacterial therapeutic agents in the 1930s (Deresinski, 2009; Sulakvelidze, 2011). After antibiotic treatments became predominant because of their advantages over phage in terms of broad-spectrum capabilities, ease of production, and stability of the resulting preparations, phage therapy continued to be pursued only in the Soviet Union and a few other countries (Summers, 2001; Stone, 2002; Sulakvelidze, 2011). Now, interest has resurfaced for alternatives for antibiotic-based treatments and there is a renewed interest in potential applications for phage therapy, including food production (Joerger, 2003; Connerton & Connerton, 2005; Huff *et al.*, 2005; Hudson *et al.*, 2005).

In this review, the potential for phage administration as well as limitations will be examined in the context of animal production systems. Although virtually no work has been done with phage treatments in organic meat production, the results from conventional food animal production will be used to illustrate possibilities for comparable food safety needs in organic meat animal production. Finally, the purpose of this review is not to provide an exhaustive catalog of all the bacteriophage therapy work that has ever been conducted in food production and other systems, as there are number of excellent reviews that already do that quite well (Sulakvelidze *et al.*, 2001; Summers, 2001; Joerger, 2003; Huff *et al.*, 2004, 2005; Hudson *et al.*, 2005; Johnson *et al.*, 2008; Letarov & Kulikov, 2009; Sulakvelidze, 2011). Instead, the purpose of this review is to examine bacteriophage biology from an ecological context in animal production systems, for understanding the limitations of phage therapy and developing potential strategies for successful implementation.

23.2 BACTERIOPHAGE BIOLOGY

Bacteriophage can be grouped in terms of genome structure by being either single stranded or double stranded, by genetic composition (DNA or RNA), and by morphology (Brüssow & Hendrix, 2002). There are also life cycle differences, namely, virulent or lytic versus temperate or lysogenic bacteriophage. Virulent or lytic phage are locked into a continuous infection–lysis cycle with their respective host, while lysogenic/temperate phage can exist either in lytic form or as a prophage (nonlytic) contained within a host bacteria (Hayes, 1968). Lysogenic phage DNA can be integrated into the bacterial host genome and replicated along with the bacterial genetic material, or at some point be induced by an environmental trigger to excise from the host's chromosome and enter into a lytic cycle (Sulakvelidze, 2011).

Not all phages are necessarily good candidates for therapy applications (Joerger, 2003). Clearly, lytic phages are preferred since they are obligated to always infect and lyse their bacterial host to perpetuate their existence. Temperate phage, on the other hand, may become integrated in the bacterial host's DNA and will not lyse the bacteria, and, more importantly, could impart immunity to the host against further infection by external phage attack (Connerton & Connerton, 2005). The basic lytic cycle of a phage consists of the following stages (Freifelder, 1983):

(i) Adsorption to specific bacterial surface receptors.
(ii) Injection of phage genetic phage material into the bacterial cell wall.
(iii) Commandeering the bacterial cell's replication system.
(iv) Production of phage components.
(v) Phage particle assembly.
(vi) Release of new phage.

23.3 POSTHARVEST APPLICATION OF BACTERIOPHAGE IN MEAT PROCESSING

Phages can be theoretically used to control bacterial populations at any stage of the food chain. For preharvest applications, phage can be introduced into food and water to prevent the spread of disease at the farm or feedlot (Sklar & Joerger, 2001; Joerger, 2003). An additional attractive feature for phage interventions is the lower development cost compared to development of a new antibiotic (Brüssow, 2005). Although much of the phage therapy research conducted thus far has been directed toward food-borne pathogen reduction via the GI tract administration, there is no reason that phage cannot be used to control bacteria in the postharvest phases of food production, i.e., clearing pathogens from the carcasses or the processing plant.

Phage therapy for organic meat production is particularly attractive because phages represent an organically acceptable intervention that can be made fairly specific for targeting only the food-borne pathogens of interest, on either the meat surface or in the rinses as well as other contact areas in the processing plant. To date, virtually no research has been done with phage and organic meats; hence, the forthcoming discussion is focused on applications in conventional meats.

Bacteriophages have been directly applied to meat surfaces to reduce specific food-borne pathogen loads. Atterbury *et al.* (2003b) demonstrated that specific bacteriophage would reduce recovery of inoculated *Campylobacter* when added to chicken skin models originating from *Campylobacter*-free chicken carcasses and they suggested that reduction could be enhanced if combined with holding the carcasses at 4°C. However, Atterbury *et al.* (2003a) observed that although *Campylobacter* phage could be isolated from frozen retail poultry, they did not survive freeze–thaw cycles particularly well. Goode *et al.* (2003)

confirmed that application of either *Salmonella* or *Campylobacter* phages resulted in lysis of the host cell on the chicken surface and was not a product of the bacterial recovery media used for enumeration. Higgins *et al.* (2005) isolated *Salmonella* Enteritidis specific lytic phages from municipal wastewater and demonstrated that they could be used to reduce *S.* Enteritidis in broiler rinse water, on broiler carcasses after a spray application, and on turkey carcasses.

In poultry products and processing plants, there has been continued interest in utilizing phage with broader host range properties to limit *Salmonella* contamination. Whichard *et al.* (2003) applied *Salmonella* Felix O1, a well-known broad-spectrum *Salmonella* bacteriophage, as a liquid inoculum to *Salmonella* Typhimurium DT104 experimentally-contaminated chicken frankfurters and achieved an approximate 2-log reduction of *S.* Typhimurium. More recently, Bielke *et al.* (2007a) demonstrated that a wide host range *Salmonella* bacteriophage when applied as a spray mist could reduce both *S.* Typhimurium and *S.* Enteritidis on commercially processed broiler carcasses that had been removed from the processing line before entering the chill tank. However, they used high levels of phage (10^9 plaque-forming unit (PFU) per carcass) based on their previous observations that high levels of the phage are required to ensure effectiveness and any attempts to lower the amount of phage administered to the carcasses decreased phage efficacy (Bielke *et al.*, 2007a; Higgins *et al.*, 2005).

For other food-borne pathogens associated with meats there has only been limited research thus far. Reduction of *E. coli* O157:H7 by specific phage application on meats has been demonstrated for steak surfaces (O'Flynn *et al.*, 2004) and when intermixed with the bacterial host in ground beef (Abuladze *et al.*, 2008). *Listeria* phage applications were initially examined for fresh produce, but more recently *Listeria* phage application research has focused on ready-to-eat foods that are typically consumed with only minimal processing and therefore, represent a high food safety risk for food-borne pathogens such as *Listeria monocytogenes* (Leverentz *et al.*, 2003; Lianou & Sofos, 2007; Guenther *et al.*, 2009; Hoelzer *et al.*, 2011). Holck and Berg (2009) demonstrated that a commercial source of *Listeria* lytic phage could be used in conjunction with a protective bacterial culture, *Lactobacillus sakei*, to achieve a 100-fold reduction of *L. monocytogenes* in cooked ham after 14–28 days of storage. When broad host range *Listeria* bacteriophages were applied to hot dogs, sliced turkey meats, smoked salmon, and seafood, over 13 days of storage at either 4°C or 20°C, *L. moncytogenes* populations exhibited as much as a 5-log decrease, while the phage retained most of their lytic activity (Guenther *et al.*, 2009). As seen with the *Salmonella* poultry product studies described earlier higher, phage doses were more effective in reducing *L. monocytogenes* than lower doses.

Bacteriophage application to meat products from a regulatory and subsequent commercialization standpoint has been a slow process, although ListShield™, a commercial *Listeria* phage-based cocktail for poultry and meat was approved by the Food and Drug Administration in 2006 (Brüssow, 2005; Sulakvelidze, 2011). Further refinements for phage delivery and application should accelerate commercialization in the meat industry. For example, immobilizing phage on modified cellulose membranes has been recently developed as a means to stabilize *L. monocytogenes* and *E. coli* O157:H7 phages and retain their infectivity prior to application to raw and ready-to-eat meats (Anany *et al.*, 2011).

In summary, it appears that bacteriophage application to meats has merits for commercialization but will require regulatory approval for additional phage products. There is no reason to believe they could not be used in organic meats as long as they are not genetically modified (USDA National Organic Program, 2008), but experiments need to be conducted with these meats to determine their effectiveness and obviously any commercial products developed would have to meet organic standards.

23.4 PREHARVEST PHAGE THERAPY

The key for preharvest use of bacteriophage control measures is not only effective oral administration of the phage, but survival of the phage until it arrives at sites in the intestine colonized by the target bacterial host. In practice, orally administered bacteriophages have yielded mixed results in animal studies. In early work using bacteriophage to treat *E. coli* diarrhea of young pigs, lambs, and calves, bacterial numbers were sufficiently reduced to be considered consistently successful (Williams Smith & Huggins, 1983; Williams Smith *et al.*, 1987a, 1987b). Similar effectiveness has been noted for treating *Clostridium perfringens* necrotic enteritis and *E. coli* septicemia in orally inoculated calves (Barrow *et al.*, 1998; Miller *et al.*, 2010). These studies suggest that that phage therapy is consistently effective on GI tract bacterial infections where the numbers of pathogens are relatively high.

In previous studies where phages have been orally administered to treat intestinal colonized food-borne pathogens, phage therapy has met with limited success (Berchieri *et al.*, 1991; Sklar & Joerger, 2001). Berchieri *et al.* (1991) used *S.* Typhimurium phage isolated from a variety of sources including chickens, experimentally infected with *S.* Typhimurium, poultry feed, and human sewage, and observed that phage administration did reduce viable levels of *S.* Typhimurium in the crop, small intestine, and ceca. Large doses (greater than 10^{10} PFU/mL) of phage, however, were more effective and the phage disappeared once *S.* Typhimurium levels dropped below a certain level. Sklar & Joerger (2001), using a range of *S.* Enteritidis bacteriophage amendments, detected measurable but generally very limited reductions in *S.* Enteritidis recovered from infected chickens. Likewise, Hurley *et al.* (2008) observed minimal impact of phage administration on *S.* Typhimurium colonization of broiler chicks. Unlike bacterial infections where bacterial numbers are fairly high, it appears that using phage to eliminate food-borne pathogens in the GI tract, where total bacterial populations are high and food-borne pathogens are only a minor component, may limit effectiveness of the phage introduced into the GI tract.

Bacteriophage therapies for reducing *E. coli* O157:H7 levels in ruminants have yielded mixed results. After comparing reduced rectal fecal levels of *E. coli* O157:H7 with prevalence of phage levels in feedlot pens and water troughs, Niu *et al.* (2009) concluded that the phage originating from these environments could reduce overall levels of the food-borne pathogen and be possible routes for administering phage. In a direct comparison, Rozema *et al.* (2009) demonstrated that oral administration via stomach tube was more effective than rectal administration for reducing *E. coli* O157:H7 shedding, although neither method eliminated shedding entirely. Rivas *et al.* (2010) evaluated the efficacy of two *E. coli* O157:H7 bacteriophage in a rumen fluid-based *in vitro* system that simulated the rumen environment. Although effective under *in vitro* conditions, the phage mixture did not decrease fecal excretion of *E. coli* O157:H7 when administered orally in cattle, and fecal levels of phage were much lower than rumen levels leading the authors to conclude that some inactivation had occurred in the GI tract (Rivas *et al.*, 2010).

This suggests that other routes of administration are more suitable in cattle. Bach *et al.* (2009) reported that *E. coli* O157:H7 bacteriophage mixtures administered orally with a stomach tube to Canadian Arcott rams reduced *E. coli* O157:H7 shedding compared to controls not receiving phage, but administered phage also declined rapidly over the 21-day period causing them to suspect GI tract inactivation of the phage. Callaway *et al.* (2008b), after administering an eight cattle *E. coli* O157:H7 phage cocktail orally to sheep, observed reductions throughout the intestinal tract with the greatest reductions occurring in the cecum. Reduced rectal *E. coli* O157:H7 levels were also observed, but complete elimination was never detected. Based on the identification of the rectal–anal junction as the primary site of

E. coli O157:H7 colonization in cattle (Naylor *et al.*, 2003) and the failure of oral phage administration to reduce *E. coli* O157:H7, Sheng *et al.* (2006) devised a successful strategy of applying bacteriophage directly to the rectal–anal mucosa and also added phage to water troughs to maintain exposure to *E. coli* O157:H7 phage over the course of the study.

In summary, it is apparent that for most oral administration studies, phage efficacy was met with limited success and the GI tract has considerable influence on the phage survival. Given the more consistent success of phage therapy with pathogen infections where high target host bacterial cell numbers are present suggests that food-borne pathogens may occur in much lower numbers in the GI tract. Consequently, the low numbers of food-borne pathogens in the presence of a highly diverse GI tract indigenous microbial population may make it much more difficult for optimum phage–target bacterial interactions.

23.5 BACTERIOPHAGE AND ANIMAL HOST RESPONSE

23.5.1 Animal host entry

Interaction of phage with the animal or human host has been of interest almost from the onset of clinical trials performed shortly after the discovery of bacteriophages (Summers, 2001; Kropinski, 2006; Deresinski, 2009). Deresinski (2009) has categorized the means for clinical administration as topical, oral, rectal, and parenteral. Once administered via one of these routes, bacteriophage has the opportunity to enter the blood stream, circulate, and enter various internal organs. This, of course, has significant implications for treatment of clinical diseases, as the ability of phage to reach organs and sustain lytic populations represents a potential means to clear out an invasive bacterial pathogen. However, phage must first translocate across the host–tissue barriers to enter the blood stream and beyond. Some routes such as rectal application and intraperitoneal injections generally lead to very efficient delivery of large numbers of phage to the blood stream (Dabrowska *et al.*, 2005; Górski *et al.*, 2006). Huff *et al.* (2002, 2003) demonstrated that aerosol administration and intramuscular injection of a phage specific to avian pathogenic *E. coli*, responsible for colibacillosis in chicks, could limit respiratory infection and mortalities, but intramuscular injection was more efficient in delivering the phage to the blood stream. For oral administration, translocation across the gut wall to the bloodstream may be somewhat more complicated. For phage in the GI tract lumen, to reach the blood, would require crossing the gut barrier and recognition by intestinal cells involved in their transport such as enterocytes, M cells, and dendritic cells (Górski *et al.*, 2006).

How much phages actually circulate in the peripheral blood stream remains unknown, although phages have been isolated from various sources of sera (Górski *et al.*, 2006). However, once in the host, phage can readily be found in internal organs such as the liver, spleen, and kidney, but how long they remain in these respective organs depends on whether host bacterial cells are there as well (Dabrowska *et al.*, 2005). Barrow *et al.* (1998) tested phage administered by intramuscular injection to control experimentally induced *E. coli* septicemia and meningitis in chickens. When 3-week-old chicks were given phage and a lethal dose of *E. coli* intramuscularly (in different muscles), no illness or mortality was observed. When the chicks were intracranially inoculated with *E. coli* and intramuscularly with phage, significant protection was observed. Lower dosages of phages (10^6 and 10^4 PFU/mL) were still effective for completely limiting mortalities in birds injected intramuscularly but only the highest phage dose (10^8 PFU) protected birds inoculated with *E. coli* intracranially.

Administration of the phage could be delayed until the onset of physical symptoms and still be protective. Phages were isolated from the spleen, blood, and brain of the birds and there was some indication of *in vivo phage* multiplication. The rapid detection of high phage titers in the brain suggested that the phage were able to cross the blood–brain barrier.

23.5.2 Animal host response to phage entry

Given that phage structure includes an outer protein coat, bacteriophage are considered antigenic and thus, are recognized by antibodies (Dabrowska *et al.*, 2005). Much like other antigenic sources, similarly related phages elicit cross-reacting antibodies and can be neutralized by reacting with these antibodies (Dabrowska *et al.*, 2005). This can have a significant influence on phage therapy. For example, Huff *et al.* (2010) detected increased phage specific IgG serum levels in birds pretreated with an intramuscular injection of a phage specific for the *E. coli* causing colisepticemia. When reacted with the antibody, the phage activity was inhibited and this was reflected in the increased mortalities in these birds indicating that the phage therapy encountered immune interference.

When bacterial infections are dealt with using bacteriophage administered to the animal host through one of the routes previously discussed, the target bacteria are theoretically removed either by direct phage lysis or via an immunostimulation of antibodies in response to the target bacterial cell lysates generated by the phage (Borysowski & Górski, 2008). As Borysowski and Górski (2008) point out, if killing is primarily due to immunostimulation, then phage application to immunocompromised host animals or humans could be highly problematic. However, when studies were done with immunocompetent mice, it became clear that only direct phage lysis of bacteria would clear out the bacterial infection. This observation is supported by Wang *et al.* (2006) where they demonstrated that mortalities from a *Pseudomonas aeruginosa* bacteremia in mice could only be reduced by live lytic phage and not heat inactivated phage, and that antibody responses were no different between the two phage. Furthermore, the clearance rates of the bacteria were closely associated with the timing of the live phage intraperitoneal dose and its subsequent rapid (within 2 hours after dosing) appearance in the blood stream. In a similar fashion, Biswas *et al.* (2002), using heat inactivated bacteriophage, also demonstrated that functional bacteriophages were necessary to save mice from a vancomycin-resistant *Enterococcus faecium* bacteremia.

In the course of conducting the immune studies with bacteriophage, it was realized that not only do phage stimulate antibody production, they also initiate an innate immune response and are rapidly cleared by the reticuloendothelial system (RES) (Dabrowska *et al.*, 2005; Kropinski, 2006). This immune recognition and response to specific phage was demonstrated by Kim *et al.* (2008) when they reported that conjugating *Salmonella* and *Listeria* bacteriophage with a nonimmunogenic monomethoxy polyethylene glycol (PEG) increased circulation half-life in mice, reduced innate cell response, and enhanced avoidance of cellular defense mechanisms. Likewise, Merril *et al.* (1996) demonstrated that serial passage by repeated intraperitoneal injections of phage in mice could be used to select for phage mutants capable of remaining in the circulatory system for much longer periods of time. Residual phage from each round of injection was recovered from collected blood and grown in an *E. coli* mutator strain to increase the incidence of mutation and subsequent selection in phage that could evade the RES. When clearance rates were compared between the mutant phages and wild-type phage at 24 hours, the mutants were still detected in the circulatory system and the wild-type phages were not. Characterization of the resulting long-circulating phage mutants revealed an alteration in the phage head protein E suggesting that it is involved in recognition and subsequent entrapment of the phage by the RES (Merril *et al.*, 1996).

23.6 OVERCOMING BARRIERS TO BACTERIOPHAGE GI TRACT THERAPY

23.6.1 Introduction

Bacteriophages have been administered to food animals and tested on laboratory animals to eliminate food-borne pathogens in the GI tract and host animal for a number of years. However, despite the extensive research conducted over this time period, outcomes and interpretations still remain somewhat unpredictable and less than successful. Typically, bacteriophage selection and targeted food-borne pathogen responses will look highly promising based on *in vitro* studies, but this same success is not repeated *in vivo* when animal experiments are conducted.

Although in theory, administering a therapeutic phage to an animal would seem to be a fairly straightforward process with a predictable outcome, this may in fact be an overly simplistic expectation. If progress is to be made to develop therapeutic phage strategies for application to both conventional and organically raised animals, current limitations need to be identified and addressed. Two limitations or barriers to optimal GI tract bacteriophage therapy results are apparent from most of the work conducted thus far with phage administration to GI tract systems. First of all, phage must be protected from the effects of the gastric compartment and transported to the intestines in order to maximize their effects. Second, oral inoculation of phage requires relatively high numbers to be effective, suggesting inefficient phage–host contact in highly microbially dense and diverse GI tract systems such as the ceca, colon, and rumen. The forthcoming sections discuss attempts to either overcome these barriers and/or highlight where future research needs to be focused.

23.6.2 Overcoming the gastric barrier

Phages that are to be used therapeutically in the GI tract must be exceptionally hardy. Upon ingestion, they have to avoid denaturation by stomach acid and proteolytic digestion by the intestinal enzymes. Consequently, for therapeutically useful bacteriophage to be more effective, they must demonstrate an ability to retain optimal activity in the presence of the hostile GI tract environment. In general, many phages are sensitive to an acidic environment (Dabrowska *et al.*, 2005). *In vivo* and *in vitro* studies have shown that phage counts rapidly decrease at a pH of 2.0, the approximate pH found in the stomach (Williams Smith *et al.*, 1987b; Ramesh *et al.*, 1999; Koo *et al.*, 2001). The proteinaceous phage coat, which is vital for the protection of the phage's genetic material and attachment to the host bacteria, is subject to denaturation by changes of pH, temperature, and ionic strength.

This susceptibility to the stomach environment may vary for different structural components of the phage. For example, recent studies by Waseh *et al.* (2010) have demonstrated that the *Salmonella* P22 phage tail spike proteins that are involved in recognition and attachment to *Salmonella* host cells were resistant to chicken GI tract fluid proteases as well as trypsin but were somewhat susceptible to chymotrypsin and completely susceptible to pepsin. However, they demonstrated that inclusion of 10% bovine serum albumin eliminated susceptibility to these enzymes as well.

To prevent acid inactivation of bacteriophage by stomach acids, neutralization by some sort of buffering agent or acid production blockers prior to phage oral administration have been examined (Koo *et al.* 2001; Dabrowska *et al.*, 2005). *In vitro* evidence supports the importance of being able to neutralize gastric pH prior to oral phage dosing. Koo *et al.* (2001) employed a commercial source of aluminum hydroxide hydrate and magnesium hydroxide to maintain nearly neutral pH and successfully protected *Vibrio vulnificus* phage during transit through the gastric and intestinal compartments of a laboratory, simulated GI tract system. Without the addition of these compounds, the phage were rapidly eliminated. Such approaches have been successfully employed for *in vivo* studies as well. Ramesh *et al.* (1999) used sodium bicarbonate to protect *Clostridium difficile* phage in the gastric compartment of hamsters after demonstrating with *in vitro* comparisons that this phage was highly susceptible to gastric pH. More recently, Loc Carrillo *et al.* (2005) used $CaCO_3$ as part of an oral suspension to administer broad-spectrum *Campylobacter* bacteriophages to broiler chickens and protect them from low pH during passage through the proventriculus and gizzard.

Multiple studies have shown that large doses of phage have to be administered orally in order for sufficient numbers to reach the intestinal tract and alter the course of a bacterial infection. In order to be therapeutically useful, phage must not only be able to survive the hostile conditions found in the stomach and intestinal tracts of the host but retain maximal pathogen targeting and killing capabilities. In cases of diarrhea and other GI tract, oriented diseases, rapid transit of bacteriophage through the GI tract can also limit their therapeutic value (Summers, 2001). What may be needed is a method to protect phages from digestion and deliver them to intestinal sites where the target host bacterium has become established. In such instances, it may be necessary to use agents that retard gut motility or use carriers that bind to the walls of the GI tract. Ideally, a carrier should be inexpensive, nontoxic, easy to handle, and readily available, as well as acceptable for organic animal production.

23.6.3 Strategies for optimizing phage therapy in the rumen

When therapeutic bacteriophages are administered to an animal to target specific food-borne pathogens in the GI tract, effectiveness is inherently dependent on not only the ability of the bacteriophages to survive these conditions, but the physiological status of the target food-borne pathogen in this ecosystem as well. Consequently, understanding the environmental conditions experienced in a microbiologically dense GI tract such as the rumen is a key factor for predicting potential success. Knowing the physiology and ecology of the target pathogen under these conditions may offer clues as to why it is difficult to optimize the interface between the infecting bacteriophage and its target bacterial host cell.

23.6.3.1 *Food-borne pathogen ecology in the rumen*

Many of the food-borne pathogen targets of bacteriophage therapy are capable of surviving in the highly competitive and diversified rumen ecosystem. For example, several *in vitro* studies have demonstrated that *Salmonella* can survive and even grow under anaerobic conditions, are fairly tolerant of short chain fatty acids (SCFA) concentrations similar to those found in GI tracts, and will actually produce SCFA (Schiemann & Shope, 1991; Jones & Falkow, 1994; Ricke, 2003a; Dunkley *et al.*, 2009a). Likewise, *E. coli* has the ability to adapt to and subsequently tolerate SCFA (Alam & Clark, 1989; Iuchi & Lin, 1993; Guilfoyle & Hirshfield, 1994; Rowbury, 1995; Guilfoyle & Hirshfield, 1996). However, the highly competitive nature of the nonpathogenic microflora in scavenging for nutrients and the minimal oxygen levels forces facultative pathogens to grow more anaerobically. This intuitively creates environmental conditions that limit pathogen growth proliferation under most circumstances in the GI tract. Exceptions, of course, include young animals where the GI tract microflora have yet to develop and adult animals experiencing radical dietary shifts (Durant *et al.*, 1999; Ricke, 2003b; Dunkley *et al.*, 2009b). However, for the most part, food-borne pathogens such as *E. coli* are a fairly minor component of the overall GI tract microbial consortia and would be expected to be fairly slow growing (Brüssow, 2005). As an illustration of this, Weiss *et al.* (2009) estimated that *E. coli* only replicated five times daily in the colon compared to a 20-minute doubling time in a laboratory setting.

Based on the nature of host–phage interactions in other ecosystems, the limited growth of food-borne pathogens in robust GI tract ecosystems would be expected to impact phage therapy administered systems. This impact may be manifested in several ways. Hurley *et al.* (2008) used a continuous culture-based mathematical modeling system to assess lytic phage versus bacterial replication as it might occur in the large intestine. From their simulations they concluded that as the growth rate of *Salmonella* was decreased in the model simulation, there would be less impact of phage on intestinal *Salmonella* population levels. This result is in line with the conclusion drawn by Bull *et al.* (2002) from mouse studies that as bacteria slow their growth they become more refractory to phage infection. How does this phage refraction occur? Hurley *et al.* (2008) speculated that some of the pathogen host cell surface receptors such as O antigens could be modulated due to environmental conditions in the intestinal environment. This is supported by Schiemann and Shope (1991) who reported that *S.* Typhimurium suppresses synthesis of a major outer membrane protein during anaerobic growth. This is also supported by Kudva *et al.* (1999), who found that the anti-*E. coli* O157:H7 phages they isolated killed optimally when administrated at high concentrations and while incubated with aeration at 37°C, whereas killing of host bacterial cells were considerably delayed under nonaerated incubation conditions. However, Rivas *et al.* (2010) did use rumen simulating conditions to optimize *E. coli* O157:H7 phage dosage but still did not observe reductions in *E. coli* O157:H7 fecal shedding when the phages were orally administrated to cattle.

23.6.3.2 *Optimizing phage–bacterial host interaction in the rumen*

Rivas *et al.* (2010) suggested that *E. coli* O157:H7 reductions could be achieved either by more continuous phage dosing with no phage replication (passive therapy) or via establishment of a self-replicating phage (active phage therapy). There is some *in vivo* evidence that a self-replication system is possible for *E. coli*. Based on the persistence of *E. coli* T7 phage in germ-free mice over a 3-week period, Weiss *et al.* (2009) suggested that some sensitive bacterial hosts are physiologically or physically protected in the GI tract and consequently continue to feed phage sensitive bacteria into the GI tract to sustain the phage population. They concluded that this coupled with previous work by Schrader *et al.* (1997) that T7 phage can grow in starved cells, may help to perpetuate the phage over time in the GI tract.

Passive therapy occurs when the initial dose of lytic phage is responsible for bacterial host removal, while active therapy occurs when bacterial lysis and release of phage is the primary means for removing target bacteria (Payne & Jansen, 2001). So, how can a successful phage therapy for ecosystems such as the rumen be predicted, let alone be designed? For nonreplicating phage therapy, Kasman *et al.* (2002) has delineated the number of potential bacteria infected by phage into four factors, namely, (1) bacterial density, (2) phage adsorption constant, (3) phage numbers, and (4) time to interact. Payne and Jansen (2001) concluded that timing of the phage dose and phage and bacterial density are important determinants and Weld *et al.* (2004) has contended that this will require a more accurate quantitative assessment of phage growth.

Better model predictions for optimizing phage–host bacterial cell interactions may also depend on a more accurate assessment of GI tract environmental impact on individual microorganisms and their ability to compete in the GI tract. For example, Coleman *et al.* (1996) developed an 11-compartment human colon model that took into account not only individual competing bacteria including food-borne pathogens but available nitrogen and carbon substrates. Although their model was more focused on the potential for probiotics to limit food-borne establishment in the GI tract, aspects of this approach could no doubt be adapted for modeling the impact of phage on the levels of food-borne pathogens predicted by modeling their ability to compete with the indigenous microflora. Given the importance of bacterial density in predicting phage therapy success, it appears that a more thorough assessment of food-borne bacterial growth and ecological dynamics in GI tract systems will also be required using models such as the one developed by Coleman *et al.* (1996).

23.7 OPTIMIZING PHAGE SOURCES FOR THERAPEUTIC APPLICATION

As discussed earlier, the specificity of phages to its host bacterium is an advantage over broad-spectrum antimicrobials since the phage is target specific and avoids harming the beneficial microorganisms. However, this can also be a disadvantage. In a mixed population of closely related bacterial strains, a phage may only infect some of the cells of that bacterial strain. Small changes in the surface proteins or lipopolysaccharide (LPS) structure can mean the difference between phage attachment and infection versus resistance (Tanji *et al.*, 2004). For example, Kudva *et al.* (1999) isolated phages that lysed *E. coli* O157:H7 from bovine and ovine fecal samples. They subsequently screened the isolates to those that specifically bound to the O157 serotype antigen and not to other common cellular receptors such as pili, fimbrae, flagella, or LPS core. Strains of *E. coli* with absent or altered O157 antigen were not infected by these phages. However, despite the focused phage selectivity for the O157 antigen, factors that altered the LPS led to phage-resistant *E. coli* O157:H7.

A different strategy to circumvent constricted specificity that results in resistant subpopulations involves the use of phage "cocktails" that contain more than one strain of phage (Kudva *et al.*, 1999; Sklar & Joerger, 2001; Sulakvelidze *et al.*, 2001; Borie *et al.*, 2008). Multiple phages that target the same bacterial species but home in on several surface receptors would greatly reduce the chance that a bacterium could escape phage infection. Kudva *et al.* (1999) noted that none of their phages could totally kill their bacterial cultures when used alone. However, when a mixture of the phages was used, total clearance was observed. Fiorentin *et al.* (2005), using a mixture of three phages isolated from free-range chickens reported a 3.5-fold reduction in *S.* Enteritidis levels in young broiler chicks, with reduced levels continuing from these birds to 25 days after treatment. Borie *et al.* (2008) developed a mixture of three phages isolated from sewage associated with chicken flocks and demonstrated effective reduction of *S.* Enteritidis in commercial chicks with the phage uncharacteristically persisting, both in the intestine and the internal organs (liver, spleen, and heart), even in the control birds that did not carry host bacterial cells. Toro *et al.* (2005) used three *Salmonella* phage isolated from poultry environmental and equine clinical samples, respectively, alone and in combination with a commercial competitive exclusion (CE) or probiotic culture and observed inconsistent benefits against *S.* Typhimurium challenged young broiler chicks from the phage treatment alone and lack of synergism with the CE culture.

An alternative is to screen the environment for phage isolates that have a much broader spectrum bacterial target range. Given the rapid pace of genetic change in bacteria, phages are under constant pressure to retain sufficient infectivity for continued dissemination. It is relatively simple to isolate new phage strains from the environment and screen them for their antibacterial effectiveness (Stone, 2002). Furthermore, it has been historically possible to isolate phages with inherent broad-host spectrum capabilities at the genus level. One well-known example is the *Salmonella* genus specific phage Felix O1, which infects the majority of *Salmonella* strains with a smooth phenotype (Felix & Callow, 1943; Whichard *et al.*, 2003). However, finding broad host range phage therapy candidates from the environment may not always be straightforward. When Higgins *et al.* (2008) screened bacteriophage isolates from commercial poultry environments for potential lytic *Salmonella* phage candidates, they noted that high phage numbers were required to lyse multiple *Salmonella* hosts and the lytic process did not involve a normal phage infection cycle followed by phage replication.

Despite these difficulties there has been some success in isolating broad host range phage across strain and species and in some cases bacterial genera, that are capable in replicating in these multiple hosts. For example, *Salmonella* bacteriophage have been isolated that are capable of replicating in nonpathogenic bacteria, *E. coli* and *Klebsiella oxytoca* (Higgins *et al.*, 2005; Bielke *et al.*, 2007a, 2007b; Santos *et al.*, 2010). Such bacterial hosts are invaluable not only because the phage they harbor have the appeal of being broad host lytically active and therefore, usable for a wide range of *Salmonella*, but also have the advantage of avoiding using a pathogen to commercially scale up phage production and risking the introduction of a phage-resistant *Salmonella* (Santos *et al.*, 2010).

A more recent screening approach for isolating broad host range phage is to target a host cell phage receptor that is fairly conserved among bacterial strains or serotypes. For example, Kim and Ryu (2011) isolated a virulent T5-like coliphage from chicken feces and internal organs that could infect either *S.* Typhimurium or *E. coli* using the outer membrane protein for vitamin B_{12} uptake as its receptor. Ricci and Piddock (2010) used this approach to isolate *S.* Typhimurium phage that targeted the outer membrane protein TolC, a component of multidrug-resistant efflux pump. Their rationale was based on the fact that TolC was required for *Salmonella* colonization in the chicken GI tract and has the advantage of being conserved across *Salmonella* serovars such that the phage could very well be effective against them as well (Baucheron *et al.*, 2005; Buckley *et al.*, 2006; Nishino *et al.*, 2006). An intriguing part of their reasoning for this phage selection strategy is that development of phage resistance would not be a concern. Any resistance to this phage by *Salmonella* would most likely be due to alterations in the TolC protein that would lead to decreased colonization by the resulting *Salmonella* mutants and thus, still accomplish the primary goal of reducing *Salmonella* infestation in poultry flocks.

23.8 CONCLUSIONS

For organic meat food safety, development of alternatives that meet the requirements for organic acceptable restrictions will be essential as consumer demand continues to increase. A possible intervention that could be employed at both the preharvest and postharvest stages of organic production involves the use of lytic bacteriophages. Bacteriophage therapy is certainly not new, having been first proposed almost simultaneously with their discovery nearly a century ago. Obviously, with the advent of antibiotics, phage therapies took a backseat for further development except in select countries such as the Soviet Union. Now as antibiotic resistance has become an ongoing concern, bacteriophage applications have come back into vogue as potential interventions to control bacterial pathogens, both of clinical and food safety concern. Consequently, in the past decade, there has been an upswing of research activities to explore the utility of phage for food safety applications in food production systems.

For retail meat applications, bacteriophage treatments appear to be reasonably straightforward and fairly effective to the point that some commercialized products are now available. However, administration of phage therapies for preharvest food animals has been much less fruitful. There appear to be several reasons for this. First of all, there are GI tract and host animal barriers such as the acidic stomach and the host immune responses that can eliminate externally delivered phages. There is also the fundamental challenge of sustaining bacteriophage in sufficient numbers to be effective against slow growing and in some cases hard to reach food-borne pathogens residing in the GI tract. Based on what is known about the complexities of host–phage interactions in other ecosystems this is not surprising. To make progress, it appears that developing an active phage therapy using phage that are more capable of self-replication in the GI tract may offer a solution. However, much more experimental work is needed to refine and optimize approaches that will allow for sustainability of phages introduced in the GI tract system. Finally, since much of the phage research conducted thus far has involved conventional food production systems, it might be assumed that phage therapies introduced into organic food production systems would behave in the same manner. However, given the complexities already known about phage–host interactions in conventional food production this may not be true and thus, warrants future research efforts in exclusively organic-based systems. The outcome of such studies may reveal specific intervention applications where they are lacking currently for organic food production.

ACKNOWLEDGMENT

This review was partially supported by a United States Department of Agriculture (USDA) Food Safety Consortium grant and USDA National Integrated Food Safety Grant #2008-51110-04339 to Steven C. Ricke and a USDA postdoctoral fellowship to Paul Hererra.

REFERENCES

Abuladze, T., M. Li, M. Y. Menetrez, T. Dean, A. Senecal, and A. Sulakvelidze. 2008. Bacteriophages reduce experimental contamination of hard surfaces, tomato, spinach, broccoli, and ground beef by *Escherichia coli* O157:H7. *Appl. Environ. Microbiol.* 74:6230–6238.

Ackerman, H. W. and M. S. Dubow. 1987. *Viruses of Prokaryotes*, vol. 1. CRC Press, Inc., Boca Raton, FL.

Alam, K. Y. and D. P. Clark. 1989. Anaerobic fermentation balance of *Escherichia coli* as observed by in vivo nuclear magnetic resonance spectroscopy. *J. Bacteriol.* 171:6213–6217.

Anany, H., W. Chen, R. Pelton, and M. W. Griffiths. 2011. Biocontrol of *Listeria monocytogenes* and *E. coli* O157:H7 in meat using phage immobilized on modified cellulose membranes. *Appl. Environ. Microbiol.* 77:6379–6387.

Atterbury, R. J., P. L. Connerton, C. E. R. Dodd, C. E .D. Rees, and I. F. Connerton. 2003a. Isolation and characterization of *Campylobacter* bacteriophages from retail poultry. *Appl. Environ. Microbiol.* 69:4511–4518.

Atterbury, R. J., P. L. Connerton, C. E. R. Dodd, C. E. D. Rees, and I. F. Connerton. 2003b. Application of host-specific bacteriophages to the surface of chicken skin leads to a reduction in recovery of *Campylobacter jejuni*. *Appl. Environ. Microbiol.* 69: 6302–6306.

Bach, R. P. S. J., Johnson, K. Stanford, and T. A. McAllister. 2009. Bacteriophages reduce *Escherichia coli* O157:H7 levels in experimentally inoculated sheep. *Can. J. Anim. Sci.* 89:285–293.

Barrow, P., M. Lovell, and A. Berchjeri. 1998. Use of lytic bacteriophage for control of experimental *Escherichia coli* septicemia and meningitis in chickens and calves. *Clin. Diagnost Lab. Immunol.* 5:294–298.

Baucheron, S., C. Mouline, K. Praud, E. Chaslus-Dancla, and A. Cloeckaert. 2005. TolC but not AcrB is essential for multidrug-resistant *Salmonella enterica* serotype Typhimurium colonization of chicks. *J. Antimicrob. Chemother.* 55:707–712.

Berchieri, A., M. A. Lovell, and P. A. Barrow. 1991. The activity in the chicken alimentary tract of bacteriophage lytic for *Salmonella typhimurium*. *Res. Microbiol.* 142:541–549.

Bielke, L. R., S. E. Higgins, A. M. Donoghue, D. J. Donohue, B. M. Hargis, and G. Tellez. 2007a. Use of wide-host-range bacteriophages to reduce *Salmonella* on poultry products. *Int. J. Poult. Sci.* 6:754–757.

Bielke, L., S. Higgins, A. Donoghue, D. Donohue, and B. Hargis. 2007b. *Salmonella* host range of bacteriophages that infect multiple genera. *Poult. Sci.* 86:2536–7540.

Biswas, B., S. Adhya, P. Washart, B. Paul, A. N. Trostel, B. Powell, R. Carlton, and C. R. Merril. 2002. Bacteriophage therapy rescues mice bacteremic from a clinical isolate of vancomycin-resistant *Enterococcus faecium*. *Infect. Immun.* 70:204–210.

Borie, C., I. Albala, P. S ànchez, M. L. Sánchez, S. Ramírez, C. Navarro, M. A. Morales, J. Retamales, and J. Robeson. 2008. Bacteriophage treatment reduces *Salmonella* colonization of infected chickens. *Avian Dis.* 52:64–67.

Borysowski, J. and A. Górski. 2008. Is phage therapy acceptable in immunocompromised host? *Int. J. Infect. Dis.* 12:466–471.

Brüssow, H. 2005. Phage therapy: the *Escherichia coli* experience. *Microbiol.* 151:2133–2140.

Brüssow, H. and R. W. Hendrix. 2002. Phage genomics: small is beautiful. *Cell* 108:13–16.

Buckley, A. M., M. A. Webber, S. Cooles, L. P. Randall, R. M. La Ragione, M. J. Woodward, and L. J. V. Piddock. 2006. The AcrAB-TolC efflux system of *Salmonella enterica* serovar Typhimurium plays a role in pathogenesis. *Cell. Microbiol.* 8:847–856.

Bull, J. J., B. R. Levin, T. DeRouin, N. Walker, and C. A. Bloch. 2002. Dynamics of success and failure in phage and antibiotic therapy in experimental infections. *BMC Microbiol.* 35:1–10.

Callaway, T. R., J. A. Carroll, J. D. Arthington, C. Pratt, T. S. Edrington, R. C. Anderson, S. C. Ricke, P. Crandall, and D. J. Nisbet. 2008a. Citrus products decrease growth of *E. coli* O157:H7 and *Salmonella* Typhimurium in pure culture and in fermentation with mixed ruminal microorganism *in vitro*. *Foodborne Pathog. Dis.* 5:621–627.

Callaway, T. R., T. S. Edrington, A. D. Brabban, R. C. Anderson, M. L. Rossman, M. J. Engler, M. A. Carr, K. J. Genovese, J. E. Keen, M. L. Looper, E. M. Kutter, and D. J. Nisbet. 2008b. Bacteriophage isolated from feedlot cattle can reduce *Escherichia coli* O157:H7 populations in ruminant gastrointestinal tracts. *Foodborne Pathog. Dis.* 5:183–191.

Callaway, T. R., J. A. Carroll, J. D. Arthington, T. S. Edrington, R. C. Anderson, M. L. Rossman, M. A. Carr, K. J. Genovese, S. C. Ricke, P. Crandall, and D. J. Nisbet. 2011a. Orange peel products can reduce *Salmonella* populations in ruminants. *Foodborne Path. Dis.* 8:1071–1075.

Callaway, T. R., J. A. Carroll, J. D. Arthington, T. S. Edrington, M. L. Rossman, M. A. Carr, N. A. Krueger, S. C. Ricke, P. Crandall, and D. J. Nisbet. 2011b. *Escherichia coli* O157:H7 populations in ruminants can be reduced by orange peel product feeding. *J. Food Prot.* 74:1917–1921.

Callaway, T. R., J. A. Carroll, J. D. Arthington, T. S. Edrington, R. C. Anderson, S. C. Ricke, P. Crandall, C. Collier, and D. J. Nisbet. 2011c. Chapter 17. Citrus products and their use against bacteria: Potential health and cost benefits. In: R. Watson, J. K. Gerald, and V. R. Preedy (eds). *Nutrients, Dietary Supplements, and Nutriceuticals: Cost analysis versus clinical benefits*. Humana Press, New York. pp. 277–286.

Chalova, V. I., P. G. Crandall, and S. C. Ricke. 2010. Microbial inhibitory and radical scavenging activities of cold-pressed terpeneless Valencia (*Citrus sinensis*) orange oil in different dispersing agents. *J. Sci. Food Agric.* 90:870–876.

Chaves, A. V., M. L. He., W. Z. Yang, A. N. Hristov, T. A. McAllister and C. Benchaar. 2008. Effects of essential oils on proteolytic, deaminative and methanogenic activities of mixed ruminal bacteria. *Can. J. Anim. Sci.* 88:117–122.

Coleman, M. E, D. W. Dreesen, and R. G. Wiegert. 1996. A simulation of microbial competition in the human colonic ecosystem. *Appl. Environ. Microbiol.* 62:3632–3639.

Connerton, P. L. and I. F. Connerton. 2005. Microbial treatments to reduce pathogens in poultry meat. In: G. C. Mead (ed) *Food Safety Control in the Poultry Industry*. Woodhead Publishing Ltd., Cambridge. pp. 414–432.

Dabrowska, K., K. Switala-Jelen, A. Opolski, B. Weber-Dabrowska, and A. Gorski. 2005. Bacteriophage penetration in vertebrates. *J. Appl. Microbiol.* 98:7–13.

Deresinski, S. 2009. Bacteriophage therapy: Exploiting smaller fleas. *Clin. Infect. Dis.* 48:1096–1101.

Dunkley, K. D., T. R. Callaway, C. O'Bryan, M. M. Kundinger, C. S. Dunkley, R. C. Anderson, D. J. Nisbet, P. G. Crandall, and S. C. Ricke. 2009a. Cell yields and fermentation responses of a *Salmonella* Typhimurium poultry isolate at different dilution rates in an anaerobic steady state continuous culture (CC). *Antonie van Leeuwenhoek. J. Gen. Mol. Microbiol.* 96:537–544.

Dunkley, K. D., T. R. Callaway, V. I. Chalova, J. L. McReynolds, M. E. Hume, C. S. Dunkley, L. F. Kubena, D. J. Nisbet, and S. C. Ricke. 2009b. Foodborne *Salmonella* ecology in the avian gastrointestinal tract. *Anaerobe* 15:26–35.

Durant, J. A., D. E. Corrier, J. A. Byrd, L. H. Stanker, and S. C. Ricke. 1999. Feed deprivation affects crop environment and modulates *Salmonella enteritidis* colonization and invasion of Leghorn hens. *Appl. Environ. Microbiol.* 65:1919–1923.

Felix, A. and B. R. Callow. 1943. Typing of paratyphoid bacilli by means of Vi bacteriophage. *Br. Med. J.* 2:127–130.

Fiorentin, L., N. D. Vieira and W. Barioni Jr. 2005. Oral treatment with bacteriophages reduces the concentration of *Salmonella* Enteritidis PT4 in caecal contents of broilers. *Avian Pathology* 34:258–263.

Freifelder, D. 1983. *Molecular Biology—A Comprehensive Introduction to Prokaryotes and Eukaryotes*. Jones and Bartlett Publishers, Inc., Boston, MA.

Friedly, E. C., P. G. Crandall, S. C. Ricke, M. Roman, C. A. O'Bryan and V. I. Chalova. 2009. *In vitro* anti-listerial effects of citrus oil fractions in combination with organic acids. *J. Food Sci.* 74:M67–M72.

Goode, D., V. M. Allen, and P. A. Barrow. 2003. Reduction of experimental *Salmonella* and *Campylobacter* contamination of chicken skin by application of lytic bacteriophages. *Appl. Environ. Microbiol.* 69:5032–5036.

Górski, A., E. Wazna, B.W. Dabrowska, K. Dabrowska, K. Switala-Jelen, and R. Miedzybrodzki. 2006. Bacteriophage translocation. *FEMS Immunol. Med. Microbiol.* 46:313–319.

Guenther, S., D. Huwyler, S. Richard, and M. J. Loessner. 2009. Virulent bacteriophage for efficient biocontrol of *Listeria monocytogenes* in ready-to-eat foods. *Appl. Environ. Microbiol.* 75:93–100.

Guilfoyle, D. E., and I. N. Hirshfield. 1994. The molecular response of *Escherichia coli* to the short chain organic acid butyrate. *Ann. N.Y. Acad. Sci.* 730:246–248.

Guilfoyle, D. E., and I. N. Hirshfield. 1996. The survival benefit of short chain organic acids and the inducible arginine and lysine decarboxylase genes for *Escherichia coli*. *Lett. Appl. Microbiol.* 22:393–396.

Hayes, W. 1968. *The Genetics of Bacteria and Their Viruses*, 2nd edn. John Wiley & Sons. New York.

Higgins, J. P., R. L. Andreatti Filho, S. E. Higgins, A. D. Wolfenden, G. Tellez, and B. M. Hargis. 2008. Evaluation of *Salmonella*-lytic properties of bacteriophages isolated from commercial broiler houses. *Avian Dis.* 52:139–142.

Higgins, J. P., S. E. Higgins, K. L. Guenther, W. Huff, A. M. Donoghue, D. J. Donoghue, and B. M. Hargis. 2005. Use of a specific bacteriophage treatment to reduce *Salmonella* in poultry products. *Poult. Sci.* 84:1141–1145.

Hoelzer, K., B. D. Sauders, M. D. Sanchez, P. T. Olsen, M. M. Pickett, K. J. Mangione, D. H. Rice, J. Corby, S. Stich, E. D. Fortes, S. E. Roof, Y. T. Grohn, M. Wiedmann, and H. F. Oliver. 2011. Prevalence, distribution, and diversity of *Listeria monocytogenes* in retail environments, focusing on small establishments and establishments with a history of failed inspections. *J. Food Prot.* 74:1083–1095.

Holck, A. and J. Berg. 2009. Inhibition of *Listeria monocytogenes* in cooked ham by virulent bacteriophages and protective cultures. *Appl. Environ. Microbiol.* 75:6944–6946.

Hudson, J. A., C. Billington, G. Carey-Smith, and G. Greening. 2005. Bacteriophages as biocontrol agents in food. *J. Food Prot.* 68:426–437.

Huff, W. E., G. R. Huff, N. C. Rath, and A. M. Donoghue. 2010. Immune interference of bacteriophage efficacy when treating colibacillosis in poultry. *Poult. Sci.* 89:895–900.

Huff, W. G., G. R. Huff, N. C. Rath, J. M. Balog, and A. M. Donoghue. 2005. Alternatives to antibiotics: Utilization of bacteriophage to treat colibacillosis and prevent foodborne pathogens. *Poult. Sci.* 84:655–659.

Huff, W. E., G. R. Huff, N. C. Rath, J. M. Balog, and A. M. Donoghue. 2004. Chapter 27—Bacteriophage: Potential role in food safety. In: R. C. Beier, S. D. Pillai, T. D. Phillips, and R. L. Ziprin (eds)

Pre-Harvest and Post-Harvest Food Safety: Contemporary Issues and Future Directions. Blackwell Publishing Professional, Ames, IA. pp. 365–374

Huff, W. E., G. R. Huff, N. C. Rath, J. M. Balog, and A. M. Donoghue. 2003. Evaluation of aerosol spray and intramuscular injection of bacteriophage to treat an *Escherichia coli* respiratory infection. *Poult. Sci.* 82:1108–1112.

Huff, W. E., G. R. Huff, N. C. Rath, J. M. Balog, and A. M. Donoghue. 2002. Prevention of *Escherichia coli* infection in broiler chickens with a bacteriophage aerosol spray. *Poult. Sci.* 81:1486–1491.

Hurley, A., J. J. Mauer, and M. D. Lee. 2008. Using bacteriophages to modulate *Salmonella* colonization of the chicken's gastrointestinal tract: Lessons learned from *in silico* and *in vivo* modeling. *Avian Dis.* 52:599–607.

Iuchi, S. and E. C. C. Lin. 1993. Adaptation of *Escherichia coli* to redox environments by gene expression. *Mol. Microbiol.* 9:9–15.

Joerger, R. D. 2003. Alternatives to anitobiotics: Bacteriocins, antimicrobial peptidies and bacteriophages. *Poult. Sci.* 82:640–647.

Johnson, R. P., C. L. Gyles, W. E. Huff, S. Ojha, G. R. Huff, N. C. Rath and A. M. Donoghue. 2008. Bacteriophages for prophylaxis and therapy in cattle, poultry and pigs. *Anim. Health Res. Reviews* 9:201–215.

Jones, B. D. and S. Falkow, 1994. Identification and characterization of a *Salmonella typhimurium* oxygen-regulated gene required for bacterial internalization. *Infect. Immun.* 62:3745–3752.

Jones, F. T. and S. C. Ricke. 2003. Observations on the history of the development of antimicrobials and their use in poultry feeds. *Poult. Sci.* 82:613–617.

Kasman, L. M., A. Kasman, C. Westwater, J. Dolan, M. G. Schmidt, and J. S. Norris. 2002. Overcoming the phage replication threshold: A mathematical model with implications for phage therapy. *J. Virol.* 76:5557–5564.

Kim, K.P., J.D. Cha, E. H. Yang, J. Klumpp, S. Hagens, W. D. Hardt, K. Y. Lee, and M. J. Loessner. 2008. PEGlyation of bacteriophages increases blood circulation time and reduces T-helper type-1 immune response. *Microb. Biotech.* 1:247–257.

Kim, M. and S. Ryu. 2011. Characterization of a T5-like coliphage, SPC35, and differential development of resistance to SPC35 in *Salmonella enterica* serovar Typhimurium and *Escherichia coli*. *Appl. Environ. Microbiol.* 77:2042–2050.

Koo, J., D. L. Marshall, and A. DePaola. 2001. Antacid increases survival of *Vibrio vulnificus* and *Vibrio vulnificus* phage in a gastrointestinal model. *Appl. Environ. Microbiol.* 67:2895–2902.

Kropinski, A. M. 2006. Phage therapy—Everything old is new again. *Can. J. Dis. Med. Microbiol.* 17:297–306.

Kudva, I. T., S. Jelacic, P. I. Tarr, P. Youderian, and C. J. Hovde, 1999. Biocontrol of *Escherichia coli* OI57 with O157 bacteriophages. *Appl. Environ. Microbiol.* 65:3767–3773.

Letarov, A. and E. Kulikov. 2009. The bacteriophages in human- and animal body-associated microbial communities. *J. Appl. Microbiol.* 107:1–13.

Leverentz, B., W. S. Conway, M. J. Camp, W. J. Janisiewicz, T. Abuladze, M. Yang, R. Saftner, A. Sulakvelidze. 2003. Biocontrol of *Listeria monocytogenes* on fresh-cut produce by treatment with lytic bacteriophages and a bacteriocin. *Appl. Environ. Microbiol.* 69:4519–4526.

Lianou, A. and J. N. Sofos. 2007. A review of the incidence and transmission of *Listeria monocytogenes* in ready-to-eat products in retail and food service environments. *J. Food Prot.* 70:2172–2198.

Loc Carrillo, C., R. J. Atterbury, A. El-Shibiny, P. L. Connerton, E. Dillon, A. Scott, and I. F. Connerton. 2005. Bacteriophage therapy To reduce *Campylobacter jejuni* colonization of broiler chickens. *Appl. Environ. Microbiol.* 71:6554–6563.

Merril, C. R., B. Biswas, R. Carlton, N. C. Jensen, G. J. Creed, S. Zullo, S. Adhya. 1996. Long-circulationg bacteriophage as antibacterial agents. *Prc. Natl. Acad. Sci.* 93:3188–3192.

Miller, R. W., J. Skinner, A. Sulakvelidze, G. F. Mathis, and C. L. Hofacre. 2010. Bacteriophage therapy for control of necrotic enteritis of broiler chickens experimentally infected with *Clostridium perfringens*. *Avian Dis.* 54:33–40.

Nannapaneni, R., V. I. Chalova, P. G. Crandall, S. C. Ricke, M. G. Johnson, and C. A. O'Bryan. 2009b. *Campylobacter* and *Arcobacter* species sensitivity to commercial orange oil fractions. *Int. J. Food Microbiol.* 129:43–49.

Nannapaneni, R., V. I. Chalova, R. Story, K. C. Wiggins, P. G. Crandall, S. C. Ricke, and M. G. Johnson. 2009a. Ciprofloxacin-sensitive and ciprofloxacin-resistant *Campylobacter jejuni* are equally sensitive to natural orange oil-based antimicrobials. *J. Environ. Sci. Health, Part B.* 44:571–577.

Nannapaneni, R., A. Muthaiyan, P. G. Crandall, M. G. Johnson, C. A. O'Bryan, V. I. Chalova, T. R. Callaway, J. A. Carroll, J .D. Arthington, D. J. Nisbet, and S. C. Ricke. 2008. Antimicrobial activity of commercial citrus-based extracts against *Escherichia coli* O157:H7 isolates and mutant strains. *Foodborne Path. Dis.* 5:695–699.

Naylor, S. W., J. C. Low, T. E. Besser, A. Mahajan, G. J. Gunn, M. C. Pearce, I. J. McKendrick, D. G. E. Smith, and D. L. Gally. 2003. Lymphoid follicle dense mucosa at the terminal rectum is the principal site of colonization of enterohemorrhagic *Escherichia coli* O157:H7 in the bovine host. *Infect. Immun.* 71:1505–1512.

Nishino, K., T. Latifi, and E. A. Groisman. 2006. Virulence and drug resistance roles of multidrug efflux systems of *Salmonella enterica* serovar Typhimurium. *Mol. Microbiol.* 59:126–141.

Niu, Y. D., T. A. McAllister, Y. Xu, R. P. Johnson, T. P. Stephens, and K. Stanford. 2009. Prevalence and impact of bacteriophages on the presence of *Escherichia coli* O157:H7 in feedlot cattle and their environment. *Appl. Environ. Microbiol.* 75:1271–1278.

O'Bryan, C. A., P. G. Crandall, V. I. Chalova, and S. C. Ricke. 2008a. Orange essential oils antimicrobial activities against *Salmonella* spp. *J. Food Sci.* 73:M264.

O'Bryan, C. A., P. G. Crandall and S. C. Ricke. 2008b. Organic poultry pathogen control from farm to fork. *Foodborne Path. Dis.* 5:709–720.

O'Flynn, G., R. P. Ross, G. F. Fitzgerald, and A. Coffey. 2004. Evaluation of a cocktail of three bacteriophages for biocontrol of *Escherichia coli* O157:H7. *Appl. Environ. Microbiol.* 69:4519–4526.

Payne, R. J. H. and V. A. A. Jansen. 2001. Understanding bacteriophage therapy as a density-dependent kinetic process. *J. Theor. Biol.* 208:37–48.

Ramesh, V., J. A. Fralick, and R. D. Rolfe. 1999. Prevention of *Clostridium difficile*-induced ileocecitis with bacteriophage. *Anaerobe*. 5:69–78.

Ricci, V. and L. J. V. Piddock. 2010. Exploiting the role of TolC in pathogenicity: identification of a bacteriophage for eradication of *Salmonella* serovars from poultry. *Appl. Environ. Microbiol.* 76:1704–1706.

Ricke, S. C. 2003a. Perspectives on the use of organic acids and short chain fatty acids as antimicrobials. *Poultry Sci.* 82:632–639.

Ricke, S. C. 2003b. The gastrointestinal tract ecology of *Salmonella* Enteritidis colonization in molting hens. *Poultry Sci.* 82:1003–1007.

Ricke, S. C., M. M. Kundinger, D. R. Miller and J. T. Keeton. 2005. Alternatives to antibiotics: chemical and physical antimicrobial interventions and foodborne pathogen response. *Poult. Sci.* 84:667–675.

Rivas, L., B. Coffey, O. McAuliffe, M. J. McDonnell, C. M. Burgess, A. Coffey, R. P. Ross, and G. Duffy. 2010. *In vivo* and *ex vivo* evaluations of bacteriophages e11/2 and e4/1c for use in the control of *Escherichia coli* O157:H7. *Appl. Environ. Microbiol.* 76:7210–7216.

Rowbury, R. J. 1995. An assessment of environmental factors influencing acid tolerance and sensitivity in *Escherichia coli*, *Salmonella* spp. and other enterobacteria. *Lett. Appl. Microbiol.* 20:333–337.

Rozema, E. A., T. P. Stephens, S. J. Bach, E. K. Okine, R. P. Johnson, K. Stanford, and T. A. McAllister. 2009. Oral and rectal administration of bacteriophages for control of *Escherichia coli* O157:H7 in feedlot cattle. *J. Food Prot.* 72:241–250.

Russell, J. B. 1992. Another explanation for the toxicity of fermentation acids at low pH—anion accumulation versus uncoupling. *J. Appl. Bacteriol.* 73:363–370.

Santos, S. B., E. Fernandes, C. M. Carvalho, S. Sillankorva, V. N. Krylov, E. A. Pleteneva, O. V. Shaburova, A. Nicolau, E. C. Ferreira, and J. Azeredo. 2010. Selection and characterization of a multivalent *Salmonella* phage and its production in a nonpathogenic *Escherichia coli* strain. *Appl. Environ. Microbiol.* 76:7338–7342.

Schiemann, D. A. and S. R. Shope. 1991. Anaerobic growth of *Salmonella typhimurium* results in increased uptake by Henle 407 epithelial and mouse peritoneal cells in vitro and repression of a major outer membrane protein. *Infect. Immun.* 59:437–440.

Schrader, H. S., J. O. Schrader, J. J. Walker, T. A. Wolf, K. W. Nickerson, and T. A. Kokjohn. 1997. Bacteriophage infection and multiplication occur in *Pseudomonas aeruginosa* starved for 5 years. *Can. J. Microbiol.* 43:1157–1163.

Shannon, E., S. R. Milillo, M. G. Johnson, and S. C. Ricke. 2011. Efficacy of cold pressed terpeneless valencia oil and its primary components on inhibition of *Listeria* species by direct contact and exposure to vapors. *J. Food Sci.* 76:M500–M503.

Sheng, H., H. J. Knecht, I. T. Kudva, and C. J. Hovde. 2006. Application of bacteriophages to control intestinal *Escherichia coli* O157:H7 levels in ruminants. *Appl. Environ. Microbiol.* 72:5359–5366.

Sirsat, S. A., A. Muthaiyan, and S. C. Ricke. 2009. Antimicrobials for pathogen reduction in organic and natural poultry production. *J. Appl. Poultry Res.* 18:379–388.

Sklar, I. B., and R. D. Joerger. 2001. Attempts to utilize bacteriophage to combat *Salmonella enterica* serovar enteritidis infection in chickens. *J. Food Safety*. 21:15–29.

Stent, G. S. and R. Calendar. 1978. *Molecular Genetics—An Introductory Narrative*, 2nd edn. W.H. Freeman and Company, San Francisco, CA.

Stone, R. 2002. Stalin's forgotten cure. *Science* 298:728–731.

Sulakvelidze, A. 2011. Safety by nature: potential bacteriophage applications. *Microbe*. 6:122–126.

Sulakvelidze, A., Z. Alavidze, and J. G. Morris. 2001. Bacteriophage therapy. *Antimicrob. Agents Chemother*. 45:649–659.

Summers, W. C. 2001. I. Bacteriophage Therapy. *Annu. Rev. Microbiol.* 55:437–51.

Tanji, Y., T. Shimada, M. Yoichi, K. Miyanaga, K. Hori, and H. Unno. 2004. Toward rational control of *Escherichia coli* O157:H7 by a phage cocktail. *Appl. Microbiol. Biotechnol.* 64:270–274.

Toro, H., S. B. Price, S. McKee, F. J. Hoerr, A. J. Krehling, M. Perdue, and L. Bauermeister. 2005. Use of bacteriophages in combination with competitive exclusion to reduce *Salmonella* from infected chickens. *Avian Dis*. 49:118–124.

USDA National Organic Program. 2008. Agricultural Marketing Service 7, Code of Federal Regulations (CFR), Part 205: the National Organic Program.

Wang, J., B. Hu, M. Xu, Q. Yan, S. Liu, X. Zu, Z. Sun, E. Reed, L. Ding, J. Gong, Q. Q. Li, and J. Hu. 2006. Use of bacteriophage in the experimental treatment of bacteremia from imipenem-resistant *Pseudomonas aeruginosa*. *Int. J. Mol. Med.* 17:309–317.

Waseh, S., P. Hanifi-Moghaddam, R. Coleman, M. Masotti, S. Ryan S, M. Foss, R. MacKenzie1, M. Henry, C. M. Szymanski, and J. Tanha. 2010. Orally administered P22 phage tailspike protein reduces *Salmonella* colonization in chickens: Prospects of a novel therapy against bacterial infections. *PLoS One*. 5:e13904.

Weiss, M., E. Denou, A. Bruttin, R. Serra-Moreno, M.-L. Dillmann, and H. Brüssow. 2009. *In vivo* replication of T4 and T7 bacteriophages in germ-free mice colonized with *Escherichia coli*. *Virology* 393:16–23.

Weld, R. J., C. Butts, and J. A. Heinemmann. 2004. Models of phage growth and their applicability to phage therapy. *J. Theoretical Biol.* 227:1–11.

Whichard, J. M., N. Sriranganathan, and F. W. Pierson. 2003. Supression of *Salmonella* growth by wild-type large-plaque variants of bacteriophage Felix Olin liquid culture and on chicken frankfurters. *J. Food Prot.* 66:220–225.

Williams Smith, H. and M. B. Huggins. 1983. Effectiveness of phages in treating experimental *Escherichia coli* diarrhoea in calves, piglets and lambs. *J. Gen. Microbiol.* 129:2659–2675.

Williams Smith, H., M. B. Huggins, K. M. Shaw. 1987a. The control of experimental *Escherichia coli* diarrhoea in calves by means of bacteriophages. *J. Gen. Microbiol.* 133:1111–1126.

Williams Smith, H., M. B. Huggins, K. M. Shaw. 1987b. Factors influencing the survival and multiplication of bacteriophages in calves and in their environment. *J. Gen. Microbiol.* 133:1127–1135.

24 The Future of Organic Meats

Ellen J. Van Loo, Steven C. Ricke, Corliss A. O'Bryan, and Michael G. Johnson

Abstract: Consumers' concerns about intensive/conventional food production and the potential effect on human health, animal welfare, and the environment have led to a rapidly growing organic meat market worldwide. Authors in this book report consumers' increasing interest in organic meat to be due to the concern of potential food safety issues, such as bovine spongiform encephalopathy (BSE) outbreaks, or other food scares. Authors also report that animal health and welfare (AHW) and environmental impact are important for the organic consumer, but research has found that organic farming does not necessarily translate into high animal welfare standards. Many aspects of organic meat production remain either undeveloped or not considered as of yet, as is the case of organic meat by-products. Consumers express interest in purchasing organic pet food, which would offer an ideal use for the by-products of the organic meat industry. Currently, there is no clear difference in the nutritional or sensory quality of organic meat, nor is there clear proof of differences in chemical contamination between conventional and organic meat. More data are required to draw unbiased conclusions. Consumers often perceive that organic meats are safer than conventional meats, but there is virtually no data to support this. This is partially due to the fact that the research results generated from studies published thus far are often inconsistent. For the organic meat market to continue growth the production base must expand to meet increasing market demands, followed by industry integration and developing more value-added and further processed organic meat products.

Keywords: organic meat production; food safety; animal health and welfare (AHW); consumer research

24.1 SYNOPSIS OF THE DIFFERENT SECTIONS

24.1.1 Economics, market, and regulatory issues

In the United States, as well as in Europe, consumers' concerns about intensive/conventional food production and their potential effect on the human health, animal welfare, and environment have created a rapidly growing organic meat market. Some authors reported the consumers' increasing interest in organic meat to be due to their concern of potential food safety issues, such as bovine spongiform encephalopathy (BSE) outbreaks, or other food scares (*Escherichia coli, Salmonella,* and *Campylobacter*) (Siderer *et al.*, 2005; Krystallis *et al.*, 2006; Winter & Davis, 2006; Onyango *et al.*, 2007). The US organic meat market

Organic Meat Production and Processing, First Edition. Edited by Steven C. Ricke, Ellen J. Van Loo, Michael G. Johnson, and Corliss A. O'Bryan.

was valued at $470 million in sales in 2010, representing 1.8% of the $26.7 billion organic food sales in 2010 (OTA, 2011). However, while the US organic food sector is booming, in Europe, the organic food market is already highly developed with approximately 13% of the organic food sales attributed to organic meat sales. Total European sales for organic food have been estimated at €18.5 billion for 2009 (Organic Monitor, 2010). Both the United States and Europe have established certification agencies and regulation to protect the consumer and regulate the organic food production. In the United States, the United States Department of Agriculture's (USDA) National Organic Program (NOP) is in charge of the organic foods regulation. In Europe, the production, control, and labeling is monitored by the EU regulation.

The organic meat industry has gained significant importance in the organic food sector. However, limited resources with detailed information are available and as of now, there are no sources that package all the information into one publication. The purpose of this book is to address specific issues associated with organic meat production and give a comprehensive overview of all the knowledge on organic meat production and processing.

24.1.2 Management issues for organically raised and processed meat animals

Animal health and welfare (AHW) (Chapter 6) and environmental impact (Chapter 7) are important for the organic consumer and research has been focused on these topics. Large variations in AHW exist between organic farms, and organic farming does not necessarily translate into high animal welfare standards. AHW depends strongly on the management approaches at the farm level and the capability of individual producers to balance a farm system with optimal living conditions for the farm animals, within the constraints of organic regulations and markets. There has been no extensive research conducted on the environmental impact of organic meat production and processing. Consequently, minimal comprehensive conclusions of any type can be drawn since mostly case studies have been performed thus far (Chapter 7). AHW, environmental impact, and meat quality should be optimized together as a multifunctional analysis since they compete with each other. For example, using slow-growing strains improves the animal welfare and meat quality but results in a lower productivity with higher unit costs and thus may increase the environmental impact. To achieve this balance, a multifunctional analysis is needed.

Many aspects of organic meat production remain either undeveloped or not considered as of yet. For example, the development of organic meat by-products (Chapter 9) for the organic industry is a sector that has considerable market potential. Consumers are interested in purchasing organic pet food and this offers an ideal use for the by-products of the organic meat industry. Creating markets for organic by-products offers an opportunity to increase the profitability of organic meat operations.

24.1.3 Processing, sensory, and human health aspects of organic meats

Processing options for organic meat were discussed in the current book (Chapter 12). Processing of organic meat is strictly regulated in the United States. Producers of organic meat often do not have USDA-inspected processing facilities available in their area that is USDA organic and/or allows small-scale processing. Mobile processing units may be a potential solution. Antimicrobials approved for use in organic meat processing are limited. They include weak organic acids, chlorine and oxidizing compounds, microbial produced antimicrobial substances, and biopreservation technologies (Chapter 13).

Nutritional and sensory quality of organic meat has been studied. Some studies report lower fat in the carcass as well as differences in the fat distribution and higher levels of the healthier *n*-3 acids (Chapter 14). There is also no clear proof of differences in chemical contamination between conventional and organic meat (Chapter 16). Limited research has been reported on sensory properties of organic meat and more data (affective as well as descriptive) are required to draw unbiased conclusions (Chapter 15).

24.1.4 The current food safety status of organic meats

Microbial safety of conventional meat production has been studied extensively over a long period of time, while the research on organic meat is much more recent and considerably less comprehensive. The consumer often has the perception that organic meats are safer; however, there is virtually no data to support this. This is partially due to the fact that the research results generated from studies published thus far are often inconsistent. It is assumed that organically raised animals may have a greater risk of causing food-borne illnesses due to their access to outdoors and the restricted use of medication. Although the organic meat sales have increased significantly in the last decade, comparative data are scarce and therefore, no clear conclusions can be made about the food safety aspects of organic beef, poultry, and pork as compared to conventional beef, poultry, and pork (Chapters 17, 18, and 19).

24.1.5 Preharvest control measures for assuring the safety of organic meats

Alternatives to conventional preharvest control methods are being evaluated for their potential use in the organic meat industry including organic acids, antimicrobial peptides, botanicals, prebiotics, probiotics, and bacteriophages. Some promising results have been reported (Chapters 20, 21, 22, and 23). Probiotics have proven to potentially have a positive effect by promoting beneficial bacteria and limiting the food-borne pathogens in the gastrointestinal (GI) tract (Chapter 20). Antibiotics can be replaced by several natural feed additives that may have a positive effect, such as prebiotics, organic acids, botanicals, and antimicrobial peptides such as bacteriocins (Chapter 21). Prebiotics selectively promote the beneficial bacteria and the most important prebiotics include fructo-oligosaccharides, mannanoligosaccharides, and other polysaccharides such as inulin. They have potential to replace subtherapeutic antibiotics. However, some inconsistencies in reported results exist (Chapter 22). Bacteriophage (Chapter 23) offers a means to remove already colonized food-borne pathogens in the GI tract but administration of this agent still requires considerable research for optimization, and narrow bacterial host specificity and pathogen resistance are problematic factors as well.

24.2 FUTURE OF THE ORGANIC MEAT INDUSTRY

Although it is difficult to predict the extent and the direction that future growth of organic markets will take, it is anticipated that the organic meat market will go the same route as the conventional meat market development. This would involve expanding the production base to meet increasing market demands, followed by industry integration and developing more value-added and further processed organic meat products (Crandall *et al.*, 2010).

24.2.1 Limitation and challenges

The organic meat market is expected to continue to grow. For the organic meat market to do so, the infusion of new marketing and policy strategies to include the interests of nontraditional organic buyers will be necessary. There are a large number of consumers who value organic farming but they do not purchase it on a regular basis. One of the main obstacles for those consumers is the price and availability (Food Marketing Institute (FMI) and American Meat Institute (AMI), 2010; Van Loo *et al.*, 2010). In order to increase the organic food purchases from those consumers, an expansion of the organic market can be a solution (Bellows *et al.*, 2008). Since more farming operations are being converted to organic farming, the supply of organic meat will increase, which could decrease the premium price. However, in the near future, price premiums for organic meats are expected to remain high since the production of these products at the farm level is increasing slower than the rapidly growing consumer demand (Oberholtzer *et al.*, 2006).

One of the key limitations holding the organic meat market back is the limited access to USDA organic certified meat processors (Conner, 2005; Seideman *et al.*, 2010). Processing according to the organic standards would be easier if these animals could be processed on separate lines from the conventional processing operations but still use the same processing facilities; however, there is often not sufficient volume for an organic processing line to be viable. USDA organic processors also face additional inspections, which increases the cost and this may not be worthwhile if only a small percentage is being processed as USDA organic (Conner, 2005). An additional issue that adds to this problem is that many of the processing facilities currently available are often unable to accommodate small-scale processing (Seideman *et al.*, 2010). Also, small farmers often face large transportation distances to reach a USDA organic processing facility since there are few small processing plants (Conner, 2005; Van Loo *et al.*, 2011). Additionally, most meat industries are highly vertically integrated, resulting in even less small-scale processing opportunities available for the small-scale farmers. Finally, the limited availability of organic feed imposes another challenge and may limit the expansion of the organic meat sector.

The organic meat market that faces competition with similarly perceived markets, such as the natural and local food markets, creates an additional barrier for the organic meat sector to continuous growth (Mintel, 2008, 2010; Martinez *et al.*, 2010). With increasing consumer concerns about the environmental issues, sustainable and local foods may become even more popular and more directly compete with the organic food sector. The natural market is also increasing in popularity and natural foods generally have more affordable prices compared to the organic food products. Both the natural and local food markets are major competitors for the organic food industry. Likewise, the competition with the conventional meat producers continues to increase since conventional producers are gaining market share in this alternative production area by producing specialty products with certain claims such as "antibiotic-free" and "all natural," as well as incorporating animal welfare factors. Consequently, the growth seen in organic markets is paralleled by a trend in the conventional meat market moving toward greener products, thereby reducing the apparent gap between conventional and organic meat markets.

24.2.2 Organic meat industry needs

The consumers who express concerns about organic farming need to be provided with more comprehensive and transparent information about organic food production practices (Bellows *et al.*, 2008). The production methods with regard to animal welfare and environmental impact of conventional and organic meat production should be compared more extensively in the future. However, comparative studies face challenges. Often, they are only case studies and these results usually cannot be generalized. More research is necessary that directly compares organic and conventional food products to obtain a more comprehensive idea of their respective quality differences. The perceived differences in food safety, nutritional quality, and sensory characteristics between conventional and organic meats must be addressed if claims are to be substantiated. Currently, little is known about the effect of the production method (conventional or one of the alternative methods) on human and animal health responses. Outdoor access and restricted use of antibiotics and clinical pharmaceuticals may create increased food safety and animal health risks (Ricke, 2010). Even though rapid growth is occurring in the organic meat market, relatively few studies have been conducted on food safety issues such as food-borne prevalence in different meat production systems; and with the research that has been conducted thus far, inconsistencies are present in their results. Further evaluation will be necessary before results can lead to conclusions about the impact on food safety and any recommendations for improvement. Similarly, for nutritional and sensory quality, there is considerable need for more extensive research before any general conclusions can be drawn.

In addition to more scientific research on organic meat, there is a need for improving the consumer awareness about specific regulations directed towards organic meat production, processing, and retail. Many consumers are not aware of the differences between organic, natural, and other similar meat products. This illustrates the need to better communicate these standards and the meaning of the labels and claims to the consumer.

Lastly, there are no worldwide standards for organic agriculture because different organizations and governments have their own definitions of organic foods and their own certification processes. This makes international trade in organic products difficult. For the development of a worldwide trade in organic products, harmonization of standards will be necessary (Siderer *et al.*, 2005).

ACKNOWLEDGMENT

Support in writing this chapter is gratefully acknowledged for a National Integrated Food Safety Initiative (NIFSI) grant (2008-51110-04339) to coauthor Steven C. Ricke.

REFERENCES

Bellows, A. C., B. Onyango, A. Diamond, and W. K. Hallman. 2008. Understanding consumer interest in organics: Production values vs. purchasing behavior. *J. Agric. Food Ind. Organ.* 6:Article 2. Available at: www.bepress.com/jafio/vol6/iss1/art2.

Conner, D. 2005. Current status of organic meat processing in Michigan. *"Growing Michigan's organic future", Michigan Organic Conference*, Eppley Center, Michigan State University, East Lansing, MI, March 5th, 2005.

Crandall, P. G., C. A. O'Bryan, S. C. Ricke, F. T. Jones, S. C. Seideman, R. Rainey, E. A. Bihn, T. Maurer, and A. C. Fanatico. 2010. Food safety of natural and organic poultry. In: S. C. Ricke, and F. T. Jones (eds). *Perspectives on food-safety issues of animal-derived foods*. University of Arkansas Press, Fayetteville, AR. pp. 289–305.

Food Marketing Institute (FMI) and American Meat Institute (AMI). 2010. The power of meat – An in-depth look at meat through the shoppers' eyes. Joint report AMI/FMI, Arlington, VA. p. 77.

Krystallis, A., I. Arvanitoyannis, and G. Chryssohoidis. 2006. Is there a real difference between conventional and organic meat? Investigating consumers' attitudes towards both meat types as an indicator of organic meat's market potential. *J. Food Products Market.* 12:47–78.

Martinez, S., M. Hand, M. Da Pra, S. Pollack, K. Ralston, T. Smith, S. Vogel, S. Clark, L. Lohr, S. Low, and C. Newman. 2010. Local food systems – Concepts, impacts, issues. ERS/USDA Economic Research Report 97, May 2010. Available at: http://www.ers.usda.gov/Publications/ERR97/ERR97.pdf.

Mintel. 2008. Organic food – US – October 2008. *Mintel database.*

Mintel. 2010. Consumer attitudes toward natural and organic food and beverage – US – March 2010. *Mintel database.*

Oberholtzer, L., C. Greene, and E. Lopez. 2006. Organic poultry and eggs capture high price premiums and growing share of specialty markets. Available at: http://www.ers.usda.gov/Publications/LDP/2006/12Dec/LDPM15001/ (accessed December, 2006).

Onyango, B. M., W. K. Hallman, and A. C. Bellows. 2007. Purchasing organic food in US food systems: A study of attitudes and practice. *Br. Food J.* 109:399–411.

Organic Monitor. 2010. The global market for organic food & drink: business opportunities & future outlook. Organic Monitor Ltd., London. p. 270.

Organic Trade Association (OTA). 2011. *Organic Trade Association's 2011 Organic Industry Survey*. December, 22nd 2010–March, 7th 2011.

Ricke, S. C. 2010. Future prospects for advancing food – safety research in food animals. In: S. C. Ricke, and F. T. Jones (eds) *Perspectives on food-safety issues of animal-derived foods*. University of Arkansas Press, Fayetteville, AR. pp. 335–350.

Seideman, S. C., T. R. Callaway, P. G. Crandall, S. C. Ricke, and D J. Nisbet. 2010. Alternative and organic beef production: Food-safety issues. In: S. C. Ricke, and F. T. Jones (eds) *Perspectives on food-safety issues of animal-derived foods*. University of Arkansas Press, Fayetteville, AR. pp. 307–321.

Siderer, Y., A. Maquet, and E. Anklam. 2005. Need for research to support consumer confidence in the growing organic food market. *Trends Food Sci. Technol.* 16:332–343.

Van Loo, E., V. Caputo, R. M. Nayga, J. F. Meullenet, P. G. Crandall, and S. C. Ricke. 2010. Effect of organic poultry purchase frequency on consumer attitudes toward organic poultry meat. *J. Food Sci.* 75:S384–S397.

Van Loo, E. J., W. Alali, S. Welander, S. C. Ricke, and P. G. Crandall. 2011. Pastured poultry in Georgia: Survey of growers' and consumers' perspective. *Poultry Science Association Annual Meeting*, St. Louis, MO, July 16th–19th, 2011.

Winter, C. K. and S. F. Davis. 2006. Organic foods. *J. Food Sci.* 71:R117–R124.

Index

Note: Page numbers with italicized *f*'s and *t*'s refer to figures and tables, respectively.

Organic Meat Production and Processing, First Edition. Edited by Steven C. Ricke, Ellen J. Van Loo, Michael G. Johnson, and Corliss A. O'Bryan.